Self-Healing Smart Materials and Allied Applications

Scrivener Publishing
100 Cummings Center, Suite 541J
Beverly, MA 01915-6106

Publishers at Scrivener
Martin Scrivener (martin@scrivenerpublishing.com)
Phillip Carmical (pcarmical@scrivenerpublishing.com)

Self-Healing Smart Materials and Allied Applications

Edited by

Inamuddin, Mohd Imran Ahamed, Rajender Boddula and Tariq Altalhi

Scrivener Publishing

WILEY

This edition first published 2021 by John Wiley & Sons, Inc., 111 River Street, Hoboken, NJ 07030, USA and Scrivener Publishing LLC, 100 Cummings Center, Suite 541J, Beverly, MA 01915, USA

For more information about Scrivener publications please visit www.scrivenerpublishing.com.

Wiley Global Headquarters
111 River Street, Hoboken, NJ 07030, USA

For details of our global editorial offices, customer services, and more information about Wiley products visit us at www.wiley.com.

Library of Congress Cataloging-in-Publication Data

ISBN 978-1-119-71015-8

Cover image: Pixabay.Com
Cover design by Russell Richardson

Set in size of 11pt and Minion Pro by Manila Typesetting Company, Makati, Philippines

Contents

Preface

There is apprehension about the undesirable material damage caused in devices by breakage, operational fatigue, volume change, abrasion, shape deformation, and cutting eventually formed during practical usage and degradation over time, resulting in deterioration of device properties. This mechanical damage of devices will reduce their reliability and shorten their lifespan. Thus, the development of suitable materials with self-healing, electrical and ionic properties to overcome the damage is very much the need of the hour. Self-healing smart materials (SHSMs) are one of the smart materials that can automatically restore some or all of a devices' functions after suffering external mechanical damage or harsh environments. In particular, because of their wide-ranging practical applications, SHSMs are having a significant impact in industry due to the self-healing capabilities of the material, which can expand the service life, operational life, longevity, and reliability of the devices; and also reduce waste, thereby conserving resources. The advancement of wearable electronic devices which use self-activating and self-adjusting systems are promising for enhancing operational safety and lifespan. Moreover, the use of SHSMs in manufacturing devices is an excellent choice to re-establish electrical and mechanical properties in case of a mechanical failure. Hence, an understanding about "self-healable technology" and all its related concepts is very essential for modern industries and research communities since these SHSMs and their composites are having a great impact on modern wearable applications in energy and environmental science.

This book describes the design, synthesis, mechanisms, characterization, fundamental properties, functions and development of SHSMs and their composites with their associated applications. It covers cementitious concrete composites, bleeding composites, elastomers, tires, membranes, and composites in energy storage, coatings, shape-memory, aerospace and robotic applications. This book is a result of the commitment of top researchers in the field with various backgrounds and expertise. Its target audience includes materials scientists, polymer industrialists, researchers, members of R&D in wearable electronics; as well as university professors, postgraduate students and academics who are working and studying in the fields of polymers, chemical technology, biology, advanced electronics, polymer engineering, aeronautical engineering. mechanical engineering, biomedicine, advanced sciences, materials sciences, flexible energy storage, and renewable energy. A summary of the information covered in the 21 chapters is given below.

Chapter 1 describes the most recent and relevant advances in the development of self-healable polymer coatings, both extrinsic (those using external healing agents) and intrinsic. Implementation of new strategies, like remote activation of the healing process, as well as the perspectives and challenges for these innovative materials, are also discussed.

Chapter 2 overviews the current progress on the synthesis of benzoxazine-based materials for self-healing and shape-memory applications. The advantages and drawbacks of these materials are also discussed. Many examples are provided regarding the design flexibility of benzoxazine chemistry and its vast potential in designing smart materials.

Chapter 3 introduces the characteristics of self-healing in elastomers and techniques for characterization of healing ability. It also reviews self-healing in particular cases of different elastomer matrices such as natural rubber (NR), styrene butadiene rubber (SBR), polybutadiene rubber (BR), bromobutyl rubber (BIIR), silicones and polyurethane (PU).

Chapter 4 discusses various promising strategies for fabricating self-healable rubber as well as methods to improve the self-recovery properties of the rubber. In addition to that, the chapter also focuses on natural rubber modification with self-healing properties, as it has become the most critical component in manufacturing self-healable tires.

Chapter 5 outlines the self-healing methods for concrete composites in general, and details about bacterial-aided, self-healing in particular. The different mechanisms of bacterial self-healing and the factors influencing it are discussed. Strategies to enhance the performance of the healing, as well as methods employed for testing are also included.

Chapter 6 focuses on an ideal and long-lasting photovoltaic device which can heal all of the damages applied to its structure to decrease maintenance costs of solar panels. The current self-healing mechanisms and recent progress in solar cells are summarized.

Chapter 7 provides additional insights into the construction of self-healable core-shell nanofiber agents and their potential application in various areas.

Chapter 8 details the state-of-the-art in intrinsic self-healing smart materials. It summarizes some of the most significant research on SHSMs.

Future challenges that need to be overcome to enable commercial exploration of systems already developed are also discussed.

Chapter 9 discusses self-healable materials in the field of catalysis. The basics and application areas of catalysts that are reported in the literature are covered. Additionally, some specific energy fields, such as oxygen evolution and hydrogen generation catalysts, are discussed in detail.

Chapter 10 details the recent advances regarding the potential applications of various types of micro/nano-carriers of healing agents for the development of smart/intrinsic self-healing anti-corrosion polymeric coatings. The SHSMs/fillers classification, various techniques used to load inhibitors into carriers/containers, and their release/corrosion-inhibition mechanisms are introduced and conceptualized.

Chapter 11 discusses the various methods of self-healing for conductive materials. Self-healing conductive materials, such as elastomers, reversible bonds, polymers, capsules, liquids and composite, are discussed. The complex conductivity which can be established even as a coating is also discussed.

Chapter 12 deals with the preparation and properties of self-healable electronic skin. Its various applications as a pressure sensor and body motion sensor are discussed. Artificial skin can also be used for electrocardiogram and electromyograph monitoring, as well as for monitoring Parkinson's tremor.

Chapter 13 overviews self-healing smart composite materials and their different self-healing mechanisms. These materials are designed based on two approaches: (a) discharge of smart self-healing agents towards the generated damage and (b) alterable crosslinking networks.

Chapter 14 overviews the stimuli-responsive self-healable materials. It provides details about the history, different types and importance of self-healable materials. The major focus is on recent examples of self-healing polymers that show response to various stimuli. The chapter ends with a discussion about aspects concerning the commercialization of SHSMs and challenges in the area.

Chapter 15 covers mechanically-induced SHSMs with novel characteristics for numerous applications. Various methods of producing these mechanically-induced SHSMs based on different types of materials are discussed.

Chapter 16 discusses SHSMs, their method of synthesis, and application in robotics. The focus of the chapter is on describing the chemistry of self-healing and the working principles. The various approaches of self-healing are also explained such as capsule-based, vascular, and intrinsic self-healing systems. The major focus is on the application of the material in the different sections of soft robots such as skin, actuator, electronics, and sensor.

Chapter 17 discusses the various methods of developing SHSMs for aerospace structural and high-temperature applications. Materials like fiber-reinforced polymers that alter the nature of the base matrix and ceramic matrix composites are described in detail along with intrinsic and extrinsic self-healing mechanisms.

Chapter 18 presents bioinspired, magnificent, nature-based, broadscope elements with numerous functions. These elements exhibit immense power to perceive, respond, and self-heal (autonomous healing). The discussion proceeds with a brief insight into SHSMs, coatings, their types, and the exploitation process used to extract their characteristic features to benefit humankind. This chapter sheds more light on examples dealing with self-healable materials with a notion to fabricate such self-healable materials with rapid exploration and a promising platform.

Chapter 19 introduces SHSMs as promising candidates for the fabrication of electronic devices, energy storage systems, and even sensors. The progress made in SHSMs research is detailed and the preparation mechanisms and properties of the self-healing processes are summarized. In terms of self-healing batteries, both macroscopic and microscopic SHSMs are introduced.

Chapter 20 focuses on self-healing in bleeding engineered composite structures. Additionally, materials with intrinsic and extrinsic self-healing properties are discussed. Moreover, various strategies like bioinspired, biomimetic and vascular networks, are described in detail. Also discussed are approaches for the synthesis of bleeding composite structures, along with their evolved properties, repairing mechanism, disadvantages, and advantages.

Chapter 21 provides a general overview of numerous materials with self-healing attributes. Greater emphasis is also placed on autonomic and non-autonomic types of SHSMs. The mechanism of action of these SHSMs are also highlighted.

Inamuddin, Mohd Imran Ahamed,
Rajender Boddula and Tariq Altalhi
March 2021

1

Self-Healing Polymer Coatings

Facundo I. Altuna* and Cristina E. Hoppe†

Instituto de Investigaciones en Ciencia y Tecnología de Materiales (INTEMA), Universidad Nacional de Mar del Plata—CONICET, Mar del Plata, Argentina

Abstract

Traditional coatings made of thermosetting polymers could not be healed or mended because of their cross-linked structure, so damage implied the end of their service life cycle. This picture has dramatically changed since the first self-healing systems reports, moment in which this "disadvantage" of thermosets has started to vanish with the continuous increase in the availability of self-healing and recyclable thermosetting polymers. These advances constitute a breakthrough, as they would avoid replacement and catastrophic failure, encourage recycling and re-processing and prevent the generation of a considerable amount of waste, with obvious environmental and economic benefits. This chapter describes some of the more recent advances in the field of self-healing thermosetting polymers with potential application as coatings. Extrinsic self-healing thermosets use an external agent to perform the healing whereas intrinsic ones require the intervention of an external trigger for repair damage. The first part of the chapter describes the work in extrinsic self-healing thermosets, and the second part in intrinsic ones, with emphasis on polymeric networks with dynamic covalent bonds (DCBs). The most common strategies for the external triggering of intrinsic self-healing polymers are also described. Finally, challenges are discussed with the aim put in attaining the so expected end-user applications.

***Keywords*:** Self-healing, thermosets, dynamic covalent bonds (DCBs), coatings, crosslinked polymers

**Corresponding author*: faltuna@fi.mdp.edu.ar; facundoaltuna@gmail.com
†*Corresponding author*: hoppe@fi.mdp.edu.ar; cristinaehoppe@gmail.com

Inamuddin, Mohd Imran Ahamed, Rajender Boddula and Tariq Altalhi (eds.) Self-Healing Smart Materials and Allied Applications, (1–38) © 2021 Scrivener Publishing LLC

1.1 Introduction

Traditionally, materials science and technology research has been a quest for improving one or several properties of any given class of materials, so that they can have an enhanced performance. With this aim, all kinds of materials from ceramics and metals to glasses and soft polymers are nowadays designed based on strong scientific knowledge. However, during their service life, materials can suffer damages with consequences that range from affecting the way in which the materials are supposed to work to catastrophically break down or, in the case of a damaged coating, leaving the substrate deprived from its protection. In such cases, besides all the relevant physicochemical, mechanical, thermal and other properties, it would be highly desirable that the material could be mended. Doing so, it would avoid its replacement and save lots of time and economic resources. It would also prevent the generation of a considerable amount of waste, with obvious environmental and economic benefits. With this purpose, the development of self-healing polymers is gaining momentum since the first works studying healing mechanisms in polymers were published about 40 years ago [1, 2]. Another leap forward followed with the design of self-healing polymeric systems containing vessels with healing agents [3, 4], which demonstrated autonomous healing ability.

Thermosetting polymers constitute one of the two great polymer groups according to the most usual polymers classification. They distinguish themselves from thermoplastic polymers—which form the other group—by their crosslinked structure, which provides them with a set of distinctive properties, such as an improved resistance to solvents and chemical reagents, good mechanical strength and thermal stability [5]. Owed to these properties their use as protective coatings, among other applications, has become very popular. Unfortunately and also because of its crosslinked structure, typical thermosetting polymers cannot be healed. This is their main disadvantage in comparison with thermoplastics, which can be not only healed but also reprocessed and recycled by applying the proper processing that often includes a thermal treatment [6, 7]. These differences, however, are starting to vanish, as proved by the continuously increasing number of systems based on self-healing and recyclable thermosetting polymers appearing in the scientific literature. The relevance of the role that self-healing materials and recyclable thermosets are called to play has been already highlighted by the World Economic Forum, which listed them among the Top 10 Emerging Technologies in 2013 and 2015 respectively [8, 9].

In this chapter, we will describe some of the more recent advances in the field of self-healing thermosetting polymers with potential application as coatings. One of the most popular classification of self-healing thermosets is between extrinsic and intrinsic healing systems [10, 11]. The category of extrinsic self-healing thermosets encompasses all those systems that make use of an external agent to perform the healing of the damage. The self-healing composites with hollow microspheres or microfibers filled with a repairing agent, usually a thermosetting polymer precursor, are the most common examples. The healing mechanism of these materials intends to mimic how nature acts to heal wounds in animals, through a vascular system that delivers the healing agent. The healing process in polymers starts when a crack propagates through it and breaks the vessels that contain the repairing agent, which is released within the crack. At this moment, the repairing agent comes in contact with a catalyst dispersed in the matrix and reacts, forming a new polymeric network that fills the crack, restoring the mechanical properties to the material. To this group belong the first developed self-healable polymers [3, 4, 12, 13]. Figure 1.1 shows schematically the concept of extrinsic self-healing thermosetting polymers.

The other large group is formed by those self-healing polymers that do not require the intervention of an external healing agent. There are some

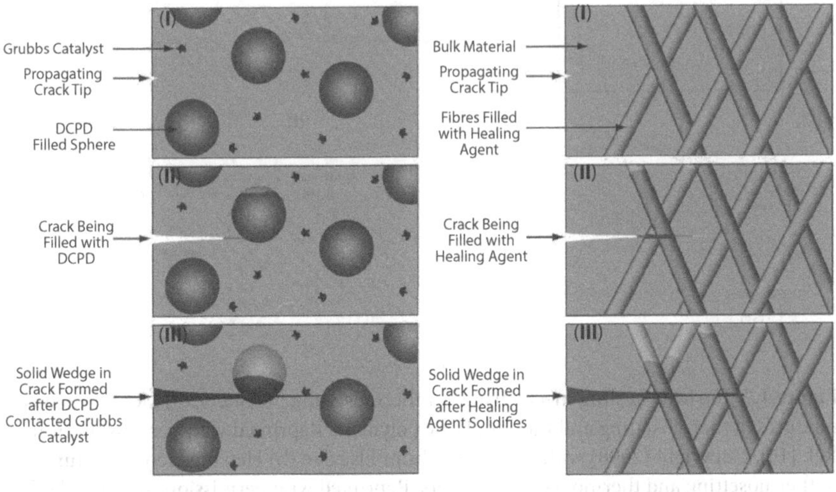

Figure 1.1 Concept of healing mechanisms in microcapsule- (left column) and hollow fiber- (right column) based self-healing composites. Extrinsic polymeric composites. Adapted from Ref. [14]; Copyright (2008) with permission from Elsevier.

examples of mixtures of thermoplastics and thermosets, either homogeneous or with a multi-phase structure, that use respectively the entanglements of dangling chains or the melted thermoplastic to mend an eventual crack [15, 16]. The majority of the reported intrinsic self-healing thermosets, however, are based on either reversible or dynamic covalent bonds. During healing, such crosslinked polymers are depolymerized or undergo a topological rearrangement, depending on whether the bonds that form the macromolecule are of reversible, or of dynamic nature [17–19]. In both cases, when the healing is produced the same polymer acts as its own healing agent. The disadvantage of most intrinsic self-healing polymers is that they need an external stimulus—such as heat, light or a mechanical stress—to induce the healing. Figure 1.2 shows schemes of the aforementioned intrinsic self-healing thermosets.

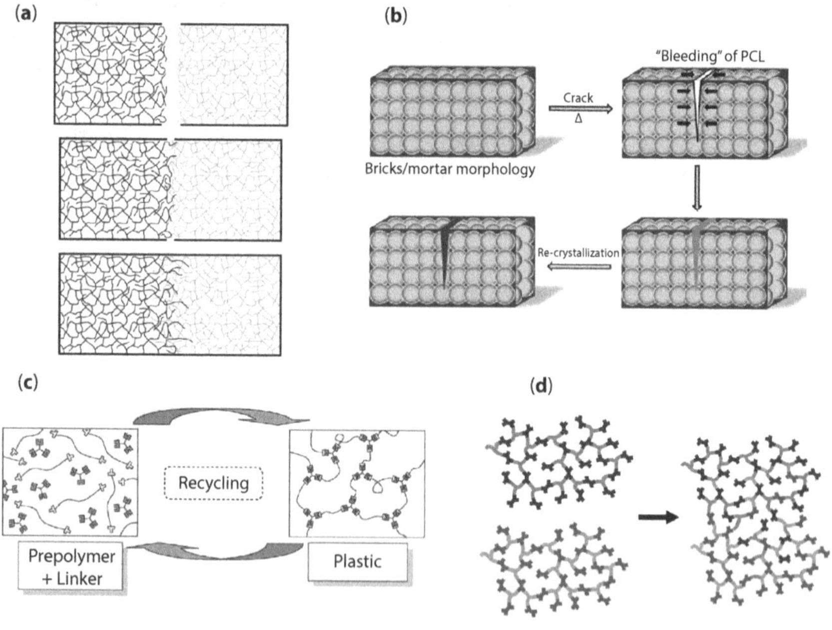

Figure 1.2 Intrinsic healing mechanisms for thermosetting polymers. (a) Homogeneous mixtures of thermosetting and thermoplastic polymers. Reprinted with permission from Ref. [14]. Copyright (2008) with permission from Elsevier. (b) Heterogeneous mixtures of thermosetting and thermoplastic polymers. Reprinted with permission from Ref. [15]; Copyright (2009) American Chemical Society. (c) Polymeric networks based on reversible covalent bonds. Reproduced with permission from Ref. [20]. Copyright (2008) John Wiley & Sons, Inc. (d) Polymeric networks based on dynamic covalent bonds. Adapted with permission from Ref. [21]. Copyright (2012) American Chemical Society.

Other characteristics of the self-healing coatings are also of critical relevance for their final applications. Perhaps the most important one regards the stimulus needed for the self-healing process to start, ranging from materials that need no stimulus to the cases where high temperatures or external stresses have to be applied. This would be another valid classification for self-healing materials, but as we will find out, very often the limit between them is somewhat blurred. In most cases extrinsic self-healing polymers are totally or nearly autonomous, and intrinsic ones need an external triggering. That is, most intrinsic self-healing coatings should not be regarded as self-healing, but instead just as healable materials. There are as well some notable exceptions to this statement that deserve to be considered.

Extrinsic self-healing polymers were developed several years before that intrinsic ones. However, nowadays intrinsic self-healing polymers are a much faster growing research field, led by polymeric networks based on dynamic covalent bonds (DCBs). We will devote the first part of this chapter to extrinsic self-healing thermosets, and a second part to intrinsic ones, with special emphasis on polymeric networks with DCBs. Finally, we will provide a perspective for future trends regarding self-healing polymeric coatings.

1.2 Extrinsic Self-Healing Polymer Coatings

The most widely explored extrinsic self-healing polymers are systems based on composites containing the healing agent into purposely designed containers that break when the material is damaged. Usually, their self-healing is autonomous, needing no other driving force than the damage itself, or in some cases the action of the surrounding environment. The key for the self-healing to take place autonomously is that the chemical precursors released can readily react at the material's operation temperature. Quite often this is room temperature, and hence these polymeric composites must be designed in order to carry the appropriate amount of catalyst to complete the curing reaction at this low temperature. Besides this, other relevant aspects of the healing mechanism need to be accounted to achieve high healing efficiencies. The vessels containing the healing agents must break, and the healing agents have to flow to fill the crack before the curing reaction produce its gelation. Hence, a good load transfer from the matrix to the vessels combined with the specific fracture toughness of the vessels is needed, and the viscosity of the healing agents must be low enough to allow them to flow into the crack. These features have also been addressed in several previous works [22].

There are several monomers and catalysts that allow reaction rates high enough to achieve self-healing at room temperature. Grubbs catalysts are one of the most widely used for autonomous self-healing composites based on microencapsulated healing agents. It was first proposed for self-healing composites by White *et al.*, to aid the crosslinking of dicyclopentadiene (DCPD) through a ring opening metathesis polymerization (ROMP) [12]. The DCPD was contained into urea microcapsules embedded within these composites, and the Grubbs catalyst was dispersed in the matrix. After healing during 48 h at room temperature, the researchers observed that the healed sample reached a load of 75% of that of the virgin sample in a fracture test. These results were very promising, and the healing efficiency values were quickly surpassed in the following years by similar systems that incorporated some improvements in their formulations and/or their processing [23, 24]. By protecting the catalyst with wax [25] a considerable raise in the healing efficiency was achieved, thanks to an enhanced stability of the catalyst, but this modification was also associated with a decrease in the damage resistance to impact. These systems were also used to study the effect of the particles' size and load on the healing efficiency, which allowed to determine that microcapsules smaller than 30 mm are efficient for small cracks (around 3 mm), but larger capsules and higher loads are needed for larger cracks [26].

However, some shortcomings of the DCPD-Grubbs catalyst systems (low long term stability due to possible side reactions with air and the polymer matrix) [27], encouraged the search for other chemistries that could overcome these difficulties. The condensation of hydroxy capped polydimethylsiloxane (HOPDMS) and polydiethoxysiloxane (PDES) with dibutyltindilaurate (DBTDL) in a vinyl ester matrix showed an improved stability, but the healing efficiencies were rather poor [27]. Much better results were observed for a polydimethylsiloxane (PDMS) elastomer filled with poly(urea-formaldehyde) (UF) microcapsules containing the same precursor and Pt initiator used to synthesize the PDMS matrix separately [28]. The authors studied the breaking mechanism of the capsule, and provided an interesting insight from the observation of the process of deformation and failure of a microcapsule within the elastomeric matrix. The sequence is shown in Figure 1.3. For stretch values below 1.5 the microcapsule is deformed along with the matrix, withstanding the stress, but for higher deformations, the failure of its wall releases the chemical precursors. This study focused on some very important aspects of the use of microcapsules for self-healing composites, such as the microcapsule rupture process, and how its presence affects some properties of the matrix. It clearly showed that the breaking of the capsules is a quite complex matter,

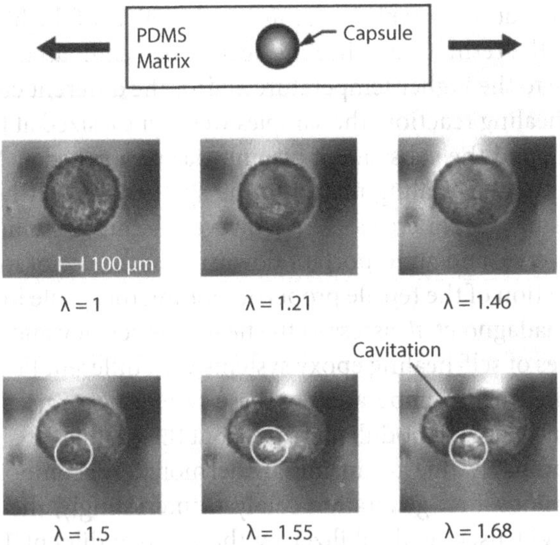

Figure 1.3 Images of a single PDMS microcapsule subjected to uniaxial tension at different values of deformation (γ). Reprinted with permission from Ref. [28]. Copyright (2007) John Wiley & Sons, Inc.

and even brittle materials (when evaluated in bulk) such as UF may need relatively high strain values to fail [28, 29]. The authors also observed an improvement on tear strength with the increase in the microcapsules concentration (up to 20 wt% of microcapsules), which is in agreement with previous results for reinforced PDMS [30, 31]. To ease the flow of the healing agent off the microcapsules and into the crack, its viscosity was reduced by using heptane as diluent [28]. Using a similar system, consisting in a silanol-terminated polydimethylsiloxane (STP) as healing agent and dibutyltin dilaurate (DTDL) as catalyst, Kim *et al.* obtained a self-healing coating that could repair itself at temperatures as low as −20 °C. The viscoelastic material produced by the healing agent was capable of protecting the coated and scratched specimens from saline solution uptake upon immersion for 48 h at −20 °C [32].

Microencapsulated epoxy precursors were also applied as external healing agent in matrices with a dispersed latent initiator [33–37]. Yin *et al.* utilized UF microcapsules containing an epoxy resin based on diglycidyl ether of bisphenol A (DGEBA), and a latent catalyst consisting in a complex of $CuBr_2$ and 2-methylimidazole (MeIm), dispersed in the matrix [33]. Their results showed very good healing efficiencies, with healed specimens showing fracture toughness values 11% higher than those of the virgin samples. As the authors explained, this is a side effect of the higher fracture toughness of the resin used as healing agent when compared to the matrix, probably due to the higher temperature and/or the different curing mechanism of the healing reaction: the samples were synthesized at temperatures up to 100 °C, and the subsequent healing was performed at 130 °C. They also applied the self-healing matrix to the synthesis of self-healing epoxy laminates, using glass fibers [33, 34]. However, the healing efficiencies were lower in this case, and an important downside of the use of microcapsules was the reduction of the tensile properties for microcapsule loads above 30 wt% [34]. Guadagno *et al.* assessed the healing efficiency and the mechanical properties of self-healing epoxy systems with different flexibilizers [36, 37]. When the curing temperature of these systems was raised from 170 to 180 °C, the mechanical moduli increased, but the healing efficiency slightly decreased, probably due to an initial thermolytic decomposition of the Hoveyda–Grubbs' first generation catalyst. Interestingly, the replacement of the Heloxy71 used as flexibilizer by the reactive diluent 1,4-butanedioldiglycidylether (BDE) leads to an increase in the compressive modulus, from around 0.7 to 1.4 GPa, with no loss in the healing efficiency (above 80%) [37]. Hia *et al.* demonstrated multiple self-healing ability in epoxy composites with alginate multicore microcapsules [38]. Three to four healing cycles were produced, though the efficiencies were around 55%, which are lower than for other systems. Moreover, the authors found that self-healing specimens have lower impact strength than the neat epoxy polymers, due to the large size of the capsules, and its high load.

Polyurethanes (PU) were explored for extrinsic self-healing systems [39, 40]. He *et al.* used isophorondiisocyanate (IPDI) encapsulated into polyurea capsules as self-healing agent for PU matrices and also for epoxy-based ones [41]. They show that complete healing (efficiencies around 100%) of both epoxy and PU matrices is achieved for capsules with diameters of 96 mm or higher. Smaller capsules produced poorer healing performances. Gil *et al.* used microencapsulated diisocyanate to improve the tensile strength of collagen [42]. They used two different diisocyanates: IPDI and 4,4'-diphenylmethane diisocyanate (MDI); the isocyanate groups react with the collagen, creating new crosslinks and mending the damages. Other healing agents employed for extrinsic self-healing coatings, including thiol-ene and azide-alkyne precursors, as well as vinyl ester and unsaturated polyesters have also been tested, as pointed out in a review by Hillewaere and Du Prez [43].

A smart approach proposes to develop self-healing systems especially designed so that the healing can be activated by the environment surrounding the material when a crack propagates. Different research groups used encapsulated isocyanate that reacts with water when released into aqueous environments, producing coatings with potential applications in offshore and other marine devices. Di Credico *et al.* also used encapsulated IPDI to provide self-healing capacity to a DGEBA-based epoxy matrix, thanks to the crosslinking reaction of isocyanate with the surrounding water [44]. Figure 1.4-I shows the damaged coating, and the healed coating after immersion for 48 h in salty water. The authors emphasized that the rough outer surface of the microcapsules played a key role improving the adhesion to the matrix, allowing the capsule to fail and release the IPDI. Wang *et al.* also used IPDI as healing agent to repair cracks in an alkyd varnish coating (AVC) [45]. Figure 1.4-II shows the aspect of the mended scratch on different substrates after different times of exposure to marine salty water. Though some seawater could penetrate into the coating and reach the substrate, the self-healing prevented a much larger damage.

Light is another possible external stimulus that can be harnessed to trigger the healing response. Sunlight was proposed by some authors to induce the self-healing of polymeric coatings. Song *et al.* designed the first

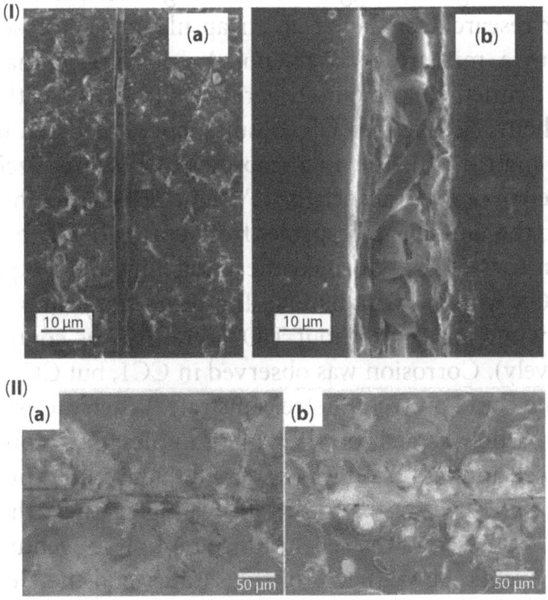

Figure 1.4 (I)—SEM micrographs of crack in a coating with PU/PUF microcapsules (a) before and (b) after immersion in salt water for 48 h. Reprinted from Ref. [44]; Copyright (2013) with permission from Elsevier. (II)—Alkyd varnish coatings on a titanium surface after 200 and 1,200 h of seawater immersion. Reprinted from Ref. [45] with permission from The Royal Society of Chemistry.

photoinduced microcapsule-based self-healing coating from tetraethyl ortosilicate and a polysiloxane, with encapsulated methacryloxypropyl-terminated polydimethylsiloxane (MAT-PDMS) and benzoinisobutyl ether (BIE) photoinitiator [46]. The conversion of the healing agent upon exposure to real sunlight was somewhat lower than when an artificial UV light was used instead. However, several tests (optical and SEM examination, and measurements of water uptake and chloride ion penetration) showed that relatively short times (4 h) of sunlight exposition are enough to produce an acceptable healing. Figure 1.5 shows the results obtained, comparing the performance of the self-healing coating with the control coating (without the encapsulated healing agent). Khalaj Asadi *et al.* successfully encapsulated a sunlight curable silicon based resin, and studied several parameters affecting some final properties of the capsules [47]. They used polyvinylpyrrolidone as emulsifier, and established the optimum amount to obtain higher yields. These capsules showed good resistance to water and xilene, but ethanol produced a rapid deterioration of the microcapsules. The authors remarked that these microcapsules can be readily used for self-healing polymers formulations.

The use of UV light to trigger the mending reactions was explored as well by other researchers. Gao *et al.* encapsulated a photosensitive resin obtained from a mixture of Bisphenol A epoxy resin diacrylate ester (BAEA) and trimethylolpropane-triacrylate (TMPTA) with 1-hydroxy-cyclohexyl-phenyl-ketone into UF capsules, and embedded them into an epoxy-amine matrix [48]. The microcapsules were synthesized containing TiO_2 in its shell in order to absorb the UV light and protect the photosensitive resin into the undamaged capsules from curing before being released. Anticorrosion tests were performed after scratching the samples and healing during 30 s with UV irradiation. The neat epoxy-amine matrix and a composite with capsules without TiO_2 were used as control (CC1 and CC2 respectively). Corrosion was observed in CC1, but CC2 and the self-healing coating could protect the steel substrate. When the experiment was repeated with a new scratch on the same samples, CC2 could no longer protect the steel, and corrosion was observed. Figure 1.6 shows the images of the samples after each test. For the self-healing coatings, the experiment was repeated 5 times, and the self-healing ability was eventually lost after 5 irradiation events. SEM images of the scratch and the electric current measured for each case confirmed the previous observations.

Zhu *et al.* also used a UV-curable healing agent into microcapsules with a rapidly degradable inner polymeric shell and an outer TiO_2 shell that can absorb UV radiation [49]. The action of the TiO_2 shell helps to degrade the inner shell, releasing the healing agent. Hence, the self-healing composite

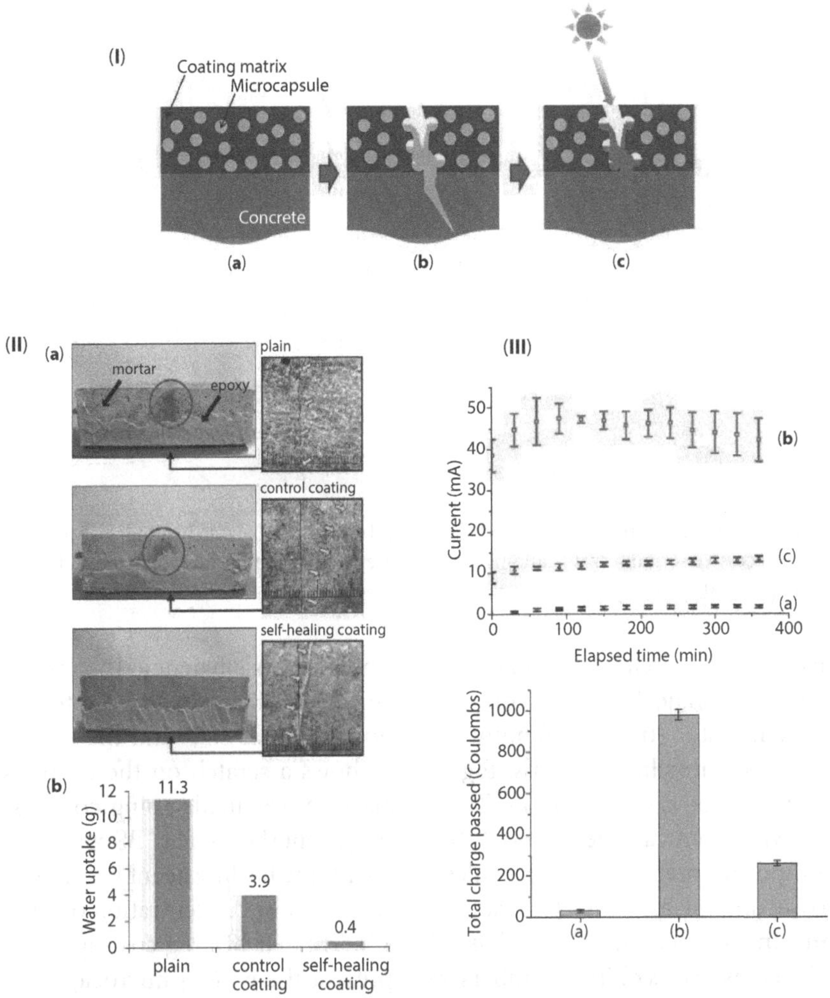

Figure 1.5 (I)—Scheme of the sunlight induced healing mechanism: the crack breaks the microcapsules and release the healing agent, which undergoes the crosslinking reaction upon exposure to sunlight. (II)—Water uptake measurements for the plain mortar, and mortars coated with the control and the self-healing coating. (III)—Chloride penetration tests. Current vs. elapsed time, and accumulated charge during 6 h for the undamaged control coating (a), scribed control coating (b) and scribed and healed self-healing coating (c). Reprinted with permission from Ref. [46]. Copyright (2013) American Chemical Society.

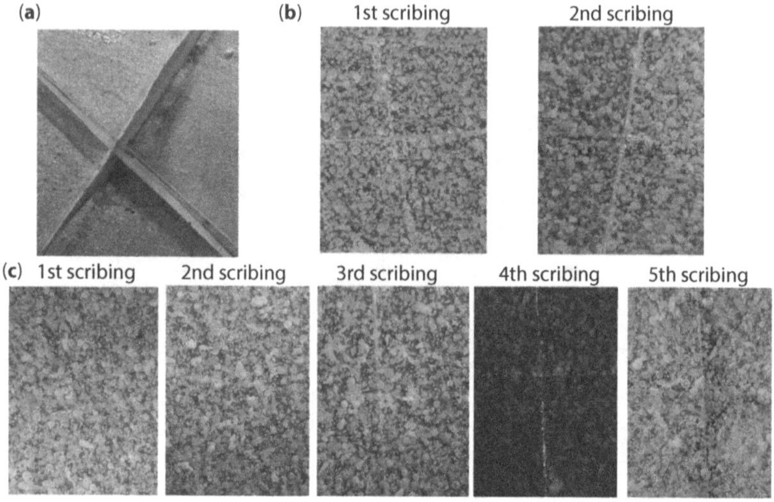

Figure 1.6 Steel substrates coated with (a) CC1, (b) CC2 and (c) self-healing coating, after successive scribing and healing sequences. Reprinted with permission from Ref. [48]; Copyright (2015) American Chemical Society.

displays a dual release mechanism that enhances its efficiency. The microencapsulated healing agent consisted in an epoxy silicone with a photosensitive initiator (triarylsulfonium hexafluorphosphate salt) and the matrix was based in silicone resins. Figure 1.7 shows a scratch on the coatings after 12 h of UV irradiation. The comparison was made using composites with microcapsules without the healing agent (labeled as "BS-xx"), and composites prepared with capsules filled with the healing agent but unable to fail and release it by UV irradiation, due to a low concentration of TiO_2 NPs in its outer shell (labeled as "CS-xx"). The self-healing coatings were labeled as "SH-xx". The numbers xx represent the wt% of microcapsules. The effect of the healing agent released within the crack is very clear, and for a microcapsules load of 60 wt% the healing seems to be excellent.

Some drawbacks of the use of microcapsules/hollow microfibers are worth to mention. Samadzadeh *et al.* [50] have mentioned some of them, including the negative side effects on the mechanical properties of the material, such as Young's modulus and ultimate stress [50, 51]. Adhesive properties can also suffer a decrease due to the presence of microcapsules [50]. In most cases a compromise between an acceptable healing with a minor deterioration of the resistance has to be reached. Additionally, there are some aspects that should not be overlooked when designing a self-healing composite based on the dispersion of microcapsules with a healing agent in a polymeric matrix. The adhesion between the capsule and

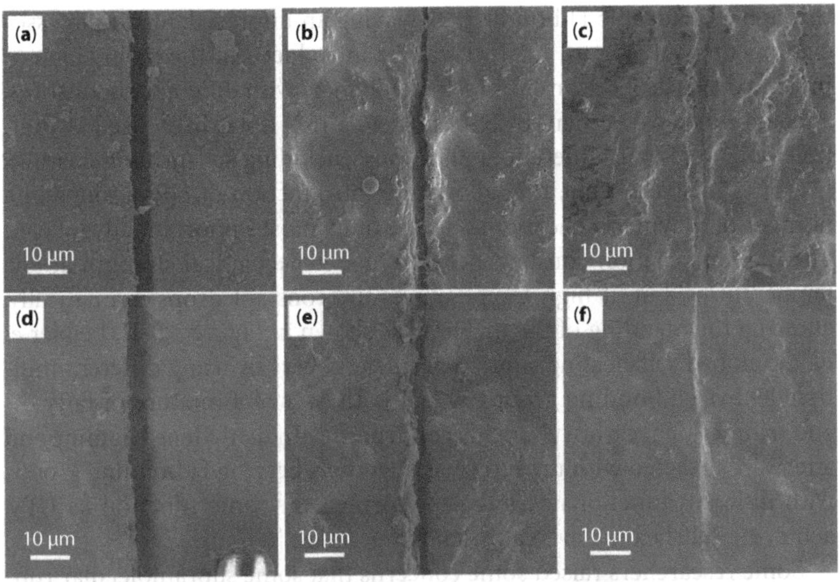

Figure 1.7 SEM images of the scratch of (a) BS-55, (b) CS-55, (c) SH-55, (d) BS-60, (e) CS-60, and (f) SH-60 after 12 h of UV irradiation (wavelength: 310 nm; power: 582 W/m2). Reprinted with permission from Ref. [49]. Copyright (2019) American Chemical Society.

the matrix plays a very important role, since it is directly related with the load transfer to the microcapsule, and to its ability to release the healing agent [22, 49]. Another disadvantage is that once the healing agent has been consumed in one or multiple repairing events, the material loses its self-healing feature. This last disadvantage is one of the most important differences in comparison with intrinsic self-healing systems, as we will show in the next section.

1.3 Intrinsic Self-Healing Polymer Coatings

The expression intrinsic self-healing polymers designates a wide variety of systems with autonomous or induced healing ability. A large portion of them are polymeric networks with either reversible or dynamic bonds, and are nowadays the most relevant kind of intrinsic self-healing thermosetting polymers with potential use as coatings. The use of reversible H-bonds to prepare self-healing supramolecular polymers received a great impulse following the first report of the synthesis and use of 2-ureido-4[1H]-pyrimidinone (UPy) for that end [52, 53]. UPy is a functional group that associates through a quadruple H bonding, displaying an unprecedented high association constant (Figure 1.8-I). It can be introduced in the main chain or in pendant ones of linear or branched polymers with different chemistries, and since its development by Sijbesma *et al.* [52] it has been used in multiple polymers with different applications, including self-healing coatings [54–56]. This versatility, that allows the introduction of UPy groups in a number of polymeric backbones, is one of its most important advantages. The high association constant provides good mechanical strength at low temperatures, but at high temperatures the constant drops dramatically, allowing the mending of the material, as shown in Figure 1.8-II. Leibler *et al.* developed self-healing supramolecular rubbers by using different multiple hydrogen bonding groups [57–59]. They used a mixture of fatty di- and trifunctional carboxylic acids condensed with diethylene triamine and afterwards reacted with urea to yield oligomers bearing H bonding groups with different functionalities that performed the same function as UPy, achieving self-healing even at room temperature [60].

Some researchers raised some concerns that some supramolecular rubbers may not display an adequate solvent resistance to be used as coatings,

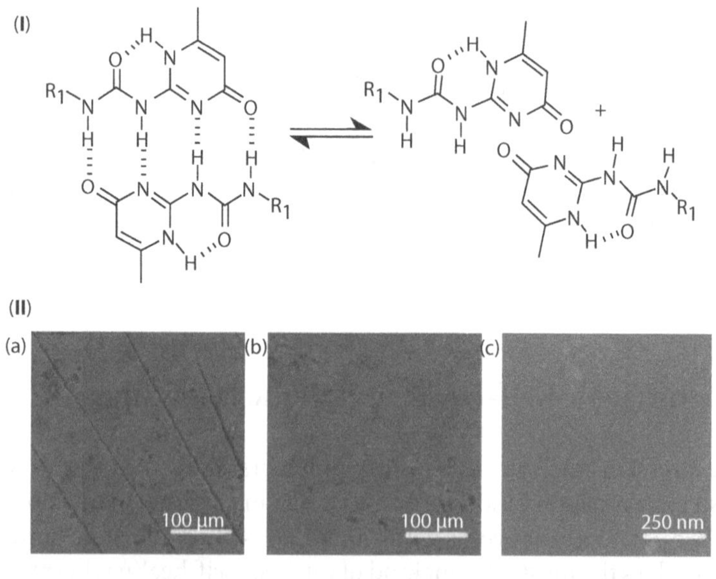

Figure 1.8 (I) 2-ureido-4[1H]-pyrimidinone (UPy), and a scheme showing its association-dissociation process. (II) Self-healing of a UPy-based coating at 80 °C for 2 min: (a) and (b) optical microscope images of the coating before and after healing, and (c) SEM image of the healed coating. Reproduced with permission from Ref. [55]; Copyright (2018) John Wiley & Sons, Inc.

and overcame this drawback by mixing supramolecular with covalent bonds [61–65]. While the introduction of covalent bonds makes a more resistant and mechanically sound material, its backside is a decrease of the self-healing capacity, needing more time and/or higher temperatures to repair a damage in the materials.

Networks built entirely from covalent bonds can also display self-healing ability, if some of the covalent bonds are of reversible nature. The use of reversible crosslinks based on the Diels–Alder (DA) chemistry (Figure 1.9-I) dates back to 1979 [66, 67]. Nonetheless, the use of DA crosslinks to produce self-healing crosslinked polymers was not explored until several years later [17, 68]. Chen and coworkers used multifunctional dienes and dienophiles to produce a reversible network that underwent depolymerization when heated to 120–150 °C [69]. Researchers from the same group improved some aspects of their materials by using a different maleimide to avoid the use of solvents and obtain colorless polymers, more suitable to use as coatings [70]. They identified the DA adducts as the responsible for the crack propagation, due to the relatively low amount of energy needed to break them [69, 70]. Therefore, it can be expected a very good degree of healing once the same covalent bonds are restored through a convenient thermal treatment. The temperature requirement to produce the retro-DA can be tuned by selecting the diene–dienophile pair, covering a wide temperature range from room temperature or mild heating

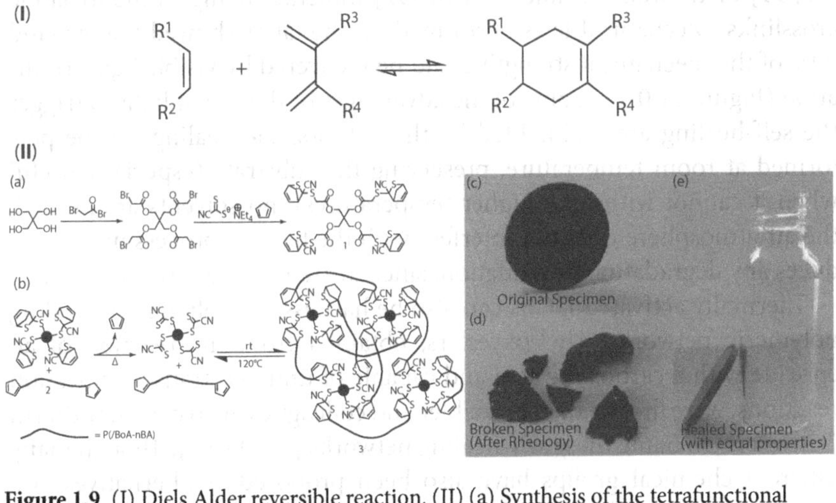

Figure 1.9 (I) Diels Alder reversible reaction. (II) (a) Synthesis of the tetrafunctional monomer (1); (b) formation of the DA-based network (3); (c)–(e) healing/recycling process. Reprinted with permission from Ref. [72]. Copyright (2014) John Wiley & Sons, Inc.

[71, 72] to above 250 °C [73, 74]. Figure 1.9-II shows an example of DA based self-healing material. One of the weaknesses of the DA-based intrinsic self-healing thermosetting polymers is that during the healing process the material loses its crosslinked structure. An alternative to overcome this drawback is the use of the dynamic equilibrium inherent to the DA adducts. Using this approach, Zhang and colleagues demonstrated plastic deformation at low temperatures in a DA-based polymer, without complete de-polymerization, but they did not offer self-healing proofs [75].

Dynamic covalent bonds (DCBs), in contrast to reversible bonds, can be used to produce self-healing intrinsic polymers that maintain their cross-linked structure even at high temperatures during the healing process [19, 76, 77]. The exchange reactions between dynamic bonds need to be activated through an external stimulus. The most usual one is heat, but some dynamic bonds respond to light, and in a lesser extent to other stimuli such as mechanical stress [78–80] or solvents [81, 82]. Light responsive DCBs based on the sulfur chemistry were extensively studied by C. Bowman's group [83–89]. The healing mechanism of these materials relies on bond exchanges through addition-fragmentation chain transfer. These polymers are capable not only of self-healing, but also of undergoing selective plastic deformation through the use of appropriate masks. This can be used to create 2D patterns on the material, as shown in Figure 1.10-I. Matyjaszewski and co-workers also used light-activated reactions to produce self-healing polymers. Their polymeric networks have either trithiocarbonate (TTC) [90, 91] or thiuram disulfide (TDS) [92] moieties acting as the dynamic crosslinks. Mechanical tests performed on the latter showed that around 90% of the mechanical strength could be recovered by visible light irradiation (Figure 11.0-II). Some of the advantages of the use of light to trigger the self-healing are highlighted by the authors: the healing can be performed at room temperature, preserving the substrate (especially useful when it cannot withstand higher temperatures); no solvents are needed; the air atmosphere does not interfere with the healing process nor it produces any degradation by oxidation, since the temperature remains low.

Thermally activated DCBs can also be harnessed to obtain self-healing polymeric networks. Hydroxyesters, obtained from the reaction of an epoxide with a carboxylic acid and capable of undergo transesterification reactions, were first studied by Leibler and colleagues, who coined the term "vitrimers" to name these self-healing networks [21, 93, 94]. Though many different chemical groups have also been proposed as alternatives [18, 19], vitrimers based on the transesterification reaction remain the most widely studied ones. A variety of epoxy and poly-carboxylic acid precursors have been used in combination with diverse catalysts. The first systems

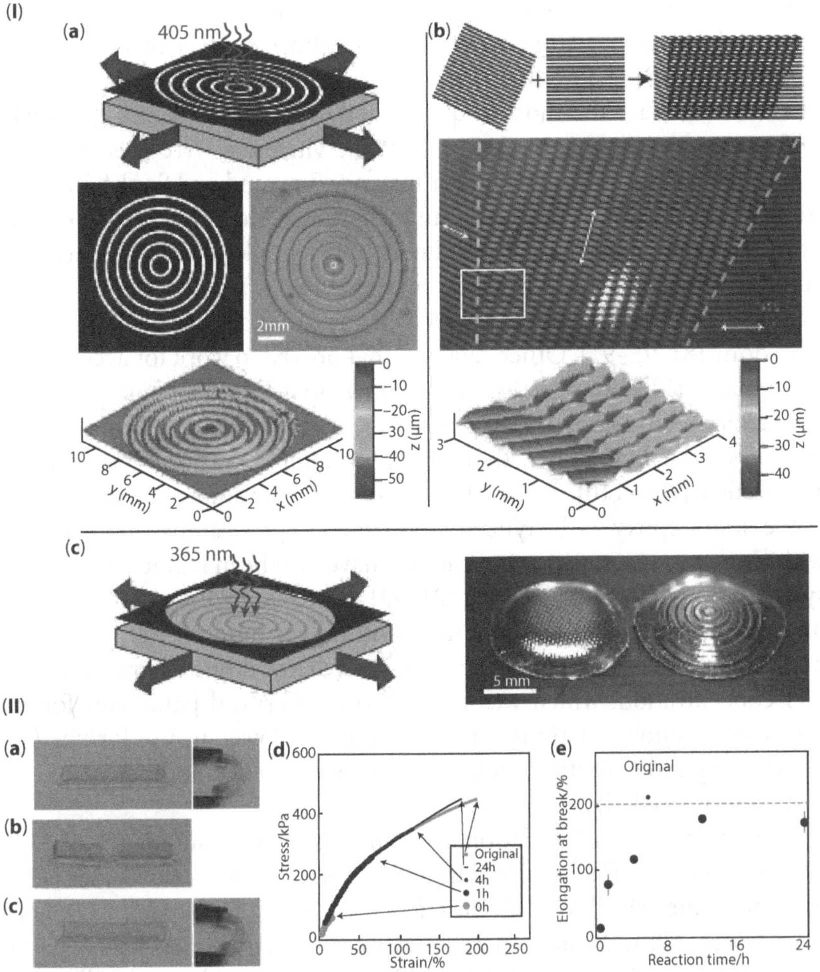

Figure 1.10 (I) Mechanopatterning of two different samples of thiol-ene based elastomers. (a) Using a mask with concentric rings. (b) Using a mask with horizontal lines in a two step process, rotating the mask 30° between the irradiation steps. Reprinted with permission from Ref. [84]. Copyright (2011) John Wiley & Sons, Inc. (II). (a)–(c) As-prepared, cut and self-healed sample of a TDS crosslinked polymer. (d) Stress–strain curves of samples with different healing times. (e) Dependence of the elongation at break with the healing time. Reprinted with permission from Ref. [92]. Copyright (2012) John Wiley & Sons, Inc.

developed by Leibler *et al.* consisted in a commercial epoxy resin (DGEBA) crosslinked with a mixture of di- and tri-functional fatty acids, with Zn^{+2} to catalyze the transesterification [21, 93]. They showed that a thermal treatment at moderate temperature (100–150 °C) produces a very good welding between two pieces of the vitrimer. Other catalysts tested by the same authors include triphenylphosphine (PPh_3) and triazobicyclodecene (TBD) [94]. TBD showed an excellent catalytic activity, similar to Zn^{+2}. TBD and Zn^{+2} are the most frequent catalysts used for epoxy-carboxylic acid vitrimers [21, 81, 93, 95–103]. These vitrimers were used to study other relevant aspects of DCBs. For instance, Legrand and Soulié-Ziakovic studied how the dynamic bonds can be harnessed to obtain a better adhesion between the matrix and glass fibers [99]. We addressed the generation of a crosslinked polymer from difunctional precursors, at the expense of leaving a fraction of soluble smaller chains [104], which several authors have done [81, 95–97]. Other catalysts that proved to work for accelerating the transesterification reactions in vitrimers include tertiary amines [105], imidazoles [106] and Sn^{+2} [107]. Also several uncatalyzed systems demonstrated self-healing ability, though higher temperatures or longer times are needed [108–110]. Figure 1.11 shows some examples of self-healing processes of epoxy-carboxylic acid vitrimers. Epoxy-anhydride networks with Zn^{+2} as transesterification catalyst have also been studied as vitrimer systems though to a lesser extent [21, 93]. Epoxy-anhydride networks have higher glass transition temperatures, and higher modulus (at room temperature) when compared to epoxy-carboxylic acid ones. However, the –OH concentration, which was proved to be a critical parameter for the transesterification reactions rate [21], is much lower in the former (the epoxy-anhydride reaction yields ester groups, hence another source of hydroxyls is needed).

A wide variety of functional groups such as disulfide [111–114], hydroxyurethanes [78], vinylogous urethanes [115–117], anhydrides [118], dioxaborolanes [119], triazolium salts [120], Schiff bases [121, 122], acylhydrazones [123], among others [124] were successfully used as dynamic crosslinks following the groundbreaking work of Leibler and colleagues, with varying degrees of healing efficiencies. Figure 1.12 shows some examples of vitrimers based on different chemistries.

An important aspect to be considered is the temperature needed to activate the exchange reactions that allows the network to flow. A characteristic temperature Tv can be defined, below which the exchange reactions are frozen, or take place only at rates low enough to be dismissed. Above Tv, the exchange reactions proceed at an appreciable rate, and the material flows, and therefore its usage should be always limited to temperatures below Tv. If Tv is low enough to enable self-healing at room temperature, then it would also be expected that the polymeric network can be easily deformed permanently at the same temperature [125]. A balance between good mechanical properties and dimensional stability at the service temperature

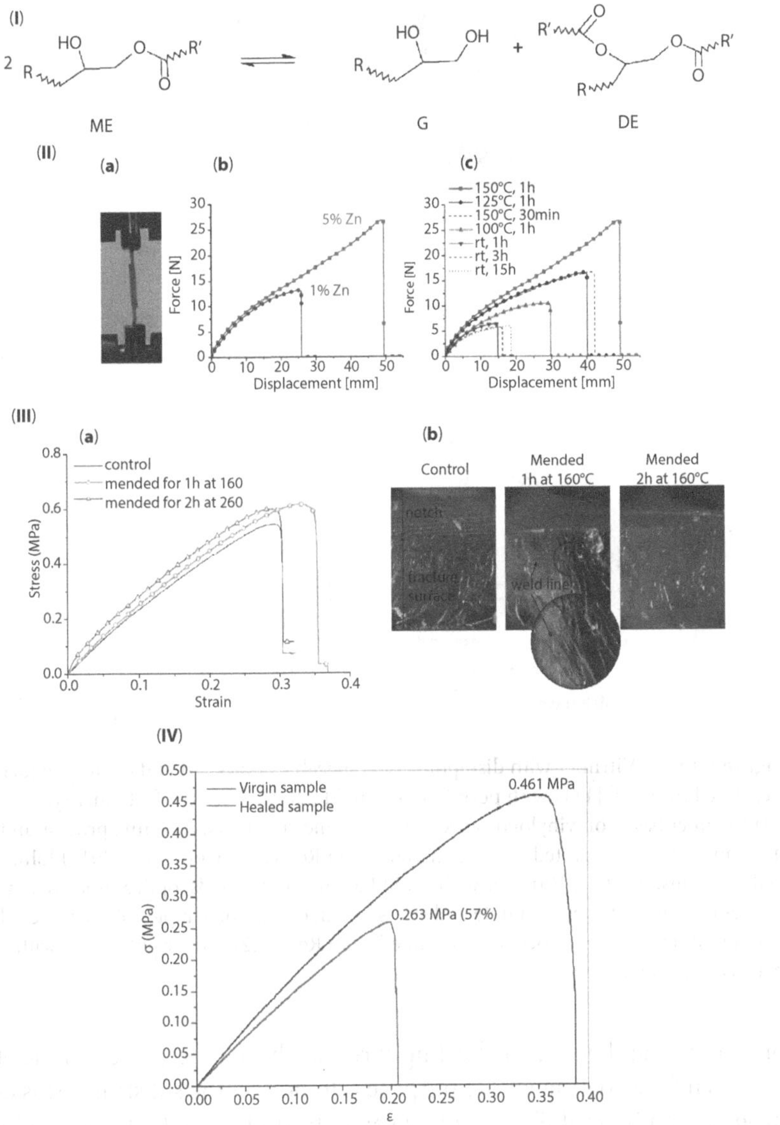

Figure 1.11 (I) Transesterification reaction between two β-hydroxyester groups. (II) Zn + 2 catalyzed DGEBA-Pripol1040 vitrimers: (a) lap-shear tests fixture; (b) force displacement curves for vitrimers with different catalyst loads. Reprinted with permission from Ref. [21]. Copyright (2012) American Chemical Society. (III) ESO-CA vitrimers without external catalyst: (a) modified lap-shear test stress–strain; (b) optical microscopy images of the fracture surfaces. Reprinted from Ref. [108] with permission from The Royal Society of Chemistry. (IV) Stress–strain curves for virgin and healed DGEBA-carboxylic acid vitrimer with tertiary amines as transesterification catalyst. Adapted from Ref. [105]; Copyright (2019) with permission from Elsevier.

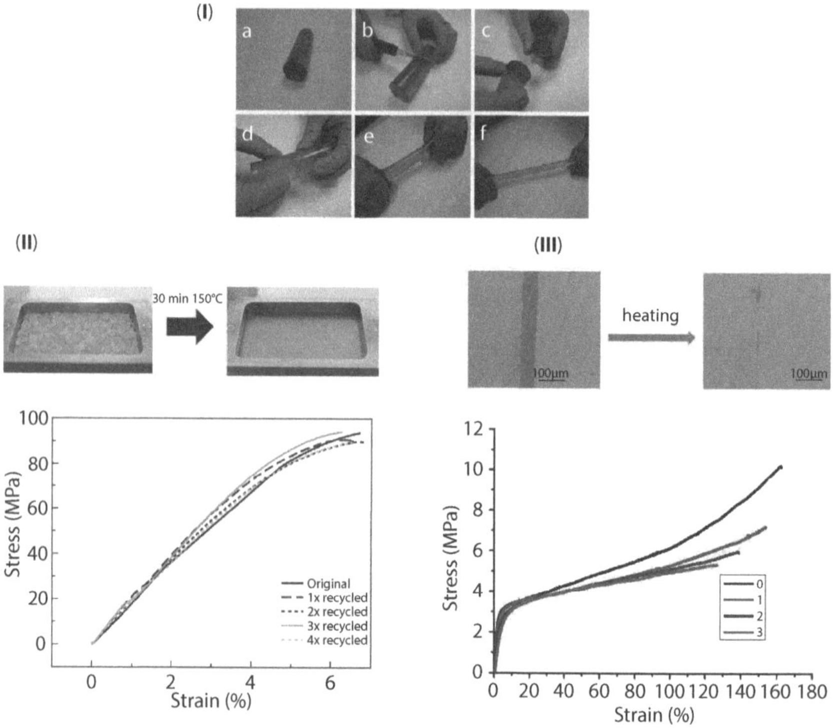

Figure 1.12 (I) Vitrimer with disulphide exchangeable bonds; cut and healing sequence. Reprinted from ref. [111] with permission from The Royal Society of Chemistry. (II) Vitrimer based on vinylogous urethane dynamic crosslinks; recycling process and mechanical tests. Reprinted with permission from Ref. [115]. Copyright (2015) John Wiley & Sons, Inc. (III) Vitrimer with Schiff base dynamic bonds; optical microscopy images of a cut and healed sample and stress–strain curves for samples of the material after multiple recycling processes. Reprinted from Ref. [122]; Copyright (2016) with permission from Elsevier.

on one side, and quick self-healing at reasonable temperatures on the other is crucial for a coating whose purpose is to protect the substrate. This compromise can be well illustrated by some disulfide based vitrimers with Tv values below room temperature [111] that also show acceptable mechanical properties. The authors claimed that the formation of quadruple H bonding between urea groups is the key to achieve such behavior, providing an additional healing mechanism, as well as crosslinks that increase the elastic modulus. Most vitrimers, however, show much higher Tv values, up to above 100 °C [18]. The activation energy of the exchange reactions (Ea) is another important parameter, which offers a measure of the sensitivity of exchange reactions rate to temperature. Typical values range from a few tens of kJ/mol to more than 120 kJ/mol. Both Tv and Ea values strongly depend on the type and amount of catalyst used in the material [94, 117].

1.4 Remote Activation of Self-Healing

Remote activation of the healing process in polymeric coatings can be performed by adding proper nanostructures to the matrix [126, 127]. Metallic nanostructures such as nanoparticles, nanorods, and nanowires [128, 129], carbon nanotubes (CNTs) [130], graphene [131] and some organic and inorganic compounds [132, 133] are known to absorb energy from electromagnetic radiation of different wavelengths and efficiently transform it into heat. This property makes them excellent candidates to produce the temperature increase needed to trigger the self-healing remotely. The mechanisms underlying the absorption of electromagnetic radiation to generate heat are out of the scope of this chapter, and will not be described here, but there are numerous articles and reviews that cover this issue, including those cited above.

Carbon nanostructures were the first to be proposed as nanoheaters. Huang *et al.* used few-layers graphene (FG) to initiate the self-healing process in thermoplastic polyurethane (TPU) [134]. The FG allowed the self-healing to be triggered by three possible methods: IR light irradiation (through the photothermal effect), an electrical current circulating through the material (resistive heating), and the application of an electromagnetic wave, in the range of the microwaves (in this case the FG act as dipoles that absorb the electromagnetic wave and generate heat through dipole distortion). The authors reported that up to 20 successive healing cycles can be obtained by IR light irradiation, with efficiencies over 99%. Recyclable composites with FG and a TPU matrix were also prepared by Fang *et al.*, who also determined that healed and recycled samples have the same conductivity as the virgin undamaged ones [135]. Graphene or graphene oxide were also incorporated to other self-healing systems such as supramolecular elastomers [136], epoxy vitrimers [137], and polyurethanes with DA reversible crosslinks [138, 139]. This permitted the usage of IR light to heat the materials, achieving not only a rapid spatially controlled self-healing, but also a spatial modulation of mechanical properties [136] and an improved mechanical performance [137]. The healing efficiencies measured for these systems were very high, typically reaching values above 90% (Figure 1.13).

CNTs were also applied to achieve indirect heating of self-healing polymeric networks. Yang and coworkers used 0.1 to 0.3 wt.% of multi-walled

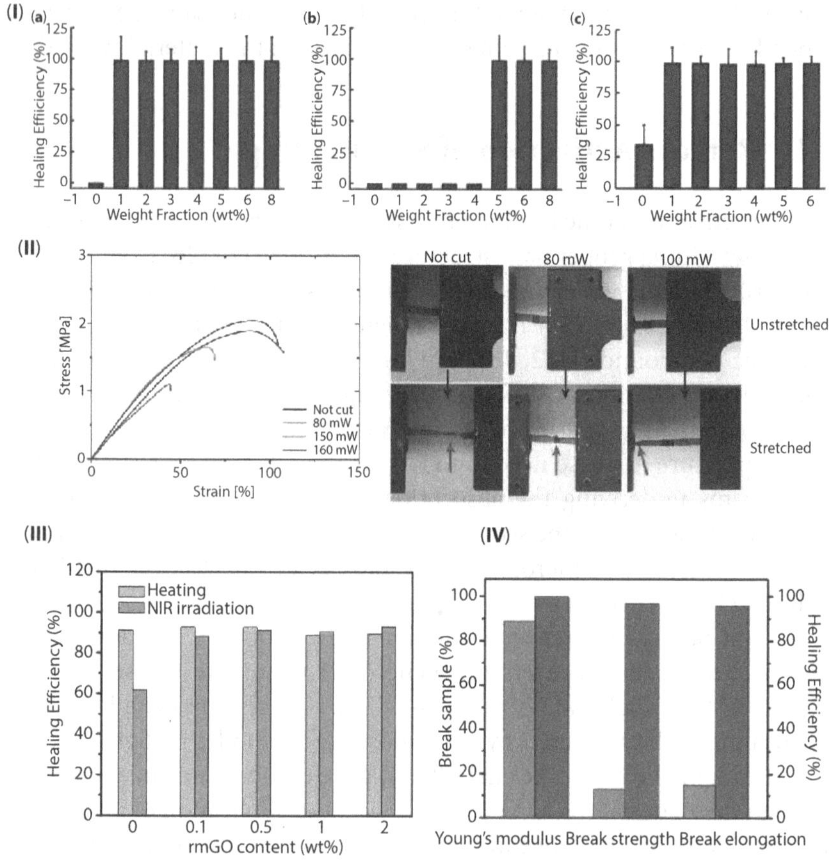

Figure 1.13 (I) Healing efficiency of a TPU with different weight fractions of FG triggered by IR light (a), an electric current (b) and an electromagnetic wave (c). Reprinted with permission from ref. [134]. Copyright (2013) John Wiley & Sons, Inc. (II) Stress-strain curves and optical photographs of tensile tests on healed supramolecular elastomers based on polyglycidols with thermally reduced graphene oxide. Reprinted with permission from ref. [136]. Copyright (2017) John Wiley & Sons, Inc. (III) Healing efficiency by direct heating and NIR irradiation on PU-graphene nanocomposites with reversible DA crosslinks. Reprinted from ref. [138]. Copyright (2018) with permission from Elsevier. (IV) Healing efficiency of a PU coating with DA reversible crosslinks with functionalized graphene nanosheets after 1 min of IR light exposure. Reprinted with permission from ref. [139]. Copyright (2019) American Chemical Society.

CNTs (MWCNTs) to trigger the self-healing process in an epoxy matrix (DGEBA with adipic acid and TBD as transesterification catalyst) with infrared light [96]. The polymeric matrix could be efficiently welded in times as short as 30 s for the highest MWCNTs content and 3 min for the lower one, with an irradiation power of 3.8 W/cm^2. MWCNTs solar light absorption was also harnessed to develop coatings with self-healing superhydrophobicity, that can be useful to generate steam and produce fresh water [140]. The coatings consist in a mixture of beewax, MWCNTs and polydimethylsiloxane (PDMS), and the superhydrophobicity can be restored thanks to the migration of beewax upon heating. The authors showed that after 20 healing cycles, the contact angle suffered only a minor decrease, from 159.3° to 155.5°. The superhydrophobicity self-healing can be achieved by direct or indirect heating—either through photothermal effect or Joule heating.

Metallic nanostructures also absorb energy from electromagnetic waves at specific wavelengths that depend on a number of variables (namely shape and size of the nanostructure, its concentration, the material of the nanostructure and that of the surroundings among others). An interesting advantage over carbon nanostructures such as CNTs and graphene is that very low concentrations of metallic nanostructures are needed. Thus transparent—most times colored—nanocomposites can be obtained, which is a very interesting feature when these materials are considered to be used as coatings. We used different amounts of gold nanoparticles (NPs) embedded in a self-healable matrix to be able to trigger the healing remotely by using a green laser ($\lambda = 532$ nm) [141]. The matrix was synthesized from epoxidized soybean oil (ESO) crosslinked with citric acid (CA), and used the β-hydroxyesters generated in the curing reaction as exchangeable bonds [108]. The Au NPs with diameters ranging from 9 to 22 nm and coated with polyvinylpyrrolidone (PVP) were added to the reacting mixture during the synthesis, and the nanocomposites showed absorption peaks centered at around 540 nm with varying intensities. Complete healing was attained when the damaged sample was irradiated with the green laser with a power density of around 1,750 mW/cm^2 for 2 h (Figure 1.14-I). An important advantage of the indirect heating through laser irradiation is that when the material is only partially fractured (i.e. there is a ligament binding both sides of the crack) the confined thermal expansion brings the crack surfaces together and contributes to an efficient healing [141, 142]. Zhang *et al.* used Au NPs embedded in a self-healing matrix of crosslinked poly ethylene oxide (PEO) [142] and in crystalline thermoplastic PEO [143] to trigger the self-healing and the shape memory function with a green laser ($\lambda = 532$ nm; up to 7,500 mW/cm^2). The addition of gold microparticles (~1,300 nm) was demonstrated as another efficient method to use visible light ($\lambda = 808$ nm) to produce the indirect heating and activate the self-healing process in a DGEBA-sebacic acid-TBD crosslinked matrix [144]. The authors observed that photothermal heating is far more efficient than direct heating to achieve a good healing (Figure 1.14-II). Similar

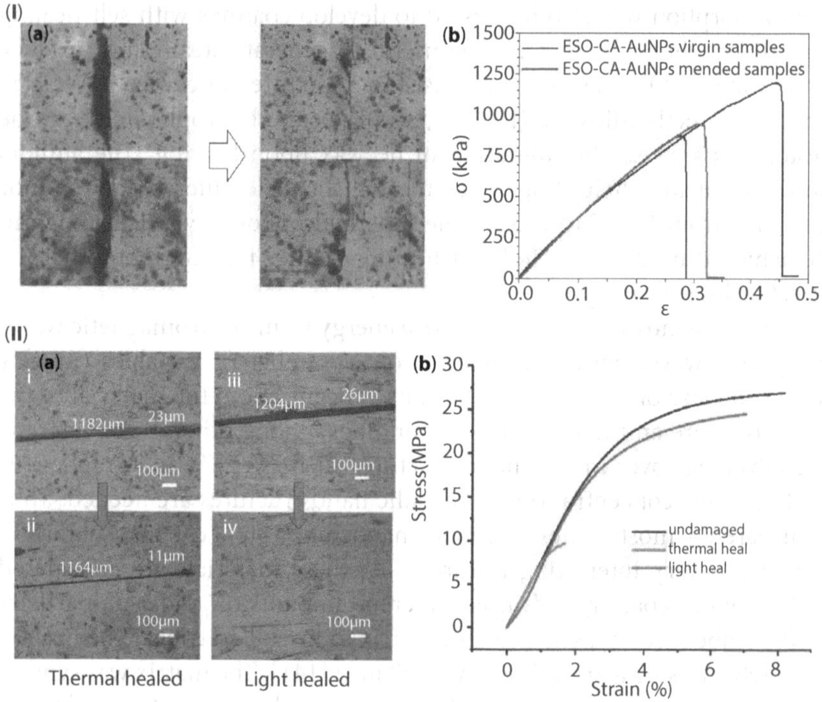

Figure 1.14 (I). (a)–(b) Optical microscopy images showing a healed sample, and stress–strain curves for the virgin and healed samples. (II) (a)-(b) Optical microscopy images of a crack healed through direct heating and by light irradiation, and the corresponding stress–strain curves for a DGEBA–sebacic acid–TBD vitrimer with gold microparticles. Reproduced with permission from Ref. [144].

findings were made in other self-healable polymers, including those with CNTs [96], showing one of the advantages of the indirect heating through photothermal effect.

Anisotropic 2- and 3-dimensional nanostructures have also been found useful to induce the needed indirect heating. These nanostructures usually display more than one surface plasmon resonance frequency, thus being capable of act as heaters when illuminated with light sources of several wavelengths [128]. Silver nanowires (NWs) were introduced within polymeric films based on polycaprolactone (PCL) and poly(vinyl alcohol) (PVA) [145]. Illumination with infrared light produced the indirect heating needed to heal the damaged film, and led to a complete recovery of its conductivity, which is also enabled by the presence of the Ag NWs. Figure 1.15 shows the experiment demonstrating the self-healing. Chen *et al.* used

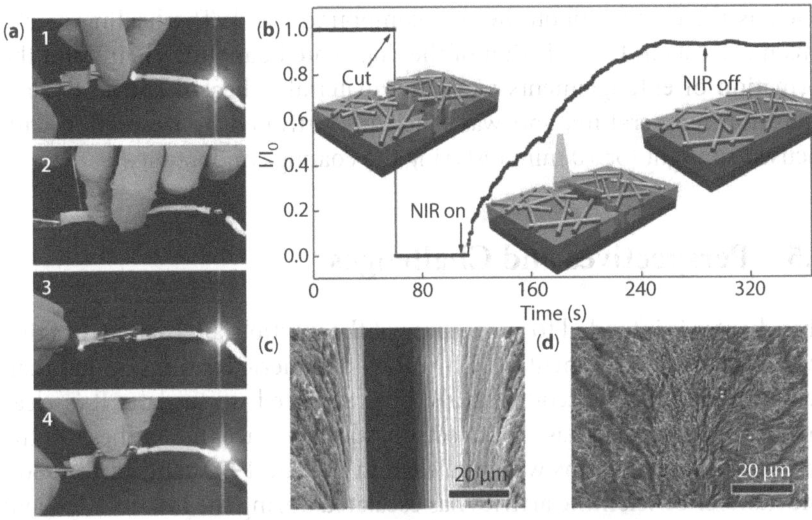

Figure 1.15 (a) Pictures of the healing process of a PCL/PLA film containing Ag NWs: (1) as-prepared film; (2) cut film; (3) film being irradiated; (4) healed film being subjected to repeated bending. (b) Current changes in the nanocomposite film during a cutting/healing process. (c, d) SEM images of a cut PCL/PLA/Ag NWs film, (c) before and (d) after being healed. Reproduced with permission from Ref. [145].

a combination of GO, Au NPs and Ag NWs to induce the self-healing process in a PCL based conductive membrane through irradiation with a green laser (λ = 808 nm) [146]. In this case, the healing was achieved thanks to the interdifussion of the electrospun PCL fibers, allowed by the heating produced by the photothermal effect, and reaching high efficiency levels (91.3% of its conductivity and 90.7% of its tensile strength were recovered after one cut-healing cycle).

Finally, several inorganic and organic compounds capable of absorbing energy in the form of light and transforming it into heat are known. Burnworth *et al.* described the self-healing process of a metallosupramolecular when irradiated with UV light thanks to the disengagement of the metal-ligand motifs [147]. In a smart approach, Chen and coworkers introduced an aniline trimer within an epoxy vitrimer matrix (DGEBA-adipic acid) and showed that it was capable of absorbing light (λ = 808 nm), increasing the temperature of the vitrimer and triggering its self-healing process [148]. Similarly, Fang and coworkers used sunlight focused to trigger the self-healing process in an epoxy resin (DGEBA) cured with a mixture of a diamine (m-xylylenediamine, MXDA) and a fluorinated monoamine (4-(heptadecafluorooctyl)aniline, HFOA), and with aniline black as the organic photothermal compound [149]. The healing mechanism consists in the diffusion of the fluorinated dangling chains and the formation of entanglements when the material is heated above its glass transition temperature, and was efficiently activated by irradiation with focused sunlight for 10 min, restoring the coating conductivity.

1.5 Perspectives and Challenges

This chapter is intended to show some of the multiple alternatives existing regarding the development of self-healing polymeric coatings, rather than being an exhaustive account of all those reported up to date. It is clear that the research field has advanced substantially since the synthesis of the first self-healing systems was published in the late 1970s and early 1980s. The number of scientific articles has escalated during the last decades, and also the number of patents filled for self-healing protective coatings and films has increased notoriously. However, commercially available products based on self-healing polymers are very scarce, probably due to some weaknesses that need to be addressed in the near future. Possibly the most important shortcoming of self-healing polymers is the need of maintaining the fracture surfaces in close contact to produce a good mending of the pieces. When these materials are used in real applications, this is an important aspect to which look after. Though it can be overcome (at least to a considerable extent) thanks to the localized indirect heating allowed by the presence of nanostructures [141, 142] or to the substrate that holds the two parts in place, it should be kept in mind that most tests on self-healing polymers are carried out in controlled environments, very different from those that the materials face during their usage.

Besides this, the changes that are introduced in the formulations in order to gain self-healing ability are very often deleterious to other properties, resulting in materials with lower Tg values and mechanical properties. Unfortunately, a high Tg value hinders a good healing efficiency in intrinsic self-healing polymers, by freezing the molecular motions that are needed to weld the polymers, whether it is achieved through interdiffusion of polymeric chains or reactions through dynamic covalent bonds. Extrinsic self-healing polymers can maintain their properties after the addition of the micro vessels containing the healing agent, but the properties of the material filling the crack are not the same as those of the original one. The cost associated to many of the raw materials needed to synthesize self-healing coatings, as well as to their processing, often complex and involving multiple steps is another disadvantage. Significant efforts will need to be devoted to find more economic and available materials, and to develop formulations with simpler processing techniques in some cases. Avoiding external catalysts is another aspect to be addressed. External catalysts can be leached from the polymeric matrix over time, preventing the self-healing response to take place and contributing to the environmental pollution which, depending on the nature of the catalyst, can be a severe problem. The inclusion of chemical structures such as cromophores or particles with functional properties—other than the photothermal effect described above—that could be used to sense and detect microcracks within the polymer would be a great leap forward. This has a special relevance considering that the healing efficiency is far higher for mending microcracks than for repairing larger failures (e.g. side-to-side cracks). Hence, if the fracture is detected at an early stage, the repairing will be much better. Some research groups have proposed to use the local increase in the resistance when a crack is generated to both detect and repair the crack through an electric current [150]. This approach still needs from an active role to screen the material for damages, but it is however a very appealing alternative to both detect and repair otherwise undetectable cracks.

Finally, the formulations for self-healing coatings should be able to be processed by some of the methods currently used at industrial level, and not only at a laboratory scale without the need for expensive transformations or new technologies. Beside its use as coatings, some of the systems described herein can find very interesting applications on other areas. For instance, self-healing polymers based on DCBs are also very good candidates to be processed through 3D printing thanks to their moldable macroscopic structure offering them the possibility to flow [81, 103]. This is a challenging aspect that also needs to be addressed, and involves tuning the material rheological properties and the final mechanical and thermal performance. Overall, in view of the most recent advances on self-healing polymeric coatings, it seems that the next years this research field will continue growing steadily, and reach end-user applications in a near future.

References

1. Wool, R.P., Material response and reversible cracks in viscoelastic polymers. *Polym. Eng. Sci.*, 18, 14, 1057–1061, 1978.
2. Jud, K., Kausch, H.H., Williams, J.G., Fracture mechanics studies of crack healing and welding of polymers. *J. Mater. Sci.*, 16, 1, 204–210, 1981.

3. Dry, C.M. and Sottos, N.R., Passive smart self-repair in polymer matrix composite materials. Proc. SPIE 1916, Smart Structures and Materials 1993: Smart Materials, 438–444, https://doi.org/10.1117/12.148501
4. Dry, C.M. and McMillan, W., Crack and damage assessment in concrete and polymer matrices using liquids released internally from hollow optical fibers. Proc. SPIE 2718, Smart Structures and Materials 1996: Smart Sensing, Processing, and Instrumentation, 448–451, 1996.
5. Pascault, J.-P., Sautereau, H., Verdu, J., Williams, R.J.J., *Thermosetting Polymers*, Marcel Dekker, New York, 2002.
6. Brazel, C.S. and Rosen, S.L., *Fundamental principles of polymeric materials*, Wiley, Hoboken, New Jersey, 2012.
7. Francis, R. (Ed.), *Recycling of Polymers: Methods, Characterization and Applications*, Wiley-VCH Verlag GmbH & Co. KGaA, Weinheim, Germany, 2016.
8. World Economic Forum's Global Agenda Council on Emerging Technologies, The top 10 emerging technologies for 2013, https://www.weforum.org/agenda/2013/02/top-10-emerging-technologies-for-2013/, 2013.
9. World Economic Forum's Meta-Council on Emerging Technologies, The top 10 emerging technologies for 2015, https://www.weforum.org/agenda/2015/03/top-10-emerging-technologies-of-2015-2/, 2015.
10. Blaiszik, B.J., Kramer, S.L.B., Olugebefola, S.C., Moore, J.S., Sottos, N.R., White, S.R., Self-Healing Polymers and Composites. *Annu. Rev. Mater. Res.*, 40, 1, 179–211, 2010.
11. Billiet, S., Hillewaere, X.K.D., Teixeira, R.F.A., Du Prez, F.E., Chemistry of Crosslinking Processes for Self-Healing Polymers. *Macromol. Rapid Commun.*, 34, 4, 290–309, 2013.
12. White, S.R., Sottos, N.R., Geubelle, P.H., Moore, J.S., Kessler, M.R., Sriram, S.R., Brown, E.N., Viswanathan, S., Autonomic healing of polymer composites. *Nature*, 409, 6822, 794–797, 2001.
13. Kessler, M.R., Sottos, N.R., White, S.R., Self-healing structural composite materials. *Compos. Part Appl. Sci. Manuf.*, 34, 8, 743–753, 2003.
14. Wu, D.Y., Meure, S., Solomon, D., Self-healing polymeric materials: A review of recent developments. *Prog. Polym. Sci.*, 33, 5, 479–522, 2008.
15. Luo, X., Ou, R., Eberly, D.E., Singhal, A., Viratyaporn, W., Mather, P.T., A Thermoplastic/Thermoset Blend Exhibiting Thermal Mending and Reversible Adhesion. *ACS Appl. Mater. Interfaces*, 1, 3, 612–620, 2009.
16. Hayes, S.A., Jones, F.R., Marshiya, K., Zhang, W., A self-healing thermosetting composite material. *Compos. Part Appl. Sci. Manuf.*, 38, 4, 1116–1120, 2007.
17. Liu, Y.-L. and Chuo, T.-W., Self-healing polymers based on thermally reversible Diels–Alder chemistry. *Polym. Chem.*, 4, 7, 2194, 2013.
18. Denissen, W., Winne, J.M., Du Prez, F.E., Vitrimers: permanent organic networks with glass-like fluidity. *Chem. Sci.*, 7, 1, 30–38, 2016.
19. Roy, N., Bruchmann, B., Lehn, J.-M., DYNAMERS: Dynamic polymers as self-healing materials. *Chem. Soc. Rev.*, 44, 11, 3786–3807, 2015.

20. Ishida, K. and Yoshie, N., Synthesis of Readily Recyclable Biobased Plastics by Diels–Alder Reaction. *Macromol. Biosci.*, 8, 10, 916–922, 2008.
21. Capelot, M., Montarnal, D., Tournilhac, F., Leibler, L., Metal-Catalyzed Transesterification for Healing and Assembling of Thermosets. *J. Am. Chem. Soc.*, 134, 18, 7664–7667, 2012.
22. Mauldin, T.C. and Kessler, M.R., Self-healing polymers and composites. *Int. Mater. Rev.*, 55, 6, 317–346, 2010.
23. Hansen, C.J., Wu, W., Toohey, K.S., Sottos, N.R., White, S.R., Lewis, J.A., Self-Healing Materials with Interpenetrating Microvascular Networks. *Adv. Mater.*, 21, 41, 4143–4147, 2009.
24. Brown, E.N., Sottos, N.R., White, S.R., Fracture testing of a self-healing polymer composite. *Exp. Mech.*, 42, 4, 372–379, 2002.
25. Patel, A.J., Sottos, N.R., Wetzel, E.D., White, S.R., Autonomic healing of low-velocity impact damage in fiber-reinforced composites. *Compos. Part Appl. Sci. Manuf.*, 41, 3, 360–368, 2010.
26. Rule, J.D., Sottos, N.R., White, S.R., Effect of microcapsule size on the performance of self-healing polymers. *Polymer*, 48, 12, 3520–3529, 2007.
27. Cho, S.H., Andersson, H.M., White, S.R., Sottos, N.R., Braun, P.V., Polydimethylsiloxane-Based Self-Healing Materials. *Adv. Mater.*, 18, 8, 997–1000, 2006.
28. Keller, M.W., White, S.R., Sottos, N.R., A Self-Healing Poly(Dimethyl Siloxane) Elastomer. *Adv. Funct. Mater.*, 17, 14, 2399–2404, 2007.
29. Keller, M.W. and Sottos, N.R., Mechanical Properties of Microcapsules Used in a Self-Healing Polymer. *Exp. Mech.*, 46, 6, 725–733, 2006.
30. Kumudinie, C. and Mark, J.E., Tearing energies for *in-situ* reinforced poly(dimethylsiloxane) networks. *Mater. Sci. Eng. C*, 11, 1, 61–66, 2000.
31. Yuan, Q.W. and Mark, J.E., Reinforcement of poly(dimethylsiloxane) networks by blended and *in-situ* generated silica fillers having various sizes, size distributions, and modified surfaces. *Macromol. Chem. Phys.*, 200, 206–220, 1999.
32. Kim, D.-M., Cho, Y.-J., Choi, J.-Y., Kim, B.-J., Jin, S.-W., Chung, C.-M., Low-Temperature Self-Healing of a Microcapsule-Type Protective Coating. *Materials*, 10, 9, 1079, 2017.
33. Yin, T., Rong, M., Zhang, M., Yang, G., Self-healing epoxy composites—Preparation and effect of the healant consisting of microencapsulated epoxy and latent curing agent. *Compos. Sci. Technol.*, 67, 2, 201–212, 2007.
34. Yin, T., Rong, M.Z., Zhang, M.Q., Self-Healing of Cracks in Epoxy Composites. *Adv. Mater. Res.*, 47–50, 282–285, 2008.
35. Pascu, O., Garcia-Valls, R., Giamberini, M., Interfacial polymerization of an epoxy resin and carboxylic acids for the synthesis of microcapsules. *Polym. Int.*, 57, 8, 995–1006, 2008.
36. Guadagno, L., Raimondo, M., Naddeo, C., Longo, P., Mariconda, A., Binder, W.H., Healing efficiency and dynamic mechanical properties of self-healing epoxy systems. *Smart Mater. Struct.*, 23, 4, 045001, 2014.

37. Guadagno, L., Raimondo, M., Vietri, U., Naddeo, C., Stojanovic, A., Sorrentino, A., Binder, W.H., Evaluation of the Mechanical Properties of Microcapsule-Based Self-Healing Composites. *Int. J. Aerosp. Eng.*, 2016, 1–10, 2016.
38. Lee Hia, I., Chan, E.-S., Chai, S.-P., Pasbakhsh, P., A novel repeated self-healing epoxy composite with alginate multicore microcapsules. *J. Mater. Chem. A*, 6, 18, 8470–8478, 2018.
39. Yang, J., Keller, M.W., Moore, J.S., White, S.R., Sottos, N.R., Microencapsulation of Isocyanates for Self-Healing Polymers. *Macromolecules*, 41, 24, 9650–9655, 2008.
40. Huang, M. and Yang, J., Facile microencapsulation of HDI for self-healing anticorrosion coatings. *J. Mater. Chem.*, 21, 30, 11123, 2011.
41. He, Z., Jiang, S., An, N., Li, X., Li, Q., Wang, J., Zhao, Y., Kang, M., Self-healing isocyanate microcapsules for efficient restoration of fracture damage of polyurethane and epoxy resins. *J. Mater. Sci.*, 54, 11, 8262–8275, 2019.
42. Brey Gil, C.S., Patricio, P.S., Oliveira, L.C., Oréfice, R.L., Improved self-healing properties of collagen using polyurethane microcapsules containing reactive diisocyanate: Improved self-healing properties of collagen. *Polym. Int.*, 65, 6, 721–727, 2016.
43. Hillewaere, X.K.D. and Du Prez, F.E., Fifteen chemistries for autonomous external self-healing polymers and composites. *Prog. Polym. Sci.*, 49–50, 121–153, 2015.
44. Di Credico, B., Levi, M., Turri, S., An efficient method for the output of new self-repairing materials through a reactive isocyanate encapsulation. *Eur. Polym. J.*, 49, 9, 2467–2476, 2013.
45. Wang, W., Xu, L., Li, X., Lin, Z., Yang, Y., An, E., Self-healing mechanisms of water triggered smart coating in seawater. *J. Mater. Chem. A*, 2, 6, 1914–1921, 2014.
46. Song, Y.-K., Jo, Y.-H., Lim, Y.-J., Cho, S.-Y., Yu, H.-C., Ryu, B.-C., Lee, S.-I., Chung, C.-M., Sunlight-Induced Self-Healing of a Microcapsule-Type Protective Coating. *ACS Appl. Mater. Interfaces*, 5, 4, 1378–1384, 2013.
47. Khalaj Asadi, A., Ebrahimi, M., Mohseni, M., Microencapsulation of a sunlight-curable silicon-based resin in the presence of polyvinylpyrrolidone. *Pigm. Resin Technol.*, 47, 3, 272–278, 2018.
48. Gao, L., He, J., Hu, J., Wang, C., Photoresponsive Self-Healing Polymer Composite with Photoabsorbing Hybrid Microcapsules. *ACS Appl. Mater. Interfaces*, 7, 45, 25546–25552, 2015.
49. Zhu, Y., Cao, K., Chen, M., Wu, L., Synthesis of UV-Responsive Self-Healing Microcapsules and Their Potential Application in Aerospace Coatings. *ACS Appl. Mater. Interfaces*, 11, 36, 33314–33322, 2019.
50. Samadzadeh, M., Boura, S.H., Peikari, M., Kasiriha, S.M., Ashrafi, A., A review on self-healing coatings based on micro/nanocapsules. *Prog. Org. Coat.*, 68, 3, 159–164, 2010.
51. Tripathi, M., Rahamtullah, Kumar, D., Rajagopal, C., Kumar Roy, P., Influence of microcapsule shell material on the mechanical behavior of

epoxy composites for self-healing applications. *J. Appl. Polym. Sci.*, 131, 15, 40572, 2014.
52. Sijbesma, R.P., Beijer, F.H., Brunsveld, L., Folmer, B.J.B., Hirschberg, J.H.K.K., Lange, R.F.M., Lowe, J.K.L., Meijer, E.W., Reversible Polymers Formed from Self-Complementary Monomers Using Quadruple Hydrogen Bonding. *Science*, 278, 5343, 1601–1604, 1997.
53. Bosman, A.W., Sijbesma, R.P., Meijer, E.W., Supramolecular polymers at work. *Mater. Today*, 7, 4, 34–39, 2004.
54. Deflorian, F., Rossi, S., Scrinzi, E., Self-healing supramolecular polyurethane coatings: Preliminary study of the corrosion protective properties. *Corros. Eng. Sci. Technol.*, 48, 2, 147–154, 2013.
55. Wang, X., Li, Y., Qian, Y., Qi, H., Li, J., Sun, J., Mechanically Robust Atomic Oxygen-Resistant Coatings Capable of Autonomously Healing Damage in Low Earth Orbit Space Environment. *Adv. Mater.*, 30, 36, 1803854, 2018.
56. Heinzmann, C., Lamparth, I., Rist, K., Moszner, N., Fiore, G.L., Weder, C., Supramolecular Polymer Networks Made by Solvent-Free Copolymerization of a Liquid 2-Ureido-4[1*H*]-pyrimidinone Methacrylamide. *Macromolecules*, 48, 22, 8128–8136, 2015.
57. Cordier, P., Tournilhac, F., Soulié-Ziakovic, C., Leibler, L., Self-healing and thermoreversible rubber from supramolecular assembly. *Nature*, 451, 7181, 977–980, 2008.
58. Montarnal, D., Tournilhac, F., Hidalgo, M., Couturier, J.-L., Leibler, L., Versatile One-Pot Synthesis of Supramolecular Plastics and Self-Healing Rubbers. *J. Am. Chem. Soc.*, 131, 23, 7966–7967, 2009.
59. Montarnal, D., Cordier, P., Soulié-Ziakovic, C., Tournilhac, F., Leibler, L., Synthesis of self-healing supramolecular rubbers from fatty acid derivatives, diethylene triamine, and urea. *J. Polym. Sci. Part Polym. Chem.*, 46, 24, 7925–7936, 2008.
60. Mynar, J.L. and Aida, T., The gift of healing. *Nature*, 451, 7181, 895–896, 2008.
61. Sordo, F., Mougnier, S.-J., Loureiro, N., Tournilhac, F., Michaud, V., Design of Self-Healing Supramolecular Rubbers with a Tunable Number of Chemical Cross-Links. *Macromolecules*, 48, 13, 4394–4402, 2015.
62. Khor, S.P., Varley, R.J., Shen, S.Z., Yuan, Q., Thermo-reversible healing in a crosslinked polymer network containing covalent and thermo-reversible bonds. *J. Appl. Polym. Sci.*, 128, 6, 3743–3750, 2013.
63. Montarnal, D., Tournilhac, F., Hidalgo, M., Leibler, L., Epoxy-based networks combining chemical and supramolecular hydrogen-bonding cross-links. *J. Polym. Sci. Part Polym. Chem.*, 48, 5, 1133–1141, 2010.
64. Liu, L., Zhu, L., Zhang, L., A Solvent-Resistant and Biocompatible Self-Healing Supramolecular Elastomer with Tunable Mechanical Properties. *Macromol. Chem. Phys.*, 219, 4, 1700409, 2018.
65. Altuna, F.I., Casado, U., dell'Erba, I.E., Luna, L., Hoppe, C.E., Williams, R.J.J., Epoxy vitrimers incorporating physical crosslinks produced by self-association of alkyl chains. *Polym. Chem.*, 11, 7, 1337–1347, 2020.

66. Kennedy, J.P. and Castner, K.F., Thermally reversible polymer systems by cyclopentadienylation. I. A model for termination by cyclopentadienylation of olefin polymerization. *J. Polym. Sci. Polym. Chem. Ed.*, 17, 7, 2039–2054, 1979.
67. Stevens, M.P. and Jenkins, A.D., Crosslinking of polystyrene via pendant maleimide groups. *J. Polym. Sci. Polym. Chem. Ed.*, 17, 11, 3675–3685, 1979.
68. Khan, N.I., Halder, S., Gunjan, S.B., Prasad, T., A review on Diels-Alder based self-healing polymer composites. *IOP Conf. Ser. Mater. Sci. Eng.*, 377, 012007, 2018.
69. Chen, X., A Thermally Re-mendable Cross-Linked Polymeric Material. *Science*, 295, 5560, 1698–1702, 2002.
70. Chen, X., Wudl, F., Mal, A.K., Shen, H., Nutt, S.R., New Thermally Remendable Highly Cross-Linked Polymeric Materials. *Macromolecules*, 36, 6, 1802–1807, 2003.
71. Reutenauer, P., Buhler, E., Boul, P.J., Candau, S.J., Lehn, J.-M., Room Temperature Dynamic Polymers Based on Diels-Alder Chemistry. *Chem.—Eur. J.*, 15, 8, 1893–1900, 2009.
72. Oehlenschlaeger, K.K., Mueller, J.O., Brandt, J., Hilf, S., Lederer, A., Wilhelm, M., Graf, R., Coote, M.L., Schmidt, F.G., Barner-Kowollik, C., Adaptable Hetero Diels–Alder Networks for Fast Self-Healing under Mild Conditions. *Adv. Mater.*, 26, 21, 3561–3566, 2014.
73. Grigoras, M. and Colotin, G., Copolymerization of a bisanthracene compound with bismaleimides by Diels–Alder cycloaddition. *Polym. Int.*, 50, 12, 1375–1378, 2001.
74. Jones, J.R., Liotta, C.L., Collard, D.M., Schiraldi, D.A., Cross-Linking and Modification of Poly(ethylene terephthalate-*co*-2,6-anthracenedicarboxylate) by Diels–Alder Reactions with Maleimides. *Macromolecules*, 32, 18, 5786–5792, 1999.
75. Zhang, G., Zhao, Q., Yang, L., Zou, W., Xi, X., Xie, T., Exploring Dynamic Equilibrium of Diels–Alder Reaction for Solid State Plasticity in Remoldable Shape Memory Polymer Network. *ACS Macro Lett.*, 5, 7, 805–808, 2016.
76. Chakma, P. and Konkolewicz, D., Dynamic Covalent Bonds in Polymeric Materials. *Angew. Chem. Int. Ed.*, 58, 29, 9682–9695, 2019.
77. Winne, J.M., Leibler, L., Du Prez, F.E., Dynamic covalent chemistry in polymer networks: A mechanistic perspective. *Polym. Chem.*, 10, 45, 6091–6108, 2019.
78. Fortman, D.J., Brutman, J.P., Cramer, C.J., Hillmyer, M.A., Dichtel, W.R., Mechanically Activated, Catalyst-Free Polyhydroxyurethane Vitrimers. *J. Am. Chem. Soc.*, 137, 44, 14019–14022, 2015.
79. Black, A.L., Lenhardt, J.M., Craig, S.L., From molecular mechanochemistry to stress-responsive materials. *J. Mater. Chem.*, 21, 6, 1655–1663, 2011.
80. Tee, B.C.-K., Wang, C., Allen, R., Bao, Z., An electrically and mechanically self-healing composite with pressure- and flexion-sensitive properties for electronic skin applications. *Nat. Nanotechnol.*, 7, 825, 2012.

81. Yang, Y., Terentjev, E.M., Wei, Y., Ji, Y., Solvent-assisted programming of flat polymer sheets into reconfigurable and self-healing 3D structures. *Nat. Commun.*, 9, 1, 1906, 2018.
82. Shi, Q., Yu, K., Dunn, M.L., Wang, T., Qi, H.J., Solvent Assisted Pressure-Free Surface Welding and Reprocessing of Malleable Epoxy Polymers. *Macromolecules*, 49, 15, 5527–5537, 2016.
83. Scott, T.F., Schneider, A.D., Cook, W.D., Bowman, C.N., Photoinduced Plasticity in Cross-Linked Polymers. *Science*, 308, 5728, 1615–1617, 2005.
84. Kloxin, C.J., Scott, T.F., Park, H.Y., Bowman, C.N., Mechanophotopatterning on a Photoresponsive Elastomer. *Adv. Mater.*, 23, 17, 1977–1981, 2011.
85. Kloxin, C.J., Scott, T.F., Bowman, C.N., Stress Relaxation via Addition–Fragmentation Chain Transfer in a Thiol-ene Photopolymerization. *Macromolecules*, 42, 7, 2551–2556, 2009.
86. Worrell, B.T., Mavila, S., Wang, C., Kontour, T.M., Lim, C.-H., McBride, M.K., Musgrave, C.B., Shoemaker, R., Bowman, C.N., A user's guide to the thiol-thioester exchange in organic media: Scope, limitations, and applications in material science. *Polym. Chem.*, 9, 36, 4523–4534, 2018.
87. Chatani, S., Kloxin, C.J., Bowman, C.N., The power of light in polymer science: Photochemical processes to manipulate polymer formation, structure, and properties. *Polym. Chem.*, 5, 7, 2187–2201, 2014.
88. Wang, C., Goldman, T.M., Worrell, B.T., McBride, M.K., Alim, M.D., Bowman, C.N., Recyclable and repolymerizable thiol–X photopolymers. *Mater. Horiz.*, 5, 6, 1042–1046, 2018.
89. Huck, W.T.S., Polymer networks take a bow. *Nature*, 472, 7344, 425–426, 2011.
90. Amamoto, Y., Kamada, J., Otsuka, H., Takahara, A., Matyjaszewski, K., Repeatable Photoinduced Self-Healing of Covalently Cross-Linked Polymers through Reshuffling of Trithiocarbonate Units. *Angew. Chem. Int. Ed.*, 50, 7, 1660–1663, 2011.
91. Amamoto, Y., Otsuka, H., Takahara, A., Matyjaszewski, K., Changes in Network Structure of Chemical Gels Controlled by Solvent Quality through Photoinduced Radical Reshuffling Reactions of Trithiocarbonate Units. *ACS Macro Lett.*, 1, 4, 478–481, 2012.
92. Amamoto, Y., Otsuka, H., Takahara, A., Matyjaszewski, K., Self-Healing of Covalently Cross-Linked Polymers by Reshuffling Thiuram Disulfide Moieties in Air under Visible Light. *Adv. Mater.*, 24, 29, 3975–3980, 2012.
93. Montarnal, D., Capelot, M., Tournilhac, F., Leibler, L., Silica-Like Malleable Materials from Permanent Organic Networks. *Science*, 334, 6058, 965–968, 2011.
94. Capelot, M., Unterlass, M.M., Tournilhac, F., Leibler, L., Catalytic Control of the Vitrimer Glass Transition. *ACS Macro Lett.*, 1, 7, 789–792, 2012.
95. Pei, Z., Yang, Y., Chen, Q., Terentjev, E.M., Wei, Y., Ji, Y., Mouldable liquid-crystalline elastomer actuators with exchangeable covalent bonds. *Nat. Mater.*, 13, 1, 36–41, 2013.

96. Yang, Y., Pei, Z., Zhang, X., Tao, L., Wei, Y., Ji, Y., Carbon nanotube–vitrimer composite for facile and efficient photo-welding of epoxy. *Chem. Sci.*, 5, 9, 3486, 2014.
97. Yang, Y., Pei, Z., Li, Z., Wei, Y., Ji, Y., Making and Remaking Dynamic 3D Structures by Shining Light on Flat Liquid Crystalline Vitrimer Films without a Mold. *J. Am. Chem. Soc.*, 138, 7, 2118–2121, 2016.
98. Demongeot, A., Mougnier, S.J., Okada, S., Soulié-Ziakovic, C., Tournilhac, F., Coordination and catalysis of Zn^{2+} in epoxy-based vitrimers. *Polym. Chem.*, 7, 27, 4486–4493, 2016.
99. Legrand, A. and Soulié-Ziakovic, C., Silica–Epoxy Vitrimer Nanocomposites. *Macromolecules*, 49, 16, 5893–5902, 2016.
100. Chabert, E., Vial, J., Cauchois, J.-P., Mihaluta, M., Tournilhac, F., Multiple welding of long fiber epoxy vitrimer composites. *Soft Matter*, 12, 21, 4838–4845, 2016.
101. Yang, Y., Peng, G., Wu, S., Hao, W., A repairable anhydride-epoxy system with high mechanical properties inspired by vitrimers. *Polymer*, 159, 162–168, 2018.
102. Yu, K., Taynton, P., Zhang, W., Dunn, M.L., Qi, H.J., Reprocessing and recycling of thermosetting polymers based on bond exchange reactions. *RSC Adv.*, 4, 20, 10108, 2014.
103. Shi, Q., Yu, K., Kuang, X., Mu, X., Dunn, C.K., Dunn, M.L., Wang, T., Jerry Qi, H., Recyclable 3D printing of vitrimer epoxy. *Mater. Horiz.*, 4, 4, 598–607, 2017.
104. Altuna, F., Hoppe, C., Williams, R., Epoxy Vitrimers: The Effect of Transesterification Reactions on the Network Structure. *Polymers*, 10, 1, 43, 2018.
105. Altuna, F.I., Hoppe, C.E., Williams, R.J.J., Epoxy vitrimers with a covalently bonded tertiary amine as catalyst of the transesterification reaction. *Eur. Polym. J.*, 113, 297–304, 2019.
106. Altuna, F.I., Hoppe, C.E., Williams, R.J.J., Shape memory epoxy vitrimers based on DGEBA crosslinked with dicarboxylic acids and their blends with citric acid. *RSC Adv.*, 6, 91, 88647–88655, 2016.
107. Brutman, J.P., Delgado, P.A., Hillmyer, M.A., Polylactide Vitrimers. *ACS Macro Lett.*, 3, 7, 607–610, 2014.
108. Altuna, F.I., Pettarin, V., Williams, R.J.J., Self-healable polymer networks based on the cross-linking of epoxidised soybean oil by an aqueous citric acid solution. *Green Chem.*, 15, 12, 3360–3366, 2013.
109. Lu, L., Fan, J., Li, G., Intrinsic healable and recyclable thermoset epoxy based on shape memory effect and transesterification reaction. *Polymer*, 105, 10–18, 2016.
110. Zhang, H. and Xu, X., Improving the transesterification and electrical conductivity of vitrimers by doping with conductive polymer wrapped carbon nanotubes. *Compos. Part Appl. Sci. Manuf.*, 99, 15–22, 2017.

111. Rekondo, A., Martin, R., Ruiz de Luzuriaga, A., Cabañero, G., Grande, H.J., Odriozola, I., Catalyst-free room-temperature self-healing elastomers based on aromatic disulfide metathesis. *Mater. Horiz.*, 1, 2, 237, 2014.
112. Martin, R., Rekondo, A., Ruiz de Luzuriaga, A., Santamaria, A., Odriozola, I., Mixing the immiscible: Blends of dynamic polymer networks. *RSC Adv.*, 5, 23, 17514–17518, 2015.
113. Ruiz de Luzuriaga, A., Matxain, J.M., Ruipérez, F., Martin, R., Asua, J.M., Cabañero, G., Odriozola, I., Transient mechanochromism in epoxy vitrimer composites containing aromatic disulfide crosslinks. *J. Mater. Chem. C*, 4, 26, 6220–6223, 2016.
114. Ruiz de Luzuriaga, A., Martin, R., Markaide, N., Rekondo, A., Cabañero, G., Rodríguez, J., Odriozola, I., Epoxy resin with exchangeable disulfide crosslinks to obtain reprocessable, repairable and recyclable fiber-reinforced thermoset composites. *Mater. Horiz.*, 3, 3, 241–247, 2016.
115. Denissen, W., Rivero, G., Nicolaÿ, R., Leibler, L., Winne, J.M., Du Prez, F.E., Vinylogous Urethane Vitrimers. *Adv. Funct. Mater.*, 25, 16, 2451–2457, 2015.
116. Lessard, J.J., Garcia, L.F., Easterling, C.P., Sims, M.B., Bentz, K.C., Arencibia, S., Savin, D.A., Sumerlin, B.S., Catalyst-Free Vitrimers from Vinyl Polymers. *Macromolecules*, 52, 5, 2105–2111, 2019.
117. Denissen, W., Droesbeke, M., Nicolaÿ, R., Leibler, L., Winne, J.M., Du Prez, F.E., Chemical control of the viscoelastic properties of vinylogous urethane vitrimers. *Nat. Commun.*, 8, 1, 14857, 2017.
118. Lawton, M.I., Tillman, K.R., Mohammed, H.S., Kuang, W., Shipp, D.A., Mather, P.T., Anhydride-Based Reconfigurable Shape Memory Elastomers. *ACS Macro Lett.*, 5, 2, 203–207, 2016.
119. Röttger, M., Domenech, T., van der Weegen, R., Breuillac, A., Nicolaÿ, R., Leibler, L., High-performance vitrimers from commodity thermoplastics through dioxaborolane metathesis. *Science*, 356, 6333, 62–65, 2017.
120. Obadia, M.M., Mudraboyina, B.P., Serghei, A., Montarnal, D., Drockenmuller, E., Reprocessing and Recycling of Highly Cross-Linked Ion-Conducting Networks through Transalkylation Exchanges of C–N Bonds. *J. Am. Chem. Soc.*, 137, 18, 6078–6083, 2015.
121. Lei, Z.Q., Xie, P., Rong, M.Z., Zhang, M.Q., Catalyst-free dynamic exchange of aromatic Schiff base bonds and its application to self-healing and remolding of crosslinked polymers. *J. Mater. Chem. A*, 3, 39, 19662–19668, 2015.
122. Li, H., Bai, J., Shi, Z., Yin, J., Environmental friendly polymers based on schiff-base reaction with self-healing, remolding and degradable ability. *Polymer*, 85, 106–113, 2016.
123. Kuhl, N., Bode, S., Bose, R.K., Vitz, J., Seifert, A., Hoeppener, S., Garcia, S.J., Spange, S., van der Zwaag, S., Hager, M.D., Schubert, U.S., Acylhydrazones as Reversible Covalent Crosslinkers for Self-Healing Polymers. *Adv. Funct. Mater.*, 25, 22, 3295–3301, 2015.

124. Zhang, Z.P., Rong, M.Z., Zhang, M.Q., Polymer engineering based on reversible covalent chemistry: A promising innovative pathway towards new materials and new functionalities. *Prog. Polym. Sci.*, 80, 39–93, 2018.
125. Dahlke, J., Zechel, S., Hager, M.D., Schubert, U.S., How to Design a Self-Healing Polymer: General Concepts of Dynamic Covalent Bonds and Their Application for Intrinsic Healable Materials. *Adv. Mater. Interfaces*, 5, 17, 1800051, 2018.
126. Altuna, F.I., Puig, J., Hoppe, C.E., Williams, R.J.J., Remote actuation of epoxy nanocomposites with functional properties. *J. Argent. Chem. Soc.—An. Asoc. Quim. Argent.*, 105, 2, 114–134, 2018.
127. Habault, D., Zhang, H., Zhao, Y., Light-triggered self-healing and shape-memory polymers. *Chem. Soc. Rev.*, 42, 17, 7244, 2013.
128. Baffou, G., Quidant, R., Girard, C., Heat generation in plasmonic nanostructures: Influence of morphology. *Appl. Phys. Lett.*, 94, 15, 153109, 2009.
129. Govorov, A.O. and Richardson, H.H., Generating heat with metal nanoparticles. *Nano Today*, 2, 1, 30–38, 2007.
130. Rance, G.A., Marsh, D.H., Nicholas, R.J., Khlobystov, A.N., UV–vis absorption spectroscopy of carbon nanotubes: Relationship between the π-electron plasmon and nanotube diameter. *Chem. Phys. Lett.*, 493, 1–3, 19–23, 2010.
131. Chen, G., Wu, K., Zhang, Q., Shi, Y., Lu, M., Dual-Responsive Shape Memory and Thermally Reconfigurable Reduced Graphene Oxide-Vitrimer Composites. *Macromol. Res.*, 27, 6, 526–533, 2019.
132. Bao, Z., Liu, X., Liu, Y., Liu, H., Zhao, K., Near-infrared light-responsive inorganic nanomaterials for photothermal therapy. *Asian J. Pharm. Sci.*, 11, 3, 349–364, 2016.
133. Jung, H.S., Verwilst, P., Sharma, A., Shin, J., Sessler, J.L., Kim, J.S., Organic molecule-based photothermal agents: An expanding photothermal therapy universe. *Chem. Soc. Rev.*, 47, 7, 2280–2297, 2018.
134. Huang, L., Yi, N., Wu, Y., Zhang, Y., Zhang, Q., Huang, Y., Ma, Y., Chen, Y., Multichannel and Repeatable Self-Healing of Mechanical Enhanced Graphene-Thermoplastic Polyurethane Composites. *Adv. Mater.*, 25, 15, 2224–2228, 2013.
135. Fang, M., Li, D., Lin, H., Luo, C., Qi, R., Peng, H., Flexible and recyclable conductive composite based on few-layered graphene with excellent self-healing capability. *Eur. Polym. J.*, 108, 536–541, 2018.
136. Noack, M., Merindol, R., Zhu, B., Benitez, A., Hackelbusch, S., Beckert, F., Seiffert, S., Mülhaupt, R., Walther, A., Light-Fueled, Spatiotemporal Modulation of Mechanical Properties and Rapid Self-Healing of Graphene-Doped Supramolecular Elastomers. *Adv. Funct. Mater.*, 27, 25, 1700767, 2017.
137. Yang, Z., Wang, Q., Wang, T., Dual-Triggered and Thermally Reconfigurable Shape Memory Graphene-Vitrimer Composites. *ACS Appl. Mater. Interfaces*, 8, 33, 21691–21699, 2016.

138. Lin, C., Sheng, D., Liu, X., Xu, S., Ji, F., Dong, L., Zhou, Y., Yang, Y., NIR induced self-healing electrical conductivity polyurethane/graphene nanocomposites based on Diels–Alder reaction. *Polymer*, 140, 150–157, 2018.
139. Wu, S., Li, J., Zhang, G., Yao, Y., Li, G., Sun, R., Wong, C., Ultrafast Self-Healing Nanocomposites via Infrared Laser and Their Application in Flexible Electronics. *ACS Appl. Mater. Interfaces*, 9, 3, 3040–3049, 2017.
140. Weng, D., Xu, F., Li, X., Li, Y., Sun, J., Bioinspired photothermal conversion coatings with self-healing superhydrophobicity for efficient solar steam generation. *J. Mater. Chem. A*, 6, 47, 24441–24451, 2018.
141. Altuna, F.I., Antonacci, J., Arenas, G.F., Pettarin, V., Hoppe, C.E., Williams, R.J.J., Photothermal triggering of self-healing processes applied to the reparation of bio-based polymer networks. *Mater. Res. Express*, 3, 4, 045003, 2016.
142. Zhang, H. and Zhao, Y., Polymers with Dual Light-Triggered Functions of Shape Memory and Healing Using Gold Nanoparticles. *ACS Appl. Mater. Interfaces*, 5, 24, 13069–13075, 2013.
143. Zhang, H., Fortin, D., Xia, H., Zhao, Y., Fast Optical Healing of Crystalline Polymers Enabled by Gold Nanoparticles. *Macromol. Rapid Commun.*, 34, 22, 1742–1746, 2013.
144. Wang, Z., Li, Z., Wei, Y., Ji, Y., Gold Nanospheres Dispersed Light Responsive Epoxy Vitrimers. *Polymers*, 10, 1, 65, 2018.
145. Li, Y., Chen, S., Wu, M., Sun, J., Rapid and Efficient Multiple Healing of Flexible Conductive Films by Near-Infrared Light Irradiation. *ACS Appl. Mater. Interfaces*, 6, 18, 16409–16415, 2014.
146. Chen, L., Si, L., Wu, F., Chan, S.Y., Yu, P., Fei, B., Electrical and mechanical self-healing membrane using gold nanoparticles as localized "nano-heaters". *J. Mater. Chem. C*, 4, 42, 10018–10025, 2016.
147. Burnworth, M., Tang, L., Kumpfer, J.R., Duncan, A.J., Beyer, F.L., Fiore, G.L., Rowan, S.J., Weder, C., Optically healable supramolecular polymers. *Nature*, 472, 7343, 334–337, 2011.
148. Chen, Q., Yu, X., Pei, Z., Yang, Y., Wei, Y., Ji, Y., Multi-stimuli responsive and multi-functional oligoaniline-modified vitrimers. *Chem. Sci.*, 8, 1, 724–733, 2017.
149. Fang, L., Chen, J., Zou, Y., Chen, S., Fang, T., Lu, C., Xu, Z., Self-Healing Epoxy Coatings via Focused Sunlight Based on Photothermal Effect. *Macromol. Mater. Eng.*, 302, 9, 1700059, 2017.
150. Pu, W., Fu, D., Wang, Z., Gan, X., Lu, X., Yang, L., Xia, H., Realizing Crack Diagnosing and Self-Healing by Electricity with a Dynamic Crosslinked Flexible Polyurethane Composite. *Adv. Sci.*, 5, 5, 1800101, 2018.

2

Smart Phenolics for Self-Healing and Shape Memory Applications

Baris Kiskan* **and Yusuf Yagci**

Istanbul Technical University, Department of Chemistry, Istanbul, Turkey

Abstract

Although there are many types of curable systems, phenolic resins are still one of the major polymers and have very large production volumes. Beside, an alternative phenolic resin emerged in the last decades as more reactive and releasing less by-products during curing compared to classical phenolics. These resins are based on 1,3-benzoxazines and their polymers are coined as polybenzoxazines. Among high performance thermosets, polybenzoxazines exhibit a range of features. In general, polybenzoxazines exhibit low shrinkage during their production, high char yield and Tg. Moreover, they have good mechanical performance with low water adsorption. Consequently, polybenzoxazines gained a vast interest in polymer science. However, the general interest for polybenzoxazines is related to their usage as composite ingredients, dielectric materials etc. On the other hand, the reactive nature of benzoxazine monomers and their huge design flexibility proved that these compounds have vast potential in designing smart materials. Although the development of these materials is in benzoxazine chemistry is premature, it is expected that there will be intensive studies to synthesize benzoxazine based self-healing, shape memory materials. In this chapter, an overview about polybenzoxazines and their current progress about the synthesis of benzoxazine-based materials for these unconventional uses are summarized.

Keywords**:** Polybenzoxazines, benzoxazine resins, ring opening polymerization, smart phenolic resins

**Corresponding author*: kiskanb@itu.edu.tr

Inamuddin, Mohd Imran Ahamed, Rajender Boddula and Tariq Altalhi (eds.) Self-Healing Smart Materials and Allied Applications, (39–64) © 2021 Scrivener Publishing LLC

2.1 Introduction

Although there are many different types of thermosets that were synthesized as high performance polymers during the last century, phenolic resins, as the first man-made plastic, have been continuously used in industry for more than 100 years. As a reflection of this interest, many papers and books were published and conference activities were successfully performed about these resins. The worldwide production of classical phenolic resins, resole and novolac, has reached to a very large volume such as 5 million tons per year and the growth of the production is parallel to the global economic growth rate [1–3]. Definitively, these resins will survive in the second century of their existence due to the interest in global markets. Accordingly, novel applications have already emerged related to phenolic resins such as porous materials by the carbonization of phenolic foams, fiber reinforced phenolics, modern composites for aircrafts, blast protection materials, phenolic resins based adhesives for honeycomb sandwich panels and more [4]. On the other hand, resole and novolac synthesis requires formaldehyde (Scheme 2.1) and the emission problem of this molecule resulted in regulations for phenolic industry. Especially, in wood products formaldehyde residues expose a serious concern of health and environment, therefore, efforts to reduce formaldehyde emissions in these resins is one of the major trend. Moreover, phenolic resins are cured either by using acids or bases before usage and the curing rate of classical phenolic resins are relatively slow compared to other resins and, besides formaldehyde emission, formation of other by-products is a fact during curing [5]. Hence, these problems gave rise to new research and development

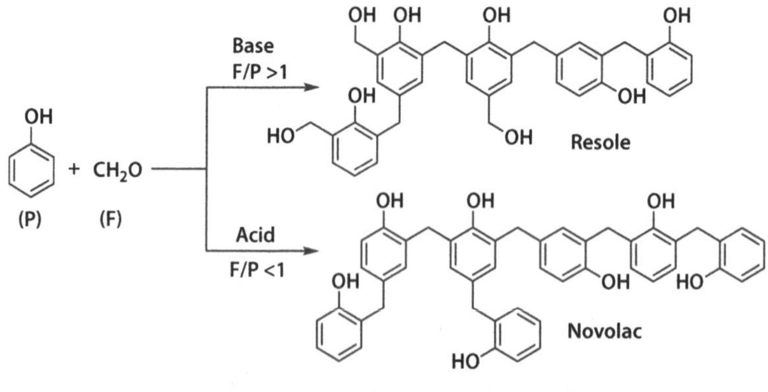

Scheme 2.1 Synthesis and representative structures for Novolac and Resole.

projects for phenolic resins with the anticipation of more reactivity and less emissions.

The benzoxazine type resins emerged with high potential as a solution for drawbacks of phenolic resins where they have not succeeded sufficiently. Polybenzoxazines (PBZs), a kind of polyaminophenol, are the curing product of 1,3-benzoxazines and are considered as a contender to novolak due to the structural resemblance (Scheme 2.2). Hence, most of the PBZs have properties both similar and also unique that are unseen in novolac resins [6]. They generally exhibit high glass transition temperatures (T_g), char yield, flame resistance, low water absorption, and good mechanical performance [6–12]. In addition, benzoxazines do not exhibit an important shrinkage during curing due to the ring-opening polymerization character [11]. As well known, shrinkage is a problem for many other curable monomer types such as acrylates. Another important superiority of benzoxazines is related to their reactive nature, and thus, by-product release is much lower compared to classical phenolic resins. Typical polymerization of a benzoxazine monomer takes place between 160 and 250 °C without any catalyst or curative (Scheme 2.2) [13–16]. The polymerization of benzoxazines proceed over a cationic ring-opening pathway, and therefore, polymerization temperatures can be reduced by using acidic compounds as catalyst [17–21]. Another advantage of benzoxazine resin chemistry is related to the synthetic simplicity of benzoxazine monomer synthesis. Any phenol and a primary amine, which can tolerate the synthesis conditions, with 2 moles of formaldehyde can yield a benzoxazine monomer (Scheme 2.2) [22–31]. Accordingly, benzoxazine chemistry has a vast design flexibility due to the presence of many different commercially available phenols and amines. The huge molecular diversity of benzoxazines unleashes many possibilities for unconventional uses ranging from self-healing to shape-memory material applications [32].

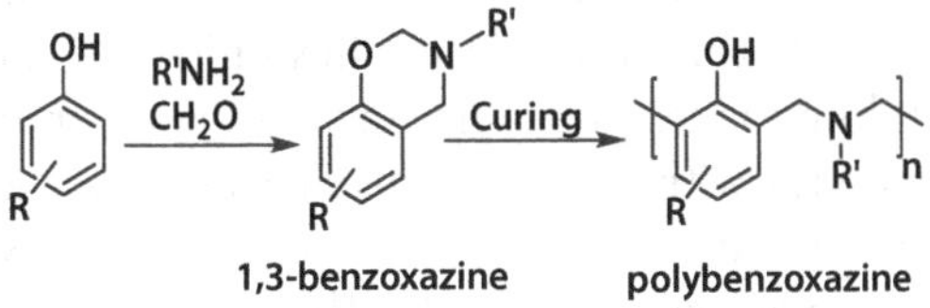

Scheme 2.2 Synthesis of a 1,3-benzoxazine and production of a polybenzoxazine therefrom.

2.2 Self-Healable Polybenzoxazines

In polymer and material science, controlling the molecular structure of polymers/materials and resulting properties has been an ongoing challenge. Among various approaches, the syntheses of responsive polymers are important in this context since their properties can be controlled by either stimulus or self-intervention of these polymers. Materials that can respond temperature, light, pH, mechanical deformation, etc. are named as smart materials and many studies have been reported about them. Among these materials, especially, self-repairing polymers gained much attention considering the opportunity to extend the lifetime by fixing damage during usage without human intervention. Excluding capsule based approaches [33], self-healing polymers can be categorized into two subbranches as "autonomous" and "stimuli-responsive" healing materials [34–37]. Stimuli-responsive materials require a triggering event and conveyed energy for repairing the damage such as electricity, thermal energy, light, and ultrasound [38–40]. By this approach, successive self-healable materials were designed based on reversible chemical reactions for multiple healings. In example, [2 + 2], [3 + 2], [4 + 2], [4 + 4] cycloadditions were used extensively [41–43]. In the case autonomic self-healing materials supramolecular attractions such as hydrogen bonding were utilized in their designs and besides chemical also physical cross-linking were established through supramolecular interaction that govern the entanglements of polymer chains. However, in both approaches, the primary mechanism can be simplified as molecular diffusions, flow of molecules interconnected with segmental mobility of polymer chains by a stimulus [44–48]. In brief, for a self-healable material a dynamic network is required with suitable functional groups capable of performing healing reactions or interactions.

As stated previously, benzoxazine chemistry has a vast design capacity to append purpose oriented functional groups to resulting polybenzoxazines. Moreover, a unique property of polybenzoxazines is related to the intra/intermolecular hydrogen bonding that contributes many interesting properties of these polymers. Those hydrogen bonding are also useful in designing materials when considering supramolecular attractions for the final material [49–51]. Thus, both design flexibility and extensive amount of hydrogen bonding makes polybenzoxazines good candidates to use as self-healable phenolics with careful molecular designs. According to the stated background, for example, polyether chains are flexible and could be appropriate in benzoxazine design for the purpose of obtaining a self-healable phenolic network. Hence, curable benzoxazine macro-monomers were synthesized from jeffamines (polyethers with primary amines at the chain ends) and a coumarine (Scheme 2.3). The obtained end-chain coumarine-benzoxazine macro-monomers were oily, which is an advantage to prepare films in Teflon molds. The molded macro-monomers were cured at *ca.* 180 °C to obtain cross-linked but soft/flexible polybenzoxazine films [52].

In this system, the coumarine functionality was selected to act as reversible reaction site based on light triggered [2 + 2] dimerization between wavelengths 300 and 350 nm. The long-chain jeffamine moieties provided the indispensable mobility for the system to ease the molecular collusions of coumarines. The self-healing efficiencies (η) of these films were calculated by comparing fracture toughness of the pristine and healed samples. The η value was found to be *ca.* 10% for the cut and healed specimen, but for cracked then healed specimen η was *ca.* 44%. Although these results are not quite sufficient for a real-life application, the system can be improved and this study is the first report about intrinsically self-healable polybenzoxazines.

In contrast to self-healing materials relied on bond reformations by external stimulus, healing with self-intervention of polymer by supramolecular attractions like hydrogen bonding are autonomous and can perform multiple healing cycles upon damage. In this context, autonomic self-healable polybenzoxazines can be obtained by carefully arranging the hydrogen bonds that naturally present on polybenzoxazines and the soft segments in these networks. Accordingly, jeffamine based main-chain polybenzoxazine precursors (PPO-Benz) were synthesized and mixed with a carboxylic acid group containing monofunctional-benzoxazine (Benz-COOH) in molds with various mass percentages prior to curing [53]. The obtained films were flexible and the material contained more hydrogen

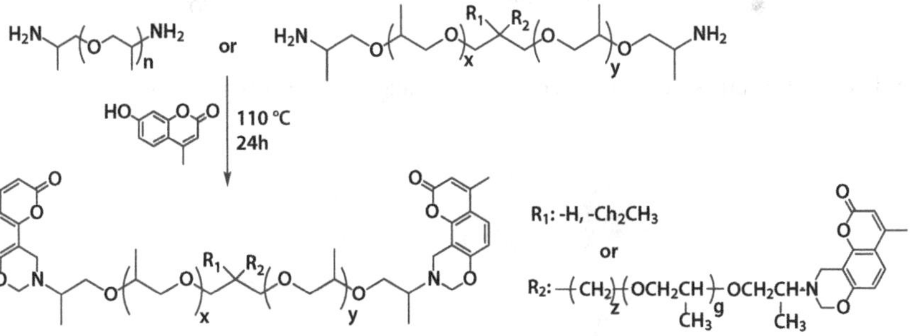

Scheme 2.3 Synthesis of end-chain coumarine functional benzoxazine macromonomers.

bonds compared to pristine polybenzoxazine due to excess phenolic –OH and –COOH functionalities originating from Benz-COOH (Scheme 2.4).

Films were prepared with different mass ratios of PPO-Benz and Benz-COOH by using solvent casting method. These films then cured at 180 °C before healing tests. Typically, the films were cut into two pieces and kept in contact for 12 h at room temperature. Then, the healing efficiencies of the healed films were found by tensile tests by calculating their toughness recovery. The extent of healing was found to be related to the added amount of Benz-COOH in the films. The cut films were able to restore themselves to certain degrees of healing (Figure 2.1). For example, healing efficiency was calculated as 96% for the polybenzoxazine film with 10% Benz-COOH but only 26% for the sample bearing 2.5% Benz-COOH. The findings clearly reveal that the number of hydrogen bonds is the major effect on recovery and the presence of Benz-COOH in curing formulations, as extra hydrogen bonding source, augments the self-mending ability.

In another strategy, supramolecular attractions and S–S bond cleavage-reformations were used to design recyclable and self-healable polybenzoxazines. In this approach, low cost chemicals were converted to self-healable materials in only 30 min with a simple process. This strategy relies on inverse vulcanization of benzoxazines, which is simply performed via mixing benzoxazine monomer or polybenzoxazine prepolymers with elemental sulfur at *ca.* 180 °C [54]. According to the proposed reaction mechanisms,

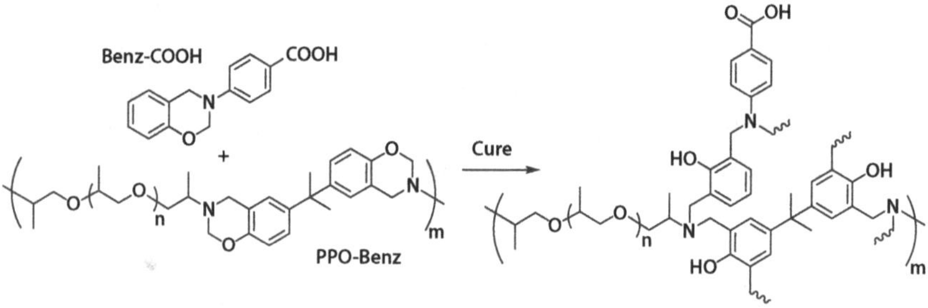

Scheme 2.4 Synthesis of polybenzoxazine with augmented hydrogen bonds.

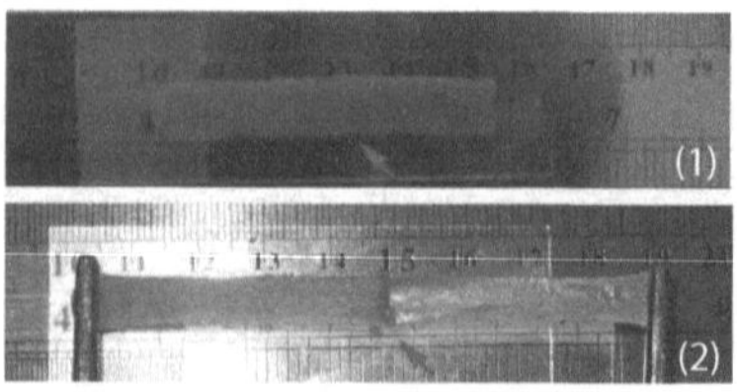

Figure 2.1 The images of cut-healed PPO-Benz/Benz-COOH films.

the reaction proceeds over a radical mediated process which was named as sulfur radical transfer and coupling (SRTC) (Scheme 2.5) [55].

Main-chain polybenzoxazine from poly(propylene oxide) (PPO-Benz), a di-ally functional benzoxazine monomer (B-ala) and sulfur was heated up to 185 °C (Scheme 2.6) for self-healable material fabrication [56]. The obtained crosslinked polymeric films was recycled and healed up to 5 cycles by heating (Figure 2.2). As observed from tensile tests, brittleness of the films increased and the toughness decreased after each thermal (180 °C) healing. The stress value of the sample was measured 877 kPa after the 1st and 2,007 kPa after 5th healings. Conversely, elongation at break reduced gradually from *ca.* 80% to *ca.* 20% after 5 healing cycles. Such toughness loss indicates the reduction of chain mobility after healings. Because S–S bond cleavage and reformation reactions in each cycle shortens the polysulfide chains. The rigidity and crosslinking density of the material increased due to these short chains and self-healing ability after a certain cycle number eventually reduced. Although healing of this system is limited, the results demonstrate that it is possible to use elemental sulfur with benzoxazines to produce recyclable and self-healable poly (benzoxazines-*co*-sulfide)s.

Obviously, previous studies revealed that benzoxazines with poly(propylene oxide)s has high potential to design different self-healing materials. In this line, for self-healable polybenzoxazines, PPO-benz were used for ring-opening polymerization and subsequent ketene formation by light for healing reactions. The oxoketenes were generated over a bisdioxinone (BisBDiox) molecule which was admixed in the benzoxazine precursor [57, 58]. Upon irradiation *ca.* 300 nm at room temperature, bisdioxinones

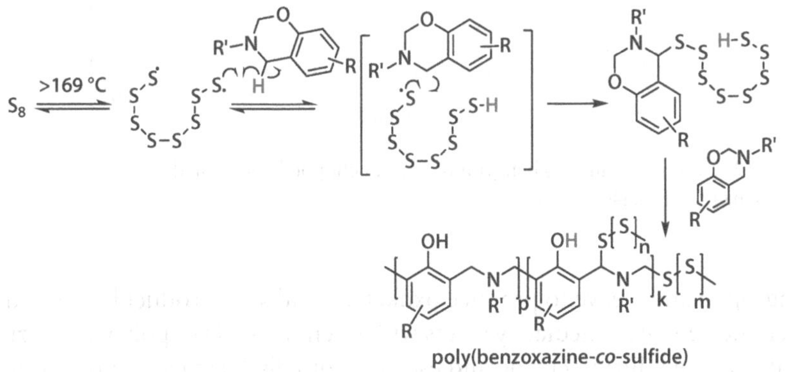

Scheme 2.5 Simplified mechanism of inverse vulcanization reaction of a benzoxazine monomer and elemental sulfur to produce a poly(benzoxazine-*co*-sulfide).

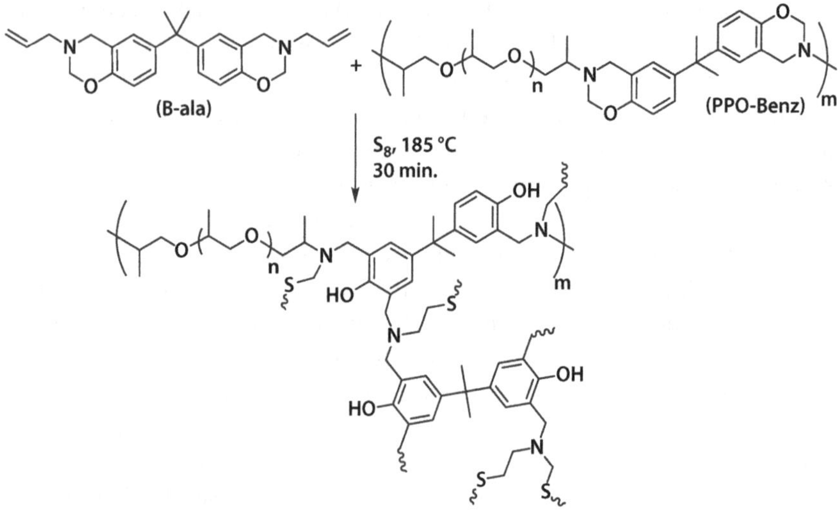

Scheme 2.6 Synthesis of recyclable and self-healable poly(benzoxazines-co-sulfide).

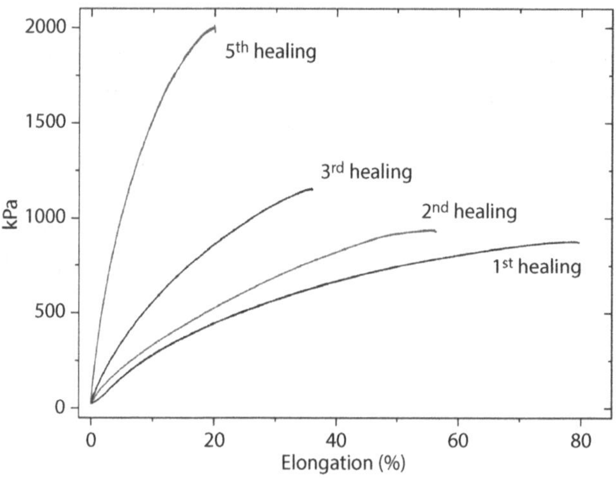

Figure 2.2 Stress–Strain (%) behavior of 5 times chopped and healed PPOBenz40%–B-ala40%–S20%.

ring-open and cleave to produce oxoketene and side-product ketone. And then the ketene immediately reacts with phenolic –OH of polybenzoxazine that occurred during curing and esters eventually form (Scheme 2.7) [59]. By this way, light triggered crosslinking was achieved and healing on surface was achieved.

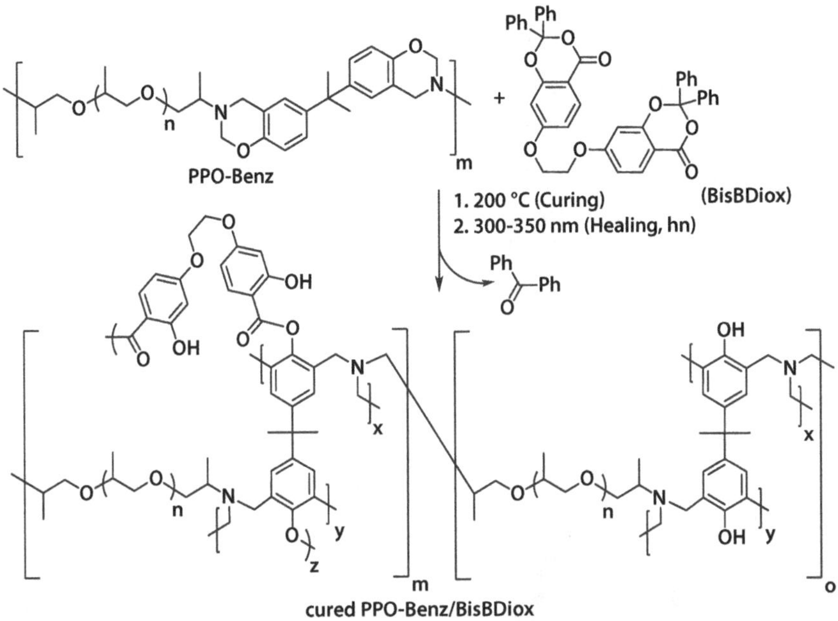

Scheme 2.7 Curing and subsequent light induced healing of PPO-Benz/BisBDiox system.

A typical self-healable film was prepared by dissolving PPO-Benz and BisBDiox (5% w/w) in tetrahydrofuran (THF) and then this solution was used to cast a thin layer on silicon wafers by spin coating technique. After drying, the obtained films had a thickness of *ca.* 300 nm. Then, these films were cured gradually by using an open-air oven at 180 °C and at 200 °C for ~30 min. To test healing, initially, the surface was scratched by using an atomic force microscopy (AFM) and its nano-indentation diamond tip. All the films were monitored by AFM in tapping mode before light exposure. Then, the films were exposed to light between 300 and 350 nm for 10 h under THF vapor at room temperature to trigger oxoketene formation. During irradiation, BisBDiox acted as bridging agent and polybenzoxazines chemically bind to each other over ester linkages and efficient healing on surface was achieved. The segmental mobility of the polybenzoxazine chains was sufficient to ship BisBDiox molecules to the damaged zone for healing reactions to take place. The amount of healing could not be monitored by classical methods, however the damaged zone was monitored by using AFM (Figure 2.3) and complete healing was observed especially in the case of THF vapor assistance. AFM results clearly reveal that using BisBDiox as self-healing agent is a convenient strategy due to its ability to react with nucleophiles and compatibility with phenolics.

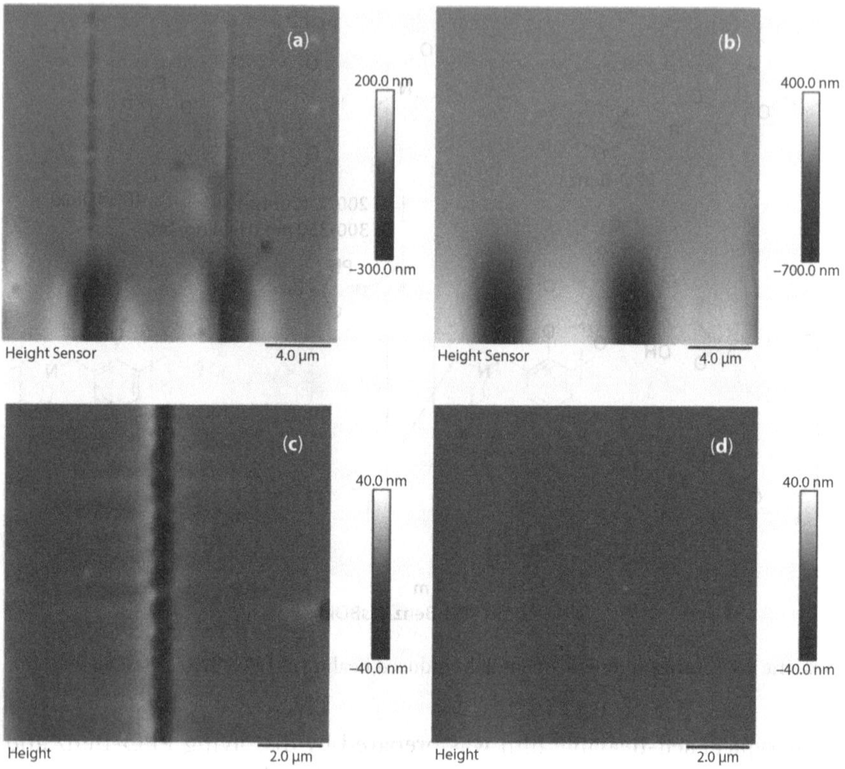

Figure 2.3 AFM pictures of PPO-Benz/BisBDiox before (a, c) and after irradiation (b, d).

In another approach, ester formation on polybenzoxazines were also used to obtain self-healable polybenzoxazines. In contrast to previous method, the healing is based on ester exchange reactions. For the purpose, succinic anhydride was admixed with bisphenol F derived benzoxazine prior to curing. The ring-opening reaction generates free phenolic OH groups that are available to react with anhydride at high temperature. By this way, the polybenzoxazine contains both phenolic OH and carboxylic groups in the network (Scheme 2.8) [60].

Therefore, the network structure is suitable for possible transesterification reactions that can be used for healing the damaged material. It was shown that such a structure can behave like thermoplastic polymer at *ca.* 140 °C with a transesterification catalyst, zinc acetate ($Zn(Ac)_2$), and exhibit self-healing even after several healing cycles. An optical microscope was used to demonstrate the autonomic repairing capability of these thermosets. Accordingly, a cut-healed sample was presented in Figure 2.4a. Initially, a cut approximately 65.8 μm wide was generated on the surface of

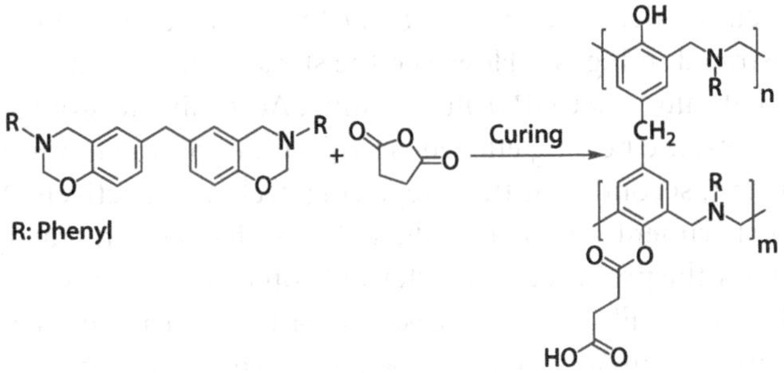

Scheme 2.8 Synthesis of polybenzoxazine-succinic anhydride thermoset.

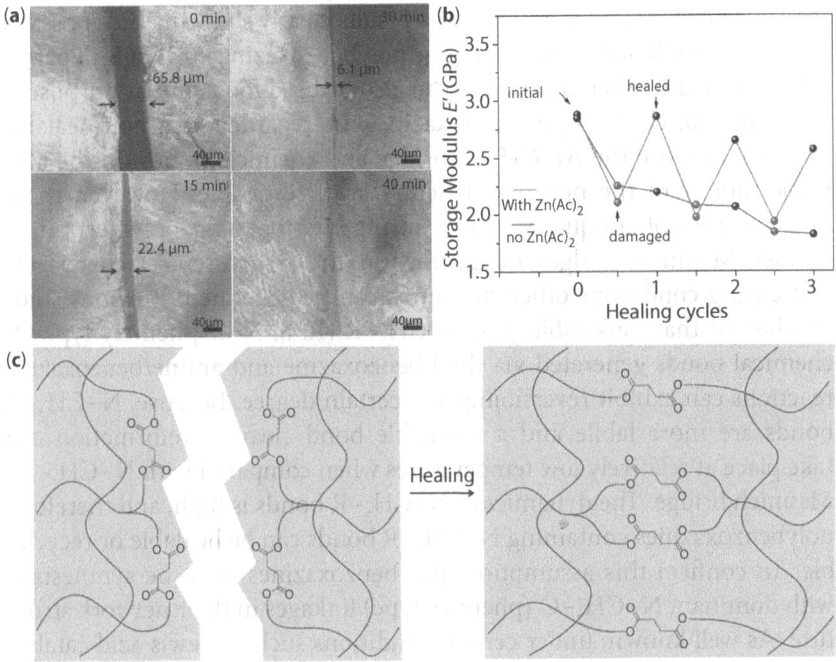

Figure 2.4 (a) Images showing the healing on the surface of sample film (b) Storage modulus for damage and healing cycles, (c) A simplified illustration showing transesterifications for healing (Copyright: https://creativecommons.org/licenses/by/4.0/).

the material. As seen in Figure 2.4, the damaged area decreased reasonably, and diameter of the damaged zone reduced to *ca.* 6.1 μm after a certain time. Moreover, the damage was almost healed after heating the material at 140 °C for 40 min. The ability of repeated recovery was also tested via mechanical measurements of the present thermoset. The storage modulus

(E′) of specimen decreased from *ca.* 2.9 to 2.1 GPa after the first damage as demonstrated in Fig. 4b. However, the storage modulus almost reached to its original value (2.86 GPa) after heating. According to storage modulus measurements, the healing efficiencies (η) of the specimen was 99, 92 and 89% after first, second, and third repairing cycles, respectively. Moreover, a control experiment was also conducted to understand the catalytic effect of $Zn(Ac)_2$ for the proposed transesterification reaction. And in contrast to catalyzed samples, E′ values decreased linearly after each damage and the samples without $Zn(Ac)_2$ could not be recovered effectively.

In conclusion, based on the strategy of ester exchange reactions, a novel self-healable polybenzoxazine system was presented by using simple chemical such as succinic anhydride and bisphenol F based commercially available benzoxazine monomer. The results clearly show that it is possible to obtain a bulk-state self-healing in polybenzoxazines through dynamic ester bonds and other anhydrides can also be used for a similar purpose.

As known, classical polybenzoxazines are regarded as a non-healable polymer, because the Ar–CH_2–N bridge and chemical bonds on the aromatic moiety of the network structure are irreversible. However, those linkages are not unique occurred during ring-opening polymerization, besides Mannich bridges depending on the structure of benzoxazibe and curing conditions other structures can be generated. Previous studies showed that, reversibility of N–CH_2–R (R: S, N or phenoxy type O) chemical bonds generated via thiol/benzoxazine and amine/benzoxazine reactions can exhibit reversibility at a certain degree. Because, N–CH_2–R bonds are more labile and a reversible bond cleavage/reformation can take place at relatively low temperatures when compared with N–CH_2–Ar Mannich bridge. The dynamics of N–CH_2–R bonds is high, and therefore, polybenzoxazines containing N–CH_2–R bonds can be healable or recyclable. To confirm this assumption, polybenzoxazines must be synthesized with dominant N–CH_2–O (phenoxy type) linkages in their network structure. As well known, under certain conditions such as Lewis acid catalyst and low curing temperatures force the ring-opening polymerization to generate more N–CH_2–O than N–CH_2–Ar bonds. Moreover, by blocking the ortho-position of a benzoxazine monomer similar results can be obtained and possible reversibility based on phenoxy type N–CH_2–O bonding can form a self-healable polybenzoxazines without complicated monomer or macromonomer designs. This possibility was investigated by curing benzoxazines with functional groups at ortho-, meta- and para- positions. After curing those benzoxazines at 210 °C for 4 h, the obtained polybenzoxazines were compressed at 180 °C under 5 MPa for 20 min. Polybenzoxazines from ortho-functional monomers (Scheme 2.9)

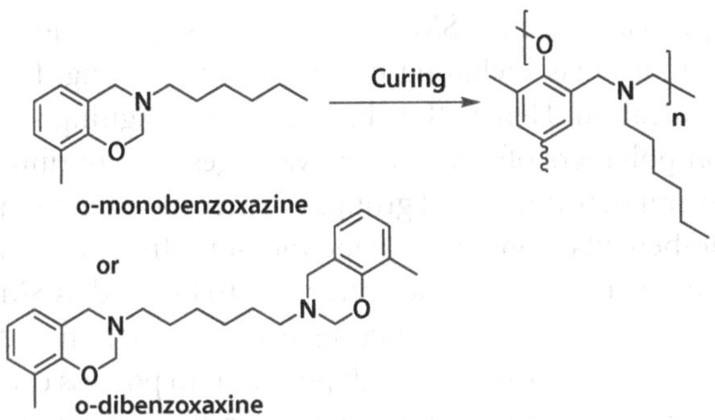

Scheme 2.9 The structures of ortho functional mono and dibenzoxazines and their curing to give phenoxy linkage containing polybenzoxazines.

exhibited healing even after crushing into pieces, which is unusual for classical polybenzoxazines [61]. The deformation and healing behaviors of the ortho-functional polybenzoxazines (Scheme 2.9) were found to be quite similar according to experiments. Further molding studies for ortho-functional polybenzoxazines also revealed that the finely chopped fragments of these polybenzoxazines can even be remolded successfully into the different shapes by using molds after compressing under 5 MPa, at *ca.* 180 °C for 2 h. Higher pressures were much beneficial since the amount defects in the molded samples were reduced. The recycled sample from ortho-difunctional benzoxazine showed 48 MPa of tensile strength with a self-healing efficiency more than 90%. In summary, this work proved that phenoxy type N–CH_2–O bonding is a dynamic covalent bond and can be generated on polybenzoxazine by using an ortho-blocked benzoxazine monomer. Moreover, the amount of phenoxy structures can be increased by controlling the curing temperature. The exchange reactions between N–CH_2–O (phenoxy) bonds is the major cause that facilitates both healing and recycling of the polybenzoxazines.

2.3 Benzoxazine Resins for Shape Memory Applications

Among smart materials shape memory polymers (SMPs) have high potential in material chemistry since SMPs can shift between different shapes and return back to their initial shape upon stimulus [62, 63]. Generally, these materials exhibit SMP property by temperature change using infrared light, a heater or simply warm water [64–66]. Due their transition between shapes via stimulus, SMPs have various application possibilities in many different areas including electronics, packaging, textiles, transportation, medication. Hence, the challenge for designing such materials with common polymers offering cost advantages is a continuous interest. Moreover, several new functional groups are incorporated in these systems to obtain side-benefits along with shape memory. In this manner, epoxies and related polymers have attracted attention to be used as SMPs, because epoxies exhibit low T_g, high tolerance against deformation and low recovery stress. Therefore, epoxies have high potential to possess excellent shape memory performance. On the other hand, pristine epoxy polymers had practical application due to requirement of high strength and stiffness besides shape memory property. Thus, addition of benzoxazine monomers into epoxies may generate SMP thermosets having better properties. The compatibility of benzoxazine and epoxies was found to be high and epoxy thermosets were produced successfully. It was reported that benzoxazine resins acted as curing compounds for epoxies and copolymers with improved properties were obtained compared to pristine epoxy polymers due to the synergism between epoxies and benzoxazine resins [67–70]. Polybenzoxazine-epoxy based SMPs were fabricated by Rimdusit *et al.* and those SMPs exhibited better thermal stability and mechanical performance [71]. Polybenzoxazine-epoxy SMPs were synthesized by using a bisbenzoxazine derived from bisphenol A and aniline (BA-a), two different epoxies EPON 826, NGDE and Jeffamine D230 (as curative) (Scheme 2.10). All these ingredients were mixed in a mold and melted at 70 °C to obtain a homogeneous mixture prior to curing in an open-air oven. Mechanical properties, shape recovery capacity and thermal properties of the obtained

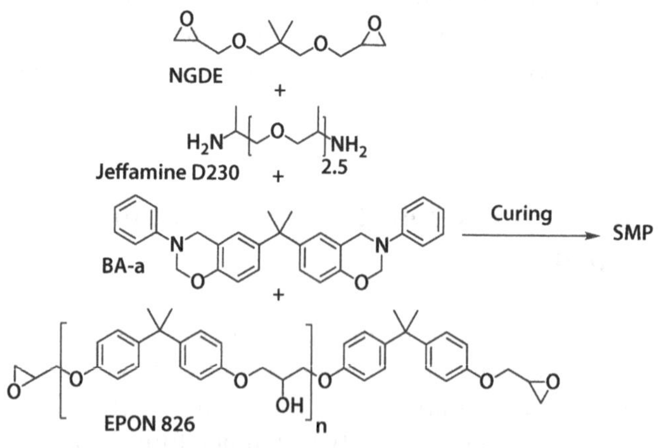

Scheme 2.10 Synthesis of polybenzoxazine-epoxy SMP.

SMPs were characterized (Figure 2.5). Accordingly, all of the deformed SMP samples are completely recovered to their original shapes after a few minutes suggesting a good shape recovery performance. However, increasing the amount of BA-a content in epoxy-benzoxazine mixture affected the shape recovery time in a negative manner. Because, polybenzoxazine moiety is hard and naturally would increase the rigidity of the materials, resulting in a slow shape memory response.

Another SMP based on poly(ε-caprolactone) (PCL) and polybenzoxazine was reported as alternative to epoxy systems [72, 73]. Hydroxyl (PCL-OH) and tosyl (PCL-OTs) terminated two different types of PCLs were used with difunctional benzoxazine monomer to prepare SMPs. The curing studies revealed that polybenzoxazines could not bind to hydroxyl terminated PCL, however, polybenzoxazines were able to bind tosyl terminated PCL by substitution reaction between phenolic –OH and tosyl groups (Scheme 2.11). By this way polybenzoxazines were covalently incorporated to PCL chains and thermally triggered shape memory behavior with high shape recovery (S_r) and fixity (S_f) were obtained for these copolymer thermostes. Unbound PCL-polyenzoxazines (PCL-(OH_2)/benzoxazine) was shown to be brittle, heterogeneous in morphology, and therefore, lacking

Figure 2.5 Photographs of polybenzoxazine-epoxy samples showing original (a), temporary (b) and recovered (c) shapes. Copyright: CC BY-NC-SA license: https://creativecommons.org/licenses/by-nc-sa/3.0/).

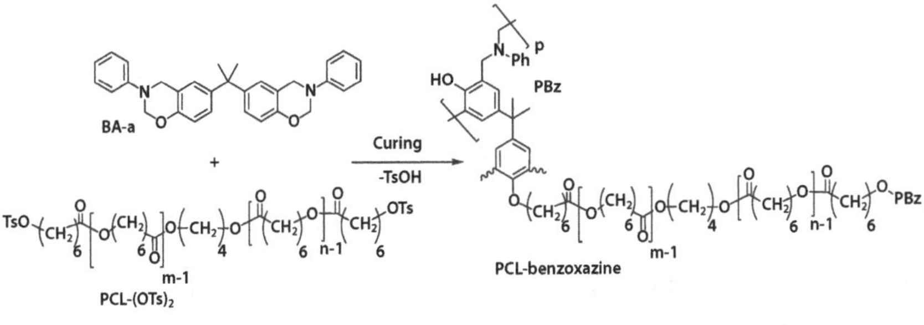

Scheme 2.11 Synthesis of crosslinked PCL-benzoxazine.

of shape memory properties. Owing to the fact that PCL and polybenzoxazines are not interconnected, the pure polybenzoxazine zones in the matrix can be considered as the reason of the brittleness. Because polybenzoxazines are known to have high crosslinking density with strong intra- and inter-hydrogen bonds.

Shape memory properties of all PCL-polybenzoxazines, PCL-(OTs_2)/benzoxazines, were studied by bending tests both at ambient temperature and above the phase transition temperatures predetermined by dynamic mechanical analysis. Accordingly, unlike PCL-(OH_2)/benzoxazine, PCL-(OTs_2)/benzoxazine samples softened at 100 °C and bending was much easier for high PCL containing copolymers (Figure 2.6). Thus, increasing PCL content in PCL-(OTs_2)/benzoxazine improved both S_f and S_r values and among all samples for PCL-(OTs_2)-80/benzoxazine-20 (80:20, w/w) was almost 100% even after several cycles. According to gel content analysis, this sample had the largest amount of polybenzoxazine bonded with PCL and the largest amount of free PCL.

In summary, interconnected PCL and polybenzoxazine chains generated hard and soft segments required for a typical SMP system. The amount of PCL was found to be crucial for the proposed system because PCL endowed strong fixing abilities via molecular interactions between the polybenzoxazine and PCL chains. Besides, increasing PCL content augmented the bending ability of the copolymer at temperatures above PCLs melting temperature.

In another approach, polybenzoxazines with both intrinsic self-healing and shape memory properties was reported. In order to synthesize this smart material initially the benzoxazine was obtained via reacting formaldehyde, bisphenol A and bis(3-aminopropyl) terminated polydimethylsiloxane (Scheme 2.12) [74]. By this way, a main-chain polybezoxazine was obtained having film formation ability and essential chain dynamics due to

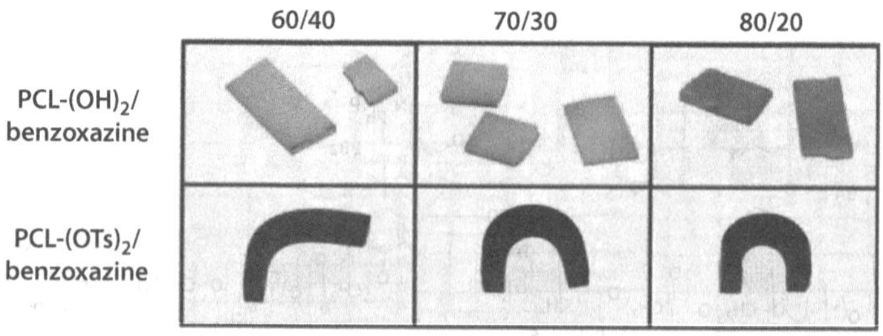

Figure 2.6 Images for PCL/benzoxazine materials; Bending tests of PCL-(OH_2)/benzoxazine samples failed at 100 °C (top), shape recovery of PCL-(OTs_2)/benzoxazine samples (bottom). (Elsevier Ltd. License number: 4790661063144).

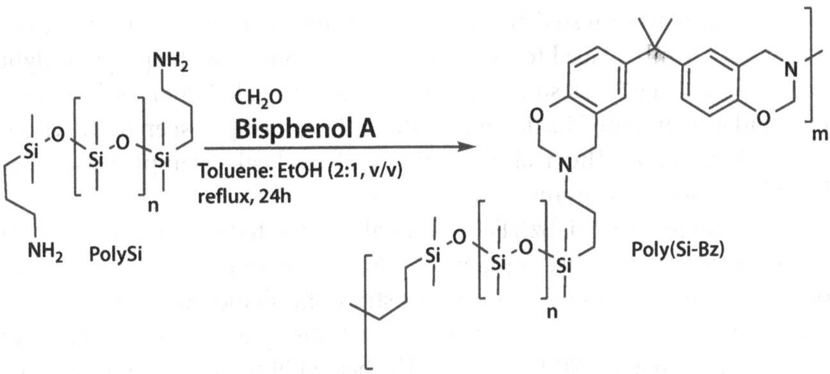

Scheme 2.12 Synthesis of Poly(Si-Bz) from bis(3-aminopropyl) terminated polydimethylsiloxane (PolySi).

the polydimethylsiloxane segments. The crosslinked films of main-chain polybenzoxazine precursor (Poly(Si-Bz)) containing 2% $FeCl_3$ by mass were prepared in Teflon molds and could be cured at relatively low temperatures (100–120 °C). Because, Lewis acids are good catalysts to reduce the ring-opening polymerization of benzoxazines, these compounds can even trigger ring-opening reaction at room temperature [15, 75].

During the studies about the effect of Lewis acid catalysts, it was well established that the CH_2–N–CH_2 bridges and phenolic –OH groups of polybenzoxazines can bind metal ions by using nitrogen and oxygen atoms [76–79]. For example, a coordination system was demonstrated between polybenzoxazine and Hg^{2+} ions and the use of polybenzoxazines for environmental issues such as heavy metal extraction from water was reported, previously [80, 81]. As known, similar metal–ligand interactions has also been shown to be beneficial for the fabrication of self-healing materials [82–86]. For example, efficient self-healing was observed for polymers having 2,6-pyridinedicarboxamide groups. In these polymers, adjacent strong and weak metal ion binding sites were located in a single ligand to generate highly dynamic metal–ligand interactions [87]. Similarly, the metal binding ability of polybenzoxazine was exploited and Fe^{3+} ligated polybenzoxazines were prepared for both self-repairing and shape memory applications simply by adding $FeCl_3$ into polybenzoxazines. In this approach, the molecular design concept utilizes two different attractions for self-healing ability such as strong inter- and intra-molecular hydrogen bonding and metal-ligand interactions of Fe^{3+}–phenolic oxygen–tertiary amine ligation. Accordingly, self-standing films of Poly(Si-Bz)/$FeCl_3$ were prepared at low curing temperatures and by this way inner hydrogen-bonding interactions were not disturbed and utilized effectively for self-healing applications. The healing of these films were tested by cutting them into two pieces and then cut edges were firmly placed to maintain efficient contact with applying slight pressure by using glass slides. The film was self-healed after 24 h of contact, and it maintained its flexibility after self-repairing as seen in the visual image (Figure 2.7). The healing efficiency of cut-healed sample was calculated by tensile measurements as *ca.* 61%.

Interestingly, Poly(Si-Bz)/$FeCl_3$ films also exhibited shape recovery (SR) property besides self-healing ability, which is another important feature of smart polymers. In order to demonstrate the shape memory property, curled or spiral shapes were given to these films by using a pin or fastener and then heated at *ca.* 90°C in water. The heated films cooled to room temperature rapidly to fix curled/spiral shapes. These samples were unwrapped by hand and deformation recovered rapidly after removal of force and the films returned back to their fixed shapes in 5–7 s (Figure 2.8).

Similar experiments were performed for the films without containing Fe^{3+} salt and SMP behavior could not be observed. Therefore, in

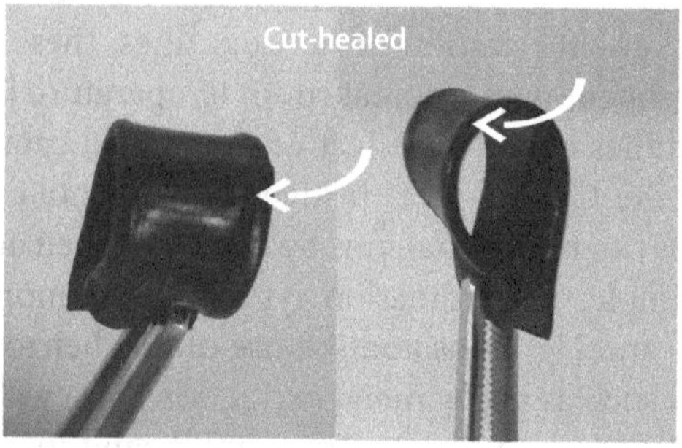

Figure 2.7 The images of cured Poly(Si-Bz)/$FeCl_3$ film after healing. Arrow indicates the contact edges.

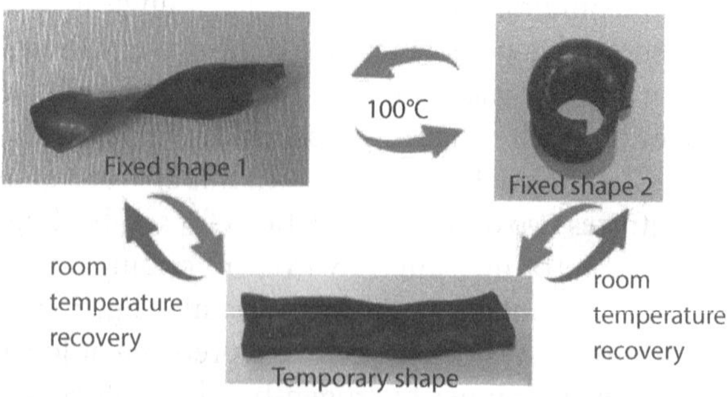

Figure 2.8 Images of shape recovery behavior of cured Poly(Si-Bz)/$FeCl_3$ film.

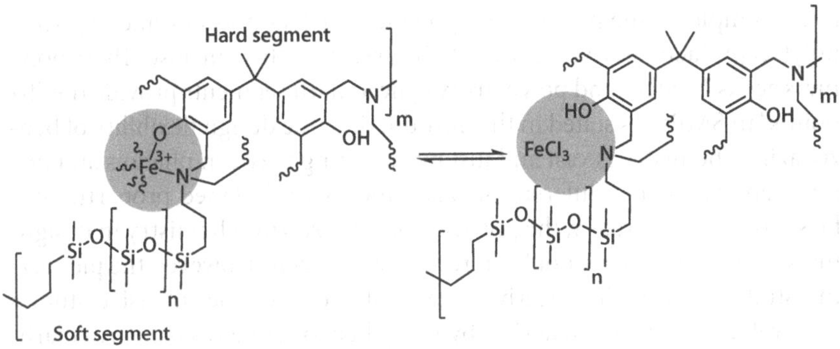

Scheme 2.13 Metal–ligand bond breakage and reforming at certain temperature changes.

the light of blank experiments, it was concluded that SMP behavior of Poly(Si-Bz)/$FeCl_3$ films may stem from dynamic metal ligand interactions between Fe^{3+} and oxygen/nitrogen atoms polybenzoxazine chain (Scheme 2.13). Possibly, the metal complex exhibits reversible weak bonding at 90–100 °C, but during rapid cooling to room temperature the given shape fixes due to reformation of stronger metal–ligand attractions. Generally, in SMP systems hard and soft segments are required and besides dynamic metal–ligand binding, polysiloxane chains act as soft segments and polybenzoxazine provides the hard segment. Accordingly, it can be deduced that polysiloxane chains and metal–ligand interactions form the "switch" and hard polybenzoxazine chains generate the "fix points" in this SMP.

2.4 Conclusion

In this chapter, self-healing and shape memory systems based on benzoxazines are reviewed. The results demonstrate that it is possible design various different smart materials by using benzoxazines. In general, polybenzoxazines with soft segments such as poly(propylene oxide)s (Jeffamines) can be exhibit self-healing ability with different levels of success. The healing mechanisms can be autonomous or stimuli induced types depending on the design of polybenzoxazines. Especially, in terms of autonomous healing systems, the hydrogen bonds of polybenzoxazines can be exploited. However, from the practical point of view, the hydrogen bonding in polybenzoxazines must augmented for successful results by additives or designing special benzoxazines. Moreover, by mixing benzoxazine resins with suitable polymers shape memory polymers can be obtained easily after a simple curing process. Soft polymers such as polysiloxanes, epoxies, poly(-caprolactone)s were proven to be useful for this purpose. These polymers act as "switch" and polybenzoxazine as hard segments provide the "fix points" in SMPs. As stated in the introduction, the design flexibility of benzoxazine chemistry is vast and just by switching between phenols and primary amines, many different benzoxazines with designed properties can be synthesized. Actually, the potential of benzoxazine chemistry in designing smart materials has not been recognized widely. However, the pioneering studies in this line clearly reveal that there will be intensive studies to develop new smart materials by using benzoxazine resins. Particularly, different healing strategies or SMPs based on benzoxazine resins would be one of the major interest in both academia and industry.

References

1. Pilato, L., *Phenolic resins: A century of progress*, Springer-Verlag Berlin, Heidelberg, 2010.
2. Iredale, R.J., Ward, C., Hamerton, I., Modern advances in bismaleimide resin technology: A 21st century perspective on the chemistry of addition polyimides. *Prog. Polym. Sci.*, 69, 1, 2017.
3. Nair, C.P.R., Advances in addition-cure phenolic resins. *Prog. Polym. Sci.*, 29, 401, 2004.
4. Kiskan, B. and Yagci, Y., The journey of phenolics from the first spark to advanced materials. *Isr. J. Chem.*, 2019, In Press 2020, 60, 20-32. https://onlinelibrary.wiley.com/doi/full/10.1002/ijch.201900086.
5. Forsdyke, K.L. and Starr, T.F., *Thermoset resins*, Rapra Technology Limited, Shawbury, UK, 2002.
6. Ghosh, N., Kiskan, B., Yagci, Y., Polybenzoxazines—New high performance thermosetting resins: Synthesis and properties. *Prog. Polym. Sci.*, 32, 1344, 2007.
7. Kiskan, B., Ghosh, N., Yagci, Y., Polybenzoxazine-based composites as high-performance materials. *Polym. Int.*, 60, 167, 2011.
8. Kim, H.J., Brunovska, Z., Ishida, H., Synthesis and thermal characterization of polybenzoxazines based on acetylene-functional monomers. *Polymer*, 40, 6565, 1999.
9. Lin, C.H., Chang, S.L., Shen, T.Y., Shih, Y.S., Lin, H.T., Wang, C.F., Flexible polybenzoxazine thermosets with high glass transition temperatures and low surface free energies. *Polym. Chem.*, 3, 935, 2012.
10. Xu, R.J., Schreiber, H.P., Huang, M., Ishida, H., Polybenzoxazine resins: Aspects of interaction and adsorption behavior. *J. Polym. Sci. Part B: Polym. Phys.*, 37, 1441, 1999.

11. Ishida, H. and Allen, D.J., Physical and mechanical characterization of near-zero shrinkage polybenzoxazines. *J. Polym. Sci. Part B: Polym. Phys.*, 34, 1019, 1996.
12. Ishida, H. and Lee, Y.H., Synergism observed in polybenzoxazine and poly(epsilon-caprolactone) blends by dynamic mechanical and thermogravimetric analysis. *Polymer*, 42, 6971, 2001.
13. Cid, J.A., Wang, Y.X., Ishida, H., Cationic polymerization of benzoxazine monomers by boron trifluoride complex. *Polym. Polym. Compos.*, 7, 409, 1999.
14. Zhang, K., Tan, X., Wang, Y., Ishida, H., Unique self-catalyzed cationic ring-opening polymerization of a high performance deoxybenzoin-based 1,3-benzoxazine monomer. *Polymer*, 168, 8, 2019.
15. Dunkers, J. and Ishida, H., Reaction of benzoxazine-based phenolic resins with strong and weak carboxylic acids and phenols as catalysts. *J. Polym. Sci. Part A: Polym. Chem.*, 37, 1913, 1999.
16. Andreu, R., Reina, J.A., Ronda, J.C., Carboxylic acid-containing benzoxazines as efficient catalysts in the thermal polymerization of benzoxazines. *J. Polym. Sci. Part A: Polym. Chem.*, 46, 6091, 2008.
17. Hamerton, I., McNamara, L.T., Howlin, B.J., Smith, P.A., Cross, P., Ward, S., Examining the initiation of the polymerization mechanism and network development in aromatic polybenzoxazines. *Macromolecules*, 46, 5117, 2013.
18. Ishida, H. and Rodriguez, Y., Curing kinetics of a new benzoxazine-based phenolic resin by differential scanning calorimetry. *Polymer*, 36, 3151, 1995.
19. Su, Y.C., Yei, D.R., Chang, F.C., The kinetics of b-a and p-a type copolybenzoxazine via the ring opening process. *J. Appl. Polym. Sci.*, 95, 730, 2005.
20. Liu, C., Shen, D., Sebastián, R.M.a., Marquet, J., Schönfeld, R., Mechanistic studies on ring-opening polymerization of benzoxazines: A mechanistically based catalyst design. *Macromolecules*, 44, 4616, 2011.
21. Liu, C., Shen, D., Sebastián, R.M., Marquet, J., Schönfeld, R., Catalyst effects on the ring-opening polymerization of 1,3-benzoxazine and on the polymer structure. *Polymer (United Kingdom)*, 54, 2873, 2013.
22. Imran, M., Kiskan, B., Yagci, Y., Concise synthesis and characterization of unsymmetric 1, 3-benzoxazines by tandem reactions. *Tetrahedron Lett.*, 54, 4966, 2013.
23. Kaya, G., Kiskan, B., Yagci, Y., Phenolic naphthoxazines as curing promoters for benzoxazines. *Macromolecules*, 51, 1688, 2018.
24. Kaya, G., Kiskan, B., Yagci, Y., Coumarines as masked phenols for amide functional benzoxazines. *Polym. Chem.*, 10, 1268, 2019.
25. Andreu, R., Espinosa, M.A., Galia, M., Cadiz, V., Ronda, J.C., Reina, J.A., Synthesis of novel benzoxazines containing glycidyl groups: A study of the crosslinking behavior. *J. Polym. Sci. Part A: Polym. Chem.*, 44, 1529, 2006.
26. Zuniga, C., Larrechi, M.S., Lligadas, G., Ronda, J.C., Galia, M., Cadiz, V., Polybenzoxazines from renewable diphenolic acid. *J. Polym. Sci. Part A: Polym. Chem.*, 49, 1219, 2011.

27. Oie, H., Sudo, A., Endo, T., Acceleration effect of n-allyl group on thermally induced ring-opening polymerization of 1,3-benzoxazine. *J. Polym. Sci. Part A: Polym. Chem.*, 48, 5357, 2010.
28. Agag, T. and Takeichi, T., Synthesis and characterization of novel benzoxazine monomers containing allyl groups and their high performance thermosets. *Macromolecules*, 36, 6010, 2003.
29. Baqar, M., Agag, T., Ishida, H., Qutubuddin, S., Methylol-functional benzoxazines as precursors for high-performance thermoset polymers: Unique simultaneous addition and condensation polymerization behavior. *J. Polym. Sci. Part A: Polym. Chem.*, 50, 2275, 2012.
30. Lyu, Y. and Ishida, H., Natural-sourced benzoxazine resins, homopolymers, blends and composites: A review of their synthesis, manufacturing and applications. *Prog. Polym. Sci.*, 99, 101168, 2019.
31. Zhang, K., Han, M., Han, L., Ishida, H., Resveratrol-based tri-functional benzoxazines: Synthesis, characterization, polymerization, and thermal and flame retardant properties. *Eur. Polym. J.*, 116, 526, 2019.
32. Kiskan, B., Adapting benzoxazine chemistry for unconventional applications. *React. Funct. Polym.*, 129, 76, 2018.
33. Brown, E.N., White, S.R., Sottos, N.R., Microcapsule induced toughening in a self-healing polymer composite. *J. Mater. Sci.*, 39, 1703, 2004.
34. Wool, R.P. and Oconnor, K.M., A theory of crack healing in polymers. *J. Appl. Phys.*, 52, 5953, 1981.
35. White, S.R., Sottos, N.R., Geubelle, P.H., Moore, J.S., Kessler, M.R., Sriram, S.R., Brown, E.N., Viswanathan, S., Autonomic healing of polymer composites. *Nature*, 409, 794, 2001.
36. Kessler, M.R., Sottos, N.R., White, S.R., Self-healing structural composite materials. *Compos. Part A: Appl. S.*, 34, 743, 2003.
37. Kumar, A., Stephenson, L.D., Murray, J.N., Self-healing coatings for steel. *Prog. Org. Coat.*, 55, 244, 2006.
38. Stuart, M.A.C., Huck, W.T.S., Genzer, J., Müller, M., Ober, C., Stamm, M., Sukhorukov, G.B., Szleifer, I., Tsukruk, V.V., Urban, M. *et al.*, Emerging applications of stimuli-responsive polymer materials. *Nat. Mater.*, 9, 101, 2010.
39. Lu, W., Le, X.X., Zhang, J.W., Huang, Y.J., Chen, T., Supramolecular shape memory hydrogels: A new bridge between stimuli-responsive polymers and supramolecular chemistry. *Chem. Soc. Rev.*, 46, 1284, 2017.
40. Basak, S., Nanda, J., Banerjee, A., Multi-stimuli responsive self-healing metallo-hydrogels: Tuning of the gel recovery property. *Chem. Commun.*, 50, 2356, 2014.
41. Liu, Y.-L. and Chuo, T.-W., Self-healing polymers based on thermally reversible diels-alder chemistry. *Polym. Chem.*, 4, 2194, 2013.
42. Zhang, M.Q. and Rong, M.Z., Intrinsic self-healing of covalent polymers through bond reconnection towards strength restoration. *Polym. Chem.*, 4, 4878, 2013.

43. Xu, W.M., Rong, M.Z., Zhang, M.Q., Sunlight driven self-healing, reshaping and recycling of a robust, transparent and yellowing-resistant polymer. *J. Mater. Chem. A*, 4, 10683, 2016.
44. Brown, E.N., Sottos, N.R., White, S.R., Fracture testing of a self-healing polymer composite. *Exp. Mech.*, 42, 372, 2002.
45. Cordier, P., Tournilhac, F., Soulié-Ziakovic, C., Leibler, L., Self-healing and thermoreversible rubber from supramolecular assembly. *Nature*, 451, 977, 2008.
46. Yuan, Y.C., Rong, M.Z., Zhang, M.Q., Chen, J., Yang, G.C., Li, X.M., Self-healing polymeric materials using epoxy/mercaptan as the healant. *Macromolecules*, 41, 5197, 2008.
47. Phadke, A., Zhang, C., Arman, B., Hsu, C.-C., Mashelkar, R.A., Lele, A.K., Tauber, M.J., Arya, G., Varghese, S., Rapid self-healing hydrogels. *Proc. Natl. Acad. Sci. U.S.A.*, 109, 4383, 2012.
48. Abdallh, M., Hearn, M.T.W., Simon, G.P., Saito, K., Light triggered self-healing of polyacrylate polymers crosslinked with 7-methacryloyoxycoumarin crosslinker. *Polym. Chem.*, 8, 5875, 2017.
49. Kim, H.D. and Ishida, H., A study on hydrogen bonding in controlled-structure benzoxazine model oligomers. *Macromol. Symp.*, 195, 123, 2003.
50. Yang, P., Wang, X., Fan, H., Gu, Y., Effect of hydrogen bonds on the modulus of bulk polybenzoxazines in the glassy state. *Phys. Chem. Chem. Phys.*, 15, 15333, 2013.
51. Shen, X., Cao, L., Liu, Y., Dai, J., Liu, X., Zhu, J., Du, S., How does the hydrogen bonding interaction influence the properties of polybenzoxazine? An experimental study combined with computer simulation. *Macromolecules*, 51, 4782, 2018.
52. Kiskan, B. and Yagci, Y., Self-healing of poly (propylene oxide)-polybenzoxazine thermosets by photoinduced coumarine dimerization. *J. Polym. Sci. Part A: Polym. Chem.*, 52, 2911, 2014.
53. Arslan, M., Kiskan, B., Yagci, Y., Benzoxazine-based thermosets with autonomous self-healing ability. *Macromolecules*, 48, 1329, 2015.
54. Arslan, M., Kiskan, B., Yagci, Y., Combining elemental sulfur with polybenzoxazines via inverse vulcanization. *Macromolecules*, 49, 767, 2016.
55. Lin, H.K. and Liu, Y.L., Sulfur radical transfer and coupling reaction to benzoxazine groups: A new reaction route for preparation of polymeric materials using elemental sulfur as a feedstock. *Macromol. Rapid Commun.*, 39, 1700832, 2018.
56. Arslan, M., Kiskan, B., Yagci, Y., Recycling and self-healing of polybenzoxazines with dynamic sulfide linkages. *Sci. Rep.*, 7, 5207, 2017.
57. Tasdelen, M.A., Kumbaraci, V., Talinli, N., Yagci, Y., Photochemically masked benzophenone: Photoinitiated free radical polymerization by using benzodioxinone. *Polymer*, 47, 7611, 2006.

58. Tasdelen, M.A. and Yagci, Y., Benzodioxinone photochemistry in macromolecular science: Progress, challenges, and opportunities. *ACS Macro Lett.*, 6, 1392, 2017.
59. Arslan, M., Motallebzadeh, A., Kiskan, B., Demirel, A.L., Kumbaraci, I.V., Yagci, Y., Combining benzoxazine and ketene chemistries for self-healing of high performance thermoset surfaces. *Polym. Chem.*, 9, 2031, 2018.
60. Fu, F., Huang, M., Zhang, W., Zhao, Y., Liu, X., Thermally assisted self-healing behavior of anhydride modified polybenzoxazines based on transesterification. *Sci. Rep.*, 8, 10325, 2018.
61. Zhang, L., Zhao, Z., Dai, Z., Xu, L., Fu, F., Endo, T., Liu, X., Unexpected healability of an ortho-blocked polybenzoxazine resin. *ACS Macro Lett.*, 8, 506, 2019.
62. Lendlein, A. and Kelch, S., Shape-memory polymers. *Angew. Chem. Int. Edit.*, 41, 2034, 2002.
63. Bilici, C., Can, V., Nöchel, U., Behl, M., Lendlein, A., Okay, O., Melt-processable shape-memory hydrogels with self-healing ability of high mechanical strength. *Macromolecules*, 49, 7442, 2016.
64. Lendlein, A., Jiang, H.Y., Junger, O., Langer, R., Light-induced shape-memory polymers. *Nature*, 434, 879, 2005.
65. Okay, O., Semicrystalline physical hydrogels with shape-memory and self-healing properties. *J. Mater. Chem. B*, 7, 1581, 2019.
66. Kurt, B., Gulyuz, U., Demir, D.D., Okay, O., High-strength semi-crystalline hydrogels with self-healing and shape memory functions. *Eur. Polym. J.*, 81, 12, 2016.
67. Kimura, H., Matsumoto, A., Hasegawa, K., Ohtsuka, K., Fukuda, A., Epoxy resin cured by bisphenol a based benzoxazine. *J. Appl. Polym. Sci.*, 68, 1903, 1998.
68. Rao, B.S., Reddy, K.R., Pathak, S.K., Pasala, A., Benzoxazine-epoxy copolymers: Effect of molecular weight and crosslinking on thermal and viscoelastic properties. *Polym. Int.*, 54, 1371, 2005.
69. Liu, Y., Gao, S., Gong, X., Xue, Q., Lu, Z., Benzoxazine-epoxy thermosets with smectic phase structures for high thermal conductive materials, *Liq. Cryst.*, 46, 11, 1686–1695, 2019.
70. Rimdusit, S. and Ishida, H., Synergism and multiple mechanical relaxations observed in ternary systems based on benzoxazine, epoxy, and phenolic resins. *J. Polym. Sci. Part B: Polym. Phys.*, 38, 1687, 2000.
71. Sarawut, R., Montha, L., Kasinee, H., Pornnapa, K., Isala, D., Shape memory polymers from benzoxazine-modified epoxy. *Smart Mater. Struct.*, 22, 075033, 2013.
72. Schäfer, H., Hartwig, A., Koschek, K., The nature of bonding matters: Benzoxazine based shape memory polymers. *Polymer*, 135, 285, 2018.
73. Schäfer, H. and Koschek, K., Effect of poly(-caprolactone) in polybenzoxazine blends and respective copolymers on morphology and mechanical properties. *Eur. Polym. J.*, 108, 582, 2018.

74. Arslan, M., Kiskan, B., Yagci, Y., Benzoxazine-based thermoset with autonomous self-healing and shape recovery. *Macromolecules*, 51, 10095, 2018.
75. Sudo, A., Hirayama, S., Endo, T., Highly efficient catalysts-acetylacetonato complexes of transition metals in the 4th period for ring-opening polymerization of 1,3-benzoxazine. *J. Polym. Sci. Part A: Polym. Chem.*, 48, 479, 2010.
76. Arslan, M., Kiskan, B., Yagci, Y., Ring-opening polymerization of 1,3-benzoxazines via borane catalyst. *Polymers*, 10, 239, 2018.
77. Low, H.Y. and Ishida, H., Improved thermal stability of polybenzoxazines by transition metals. *Polym. Degrad. Stabil.*, 91, 805, 2006.
78. Chaisuwan, T., Komalwanich, T., Luangsukrerk, S., Wongkasemjit, S., Removal of heavy metals from model wastewater by using polybenzoxazine aerogel. *Desalination*, 256, 108, 2010.
79. Aydogan, C., Kiskan, B., Hacioglu, S.O., Toppare, L., Yagci, Y., Electrochemical manipulation of adhesion strength of polybenzoxazines on metal surfaces: From strong adhesion to dismantling. *RSC Adv.*, 4, 27545, 2014.
80. Akay, S., Kayan, B., Kalderis, D., Arslan, M., Yagci, Y., Kiskan, B., Poly(benzoxazine-co-sulfur): An efficient sorbent for mercury removal from aqueous solution. *J. Appl. Polym. Sci.*, 134, 45306, 2017.
81. Taskin, O., Kiskan, B., Aksu, A., Balkis, N., Weber, J., Yagci, Y., Polybenzoxazine: A powerful tool for removal of mercury salts from water. *Chem.—Eur. J.*, 20, 10953, 2014.
82. Balkenende, D.W.R., Coulibaly, S., Balog, S., Simon, Y.C., Fiore, G.L., Weder, C., Mechanochemistry with metallosupramolecular polymers. *J. Am. Chem. Soc.*, 136, 10493, 2014.
83. Bode, S., Bose, R.K., Matthes, S., Ehrhardt, M., Seifert, A., Schacher, F.H., Paulus, R.M., Stumpf, S., Sandmann, B., Vitz, J. *et al.*, Self-healing metallopolymers based on cadmium bis(terpyridine) complex containing polymer networks. *Polym. Chem.*, 4, 4966, 2013.
84. Bode, S., Zedler, L., Schacher, F.H., Dietzek, B., Schmitt, M., Popp, J., Hager, M.D., Schubert, U.S., Self-healing polymer coatings based on crosslinked metallosupramolecular copolymers. *Adv. Mater.*, 25, 1634, 2013.
85. Sandmann, B., Happ, B., Kupfer, S., Schacher, F.H., Hager, M.D., Schubert, U.S., The self-healing potential of triazole-pyridine-based metallopolymers. *Macromol. Rapid Commun.*, 36, 604, 2014.
86. Zheng, P., Arantes, G.M., Field, M.J., Li, H., Force-induced chemical reactions on the metal centre in a single metalloprotein molecule. *Nat. Commun.*, 6, 7569, 2015.
87. Li, C.-H., Wang, C., Keplinger, C., Zuo, J.-L., Jin, L., Sun, Y., Zheng, P., Cao, Y., Lissel, F., Linder, C. *et al.*, A highly stretchable autonomous self-healing elastomer. *Nat. Chem.*, 8, 618, 2016.

3

Self-Healable Elastomers

Mariajose Cova Sánchez[1], Daniela Belén García[2], Mariano Martin Escobar[2,3]* and Marcela Mansilla[3]

[1]Departamento de Asistencia a la Industria del Caucho, Instituto Nacional de Tecnología Industrial, Buenos Aires, Argentina
[2]Departamento de Diseño de materiales, Instituto Nacional de Tecnología Industrial, Buenos Aires, Argentina
[3]INTI, CONICET, Departamento de Diseño de Materiales, Buenos Aires, Argentina

Abstract
The chapter is divided in three sections: the first one is an introduction about the characteristics of self-healing in elastomers; the second one included the characterization methods of healing ability. The last section consists on a review of self-healing in particular cases of different elastomer matrices, such as: natural rubber (NR), styrene butadiene rubber (SBR), polybutadiene rubber (BR), bromobutyl rubber (BIIR), silicones and polyurethane (PU).

Keywords: Healing, rubber, disulfide, hydrogen bonding, diels–alder

3.1 Introduction

Elastomers are elastic polymers that are highly flexible. They are used in a huge number of applications, such as tires, seals, biomaterials, soft robotics, wearable electronic devices and sensors. Elastomer compounds present excellent mechanical properties due to the formation of a stable three dimensional network of covalent bonds in the vulcanization process, but make the material not or almost not self-healable due to the irreversibility of the covalent bonds. Actually, the rubber industry looks for new challenges to contribute to sustainable development. Consequently, it is necessary to develop sustainable materials and methods in order to extend the

**Corresponding author*: mescobar@inti.gob.ar

Inamuddin, Mohd Imran Ahamed, Rajender Boddula and Tariq Altalhi (eds.) Self-Healing Smart Materials and Allied Applications, (65–98) © 2021 Scrivener Publishing LLC

life cycle of vulcanized elastomers [1]. An alternative to reach this goal is the concept of self-healable elastomeric compounds.

Self-healable materials can be defined as a smart material that responds to stimulus to extend its service life by restoring mechanical integrity after damage. The capacity for self-healing is usually observed in animals and plants. Any industrial component with this capacity would be beneficiated in the presence of defects in service, such as micro cracks [2, 3]. The goal of developing self-healable materials is to extend the long-term durability by effectively removing any local damage that occurs during its service life [4–8]. Another advantage of self-healable rubber is that can prevent catastrophic failure [9].

Self-healable materials can be classified as autonomous (do not need a stimulus to conduct the self-healing process), or non-autonomous, in which a stimulus is required to trigger the self-healing mechanism. Some of the most common stimuli are light, mechanical, chemical or heat (Figure 3.1).

Other possible classifications of self-healing materials are as extrinsic and intrinsic types, according to the healing mechanism [10, 11]. The first one is characterized by the storage of the healing agents in nanotubes, microcapsules, or other micro-containers, which can be released after the matrix suffers damage [12, 13]. Intrinsic type refers to those polymers containing covalent or non-covalent either reversible bonds, which activate by external stimulation after damage. Dynamic covalent bonds can be disulfide bonding [14] and reversible Diels–Alder reaction (DA) [15], whereas dynamic non-covalent bonds can be metal–ligand coordination

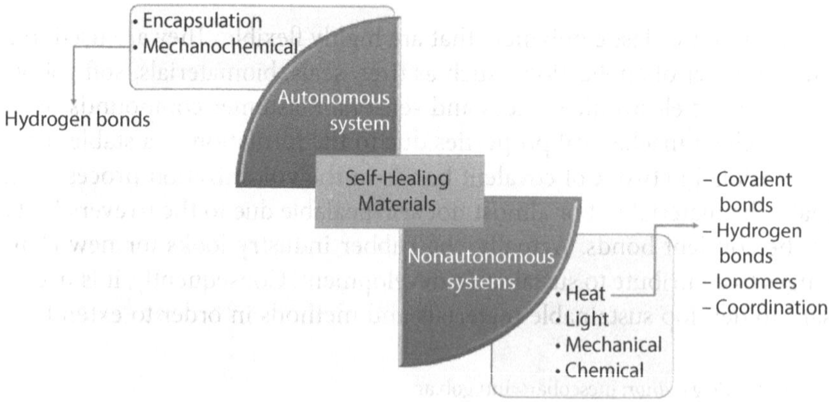

Figure 3.1 Classification scheme of self-healing systems.

[16, 17], hydrogen bonds [18, 19], hydrophobic and π–π interactions [20–22]. When the self-healing mechanism is due to dynamic bonds, the mechanical strengths are weaker than those systems with permanent covalent bonds. To overcome this drawback, some authors proposed the incorporation of crystalline domains within the polymer matrix to increase the toughness of the system [23–26].

Covalent bonds are also capable to self-cure partially when the compound is formulated in a way to obtain high content of di- and polysulfides bridges and a low crosslinks density that permits the rearrangement of the bonds that had been broken at the healed interface. Sometimes, this kind of healing mechanism is preferred over hydrogen bonds, because the last one breaks easily when a mechanical effort is applied [2].

When thermal activation for self-healing is used, systems based on DA reaction are the most common. This reaction consists in a [4 + 2] cycloaddition between an electron poor dienophile and an electron rich diene, producing a cyclohexene adduct which could be reverted via retro-DA reaction (rDA) using high temperature conditions (see Figure 3.2). This reaction presents fast kinetics and soft reaction conditions [27].

Amide is another important bond type for self-healing applications, but it requires extreme conditions to be reversed. A reversion method includes the use of bulky substituents that promotes the breakage of the amine bond.

Wool *et al.* [28] analyzed the mechanism of intrinsic self-healing polymers and summarized that the process takes five phases: (i) surface rearrangement, (ii) surface approach, (iii) wetting, (iv) diffusion and (v) randomization during which the longitudinal chain diffusion is responsible for crack healing.

3.2 Self-Healing in Elastomers

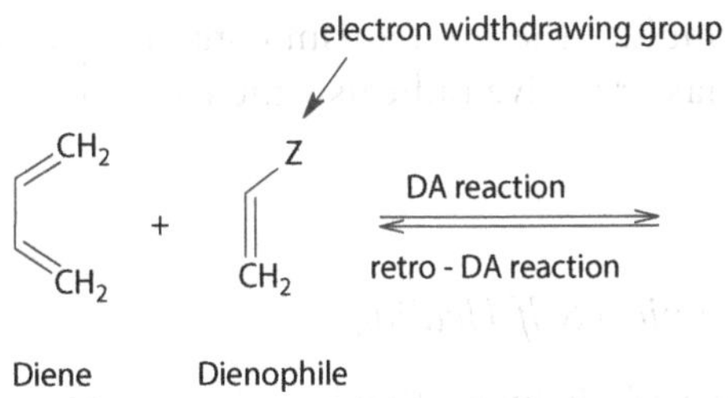

Figure 3.2 Schematic representation of thermally reversible Diels–Alder (DA) reaction of a diene with a dienophile.

3.2.1 Self-Healing Mechanism

There are four mechanisms to trigger the self-healable process in materials: heat induced, light induced, mechanochemical and encapsulation (Figure 3.3). Mechanochemical self-healing and encapsulation do not need a stimulus and the sample can be healed in a variety of conditions. The non-autonomous heat and light triggered self-healing systems have been more developed [2]. Each mechanism is described briefly.

3.2.1.1 Heat Stimulated Self-Healing

Heat induced self-healing can be performed in reversible or dynamic chemical bonds, such as DA reaction, radical based systems or supramolecular interactions: ionic interactions, halogen or hydrogen bonds, metal–ligand interactions, π–π interactions or host-guest interactions [29].

The DA reaction is the most commonly used mechanism when it comes to thermo-reversible self-healing based on covalent bonds [30, 31]. Authors describe the healing process in four steps: a) the material is heated up to break the DA bonds, b) part of thermal energy becomes in kinetic energy that increases the mobility of chains, promoting the filling of the voids; c) cool the sample, allowing the formation of DA bonds; and d) the mechanical properties of the material are recovered.

3.2.1.2 Light Stimulated Self-Healing

There are different methods to promote the self-healing process induced by light. The first one consists of using a photo crosslinking reaction between molecular parts that are photo reactive. Another method uses rebuild that links up polymer chains on both side of the cracks. Covalent bonds like disulfides, allyl sulfides or trithiocarbonates are susceptible to this stimulus. The mechanism consists of breaking these covalent bonds and make reactive radicals which are combined to form new bonds.

3.2.1.3 Mechanochemical Self-Healing

This mechanism is particularly interesting in poly-urea and polyurethane urea based systems, in which the amine bonds is present [32]. The reversion of the amine bond can form ketene, which is highly reactive for the reversible interaction. Therefore, the isocyanate can be used as a reactant with an amine group to form a urea bond. This type of bonds is stronger compared to hydrogen bonding.

Rekondo *et al.* [33] developed a self-healing poly (urea urethane) that contains both amino and quadruple hydrogen bonds. A sample with those characteristics can be completely separated and bonded by physical contact, exhibiting a rapid self-healing ability at room temperature due to the combination of both bonds type.

3.2.1.4 *Encapsulation*

There are different strategies to contain the healing agents inside the material such as hollow fibers or a microvascular network filled with a crosslinking agent, microspheres or microcapsules of crosslinking agent and superparamagnetic nanoparticles, as is shown in Figure 3.3 [34]. The mechanism to heal the sample is similar in cases of hollow fiber, microspheres or microcapsules and microvascular network: once the crack propagates, the container is broken and releases the crosslinking agents; they fill the crack and then repair the structure. In the case of super paramagnetic nanoparticles dispersed in the matrix, the self-healing mechanism is activated through the application of an external magnetic field that induces a vibration in the particles. Therefore, the temperature increases and facilitates the self-healing.

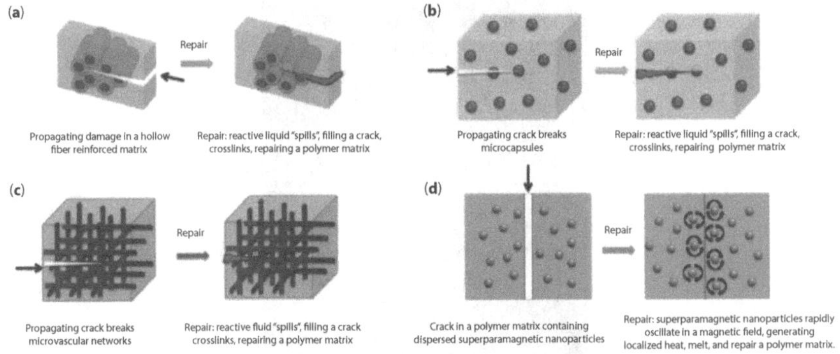

Figure 3.3 Self-healing illustration of common encapsulation methods, (a) hollow fibers inside a polymer, (b) Microspheres/microcapsules, (c) microvascular network, and (d) superparamagnetic nanoparticles (Reprinted from Yang *et al.* [34], open access).

3.2.2 Characterization of Healing Process

The healing efficiency can be calculated as the ratio between the tensile strength results for the healed and original sample according to Equation (3.1) [35–38]:

$$healing\ efficiency\,[\%]=\frac{\sigma_b^{healed}}{\sigma_b^{pristine}}\times 100 \tag{3.1}$$

This relation also works for others mechanical properties such as failure stress, failure strain, energy at break or J-integral at crack initiation [36].

Raman Spectroscopy can be used following the evolution of the three main spectral bands: and C–C stretching (υ C–C = 1,590 cm^{-1}), C–S stretching (υ C–S = 650 cm^{-1}), S–S stretching (υ S–S = 500 cm^{-1}) [35], which are related to the chemical structure of the crosslinked material.

In addition, there are optical methods to characterize the healed sample. Optical microscope and digital cameras are used to analyze the healing process [39] and also scanning electron microscopy (SEM) to study the

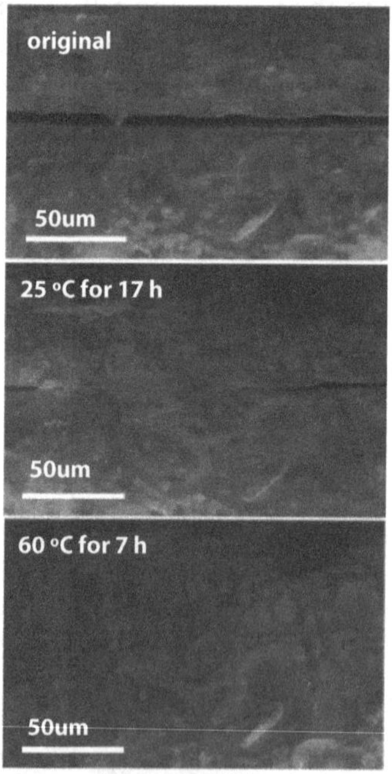

Figure 3.4 SEM images to trace the healing process of the scratch. (Reused with permission and modified after Peng *et al.* [40]).

healing of the scratches [40] as can be seen in Figure 3.4. The figure shows two samples of a compound, which were cut and submitted at different conditions to recover the mechanical properties: room temperature (that take 17 h) and 60 °C (that take 7 h).

3.3 Particular Cases in Different Elastomers

3.3.1 Natural Rubber (NR)

Natural Rubber (NR) is obtained from rubber plants through coagulation process being the most used type the *Hevea brasiliensis*. NR is chemically composed by cis-1,4-polyisoprene which presents exceptional properties like superior tear resistance, high resilience and fatigue resistance.

Some authors have studied the reprocessability of epoxidized natural rubber (ENR), adding different fillers. Cao *et al.* reported the preparation and characterization of ENR modified by a 2,2,6,6-tetramethylpiperidine-1-oxyl (TEMPO) oxidized cellulose nanocrystals (TOCNs), generating free carboxyl groups, which serve as reinforcing fillers and also as cross-linking agents via epoxy-acid reaction [41]. The linkages between the rubber chains and the reinforcement particles consisted in β-hydroxyl ester, as can be seen in Figure 3.5. The interfacial ester bonds were formed between epoxy groups and carboxyl groups and that was probed by FTIR results.

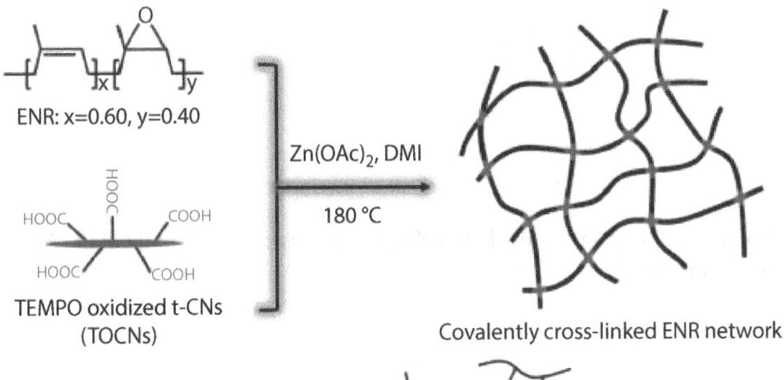

Figure 3.5 Illustration of the TOCNs cross-linked ENR network based on epoxy-acid reaction, using zinc acetate (Zn(OAc)2) and 1,2-dimethylimidazole (DMI) as additives (Reprinted with permission from Cao *et al.* [41]).

The covalent rubber network presented high fracture strain (>600%) and tensile stress (>5 MPa). Furthermore, the network with exchangeable β-hydroxyl ester bonds at rubber–TOCNs interface can change the network topology via trans-esterification reactions. The samples achieved up to 80% self-healing efficiency.

Xu *et al.* developed recyclable and self-healable ENR/citric acid-modified bentonite (CABt) composites (Figure 3.6) [42]. CABt presents numerous carboxyl groups on surface, which react with ENR through exchangeable β-hydroxyl ester linkages, as can be seen in Figure 3.7. Meanwhile, the inherent stickiness of ENR matrix and the low crosslinked network facilitate transesterification reactions of β-hydroxyl ester linkages and chains diffusion, which make ENR/CABt composites recyclable and healable.

Some studies have focused on the DA reactions because of its reversible crosslinking property. Trovatti *et al.* [43] report the furan-modified NR reversible crosslinking. A comparison between the furan-modified NR 1H-NMR spectra and its retro-DA de-crosslinked rubber gives evidence of the reversibility of the reaction. Tanasi *et al.* [27] studied the use of a DA reaction to obtain a NR reversibly crosslinked and with self-repair capacity. NR was crosslinked via DA reactions following the procedure shown in

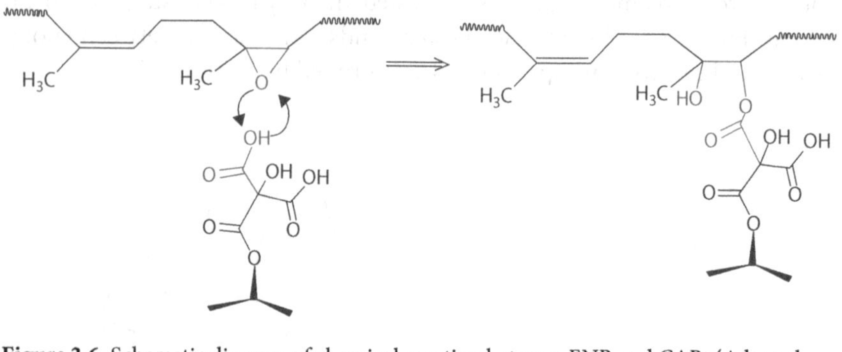

Figure 3.6 Schematic diagram of chemical reaction between ENR and CABt (Adapted with permission from Xu *et al.* [42]).

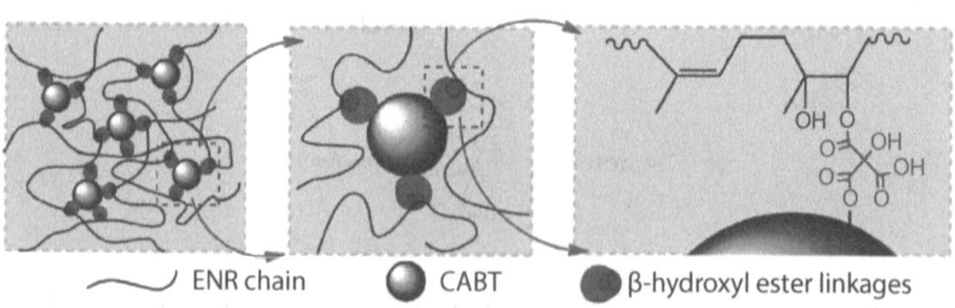

Figure 3.7 Schematic diagram of hybrid rubber network in cured ENR/CABt composite CABt (Reprinted with permission from Xu *et al.* [42]).

Figure 3.8: i) NR reacted with maleic anhydride to produce NR-g-MA; ii) furan moieties were grafted to the rubber generating NR-g-furan; iii) pending furans were crosslinked with a bismaleimide, obtaining thermoreversible bridges which crosslinked NR matrix (NR-DA). Authors explained through phenomenological model that the reversibility of the crosslinks could be achieved by dynamic mechanical analysis complemented with chemical analysis. The results show that the rubber compound present mechanical

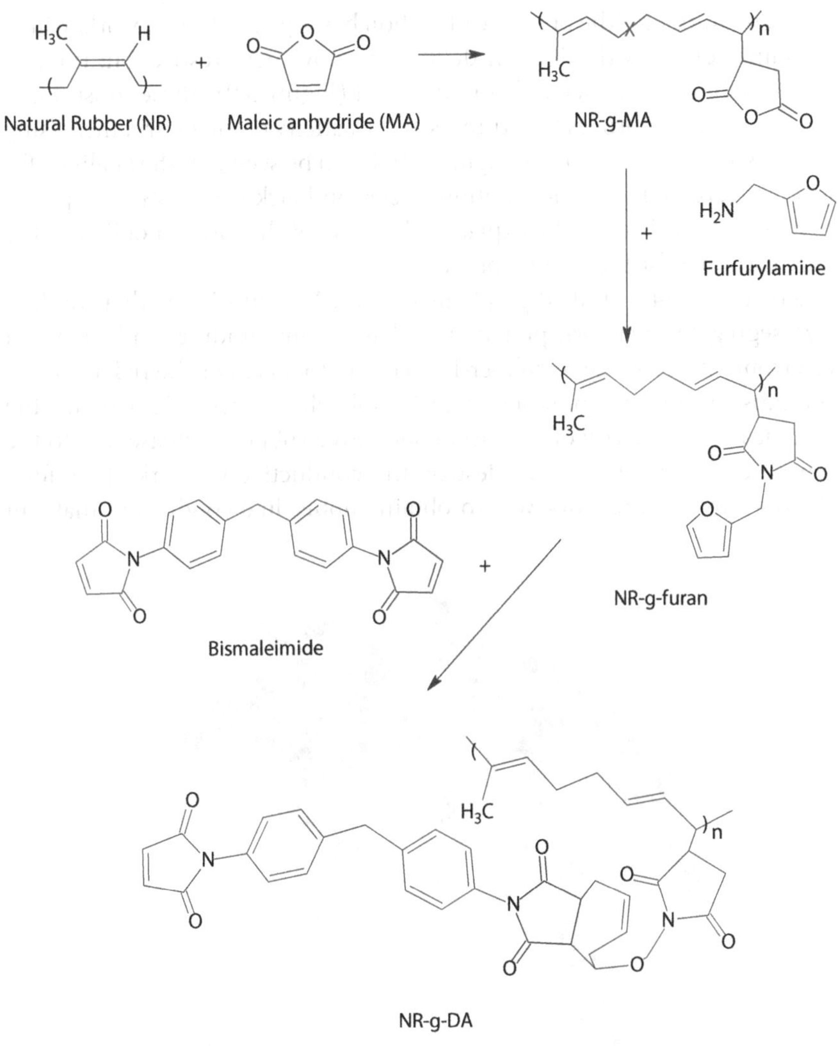

Figure 3.8 Schematic representation of the chemical reaction between: (a) NR and MA, (b) NR-g-MA and FFA, (c) NR-g-furan and bismaleimide (Adapted with permission from Tanasi *et al.* [27]).

properties and a crosslinks density comparable whit those of a vulcanized NR with low sulfur content, that represent a healing efficiency greater than 80% at low strain.

More recently, Khimi *et al.* studied the influence of carbon black on the self-healing efficiency of NR and investigated an intrinsic self-healing NR by ionomeric interactions [9]. Compounds using NR were mixed in a conventional laboratory two roll mill, combining with Zinc Oxide, Stearic acid, Zinc thiolate, dicumyl peroxide (DCP) and Carbon Black grade N330 in different amounts. The presence of carbon black generate a network, which contains rubber with different degrees of movement restriction: trapped rubber, occluded rubber and bound rubber (Figure 3.9). These constrained rubber region improves the stiffness and elasticity of the matrix increasing tensile strength. However, in Figure 3.10 it can be seen that the healing efficiency decreased with the addition of carbon black from 98% to approximately 40%. This could be explained because of the lower mobility of the chains rubber for the healing process.

Zhan *et al.* [44] studied graphene (GE)/NR composites with a conductive segregated network prepared by latex compounding, with focus in electronic devices due to rubber-like conductors are considered as one of the most important components in flexible electronics [45]. Usually, the electrical conductivity of a typical conductive rubber decrease due to the tensile cycles in service can destroy the conductive network. Therefore, the objective of the work was to obtain rubber-like conductors that can

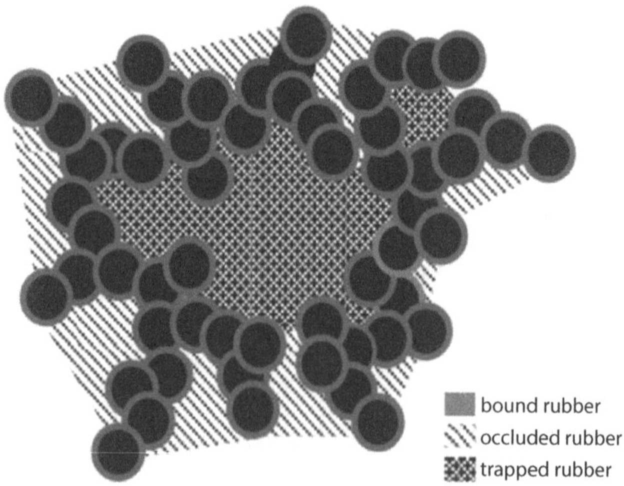

Figure 3.9 Schematic presentation of carbon black filler network (Reprinted with permission from Khimi *et al.* [9]).

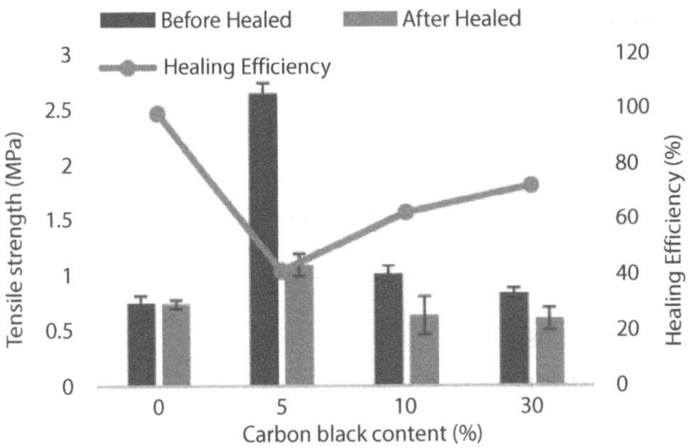

Figure 3.10 Healing efficiency of tensile strength with different carbon black content (Reprinted with permission from Khimi *et al.* [9]).

recover its original properties by a post-treatment. Samples with 10 phr of GE were made and an electrical conductivity of 2.7 S/m was obtained, with a relatively good flexibility. Then, after 4 tensile cycles and subsequent thermal treatment, the electrical conductivity of the sample increased nearly 2 times than that without treatment, indicating that the network destroyed during the tensile cycles can be healed during the post thermal treatment.

Hernandez *et al.* [35] analyzed the influence of the disulfide and polysulfide bonds on the healing ability in NR with a conventional accelerator/sulfur ratio fixed in 0.2. The samples were vulcanized at two conditions: t_{90} (optimum time of cure) and t_{50} (partially vulcanized), obtaining a recovery of the mechanical properties in the last condition after a treatment at 70 °C for 7 h. The result highlight the compromise between healing capability and mechanical performance, in which the crosslinks density, the amount of sulfur, and the amount of labile disulfide and polysulfide crosslinks are key parameters. Through electron spin resonance it was stated that the healing mechanism is based on free sulfur radicals that are thermally induced.

Recently, Utrera-Barrios *et al.* [46] reported the development of an ENR with thermally reduced graphene oxide (TRGO) nanocomposite vulcanized with dicumyl peroxide (DCP). They found that the mechanical performance improved with the incorporation of TRGO compared with pristine ENR in more than 100% and promotes the hydrogen bonding interactions. In addition, the presence of TRGO gives self-healing capability to the system.

During the vulcanization of with organic peroxides there are two possible reactions: (1) the crosslinks can be generated by the abstraction of the adjacent hydrogen to the double bond to form the C–C bond and (2) the rupture of the double bond results in the chain growing. When ENR is

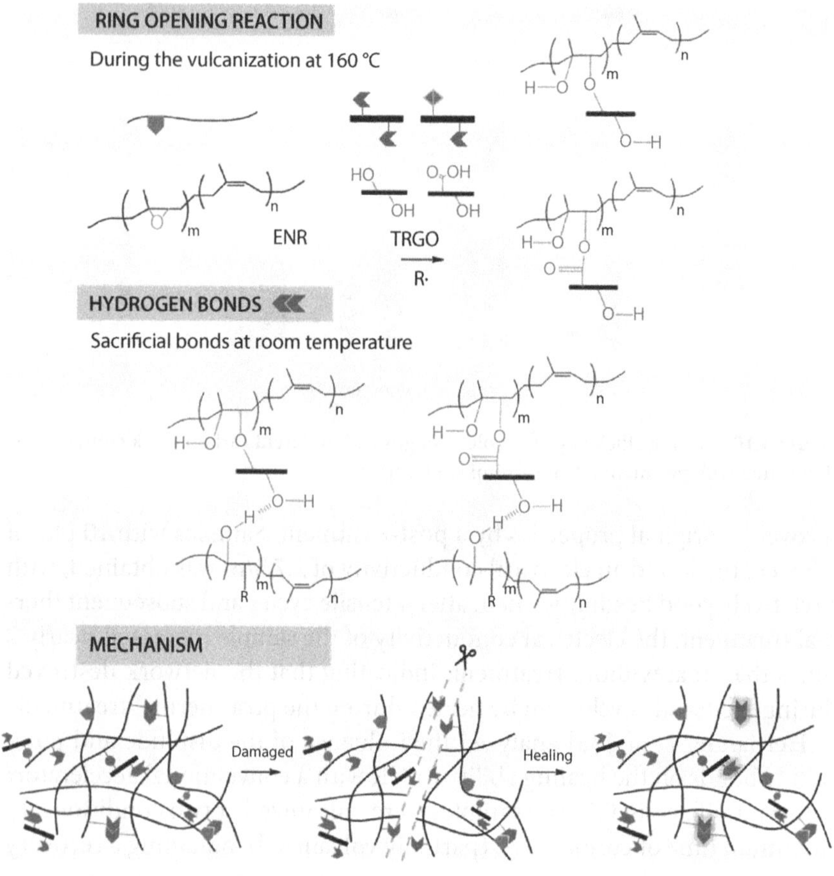

Figure 3.11 Proposed scheme of self-healing mechanism in epoxidized natural rubber and thermally reduced grapheme oxide composites (ENR/TRGO) (Reprinted with permission from Utrera-Barrios *et al.* [46]. Copyright 2020 American Chemical Society).

involved, another reaction can take place, which is shown in Figure 3.11: the ring opening of epoxy groups and consequently form of hydroxyl groups (–OH), creating in the matrix a thermo-reversible supramolecular network [47–49]. Therefore, a high concentration of epoxy groups promotes the formation of hydroxyl groups and promotes the formation of hydrogen bonds, and consequently the self-healing capability.

3.3.2 Styrene Butadiene Rubber (SBR)

Styrene butadiene rubber is the most used system among the synthetic rubbers. Due to the huge amount of scrap from tires, some researchers focused its interest on the self-healing behavior, which will allow to increases the time of use and the development of high performance smart tires.

Hernandez Santana *et al.* analyzed the self-healing behavior in SBR compounds containing ground tire rubber (GTR) particles and the coupling agent bis[3-(trietoxysilyl)propyl] tetrasulfide (TESP) [50]. Due to healing process requires chain mobility, authors prepared samples varying the accelerant/sulfur ratio (A/S) in order to evaluate the influence of the density and type of crosslinks: mono, di and polysulfides [51]. The sulfur amount was fixed in 0.7 phr.

The healing efficiency was evaluated through tensile tests: once the test was performed, the two pieces were repositioned together at 70 °C in a press at 10 bar for 7 h. Afterwards, the sample was tested again and the obtained results are shown in Figure 3.12. It was observed that systems with the lowest tensile strength exhibit the higher healing efficiency, corresponding to the ratios A/S = 0.2 and 1. The healing efficiency is attributed to the chain entanglement between the dangling chains on each piece, followed by thermal scission of the di and poly-sulfide bonds, which increases the chain mobility in the broken area. Also was confirmed that the healing efficiency is strongly reduced if the chain mobilization decreases due to a denser crosslinking network or by the use of reinforcement particles.

Araujo-Morera *et al.* [52] incorporated different amounts of GTR (10, 20 and 30 phr) into an SBR compound. The samples were submitted to 20 stretch cycles up 70 % strain in each cycle. After that, each sample was stored 12 h at room temperature and 12 h at 70 °C. Through dynamic mechanical analysis (DMA) it was stated that the elastic modulus (E')

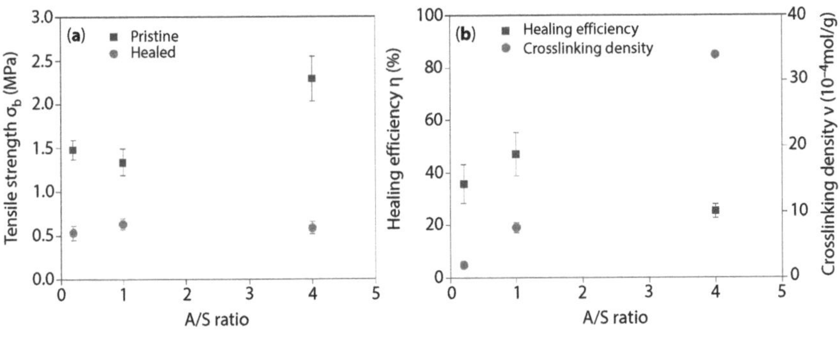

Figure 3.12 SBR vulcanized with different values of the ratio accelerant/sulfur (A/S): (a) Tensile strength of pristine and healed samples and (b) Healing efficiency and crosslink density as function of A/S (Reprinted with permission from Hernandez Santana *et al.* [50]).

diminishes in all the samples once the damage process was applied. That behavior is attributed to the Mullins effect, which is a result of different contributions like: chain scission, breaking of crosslinks, chains disentanglement, filler deagglomeration and softening of the chains that were near the filler surface (bound rubber) [53]. Once the healing protocol was applied, an efficiency of 110 % was obtained in the unfilled sample, which is attributed principally to the entanglement of the dangling chains and to the re-crosslinking of the damaged poly and disulfide crosslinks. Same authors used dielectric spectroscopy to analyze the type of interaction within the rubber composites. Figure 3.13 shows dielectric spectrum of different samples at −25 °C. In the case of the unfilled sample, either in the virgin and damage states, a similar spectrum was obtained, which indicate that the relaxation processes of the sample were exactly the same (Figure 3.13a). After the healing protocol, the spectrum presents a slight widening at high frequencies, indicating a higher number of entanglements promoted by the thermal healing treatment. In the case of the compound with 10 phr of GTR, a particular behavior was observed in Figures 3.13b and c, since the virgin sample indicates the presence of a percolated network that is broken with the cycled deformation, which cannot be recovered after the healing treatment.

Kuang *et al.* [38] crosslinked furfuryl grafted SBR (SBR-FS) with bismaleimide (M_2) via DA reaction at different molar ratios (1/1, 2/1 and 3/1). After the tensile test, the broken samples were pressed and healed at different thermal conditions (70 °C for 30 min, 100 °C for 30 min and 100 °C for 5 h) to induce the healing behavior giving by the reversibility of the DA/rDA reaction. In addition, author included carbon nanotubes functionalized with furfuryl groups (MWCNT-FA) were included in the compound.

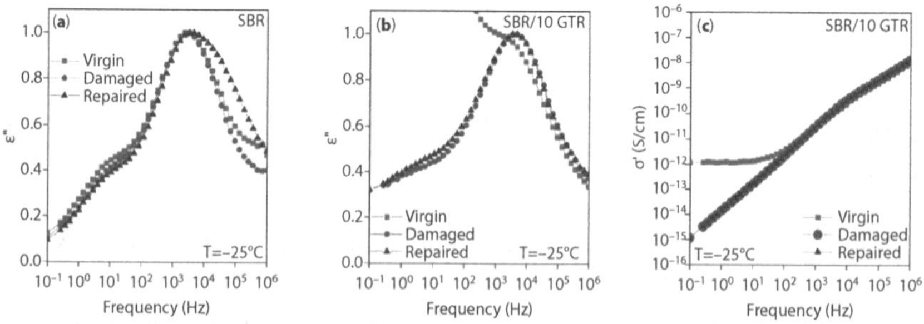

Figure 3.13 Dielectric parameters (e" and s") as function of frequency at −25 °C: (a) unfilled SBR and (b, c) SBR/10GTR compound (Reprinted from Araujo-Morera *et al.* [52], open access).

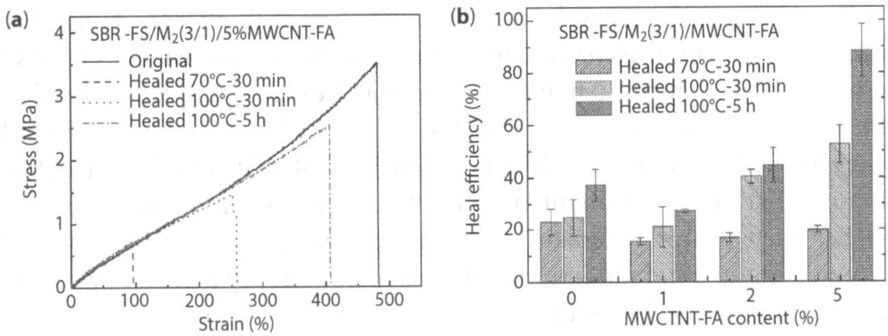

Figure 3.14 (a) Stress–strain curves of SBR-FS/M2 = 3/1 with 5% of MWCNT-FA and (b) Healing efficiency as a function of MWCNT-FA; both for the original sample and healed samples at the specified healing conditions (Adapted with permission from Kuang *et al.* [38]).

Figure 3.14(a) exhibits the stress–strain curves of samples whose molar ratio was 3/1 of furan/maleimide and 5% of MWCNT-FA. It can be observed that the stress–strain curves with the 3 different conditions after the healing process are overlapped. Moreover, the healing efficiency increases with temperature and time during which the temperature is applied. In Figure 3.14(b) is observed that a small addition of 1 phr MWCNT reduces the healing efficiency.

3.3.3 Polybutadiene Rubber

The other important synthetic rubber in the tire industry is polybutadiene rubber (BR) due to his high resistance to wear and low rolling resistance.

Xiang *et al.* analyzed samples of BR compounded with a liquid polysulfide that provides disulfide bonds and thiol end groups [54]. The curing (vulcanization) and self-healing processes was promoted through UV radiation at room temperature. The authors emphasize that a photo induction method requires less energy than that based on thermal activation. A disulfide metathesis by including diethyl disulfide (DEDS) and dibutyl disulfide (DBDS) was used due to this mechanism is promoted by the UV radiation and the metathesis was characterized at different irradiation intensities and time intervals. The amount of the liquid polysulfide was varied between 50, 100 and 150 g in 100 g of rubber, naming the compounds B1S5, B1S10 and B1S15 respectively. Through gas chromatography–mass spectroscopy it was observed that increasing the UV intensity the time require to reach the metathesis equilibrium diminishes and the optimum value for curing was determined in 20 mW/cm^2.

The self-healing was evaluated through tensile tests: after the failure the samples were put in physical contact and a radiation intensity of 20 mW/cm^2 was applied during 3 h. It was found that the compound B1S10 reaches the highest healability with an efficiency of ~60 %. Then, in this compound were applied different intensities between 5 and 40 mW/cm^2 for 2 h, determining that the optimum intensity is 30 mW/cm^2 as can be seen in Figure 3.15a. Then with that radiation amplitude the exposure time was varied from 0.5 up to 3 h, founding that the healing efficiency increases with the time, reaching a 97% (Figure 3.15b). Due to the reversibility of the disulfide metathesis the healing mechanism can be applied more than once. As can be seen in Figure 3.15c, efficiency higher than 90% is maintained even after the third cycle of healing.

Given the outstanding obtained results, the authors advanced in applying the optimum self-healing protocol (30 mW/cm^2, 3 h) in a grounded sample of the B1S10 compound in order to reprocess it and enabling to obtain a material with a different shape as is schematized in Figure 3.16. The recycle efficiency (the recovery degree in the mechanical properties after one

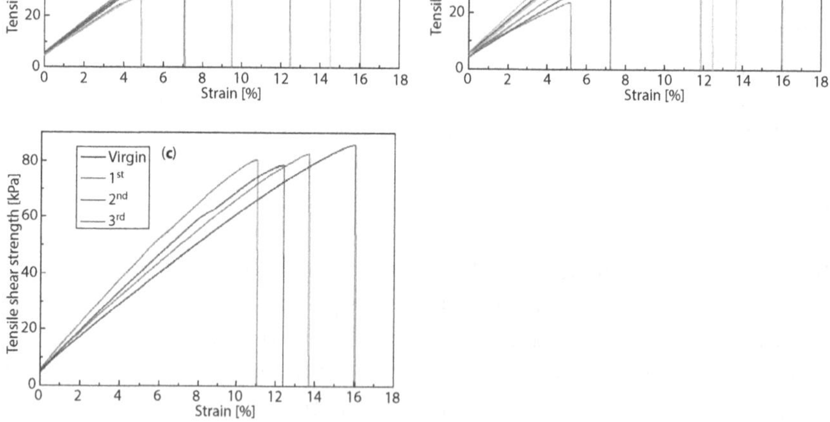

Figure 3.15 Tensile strain measurements to characterize the self-healing process induced by UV-radiation varying the following conditions: (a) intensity, (b) interval time. (c) Repeated self-healing when the sample was irradiated with 30 mW/cm2 during 3h (Adapted with permission from Xiang *et al.* [54]).

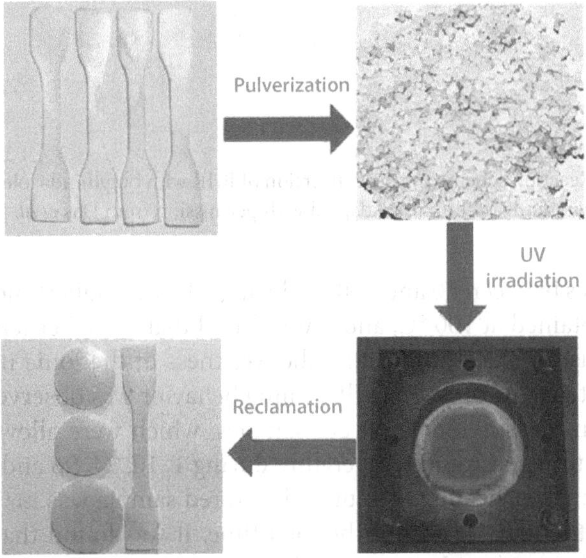

Figure 3.16 Process followed to the sample reclamation (Adapted with permission from Xiang *et al.* [54]).

treatment) decreases with the process: an efficiency of 82% is obtained when the reclamation is made once, but diminishes up to 64% after the third reclamation cycle. This process also depends on the mesh size due to when the particle is smaller the available surface to react is higher.

3.3.4 Bromobutyl Rubber

The principal characteristics of bromobutyl rubber (BIIR) are its high air impermeability and resistance to chemical agents. BIIR is obtained from bromination of butadiene rubber, enhancing the reactivity with amines and, therefore, allowing various functionalities to BIIR.

The above mentioned mechanism was used by Das *et al.* to introduce ionic functional groups in a rubber compound by using butylimidazole to increases the ionic interaction with the aim to enhance the self-healing, without no extra cross-links agent required [55]. Authors compounded BIIR and butylimidazole in an internal mixer to form amidazolium bromide groups (Figure 3.17), previously an isomerization of the bromide group is induced to obtain the reactive species. Then, the elastomer network is produced by ionic crosslinking and the obtained samples are soluble, which can be seen as an advantage from the environmental point of view. The crosslinking reaction was characterized in a rheometer at

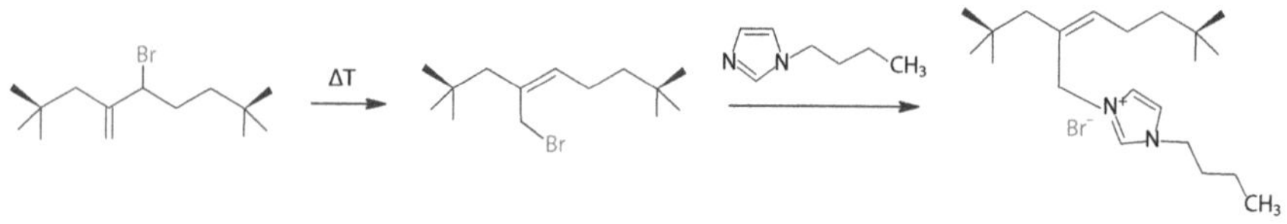

Figure 3.17 Reaction scheme for the conversion of BIIR with butylimidazole for posterior self-healing by ionic cross-linking (Adapted with permission from Das *et al.* [55]).

temperatures between 80 and 160 °C during 1 h. The highest modulus values were obtained at 100 °C, and it was found that at higher temperatures the modulus decreases, indicating the weakness of the ionic interactions at elevated temperatures. The self-healing behavior was observed through stress–strain measurements on cut samples, which were allowed to heal at different times: at room temperature during 1, 10, 24, 96 and 196 h and to explore a faster healing the cut and restored sample was maintained at 100 °C for the first 10 min of the total time. It was found that the healing effect increases with time and the obtained stress–strain curves are similar to that one from the uncut sample, as can be seen in Figure 3.18. A good self-healing is obtained due to the reversible characteristic of the ionic bonds and to the good mobility of the isobutylene chains (which is improved at higher temperatures). They did not obtain self-healing in the compound crosslinked with sulfur.

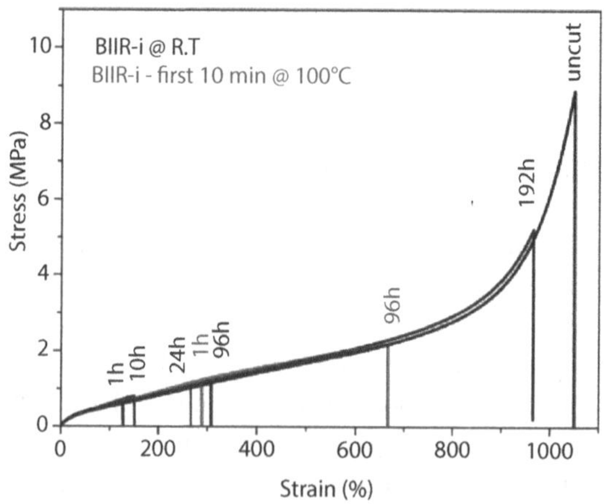

Figure 3.18 Stress–strain measurements to characterize self-healing behavior in BIIR samples modified with imidazolium (BIIR-i) when the healing was made at room temperature at different length times (blue lines) and when the first 10 minutes of heal was made at 100 °C (red lines), compared with an uncut sample (black line) (Adapted with permission from Das *et al.* [55]).

Lee and coworkers evaluated the self-healing property in pristine BIIR when is used as a coating to prevent corrosion, demonstrating that the healing process is highly stimulated by temperature by using a copper nanofiber as heater [56]. Kim *et al.* proposed a bilayer of BIIR over another one of BIIR with carbon nanotubes (in a concentration between 7 and 10 wt.%). The latter was used as heater to improve the self-healing efficiency [57]. The bi-layer was prepared through different methods: casting, painting and spraying. The angle between the applied current and the performed linear crack was varied (0, 45 and 90°) in order to study the influence of the current direction in the self-healing process. In the case of a current parallel to the crack orientation, the healing process took 2.5 h, while in the case that the current and the crack direction are parallel just 1 h is required. Then a repeating healing was performed, founding that up to a third cycle the material self-healed successfully when the sample is heated during 1 h in each cycle, as can be seen in Figure 3.19. Finally, they performed corrosion tests by applying the bi-layer in a steel substrate. To this aim, a deep scratch was made in 2 samples which were submerged in saline water and only one of the samples was heated during 1 h. After 72 h under the saline bath, the bi-layer was removed from the samples and the steel surface was observed, founding rust in the sample that was not heated (Figure 3.20).

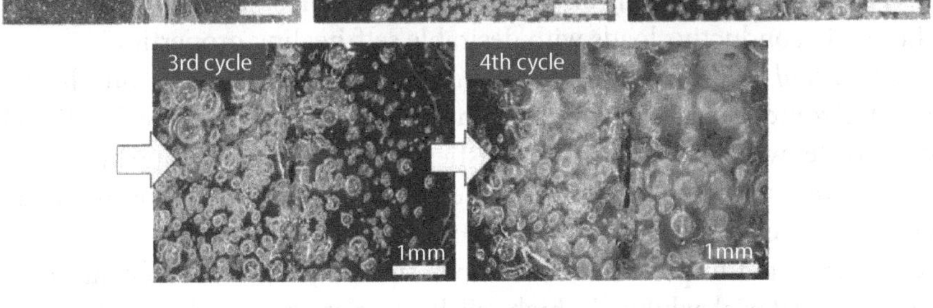

Figure 3.19 Bi-layer of BIIR and BIIR-CNT. Repeated healing test under water, the electrical current is perpendicular to the crack direction and is applied during 1 h in each cycle (Reprinted with permission from Kim *et al.* [57]).

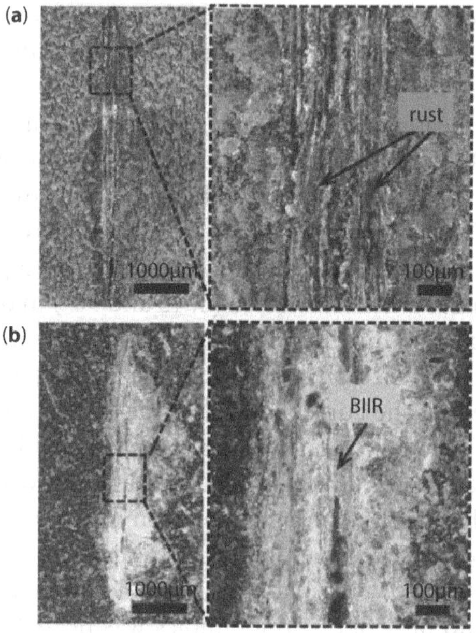

Figure 3.20 Bi-layer of BIIR and BIIR-CNT in a steel substrate to evaluate corrosion test in saline water during 72 h. (a) without heating, (b) with heating (Reprinted with permission from Kim *et al.* [57]).

3.3.5 Silicones

The silicone is characterized for its insulating properties, chemical and thermal stability, outstanding weatherability and transparency [58–60]. Some of the most common applications are automobiles, electronics, medical implants, sportswear and shoes, among others.

In some cases, it is necessary that the insulating material needs to transmit the heat generated, for example, in electronic circuits. Several authors [61] are currently working on the development of insulating materials with thermally conductive loads with desirable self-healing properties.

Zhao *et al.* [58] used thermally conductive composites based on silicone for electronic packaging materials. The challenge is to fabricate functional composites with two main characteristics: self-healing ability and high thermal conductivity. Self-healing silicone was formulated with boron nitride (BN) in order to induce the DA reaction. After the tensile test, samples were submitted to pressure and temperature to join the broken surfaces. The final material exhibited a high self-healing efficiency (almost 90%) and an increase in thermal conductivity about 500% with just 50 wt.% of BN. From Figure 3.21, it is possible to compare the stress–strain curves before

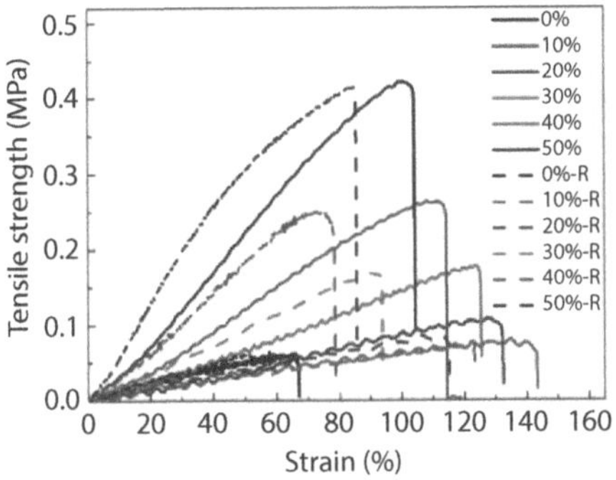

Figure 3.21 Stress–strain curves of the silicone elastomer and composites before (solid line) and after healing treatment (named with "R" and in dotted line) (Adapted with permission from Zhao *et al.* [60]).

and after healing process of the silicone elastomer and composites, observing a high structural recovery with 50 wt.% of BN. Authors explain that, due to the hindrance of the BN nanosheets, the mobility of low molecular oligomer is reduced, so it is necessary to apply certain external force (for example pressure) to facilitates the chain diffusion in the damaged interface of the composite to produce an optimal self-healing process.

Xiang *et al.* [14] studied a reversibly cured silicone elastomer which is obtained by a condensation reaction between α,ω-dihydroxyl polydimethylsiloxane and disulfide bond that contains silane coupling agent as crosslinker. This work explores the self-healing composite under sunlight by using microcontainers (for example capsules and glass capillaries), with liquid healing agents that are released when the material is damaged. The self-healing capacity was determined with a tensile test, in which samples were broken and, then, recombined under pressure to be healed by xenon exposure or natural sunlight exposure for a certain period of time. Authors observed that the healing efficiency depends on the exposure time, reaching a maximum after 48 h. Also, a misalignment of the fractured surfaces is an important variable to obtain a better self-healing. The crosslinks density was also evaluated due to a high crosslinks density reduces the wettability and diffusion of dangling chains on the fracture surface, affecting the self-healing capacity.

As it was explained before, the vulcanization agents play an important role during the dynamic reversible of disulfide bonds since they promote a higher healing efficiency. Figure 3.22 shows 80% recovery in mechanical test for healed and recycle samples, being the last one obtained by cutting the silicone sheets in small chips that were pulverized into powder and then compressed and molded.

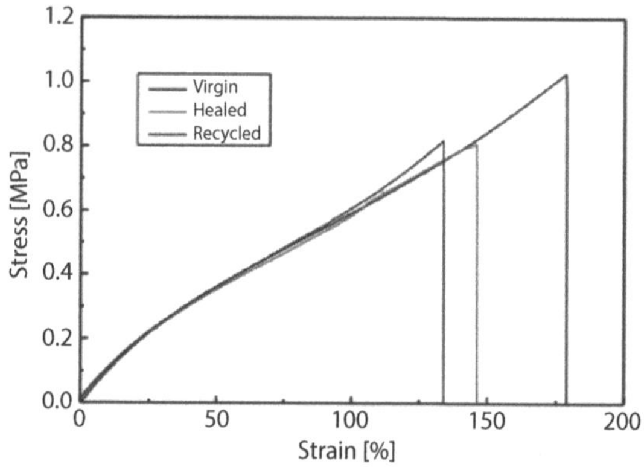

Figure 3.22 Tensile stress–strain curves of virgin, healed and recycled silicone compounds (Adapted with permission from Xiang *et al.* [14]).

Another interesting application is silicone foams, which are characterized by thermal insulation, good lightweight performance, good flexibility and excellent shock absorbent. Zhao *et al.* [60] developed silicone foam composite with self-healing properties. The concept consists in a material of acquiring different shapes without any fracture. Authors formed hybrid foam with shear stiffening gel (STG) and Methyl vinyl silicone rubber (MVQ) which maintained its shape. Expandable microspheres were used as blowing agent. These microspheres have two structures: a low-boiling volatile hydrocarbon core wrapped in the thermoplastic shell that when heated permits to the liquid inside the core to evaporate. During this process, the thermoplastic shell swells, expanding almost to hundred times. The authors evaluated the self-healing capacity after performing cuts on the samples and then bonding the surfaces together. STG samples were able to stick together, evidencing a very good viscous state. Through tensile tests, a recovery of only 11.1% was determined after healing. In the case of the elongation at break, the recovery was about 34 %. The mechanical properties of the self-healing samples are shown in Figure 3.23 (samples were named considering different MVQ proportions).

Chen *et al.* [61] synthetized a poly(dimethylsiloxane) (PDMS) elastomer with sacrificial hydrogen bonds and dynamic imine bonds. The self-healing mechanism was characterized through FT-IR, tensile test and dynamic mechanical analysis. The experimental processes and synthesis of the specimens were made in a two-step approach, where segmented PDMS products (named as PDMS-U) was obtained by reacting aminopropyl terminated polydimethylsiloxane (AP-PDMS) and 4,4'-methylenebis-(cyclohexylisocyanate) (HMDI). Then, isophthalaldehyde (IPAL) was

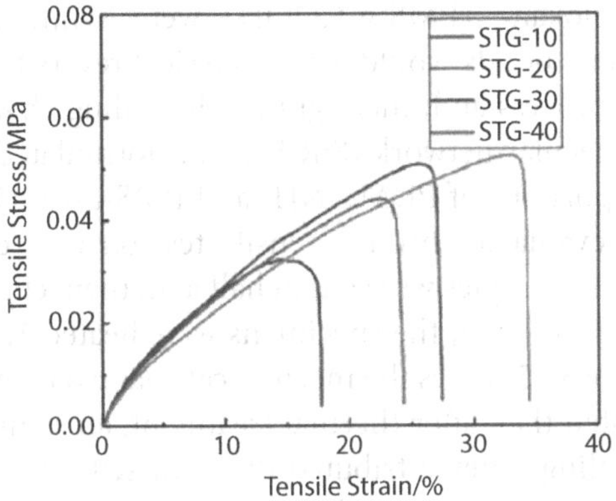

Figure 3.23 Tensile stress–strain curves of self-healed composite foams with different amounts of shear stiffening gel (Adapted with permission from Zhao *et al.* [60]).

added to form imine groups and obtain the self-healing materials with urea and imine groups (PDMS-UI). The samples varies in the molar ratio of urea to imine groups, and the samples are designated as PDMS-UI-x (x = 1–4), being x proportional to the ratio value.

Comparing the self-healing capacity of the samples, there is a decrease in the property with the increment of urea (Figure 3.24). Authors determined that the self-healing capacity depends on the concentration of imine groups.

Sun *et al.* [62] worked with dielectric elastomers to simulate artificial muscles. Amino terminated polydimethylsiloxane (PDMS-NH_2) and a

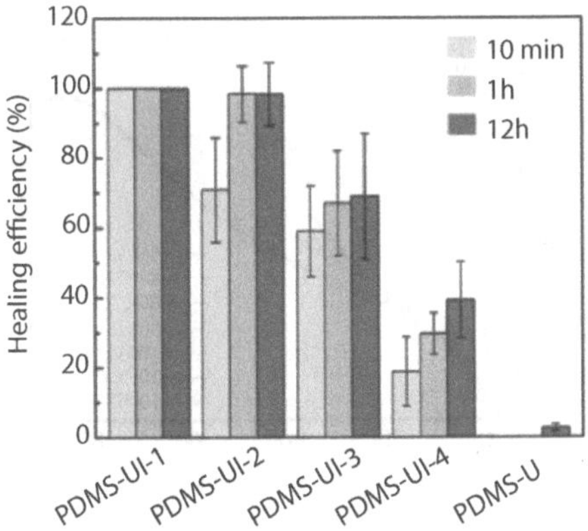

Figure 3.24 Evaluation of self-healing efficiency of PDMS-UI-x composites for different healing times (x varies between 1 and 4, being a value proportional to the molar ratio of urea to imine groups) (Adapted with permission from Chen *et al.* [62]).

polymethylvinylsiloxane (PMS-g-COOH) were synthetized, in which hydrogen and ionic bonds would act as physical reversible bonds during the crosslinking process and, then, grant self-healing characteristics. The silicone supramolecular network (SiR-SN) was formulated by incorporating different proportions of PDMS-NH_2 and PMS-g-COOH. Self-healing properties were evaluated through tensile test on thin elastomer films. Previous to the test, samples were cut in half and, then, clipped together to reattach the surfaces. Then, the specimens were heated 1, 2 and 5 h at 80 and 100 °C. Figure 3.25 shows the mechanical characterization of the samples. Authors stated that, after thermal treatment, all samples showed the ability of self-healing. They attributed different self-healing mechanisms according to the temperature used: at 80 °C leads to the reformation of hydrogen bonds, while at 100 °C there is a conversion of the hydrogen bonds into ionic bonds. In this case, there is a big difference between the

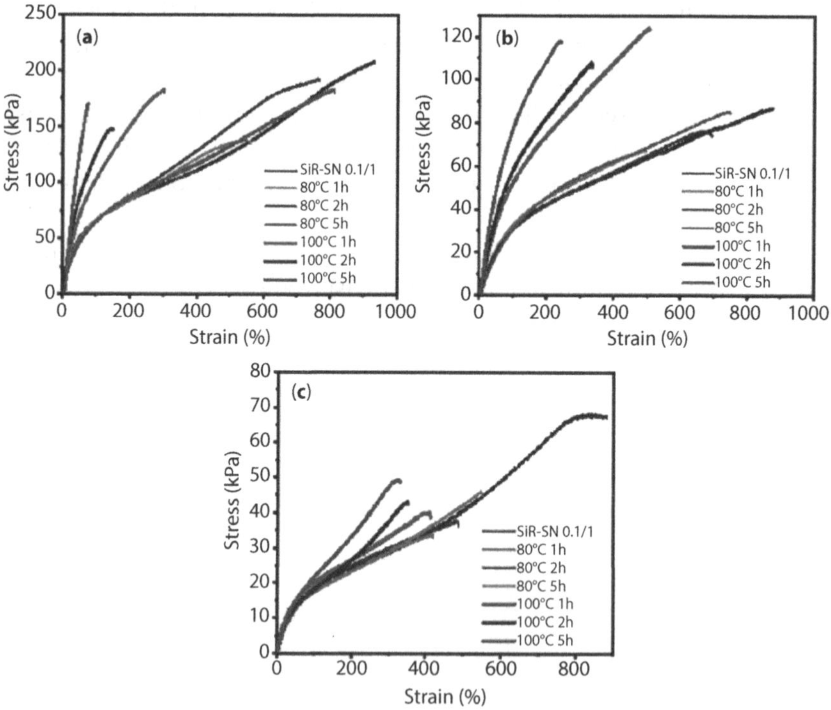

Figure 3.25 Stress–strain curves for a silicone supramolecular with different PMS-g-COOH to PDMS-NH2 ratio: (a) SiR-SN 0.1/1; (b) SiR-SN 0.2/1; (c) SiR-SN 0.5/1 before and after healing under different conditions (Reprinted with permission from Sun *et al.* [62]).

hydrogen bond proportions of the composites. A better tensile strength is observed for composites with more hydrogen bonds, because it allows to more links to convert into ionic ones. Finally, an electric field is applied in the samples in order to simulate the final application conditions. In those samples self-healed at 80 °C a mechanical recovery of 100% is obtained when an electric field is applied. Even so, they explain that the samples treated at 100 °C show a greater hardness but without reaching a 100% recovery.

3.3.6 Polyurethanes

Thermoplastic polyurethane (TPU) is a block copolymer consisting of alternating sequences of hard and soft segments, characterized by a high resilience and resistance to impacts, abrasion and tear. The TPU formulation includes diisocyanates, polyols, and chain extenders. The type and proportion of each of them have a significant influence on the final properties of the material. This system exhibited self-healing ability if dynamic disulfide bonds [63–65] and hydrogen bonding [66, 67] were introduced in its structure.

Based on these characteristics, several self-healing TPU elastomers have been designed by introducing dynamic covalent bonds and non-covalent interactions into the chains. Mainly, the most used mechanisms for self-healing TPU are based on dynamic disulfide bonds [68–70] and hydrogen bonding [71, 72].

Rekondo *et al.* [73] stated that the dynamic chemistry of disulfide bonds plays a vital role in self-healing processes. Metathesis of aromatic disulfides at room temperature is possible through the use of a tertiary amine as the catalyst. In this case bis(4-aminophenyl) disulfide was used, which is a diamine molecule characterized by an aromatic disulfide (Figure 3.26)

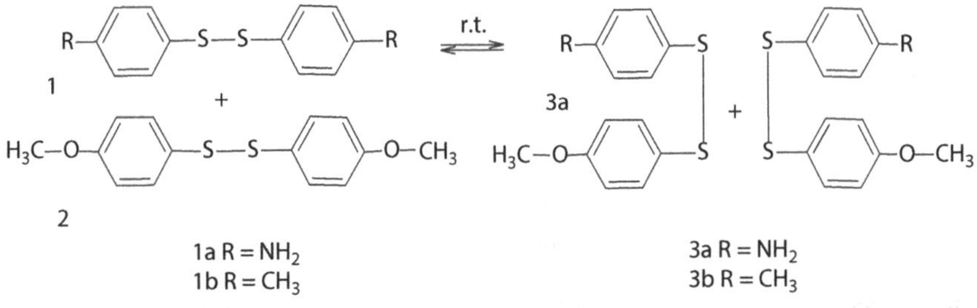

Figure 3.26 Reversible metathesis reaction of aromatic disulfides 1a–b and 2 (Adapted with permission from Rekondo *et al.* [68]).

as a dynamic crosslinking agent in poly(urea–urethane) systems. Author analyzed the use of equimolar amounts of (bis(4-aminophenyl) disulfide (1a) and bis(4-methoxyphenyl) disulfide (2) in deuterated DMSO (Figure 3.26). It was tested with 0.1 equivalents of tertiary amine (NEt_3), and the equilibrium was achieved in few minutes. If the reaction is tested without NEt_3, metathesis initiated in few minutes and the equilibrium is achieved in 22 h (Figure 3.26). With the aim to test that the reaction is catalyzed by primary aromatic amines present in 1a, the tests were carried out taking into account 1b and 2, without NEt_3. The system reaches equilibrium in 24 h if the same amount of moles of 1b and 2 are used, confirming that a catalyst is not necessary for the exchange reaction to occur.

Nevejans *et al.* [74] proposed the self-healing concept in coatings based on waterborne poly(urethane-urea) containing aromatic disulfide dynamic bonds. Authors expressed that self-healing coating material should present two opposite properties: i) high strength and ii) molecular mobility. The influence of the flexibility of the two aromatic disulfides and its concentration on self-healing efficiency was explored. The two aromatic disulfides are shown in Figure 3.27 being $S_2(Ph(CH_2)_3OH)_2$ designated as S3 and the more flexible alternative $S_2(Ph(CH_2)_6OH)_2$ is named S6.

The mobility of the different compounds was characterized through the relaxation time (in dynamic experiments) and the self-healing ability through the time needed for the scratch closure. It was concluded that, replacing S3 by S6 in the formulation, the mobility increase. By the other hand, it is possible to include a trifunctional amine as chain extender (instead of a difunctional one, which decreased the mobility) and even keep the self-healing ability.

Another strategy was presented by Zhang *et al.* [75], proposing the use of C-ON bonds in alkoxyamine to promote the self-healing action

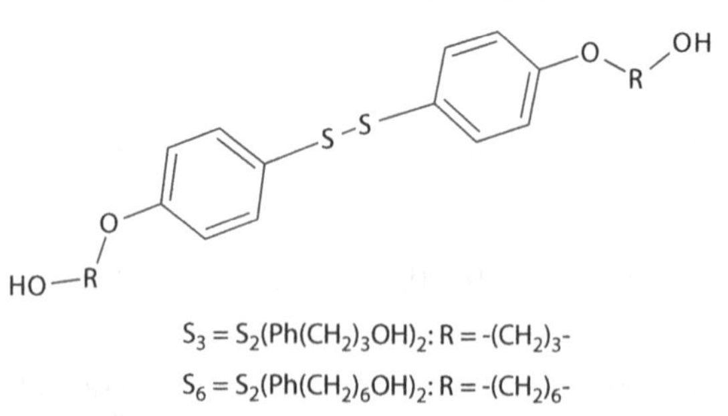

Figure 3.27 Chemical structure of bis[4-(3-hydroxypropyloxy)phenyl]disulfide (S3) and bis[4-(6-hydroxyhexoxy)phenyl]disulfide (S6) (Reprinted with permission from Nevejans *et al.* [69]).

in PU composites. Based on this idea, 5-hydroxy-2-(4-hydroxy-2,2,6,6-tetramethylpiperidin-1-yloxy)-2-methyl pentanenitrile (CTPO) was proposed as a dynamic reversible molecule that contain a nitrile group bonded to C–ON moieties. Due to nitrile has the ability to absorb electrons and stabilize carbon radicals, C-ON bonds in CTPO can be activated at a lower temperature.

Polyurethane was obtained through the copolymerization of CTPO with 3-isocyanatomethyl-3,5,5-trimethylcyclohexylisocyanate (IPDI) and polyethylene glycol (PEG), where PEG represent the soft block, while IPDI and CTPO are the hard segments. Figure 3.28 presents a scheme of the healing mechanism. The self-healing efficiency was measured through notched impact test. The same sample was damaged and healed up to three times, obtaining an average healing efficiency of 94.6, 91.7 and 89.8% for the first, second and third time, respectively.

Hu *et al.* [76] grafted a hard segment of TPU with 2-ureido-4[1H]-pyrimidione (UPy), embedding the disulfide bonds in the principal chain. UPy groups exhibits a dimerization trend that includes four hydrogen bonds in a donor–donor–acceptor–acceptor array, with rapid kinetic and high dimerization energy.

The robust interaction between quadruple H-bonding of the UPy side groups acts as supramolecular crosslinking points allowing the improvement of mechanical properties in the TPU elastomer. The self-healing abilities was characterized by a dumbbell-shaped sample, who was fully cut

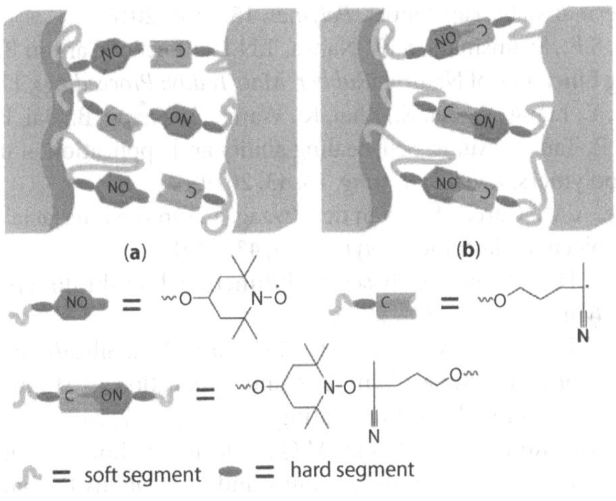

Figure 3.28 Schematic diagram of the healing reaction in PU (Adapted with permission from Zhang *et al.* [70]).

in half and cured at 100 °C for 2 h. After the applied healing procedure, it was not possible to find the original crack, suggesting that the structural integrity was recovered.

References

1. Imbernon, L. and Norvez, S., From landfilling to vitrimer chemistry in rubber life cycle. *Eur. Polym. J.*, 82, 347, 2016.
2. Backman, L., *Self-healing elastomers Comparison of methods*, pp. 9–42, Arcada University of Applied Sciences. Helsinki Finland, 2018..
3. Hu, J., Mo, R., Jiang, X., Sheng, X., Zhang, X., Towards mechanical robust yet self-healing polyurethane elastomers via combination of dynamic main chain and dangling quadruple hydrogen bonds. *Polymer*, 183, 1–2, 2019.
4. Hernández, M., Grande, A.M., Dierkes, W., Bijleveld, J., Zwaag, S., García, S.J., Turning Vulcanized Natural Rubber into a Self-Healing Polymer: Effect of the Disulfide/Polysulfide Ratio. *ACS Sustainable Chem. Eng.*, 4, 5776, 2016.
5. White, S.R., Sottos, N.R., Geubelle, P.H., Moore, J.S., Kessler, M.R., Sriram, S.R., Brown, E.N., Viswanathan, S., Autonomic healing of polymer composites. *Nature*, 409, 794, 2001.
6. Jin, H.H., Mangun, C.L., Griffin, A.S., Moore, J.S., Sottos, N.R., White, S.R., Thermally stable autonomic healing in epoxy using a dual-microcapsule system. *Adv. Mater.*, 24, 282, 2014.
7. Toohey, K.S., Sottos, N.R., Lewis, J.A., Moore, J.S., White, S.R., Self-healing materials with microvascular networks. *Nat. Mater.*, 6, 581, 2007.
8. Fang, Y., Li, J., Du, X., Dua, Z., Cheng, X., Wang, H., Thermal- and mechanical-responsive polyurethane elastomers with selfhealing, mechanical-reinforced, and thermal-stable capabilities. *Polymer*, 158, 166, 2018.
9. Khimi, S.R., Syamsinara, S.N., Najwa, T.N.L., Effect of Carbon Black on Self-healing Efficiency of Natural. *Rubber. Mat. Today: Proceedings*, 17 1064, 2019.
10. Denga, Y., Liang, X., Pei, X., Zhai, K., Wang, C., Zhang, B., Bai, B., Zhang, Y., Wang, P., Tan, Y., Xu, K. Self-healing ability and application of impact hardening polymers. *Polymer Testing*, 76, 43, 2019.
11. Grande, A.M., Garcia, S.J., Van der Zwaag, S., On the interfacial healing of a supramolecular elastomer. *Polymer*. 56, 435, 2015.
12. Shchukin D.G., Container-based multifunctional self-healing polymer coatings. *Polym. Chem.* 4, 4871, 2013.
13. Li, G.L., Zheng, Z.L., Mohwald, H., Shchukin, D.G., Silica/polymer double-walled hybrid nanotubes: Synthesis and application as stimuli-responsive nanocontainers in self-healing coatings. *ACS Nano*, 7, 2470, 2013.
14. Xiang, H.P., Rong, M.Z., Zhang, M.Q., A facile method for imparting sunlight driven catalyst-free self-healability and recyclability to commercial silicone elastomer. *Polymer*, 108, 339, 2017.

15. Zhao, J., Xu, R., Luo, G., Wu, J., Xia, H., A self-healing, re-moldable and biocompatible crosslinked polysiloxane elastomer. *J. Mater. Chem. B*, 4, 982, 2016.
16. Jin, B., Liu, M., Zhang, Q., Zhan, X., Chen, F., Silicone oil swelling slippery surfaces based on mussel-inspired magnetic nanoparticles with multiple self-healing mechanisms. *Langmuir*, 33, 10340, 2017.
17. Martin, R., Rekondo, A., Echeberria, J., Cabanero, G., Grande, H.J., Odriozola, I., Room temperature self-healing power of silicone elastomers having silver nanoparticles as crosslinkers. *Chem. Commun.*, 48, 8255, 2012.
18. Wittmer, A., Brinkmann, A., Stenzel, V., Hartwig, A., Koschek, K., Moisture-mediated intrinsic self-healing of modified polyurethane urea polymers. *J. Polym. Sci. Pol. Chem.*, 56, 537, 2018.
19. Chen, S., Zhang, B., Zhang, N., Ge, F., Zhang, B., Wang, X., Song, J., Development of selfhealing D-gluconic acetal-based supramolecular ionogels for potential use as smart quasisolid electrochemical materials. *ACS Appl. Mater. Interfaces*, 10, 5871, 2018.
20. Gulyuz, U. and Okay, O., Self-Healing Poly (acrylic acid) Hydrogels with Shape Memory Behavior of High Mechanical Strength. *Macromolecules*, 47, 6889, 2014.
21. Okay, O., Self-healing hydrogels formed via hydrophobic interactions, in: *Supramolecular Polymer Networks and Gels*, pp. 101–142, Springer International Publishing, Switzerland, 2015.
22. Fox, J., Wie, J.J., Greenland, B.W., Burattini, S., Hayes, W., Colquhoun, H.M., Mackay, M.E., Rowan, S.J., High-strength, healable, supramolecular polymer nanocomposites. *J. Am. Chem. Soc.*, 134, 5362, 2012.
23. Ling, S., Chend, W., Fane, Y., Zhenga, K., Jinb, K., Yud, K., Buehlerb, M.J., Kaplanc, D.L., Biopolymer nanofibrils: Structure, modeling, preparation, and applications. *Prog. Polym. Sci.*, 85, 1, 2018.
24. Liu, J., Ma, X., Tong, Y., Lang, M., Self-healing polyurethane based on ditelluride bonds. *Appl. Surf. Sci.*, 455, 318, 2018.
25. Xu, C., Cao, L., Lin, B., Liang, X., Chen, Y., Design of Self-Healing Supramolecular Rubbers by Introducing Ionic Cross-Links into Natural Rubber via a Controlled Vulcanization. *ACS Appl. Mater. Interfaces*, 8, 27, 17728, 2016.
26. Zhang, D.D., Ruan, Y., Zhang, B.Q., Qiao, X., Deng, G., Chen, Y., Liu, C.Y., A self-healing PDMS elastomer based on acylhydrazone groups and the role of hydrogen bonds. *Polymer*, 120, 189, 2017.
27. Tanasi, P., Hernández Santana, M., Carretero-González, J., Verdejo, R., López-Manchado, M.A., Thermo-reversible crosslinked natural rubber: A Diels–Alder route for reuse and self-healing properties in elastomers. *Polymer*, 175, 15, 2019.
28. Wool, R.P., Welding of Polymer Interfaces. *Polym. Eng. Sci.*, 9, 1340–1367, 1989.

29. Zechel, S., Geitner, R., Abend, M., Siegmann, M., Enke, M., Kuhl, N., Klein, M., Vitz, J., Gräfe, S., Dietzek, B., Schmitt, M., Popp, J., Schubert, U., Hager, M., Intrinsic self-healing polymers with a high E-modulus based on dynamic reversible urea bonds. *NPG Asia Mater.*, 9, e420, 2017.
30. Terryn, S., Brancart, J., Lefeber, D., Van Assche, J., Vanderborght, B., Self-healing soft pneumatic robots. *Sci. Rob.*, 63, 28, 2017.
31. Syrett, J.A., Becer, C.R., Haddleton, D.M., Self-healing and self mendable polymers. *Polym. Chem.*, 1, 978, 2010.
32. Ying, H., Zhang, Y., Cheng, J., Dynamic urea bond for the design of reversible and self-healing polymers. *Nat. Commun.*, 5, 3218, 2014.
33. Rekondo, A., Martin, R., Ruiz de Luzuriaga, A., Cabañero, G., Grande, H.J., Odriozola, I., Catalyst-free room-temperature self-healing elastomers based on aromatic disulfide metathesis. *R. Soc. Chem.*, 1, 237, 2014.
34. Yang, Y., *Molecular Level Design of Self-Healing Polyurethanes.* Tiger prints, All dissertations, vol. 8, p. 1685. Clamson University. Clemson, South Carolina, 2016.
35. Hernández, M., Grande, A.M., Dierkes, W., Bijleveld, J., Van der Zwaag, S., García, S.J., Turning Vulcanized Natural Rubber into a Self-Healing Polymer: Effect of the Disulfide/Polysulfide Ratio. *ACS Sustainable Chem. Eng.*, 4, 5776, 2016.
36. Grande, A.M., Garcia, S.J., Van der Zwaag, S., On the interfacial healing of a supramolecular elastomer. *Polymer*, 56, 435, 2015.
37. Kanu, N.J., Gupta, E., Kumar Vates, U., Kumar Singh, G., Self-healing composites: A state-of-the-art review. *Composites Part A*, 121, 474, 2019.
38. Kuang, X., Liu, G., Dong, X., Wang, D., Enhancement of Mechanical and Self-Healing Performance in Multiwall Carbon Nanotube/Rubber Composites via Diels–Alder Bonding. *Macromol. Mater. Eng.*, 301, 535, 2016.
39. Fang, Y., Li, J., Du, X., Dua, Z., Chenga, X., Wang, H., Thermal- and mechanical-responsive polyurethane elastomers with selfhealing, mechanical-reinforced, and thermal-stable capabilities. *Polymer*, 158, 166, 2018.
40. Peng, Y., Yang, Y., Wu, Q., Wang, S., Huang, G., Wu, J., Strong and tough self-healing elastomers enabled by dual reversible networks formed by ionic interactions and dynamic covalent bonds. *Polymer*, 157, 172, 2018.
41. Cao, L., Fan, J., Huang, J., Chen, Y., A robust and stretchable cross-linked rubber network with recyclable and self-healable capabilities based on dynamic covalent bonds. *J. Mater. Chem. A*, 7, 9, 4922, 2019.
42. Xu, C., Cui, R., Fu, L., Lin, B., Recyclable and heat-healable epoxidized natural rubber/bentonite composites. *Compos. Sci. Technol.*, 167, 421, 2018.
43. Trovatti, E., Cunha, A.G., Carvalho, A.J.F., Gandini, A., Furan-modified natural rubber: A substrate for its reversible crosslinking and for clicking it onto nanocellulose. *Int. J. Biol. Macromol.*, 95, 762, 2017.
44. Zhan, Y., Meng, Y., Li, Y., Electric heating behavior of flexible graphene/natural rubber conductor with self-healing conductive network. *Mater. Lett.*, 192, 115, 2017.

45. Zhan, Y., Lavorgna, M., Buonocore, G., Xia, H., Enhancing electrical conductivity of rubber composites by constructing interconnected network of self-assembled graphene with latex mixing. *J. Mater. Chem.*, 22, 10464, 2012.
46. Utrera-Barrios, S., Hernández, S.M., Verdejo, R., López-Manchado, M.A., Design of Rubber Composites with Autonomous Self-Healing Capability. *ACS Omega*, 5, 1902, 2020.
47. Krainoi, A., Kummerlöwe, C., Nakaramontri, Y., Vennemann, N., Pichaiyut, S., Wisunthorn, S., Nakason, C., Influence of critical carbon nanotube loading on mechanical and electrical properties of epoxidized natural rubber nanocomposites. *Polym. Test.*, 66, 122, 2018.
48. Cao, L., Yuan, D., Xu, C., Chen, Y., Biobased, self-healable, high strength rubber with tunicate cellulose nanocrystals. *Nanoscale*, 9, 15696, 2017.
49. Xu, C., Nie, J., Wu, W., Zheng, Z., Chen, Y., Self-healable, recyclable and strengthened epoxidized natural rubber/carboxymethyl chitosan bio-based composites with hydrogen bonding supramolecular hybrid network. *ACS Sustainable Chem. Eng.*, 7, 15778, 2019.
50. Hernández Santana, M., Huete, M., Lameda, P., Araujo, J., Verdejo, R., López-Manchado, M.A., Design of a new generation of sustainable SBR compounds with good trade-off between mechanical properties and self-healing ability. *Eur. Polym. J.*, 106, 273, 2018.
51. Marzocca, A.J. and Mansilla, M.A., Analysis of network structure formed in styrene-butadiene rubber cured with sulfur/TBBS system. *J. Appl. Polym. Sci.*, 103, 1105, 2007.
52. Araujo-Morera, J., Hernández Santana, M., Verdejo, R., López-Manchado, M.A., Giving a Second Opportunity to Tire Waste: An Alternative Path for the Development of Sustainable Self-Healing Styrene–Butadiene Rubber Compounds Overcoming the Magic Triangle of Tires. *Polymers*, 11, 2122, 2019.
53. Balasooriya, W., Schrittesser, B., Pinter, G., Schwarz, T., Conzatti, L., The Effect of the Surface Area of Carbon Black Grades on HNBR in Harsh Environments. *Polymers*, 11, 61, 2019.
54. Xiang, H., Yin, J., Lin, G., Liu, X., Rong, M., Zhang, M., Photo-crosslinkable, self-healable and reprocessable rubbers. *Chem. Eng. J.*, 358, 878, 2019.
55. Das, A., Sallat, A., Böhme, F., Suckow, M., Basu, D., Wießner, S., Stöckelhuber, K.W., Voit, B., Heinrich, G., Ionic Modification Turns Commercial Rubber into a Self-Healing Material. *ACS Appl. Mater. Interfaces*, 7, 20623, 2015.
56. Lee, M.W., Jo, H.S., Yoon, S.S., Yarin, A.L., Thermally driven self-healing using copper nanofiber heater. *Appl. Phys. Lett.*, 111, 011902, 2017.
57. Kim, H., Yarin, A.L., Lee, M.W., Self-healing corrosion protection film for marine environment. *Compos. Part B—Eng.*, 182, 1075987, 2020.
58. Zhao, L., Shi, X., Yin, Y., Jiang, B., Huang, Y., A self-healing silicone/BN composite with efficient healing property and improved thermal conductivities. *Compos. Sci. Technol.*, 186, 107919, 2020.

59. Ping, H., Zhi, M., Qiu, M., A facile method for imparting sunlight driven catalyst-free self- healability and recyclability to commercial silicone elastomer. *Polymer*, 108, 339–347, 2017.
60. Zhao, X. and Zhang, J., A novel composite silicone foam with enhanced safeguarding performance and self-healing property. *React. Funct. Polym.*, 138, 114–121, 2019.
61. Chen, G., Sun, Z., Wang, Y., Zheng, J., Wen, S., Zhang, J., Wang, L., Hou, J., Lin, C., Yue, Z., Designed preparation of silicone protective materials with controlled self-healing and toughness properties. *Prog. Org. Coat.*, 140, 105483, 2020.
62. Sun, H., Liu, X., Liu, S., Yu, B., Ning, N., Tian, M., Zhang, L., Silicone dielectric elastomer with improved actuated strain at low electric field and high self-healing efficiency by constructing supramolecular network. *Chem. Eng. J.*, 384, 123242, 2020.
63. Abdolah Zadeh, M.A., Esteves, A.C.C., Van der Zwaag, S., Garcia, S.J., Healable dual organicinorganic crosslinked. Sol–Gel based polymers: Crosslinking density and tetrasulfide content effect. *J. Polym. Sci. A*, 52, 1953, 2014.
64. Canadell, J., Goossens, H., Klumperman, B., Self-healing materials based on disulfide links. *Macromol.*, 44, 2536, 2011.
65. Michal, B.T., Spencer, E.J., Rowan, S.J., Stimuli-responsive reversible two-level adhesion from a structurally dynamic shape-memory polymer. *ACS Appl. Mater. Interfaces*, 8, 11041, 2016.
66. Wei, M. *et al.*, Novel Poly(tetramethylene ether)glycol and Poly (ε -caprolactone) Based Dynamic Network via Quadruple Hydrogen Bonding with Triple-Shape Effect and Self-Healing Capacity. *Appl. Mater. Interfaces*, 7 (4), 2585–2596, 2015.
67. Jiang, Z., Xiao, Y.Y., Kang, Y., Pan, M., Li, B.J., Zhang, S., Shape Memory Polymers Based on Supramolecular Interactions. *Appl. Mater. Interfaces*, 9, 20276, 2017.
68. Abdolah Zadeh, M.A., Esteves, A.C.C., Van der Zwaag, S., Garcia, S.J., Healable dual organicinorganic crosslinked. Sol–Gel based polymers: Crosslinking density and tetrasulfide content effect. *J. Polym. Sci. A*, 52, 1953, 2014.
69. Canadell, J., Goossens, H., Klumperman, B., Self-healing materials based on disulfide links. *Macromol.*, 44, 2536, 2011.
70. Michal, B.T., Spencer, E.J., Rowan, S.J., Stimuli-responsive reversible two-level adhesion from a structurally dynamic shape-memory polymer. *ACS Appl. Mater. Interfaces*, 8, 11041, 2016.
71. Wei, M. *et al.*, Novel Poly(tetramethylene ether)glycol and Poly (ε -caprolactone) Based Dynamic Network via Quadruple Hydrogen Bonding with Triple-Shape Effect and Self-Healing Capacity. *Appl. Mater. Interfaces*, 7 (4), 2585–2596, 2015.

72. Jiang, Z., Xiao, Y.Y., Kang, Y., Pan, M., Li, B.J., Zhang, S., Shape Memory Polymers Based on Supramolecular Interactions. *Appl. Mater. Interfaces*, 9, 20276, 2017.
73. Rekondo, A. and Martin, R., Catalyst-free room-temperature self-healing elastomers based onaromatic disulfide metathesis. *Mater. Horiz.*, 1, 237, 2014.
74. Nevejans, S., Ballard, N., Fernández, M., Reck, B., Asua, J., Flexible aromatic disulfide monomers for high-performance self-healable linear and cross-linked poly(urethane-urea) coatings. *Polym.*, 166, 229, 2019.
75. Zhang, Z.P., Rong, M.Z., Zhang, M.Q., Yuan, C.E., Alkoxyamine with reduced homolysis temperature and its application in repeated autonomous self-healing of stiff polymers. *Polym. Chem.*, 4, 4648, 2013.
76. Hu, J., Mo, R., Jiang, X., Sheng, X., Zhang, X., Towards mechanical robust yet self-healing polyurethane elastomers via combination of dynamic main chain and dangling quadruple hydrogen bonds. *Polym.*, 183, 121912, 2019.

4

Self-Healable Tires

Norazlianie Sazali[1,2], Mohamad Azuwa Mohamed[3]* and Zul Adlan Mohd Hir[4]

[1]*Faculty of Mechanical & Automotive Engineering Technology, University Malaysia Pahang, Pekan, Malaysia*

[2]*Centre of Excellence for Advanced Research in Fluid Flow (CARIFF), Universiti Malaysia Pahang, Lebuhraya Tun Razak, Kuantan, Malaysia*

[3]*Department of Chemical Sciences, Faculty of Science and Technology, Universiti Kebangsaan Malaysia, Bangi, Malaysia*

[4]*Faculty of Applied Sciences, Universiti Teknologi MARA Pahang, Bandar Tun Abdul Razak Jengka, Malaysia*

Abstract

Self-healable rubber is expected to contribute to the advanced manufacturing of tires which are applicable to the automobile and aviation industry. Natural Rubber (NR) is among the nature's best products and becomes the main ingredient for tire manufacturing. It is extracted through the coagulation of sap from a few trees which produce rubber. NR structure consisting of a long chain molecule, namely cis-1, 4-polyisoprene, and the outstanding properties make it unique in the industry. NR is the favored polymer compared to other synthetic rubbers because of its chosen structure building due to excellent mechanical properties. These days, indeed, the sector of tire manufacturing made from rubber is facing several obstacles to meet sustainability requirements. Thus, further renewable substances and alternative ways of prolonging the life span of vulcanized rubbers are urgently required. One potential recycling technique and/or reuse of vulcanized elastomers is through the formation of reversible networks. Specifically, the use of rubber that possess self-healing properties in the making of tires for the automotive industry is very challenging since it tends to crack when a specific force is applied. Thus, this chapter focuses on the various strategies for producing self-healing rubber and enhancing its properties.

***Keywords*:** Self-healing material, latex, self-recovery tires, rubber modification

**Corresponding author*: mazuwa@ukm.edu.my

Inamuddin, Mohd Imran Ahamed, Rajender Boddula and Tariq Altalhi (eds.) Self-Healing Smart Materials and Allied Applications, (99–122) © 2021 Scrivener Publishing LLC

4.1 Introduction

Rubbers are widely employed in numerous sectors, particularly in the automotive industry [1]. Given the escalating requests of customers, the full use of rubber worldwide exceeded 29 million tons in 2018. The chemical rubber chain cross-link prepared by either peroxide curing or sulfur vulcanization is essential to preserve the hardness, rigidity, elasticity, and mechanical robustness of the materials made from rubber required for professional usage [2, 3]. The vulcanization of rubber involved interlinked of sulfur or peroxide, which can be transformed into chemical carcinogenic substances. The rubber manufacturing industry was identified by the International Agency for Research on Cancer (IARC) as carcinogenic by the World Health Organization in 2017. Besides, the networks are irreversible interlinks formed by sulfur and peroxide, preceding to the unfeasibility to process these products of rubber [4–6]. Vast quantities of unused rubber are thrown out and become a significant concern for the environment has become the consequences [7]. By then, though, most of the rubbers rejected were destroyed for energy recovery in a low-grade form without acquired the environmental pollution affecting air and water [8].

Mixing with pure material ensued by cross-linking is another recycling technique to discard the rubber. However, the weak interface between reused rubber and pure substance significantly disintegrate the wide properties of the processed materials, thereby reducing the potential alternative to this approach. Figure 4.1 shows the structure of polymer–rubber composites obtained from waste tires [1]. A renewable solution is needed to reuse the product made from the rubber. The reprocessed of products made from rubber for renewable recycling has been invested with significant effort [9–11]. Regrettably, this method will cause considerable separation

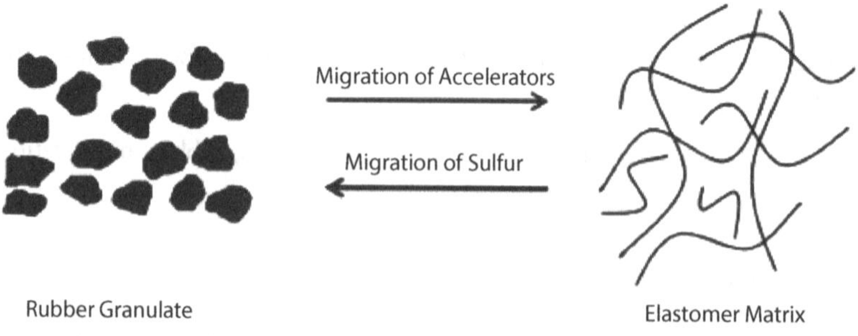

Figure 4.1 Structure of polymer–rubber composites obtained from waste tires [1].

of the rubber chain to be energy concentrated and an ecologically polluted procedure, resulting in the quality disruption to the reused rubber.

New approaches have recently been proposed for the synthesis of eco-friendly rubber through reversible non-covalent and covalent cross-linking [12–14]. The properties of rubbers made from reversible non-covalent cross-linking such as ionic strength, hydrogen bonding, and crystallinity are low due to its weak interactions [15, 16]. It is, therefore, more beneficial to use heavy covalent cross-linked, which can provide rubbers with higher mechanical strength. Reversible covalent cross-linkers were systematically investigated, including imine bond, boronic esters, and disulfide. The most commonly used method was the thermo-reversible Diels–Alder (DA) reaction to produce reversible infrastructure in rubbers. Bai *et al.* (2019) documented polybutadiene (PB) elastomer, which was thermally recyclable with cross-linked arrangements developed through the reaction of DA [17]. The PB obtained from the vulcanization process was found to have strong solvent resistance up to 100 °C, and the PB properties could be modified with several cross-linkers, like furan and maleimide. Previously, thermo-reversible crosslinked ethylene-propylene rubber was produced through the reaction of DA coupling through primary rubber grafting chains with groups of furans and then using bismalermide as a cross-linker [18–21]. Outcomes presented that cross-links may broke at temperature more than 150 °C via DA reaction and self-assembled at temperature ranging from 50 to 70 °C.

Jia *et al.* [22] demonstrated the probability of developing thermo-reversible elastomeric systems using PB with distinct molar mass relying on the Furan–Bismalermide of DA reaction. Re-molding the used materials without loss of mechanical properties has shown sufficient recycling capability of the resultant vulcanization of rubber. But only small molecular organic compounds are extracted from those previous works, such as bismalermide, which have been used for DA reaction as a dienophilic. Furthermore, no rigid filler was applied as a reinforcing agent to these thermo-reversible crosslinked rubber systems, or even as a dienophile in the reaction of DA or cross-linker for the thermo-reversible rubber system, which could significantly limit the functional usage of product made from rubber.

Including cross-linking, silica rubber fillers such as carbon black were crucial for a handful of products made from rubber [9, 23]. Fillers usage is the major approach implemented to boost rubber properties by improving mechanical durability and robustness towards heat as well as enhancing the resistance towards UV degradation, thermal aging, fatigue, and abrasion [18, 24]. Likewise, adding fillers can also effectively reduce the rate of products made from rubber [25, 26]. Moreover, both academic and functional value will be achieved by incorporating the benefits of thermo-reversible crosslinking with rigid filler in rubber vulcanizates.

4.2 Self-Healable Rubber

Self-healing rubber or self-recovering rubber is among the resurgences of intelligent material classes in engineering materials capable of recovering themselves and compatibility when degraded without having to detect or repair any manual intervention. Nevertheless, due to the rubber properties that can withstand great ricochets with insignificant or no permanent distortion, rubbers frequently struggle within fracture and fatigue methods. Developing self-recovering or healing rubber allows for the initial detention of crack propagation, thus preventing catastrophic failure. Many ways have been explored to incorporate the self-healing principle in artificial rubbers, and several studies have reported in this field [27, 28]. The end-use test results are encouraging, but no self-recovering rubbers develop a precise capability for definitive implementations. This might be attributed to constraints on artificial rubbers contrasted to unprocessed rubber. Unprocessed rubber has an enormous fault strain of any rubber. To generate widely varying properties, it may be compounded with different additives. There is minimal research devoted to investigating the competence of natural rubber to be produced as self-recovering components, as per the author's findings. Natural rubber is rubber with non-polar, which does not have any available practical categorizes, some of the reasons for deficiency of self-recovering tested using unprocessed rubber [29, 30]. Scientists focused primarily on epoxidized natural rubber (ENR) in an initial effort to improve natural rubber self-healing, gaining benefits of the oxirane polar categories found in the links of rubber. It has been documented that at high temperatures, the constituents displayed a self-healing manner in which the alterable hydrogen can be linked by interdiffusion of a molecule and successful interactions when contact is made with the broken ENR surface [31].

Nevertheless, the strength of non-covalent hydrogen exchanges is quite low. At ambient temperature, the bonding of hydrogen has a low resistance of creep and be deficient in long-term stability. Most recently, the use of groups of ionic in rubbers is a method in producing unprocessed rubber, which is self-healing. The reversibility of the ionic cross-linking amongst molecular chains of rubber activates the recovery process so that a broken sample can recover its prime characteristics. Marketable rubbers are non-polar, and no functional group of ionic is available. The production of ionic groups in chains of molecular rubber could provide the ability for the introduction of self-healing mechanisms in natural rubber [32].

The mechanical reliability of raw rubber through vulcanization is rather low, so complex rubber with filler reinforcement is a requirement for high-performance products [29]. This applies to rubbers which heal themselves. On what point the strengthening of carbon black on rubber's properties of self-recovering has yet fully explored. In this study, a recent crop of inherent self-healing natural rubber was produced throughout interactions of ionomeric, and the influence of filler reinforced carbon black on natural rubber's properties, and self-recovering efficiency was studied to grasp the maximum ability of the substances for manufacturing utilization.

4.3 Promising Strategy for Self-Healing Rubber-Based Material

Chitin is a long-chain biopolymer that is plentiful throughout the natural world in many animals and plants. Chitin nanocrystals (CNCs) are a group of crystalline particles that are positively charged, produced from crude chitin [33–36]. The previous study identified the use of CNCs as strengthening fillers in composites constructed on plastics in 2004, which showed effective improvements in mechanical performance through interactions with polymer chains that bind to hydrogen [37–39]. Afterward, CNCs were chemically changed to enhance their scattering in different plastics. A compound has developed by greasing a surface reformed CNC (L-lactide) onto poly through various previous studies [40–42]. Mechanical properties and poly (L-lactide) cell stability has been dramatically improved. The method of altering CNCs, however, is very complicated, thus significantly restricting the use of chitin nanocrystal as a filler in numerous definite manufactures [43–46].

In comparison to high-module plastics, soft rubbers with more molecular suppleness reinforce raw CNCs derived from sustainable biological resources without any alterations of chemicals. Because of its significant biodegradability, biocompatible design, outstanding water dispersion, high aspect ratio, high modulus, and exceptional polymer reinforcement capability, CNCs have theoretically become more of an eco-friendly and tempting nano-filler, rather than traditional carbon black filler used in the rubber industry. Stephen *et al.* [47] concluded that carbon nano chitin has a significant impact on improvements of carboxylated styrene-butadiene rubber (XSBR) composites in the mechanical properties.

Liu and co-workers had manufactured a composite of NR or CNCs where the mechanical properties were significantly upgraded before the CNC had been filled [48]. Nevertheless, the creation and propagation of cracks in rubber components have a significant impact on their lifetime during practical use. Recently, composites of rubber with self-healing purposes have been a substantial issue in the area of materials to fix this problem. In this case, hydrogen bonding via reversible non-covalent interaction is preferred to retain self-healing properties. Previous researchers developed a new self-recovering rubber made of thiolene functionalized with PB through a variety of cross-linking approaches in which effective self-recovery had to be activated at high temperatures [49–52].

Song and groups have recently prepared the supramolecular rubber by chemical cross-linking constitute hydrogen bonding that can be rapidly restored at room temperature [53]. Unfortunately, the effectiveness of self-healing diminished as chemical cross-linking elevated to some degree. Das *et al.* proportioned bromobutyl rubber bromide (BIIR) with self-recovery characteristics through the incorporation of ionic bromide imidazolium groups [54]. Few studies defined a self-healing characteristic that can be achieved by incorporating immense cross-linker such as zinc dimethacrylate polymerization (ZDMA) [55–58]. Regrettably, the product formed retained weak self-healing and mechanical properties, because the chains of rubber must maintain enough suppleness and flexibility to fulfill the process of self-healing. Alternatively, shape memory polymers (SMP) are one of the potential materials that could switch into temporary shapes and return to its original form as a self-healing product [56]. Figure 4.2(a) illustrating the typical measurement of shape memory effect where the heat is applied to trigger the switching temperatures (Tsw) of the polymer. In the case of shape memory rubber, the ethylene–vinyl acetate copolymer acted as the thermal-control to switch the deformation phase of rubber particles via the ZDMA-enhanced interface. The SMP approach could be beneficial to improve the robustness in automobile tires and to prolong the tire's life span where the sharp object can no longer able to flat the tires. However, the details on this enhancement are still limited in the open literature.

The supplement of suitable fillers is an achievable approach for enhancing rubber mechanical properties [59]. In several cases, fillers are not part of the self-recovering process that can inhibit the chains of rubber and enormously degrade the self-recovering effect on composites of rubber. With this in mind, as suggested, a self-recoverable composite of ENR

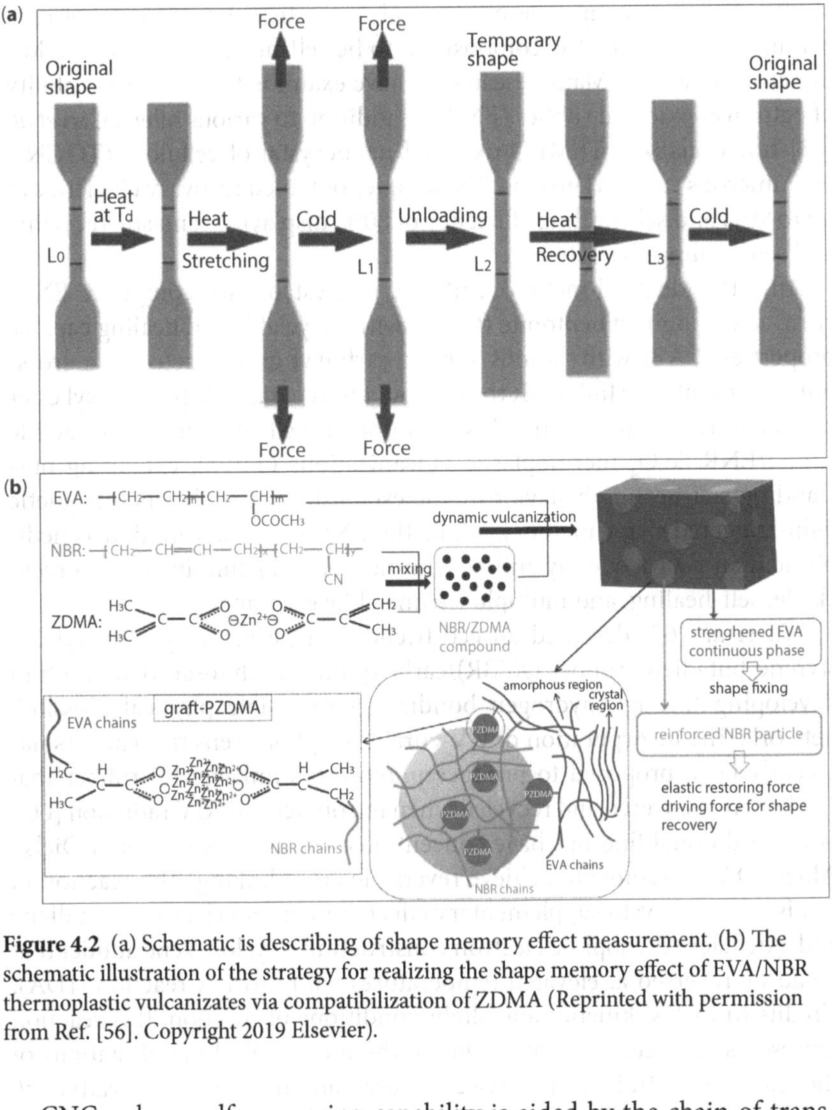

Figure 4.2 (a) Schematic is describing of shape memory effect measurement. (b) The schematic illustration of the strategy for realizing the shape memory effect of EVA/NBR thermoplastic vulcanizates via compatibilization of ZDMA (Reprinted with permission from Ref. [56]. Copyright 2019 Elsevier).

or CNCs whose self-recovering capability is aided by the chain of transmission and bonding of hydrogen integrated. Its amazing interdiffusion of molecular has selected ENR at broken interfaces, which take part as a vital function in helping rubber composite self-recovery [60–62]. The presence of the hemiacetal hydroxyl group in the CNCs structure possibly will interact with the epoxy groups in the ENR structure, thus initiating CNCs to comprise supramolecular networks throughout the creation. The hydrogen bonds developed among ENR's newly formed oxygen groups and CNC's hydroxyl groups are of the most significant importance, not just to efficiently supporting the mechanical properties of composites of rubber and also enabling the composition to be self-healed deprived of additional stimulations. Many researchers have examined the reprocess ability of natural epoxidized rubber (ENR) by addition to various fillers. Cao *et al.* [63] functionalized TEMPO-oxidized nanocrystal of cellulose (TOCNs) by numerous carboxyl groups. The samples obtained recovery efficiency of up to 90% and self-healing efficiency of 80%, displaying fantastic recycling and self-healing characteristics.

After the identical method, Xu *et al.* [64] established composite ENR/ citric acid-modified bentonite (CABt) with recyclable and healing capable properties. CABt with various surface carboxyl groups acted as a cross-linker to covalently link ENR throughout interchangeable β-hydroxyl ester connections and an effective ENR reinforcer. Self-recovering polylactide (PLA)/ENR/Fe_3O_4 thermoplastic vulcanizations (TPVs), exhibiting outstanding self-healing behavior by the establishment of thermal/magnetic light-triggered shape memory. While the ENR interacts with dodecanedioic acid and subsequently aniline trimmer (ACAT) culminated in a recyclable, self-healing, and multiple vitrimer-like elastomers.

Wu *et al.* [65] designed an eco-friendly shape memory carboxylated styrene-butadiene rubber (XSBR)/carboxy methyl chitosan (CMCS) film developing through hydrogen bonding, a reversible physical crosslink network. The incorporation of UV-curable or photo-sensitive rubbers has recently been proposed to produce photo-cross linkable PB rubber that can be self-recovered and recycled when in contact with UV radiation [66]. Some additional findings have concentrated on using reactions of Diels–Alder (DA) as remote to achieve reversible cross-linking. The reaction of DA is a [4 + 2] cyclo supplementary effect amongst an electron-rich diene and a deficient dienophile electron constructing a cyclohexene adduct that could be reversed at elevated temperatures via retro-DA reaction (rDA). Credits to its fast kinetics and slight conditions of reaction, this reaction represents an accepted choice. One of the most typical manifestations of the reaction of DA is within classes of furan and maleimide. Trovatti *et al.* researched the thermo-reversible of DA reaction as a practical method for modifying PB rubber recycling capability [67, 68], combined with a bismaleimide and furan.

Polgar *et al.* continued to work with rubber propylene-ethylene (EPM) [18, 19]. By grafting maleic anhydride (MA) onto the rubber structure, they increased EPM's reactivation. Their approach was focused on the incorporation of a furan moiety into the malleated rubber to engage as an electron-rich diene in the DA reaction. DA reaction offers the possibility of cross-linking butyl rubber (IIR). Nonetheless, according to the imperfect reactions and mixes of the products acquired through the research of the diene, and even to the unoptimized reaction parameters for polymer grafting, the implementation of the chemistry of DA to IIR has been subject to restrictions [12]. Brink and groups attempted to solve such drawbacks by introducing poly (ethylene oxide) or polystyrene (PS) and simultaneously adding carboxylic acid moieties alongside the backbone of the polymer [69]. They illustrated how butyl rubber graft copolymers could be prepared using the chemistry of DA to change their properties of tensile and their reversibility of thermal.

Trovatti *et al.* [68] even mentioned about the reversible cross-linking of natural rubber (NR) reformed by furan. They present support of this reversibility by contrasting the furan-modified NR's 1HNMR spectra with its de-crosslinked rubber retro-DA. The reaction of DA has presented assurance in preparing thermoset polymers for self-healing. DA moieties permit the self-healing abilities to be implemented in established networks of polymeric, not including substantially give in the initial properties.

The Murphy and Wudl group reported the first self-recovering polymer that was relied on the thermo-reversible reaction of DA [70]. Their pioneering research suggested a cross-linked thermoset created by multiple furan and maleimide monomer through DA cycloaddition. The corresponding network of covalent cross-linked demonstrated reversibility for 15 min when heating to 150 °C, resulting in cleavage of around 25% of the addition of DA. After cooling down to 80 °C and subsequent treatment at this temperature, the broken links were permitted to form an interconnection with the sample. Self-recovery tests were performed by splitting the sample into half. Upon injury, the two cut surfaces were taken into close interaction, cured at 150 °C to let the reaction of retro-DA, and gradually cooled to room temperature. The mechanical properties of the restored crack surface were examined by scanning electron microscopy (SEM), which showed near-absolute healing. The average productivity of healing was discovered to be about 50% of the initial energy of breakage.

Next, researchers enlarged the productivity of healing to 80%, with low-melting-point monomers by providing greater flexibility in the network [71]. Although it was not the primary occasion that chemistry of DA was implemented to accomplish reversible cross-linking, the remarkable recovery success had transformed Wudl's research into a landmark that has later sparked increased attention in self-recovery or healing of polymers [72, 73]. A comprehensive synthesis process of reversible cross-linked NR by a reaction of DA with self-recovery ability is further discussed in relative to Trovatti's research team. It is including conceptual enlightenment of the cross-link reversibility employing vibrant mechanical study under tension and shear mode and comprehensive analysis of chemical. Regional damage restoration is also practiced at various observational levels [74]. The findings suggest that the subsequent compound of rubber does have a density of cross-linking and its mechanical robustness equivalent to those of a vulcanized NR with a low sulfur loading. By captivating full gain of this reversible network of cross-linked, vulcanized rubber can be self-recovered and reprocessed for many times. Therefore, the life cycle would be prolonged, whereas discarded resources would be cut.

Natural and synthetic rubbers are polymers commonly spent in a variety of areas owing to their unusual mechanical robustness and elasticity. Rubbers had to be vulcanized or cross-linked with the creation of an irreversible network of three-dimensional (3D) for these critical applications. Nonetheless, vulcanization has always been considered a legitimate, time-consuming, and energy-intensive method [75]. Relative to the thermally generated cross-linking product, the photo-crosslinking industry is growing rapidly due to its distinctive benefits, such as low energy usage, rapid curing time, high efficiency, and wide-ranging adaptableness [74, 76]. Hence, when the photo-crosslinking technology could be extended to the industry of rubber, energy utilization, and minimal effectiveness would be easily overcome. The primary approach to producing photo-cross linkable rubber or elastomer is now the introduction in rubber chains of UV-curable or photosensitive groupings. Like a well-known photo-sensitive group, acrylate was grafted on natural rubber through acylating natural epoxidized rubber with acrylic acid to create natural acrylic rubber [77].

Besides, when acrylate monomers are introduced, the dormant double bonds present in rubber structures are also photo-polymerized to construct 3D networks [78–80]. Conversely, cinnamoyl-altered natural rubber and butyl rubber can also forcibly healed by [2 + 2] cycle, adding complex mechanism upon exposure to UV light. Then, UV-induced thiolene photo-polymerization is often employed as crosslinking inhibitors for the photo-crosslinking of natural rubber, isoprene rubber, between double bonds, carboxylated nitrile butadiene rubber, PB rubber, styrene/butadiene/styrene elastomer, butadiene–acrylonitrile copolymers, or norborne-functional poly(glycerol sebacate) elastomer, and multithiols [81–84].

Also, thiolene photopolymerization is an even more effective approach of crosslinking elastomer than double bond copolymerization with acrylate monomers [81, 83]. The previous work stated that the mechanical properties of photocured rubbers depended primarily upon cross-linking agent dosages. Still, there was no direct disclosure of thiolene photopolymerization kinetics at the site. The irreversible cross-linked networks provide excellent properties for rubber products, but until then, rubbers are also considered impenetrable and not feasible. Thus, the rubbers cross-linked are challenging to reprocess as thermoplastics. Post-consumer rubbers were primarily to be discarded as landfills, burned like coal, or reused as fillers or additives to manufacture essential components [85]. Such regulations, however, may lead to a substantial conservational problem and significant value deficit. So, a practical resolution is immediately needed for recovering the products of waste rubber.

Constitutional lively chemistry comprises structures of molecular developing dynamic networks of cross-linked relying on reversible physical contact and reaction of chemical in which cross-links were able to break and reformed reversibly, under balance influence. Grasping the benefits of such extra characteristics, self-repair and regeneration, able to control decrease stress or change of shape and to recover of used thermosets [86–89]. Different constitutional dynamic chemistries have been integrated into networks of rubber or elastomer, such as the reaction of Diels–Alder (DA), alcoxyamine moieties, disulfide metathesis, the reaction of transesterification, olefin metathesis, the association of ionic, chemistry of supramolecular, and metal-ligand interaction, providing them with inherent self-recovering and recycling properties through the rearrangement of the links amid the contact boundaries. Indeed, whether the self-recovering or recyclable rubber synthesis is repurposed and somewhat complicated, or the self-recovering or recycling cycle is most often triggered by less economical heat than light. However, this dimension has to be received scant attention.

On the other hand, the self-healing rubber based material also can be fabricated using ionic modification without using conventional crosslinking or vulcanizing agents [90]. It has been reported that the bromobutyl rubber (BIIR) with bromine functionality had led to the ionic imidazolium bromide group transformation. As a result, the formation of reversible BIIR exhibits physical cross-linking capability and extraordinary self-healing characteristics, as can be seen in Figure 4.3.

In the current study, a simple technique is developed for producing rubbers which can undergo vulcanization processed, self-recovered, and recycled by the exposure to UV radiation [91]. The rubbers are designed via thiolene photopolymerization from ordinarily existing PB rubber and liquid polysulfide, which includes both disulfide bonds and thiol end groups. The kinetics of photo-polymerization is characterized by *in situ* real-time infrared spectra (RT-IR), photo-differential calorimeter scanning (photo-DSC), and photo-dynamic rheology. Besides, there is also a critical study on the dynamic reversibility and kinetics of photo-activated by disulfide metathesis. The resulting rubbers will rearrange themselves in the cross-linked networks because of the introduction of UV radiation on the interface of disulfide metathesis, allowing for immense self-repair, reusing, and recovering capacities. The effect of the disulfide linkage material, intensity of irradiation, and time on the ability to self-heal, reprocess, and recycle are analyzed. The photo-cross linkable, self-healable, and re-processable rubbers are assumed to be conducive to saving energy, increasing the lifespan, and the waste of rubber products. In other development, the self-healable rubber can be realized via UV-induced disulfide metathesis on the interface where it can rearrange the crosslinked networks, as can be seen in Figure 4.4(a) with various shape can be obtained by UV irradiation [92].

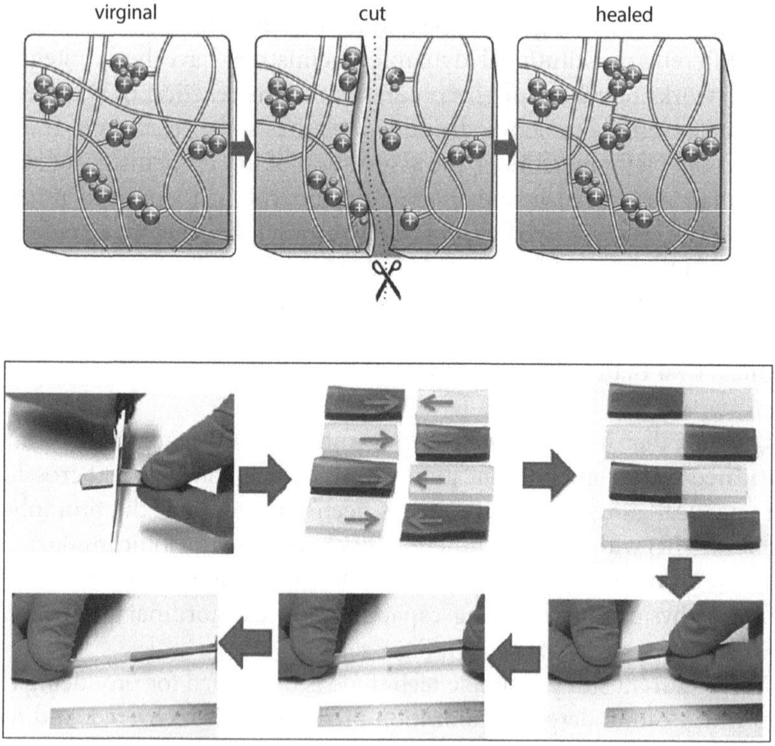

Figure 4.3 Ionic association and self-healing. (Top) Schematic representation of self-healing by the reformation of ionic associates in an ionic rubber network. (Bottom) A sample of imidazolium-modified bromobutyl rubber was cut into two pieces. After reassembling and 18 h of healing time, the sample was stretched to breakdown. For better visualization, one part of the cut sample was colored (Reprinted with permission from Ref. [90]. Copyright 2015 American Chemical Society).

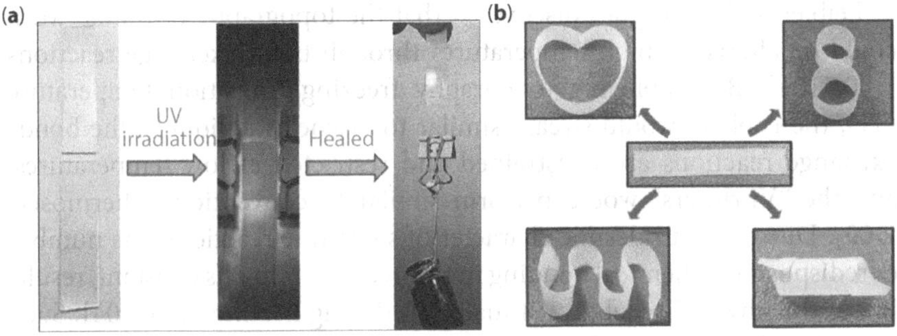

Figure 4.4 (a) Photos illustrating the healing processes of the UV-cured rubbers and (b) Rapid manufacturing of different shapes of rubber products by UV irradiation under the illumination of a 2-kW medium pressure mercury lamp within 90 s at room temperature (Reprinted with permission from Ref. [92]. Copyright 2019 Elsevier).

Ecological defense and resource conservation have to turn out to be an increasingly severe struggle in mass production, mostly in the industry of rubber [30]. Traditional vulcanized rubbers have outstanding mechanical properties because of the development of a covalently cross-linked, stable 3D network. Nonetheless, the irreversible cross-links from the vulcanization of sulfur and the cross-linking of peroxides often transform rubbers into thermosets with the inability of reusing, reutilizing, and self-recovering. For example, it is a critical problem, hundreds-over thousands of tons of products made from rubber are dumped each day in years in the automotive industry [4], building and our livings, and seriously endanger our resources and climate. For the reusing and self-recovery of cross-linked rubbers, hence, sustainable growth of the rubber industry is essential. To endow rubbers with these exciting features, the design of reversible networks in rubbers has been put into tremendous effort [88]. It is feasible for rubbers to create a constant supramolecular network centered on weak connections. Many supramolecular networks, though, will not adequately suffer reasonable stress, which prevents their functions [93, 94]. The usage of more robust covalent networks of reversible bonds of rubber would lead to disseminated mechanical properties [95]. Such networks can be repositioned in external stimuli; for example, the temperature performs self-recovering and rubber restructuring [92, 96–99]. In most cases, this method requires the design and synthesis of new rubbers instead of the modification of used rubbers since most industrial rubbers do not have suitable reactive feature groups. Furthermore, in material science, a new group of healable thermosets with captivating ductility features, named "Vitrimer" appears.

Leibler and colleagues discovered that the topography of the network could be altered at high temperatures through bond-exchange reactions [100, 101]. By employing topography freezing transition temperature (Tv), the rubbers would stream similar to viscoelastic liquids, the bond-exchange reactions are constrained and restricted at low temperatures, and the "Vitrimers" would perform similar to conventional thermosets [102]. Due to the isodesmic character of substitute reactions, the number of redisposition chemical bonding in the network remains constant, resulting in the physical reliability of materials during heating [103, 104]. Such marvelous Vitrimers performances were showed in a system of epoxy-acid that a Zn (II) transesterification catalyst was introduced into the polyester epoxy resins [105, 106]. Until now, successful technologies in eco-friendly and self-healing materials have been explored based on Vitrimers studies [104, 107–113].

The incorporation of Vitrimer-like development into networks of rubber is a practicable process for endowing rubbers with the capability to reprocess, recycle, and self-heal. Since the Vitrimer materials could be effectively produced via transesterification reaction in an epoxy–acid system, epoxidized natural rubber (ENR), resulting from NR epoxidation, is a form of commercial rubber suitable for Vitrimer-like behavior via reaction of transesterification in a network of rubber proposed to contain esters [114, 115].

Norvez and Leibler *et al.* successfully synthesized the ENR by dicarboxylic acids [116, 117], where the groups of carboxyl opened up the groups of oxirane and created β-hydroxy esters alongside the chains of rubber, creating an interlinked network. Recently, the introduction of a catalyst for transesterification, namely zinc acetate (Zn $(Ac)_2$) into the dodecanedioic acid cross-linked ENR network, could relieve the strain at high temperatures and exhibit some Vitrimer-like properties. Guo and Zhang *et al.* researched on ENR Vitrimer by integrating carboxyl group-functionalized carbon nanodot into the ENR [113]. Their ENR Vitrimer could modify the topography of the network by promoting strain relaxing through transesterification reaction under the $Zn(Ac)_2$ acceleration, which has allowed it to recycle, reshape and weld [118]. Assuming that $Zn(Ac)_2$ tends to establish to the ENR surface as time goes by and rises the risk of conservation contamination, as mention throughout this work that an ENR composite that is cross-linked by citric acid (CA)-modified bentonite (CABt) via β-hydroxyl ester in the absence of $Zn(Ac)_2$ [119, 120]. Bentonite (Bt), an available natural Earth abundance, can be a renewable substitute for replacing petroleum-based rubber strengthening agents. CABt consisting of numerous carboxyl groups is used as a cross-linker for crosslinking ENR through β-hydroxyl ester connections and an effective ENR reinforcer. Even though the cross-link degrees of ENR/CABt composites deprived of $Zn(Ac)_2$ were discovered to be low, at high temperatures, the network was still able to have less strain and show some Vitrimer-like features. The composites of ENR or CABt were assumed as biodegradable and curative at high temperatures.

4.4 Conclusion

In summary, various approaches can be employed to fabricate self-healable rubber as well as the strategy to improve the self-recovery properties of the rubber. Up until today, researchers all over the world have managed to produce rubber-based materials that exhibit self-healing properties, without sacrificing strength or elasticity. In principle, the self-healing characteristic of rubber-based tires could be ascertained by altering the rubber processing technique. Self-healable tires can be globally commercialized with the aims of the manufactured tires can be resilient towards punctured by the shape object. Various research discoveries were highlighting the self-healing phenomena rubber-based material where a cut on its surface can be cured even at room temperature. However, the low-cost and straightforward self-healable rubber-based material for tires application is more crucial for more practical and industrial application.

References

1. Sienkiewicz, M., Janik, H., Borzędowska-Labuda, K., Kucińska-Lipka, J., Environmentally friendly polymer–rubber composites obtained from waste tyres: A review. *J. Cleaner Prod.*, 147, 560–71, 2017.
2. Nisar, J., Ali, G., Ullah, N., Awan, I.A., Iqbal, M., Shah, A. *et al.*, Pyrolysis of waste tire rubber: Influence of temperature on pyrolysates yield. *J. Environ. Chem. Eng.*, 6, 3469–73, 2018.
3. Chauhan, B.S., Singh, R.K., Cho, H.M., Lim, H.C., Practice of diesel fuel blends using alternative fuels: A review. *Renewable Sustainable Energy Rev.*, 59, 1358–68, 2016.
4. Vallero, D., Chapter 31—Controlling Air Pollution from Sources, in: *Fundamentals of Air Pollution*, Fifth Edition, D. Vallero, (Ed.), pp. 881–925, Academic Press, Boston, 2014.
5. Makaruk, A., Miltner, M., Harasek, M., Biogas desulfurization and biogas upgrading using a hybrid membrane system—Modeling study. *Water Sci. Technol.*, 67, 326–32, 2013.

6. Sazali, N., Salleh, W.N.W., Ismail, A.F., Carbon tubular membranes from nanocrystalline cellulose blended with P84 co-polyimide for H2 and He separation. *Int. J. Hydrogen Energy*, 42, 9952–7, 2017.
7. White, J.L., Development of Internal-Mixer Technology for the Rubber Industry. *Rubber Chem. Technol.*, 65, 527–79, 1992.
8. Jha, A. and Bhowmick, A.K., Thermal degradation and ageing behaviour of novel thermoplastic elastomeric nylon-6/acrylate rubber reactive blends. *Polym. Degrad. Stab.*, 62, 575–86, 199.
9. Tan, K., Li, C., Meng, H., Wang, Z., Preparation and characterization of thermoplastic elastomer of poly(vinyl chloride) and chlorinated waste rubber. *Polym. Test.*, 28, 2–7, 2009.
10. Miranda, M., Pinto, F., Gulyurtlu, I., Cabrita, I., Pyrolysis of rubber tyre wastes: A kinetic study. *Fuel*, 103, 542–52, 2013.
11. Li, H., Jiang, X., Cui, H., Wang, F., Zhang, X., Yang, L. *et al.*, Investigation on the co-pyrolysis of waste rubber/plastics blended with a stalk additive. *J. Anal. Appl. Pyrolysis*, 115, 37–42, 2015.
12. Bounaceur, R., Berger, E., Pfister, M., Ramirez Santos, A.A., Favre, E., Rigorous variable permeability modelling and process simulation for the design of polymeric membrane gas separation units: MEMSIC simulation tool. *J. Membr. Sci.*, 523, 77–91, 2017.
13. Matteucci, S., Yampolskii, Y., Freeman, B.D., Pinnau, I., Transport of Gases and Vapors in Glassy and Rubbery Polymers, in: *Materials Science of Membranes for Gas and Vapor Separation*, pp. 1–47, John Wiley & Sons, Ltd. The Atrium, Southern Gate, Chichester, West Sussex PO19 8SQ, England, 2006.
14. Ismail, N.H., Salleh, W.N.W., Sazali, N., Ismail, A.F., Effect of intermediate layer on gas separation performance of disk supported carbon membrane. *Sep. Sci. Technol.*, 52, 2137–49, 2017.
15. Okada, O., Keskkula, H., Paul, D.R., Mechanical properties of blends of maleated ethylene–propylene rubber and nylon 6. *Polymer*, 42, 8715–25, 2001.
16. Li, G.-Y. and Koenig, J.L., A Review of Rubber Oxidation. *Rubber Chem. Technol.*, 78, 355–90, 2005.
17. Bai, J., Li, H., Shi, Z., Yin, J., An Eco-Friendly Scheme for the Cross-Linked Polybutadiene Elastomer via Thiol-Ene and Diels–Alder Click Chemistry. *Macromolecules*, 48, 3539–3546, 2015.
18. Polgar, L.M., Fortunato, G., Araya-Hermosilla, R., van Duin, M., Pucci, A., Picchioni, F., Cross-linking of rubber in the presence of multi-functional cross-linking aids via thermoreversible Diels–Alder chemistry. *Eur. Polym. J.*, 82, 208–19, 2016.
19. Polgar, L.M., Kingma, A., Roelfs, M., van Essen, M., van Duin, M., Picchioni, F., Kinetics of cross-linking and de-cross-linking of EPM rubber with thermoreversible Diels–Alder chemistry. *Eur. Polym. J.*, 90, 150–61, 2017.
20. Araya-Hermosilla, R., Fortunato, G., Pucci, A., Raffa, P., Polgar, L., Broekhuis, A.A. *et al.*, Thermally reversible rubber-toughened thermoset networks via Diels–Alder chemistry. *Eur. Polym. J.*, 74, 229–40, 2016.

21. Fan, Y., Cao, H., van Mastrigt, F., Pei, Y., Picchioni, F., Copper-mediated homogeneous living radical polymerization of acrylamide with waxy potato starch-based macroinitiator. *Carbohydr. Polym.*, 192, 61–8, 2018.
22. Jia, Z., Zhu, S., Chen, Y., Zhang, W., Zhong, B., Jia, D., Recyclable and self-healing rubber composites based on thermorevesible dynamic covalent bonding. *Compos. Part A: Appl. Sci. Manuf.*, 129, 105709, 2020.
23. Choudhury, A., Balmurulikrishnan, A., Sarkhel, G., Polyamide 66/EPR-g-MA blends: Mechanical modeling and kinetic analysis of thermal degradation. *Polym. Adv. Technol.*, 19, 1226–35, 2008.
24. Fan, Y., Fowler, G.D., Zhao, M., The past, present and future of carbon black as a rubber reinforcing filler—A review. *J. Cleaner Prod.*, 247, 119115, 2020.
25. Roy, K., Debnath, S.C., Pongwisuthiruchte, A., Potiyaraj, P., Up-to-date review on the development of high performance rubber composites based on halloysite nanotube. *Appl. Clay Sci.*, 183, 105300, 2019.
26. Zhang, H., Xing, W., Li, H., Xie, Z., Huang, G., Wu, J., Fundamental researches on graphene/rubber nanocomposites. *Adv. Ind. Eng. Polym. Res.*, 2, 32–41, 2019.
27. Mohajerani, A., Burnett, L., Smith, J.V., Markovski, S., Rodwell, G., Rahman, M.T. *et al.*, Recycling waste rubber tyres in construction materials and associated environmental considerations: A review. *Resour. Conserv. Recycl.*, 155, 104679, 2020.
28. Medina, N.F., Garcia, R., Hajirasouliha, I., Pilakoutas, K., Guadagnini, M., Raffoul, S., Composites with recycled rubber aggregates: Properties and opportunities in construction. *Constr. Build. Mater.*, 188, 884–97, 2018.
29. Li, Y., Zhang, X., Wang, R., Lei, Y., Performance enhancement of rubberised concrete via surface modification of rubber: A review. *Constr. Build. Mater.*, 227, 116691, 2019.
30. Siddika, A., Mamun, M.A.A., Alyousef, R., Amran, Y.H.M., Aslani, F., Alabduljabbar, H., Properties and utilizations of waste tire rubber in concrete: A review. *Constr. Build. Mater.*, 224, 711–31, 2019.
31. Sazali, N., Salleh, W.N.W., Ismail, A.F., Kadirgama, K., Othman, F.E.C., Ismail, N.H., Impact of stabilization environment and heating rates on P84 co-polyimide/nanocrystaline cellulose carbon membrane for hydrogen enrichment. *Int. J. Hydrogen Energy*, 44, 20924–20932, 2018.
32. Zhang, Z.P., Rong, M.Z., Zhang, M.Q., Polymer engineering based on reversible covalent chemistry: A promising innovative pathway towards new materials and new functionalities. *Prog. Polym. Sci.*, 80, 39–93, 2018.
33. Calvino, C., Macke, N., Kato, R., Rowan, S.J., Development, processing and applications of bio-sourced cellulose nanocrystal composites. *Prog. Polym. Sci.*, 103, 101221, 2020.
34. Younas, M., Noreen, A., Sharif, A., Majeed, A., Hassan, A., Tabasum, S. *et al.*, A review on versatile applications of blends and composites of CNC with natural and synthetic polymers with mathematical modeling. *Int. J. Biol. Macromol.*, 124, 591–626, 2019.

35. Fernandes, S.N., Lopes, L.F., Godinho, M.H., Recent advances in the manipulation of circularly polarised light with cellulose nanocrystal films. *Curr. Opin. Solid State Mater. Sci.*, 23, 63–73, 2019.
36. Wang, J., Zhang, X., Jiang, L., Qiao, J., Advances in toughened polymer materials by structured rubber particles. *Prog. Polym. Sci.*, 98, 101160, 2019.
37. Saïd Azizi Samir, M.A., Alloin, F., Paillet, M., Dufresne, A., Tangling Effect in Fibrillated Cellulose Reinforced Nanocomposites. *Macromolecules*, 37, 4313–6, 2004.
38. Sazali, N., Salleh, W.N.W., Ismail, A.F., Nordin, N.A.H.M., Ismail, N.H., Mohamed, M.A. *et al.*, Incorporation of thermally labile additives in carbon membrane development for superior gas permeation performance. *J. Natl. Gas Sci. Eng.*, 49, 376–84, 2018.
39. Sazali, N., Salleh, W.N.W., Ismail, A.F., Wong, K.C., Iwamoto, Y., Exploiting pyrolysis protocols on BTDA-TDI/MDI (P84) polyimide/nanocrystalline cellulose carbon membrane for gas separations. *J. Appl. Polym. Sci.*, 136, 46901, 2018.
40. Miao, C. and Hamad, W., *In-situ* Polymerized Cellulose Nanocrystals (CNC)–Poly(l-lactide) (PLLA) Nanomaterials and Applications in Nanocomposite Processing. *Carbohydr. Polym.*, 153, 549–558, 2016.
41. Grząbka-Zasadzińska, A., Odalanowska, M., Borysiak, S., Thermal and mechanical properties of biodegradable composites with nanometric cellulose. *J. Therm. Anal. Calorim.*, 138, 4407–4416, 2019.
42. Zong, E., Jiang, J., Liu, X., Fu, S., Xu, Y., Chu, F., Combination of lignin and L-lactide towards grafted copolymers from lignocellulosic butanol residue. *Int. J. Biol. Macromol.*, 86, 80–88, 2016.
43. Kubies, D., Ščudla, J., Puffr, R., Sikora, A., Baldrian, J., Kovářová, J. *et al.*, Structure and mechanical properties of poly(l-lactide)/layered silicate nanocomposites. *Eur. Polym. J.*, 42, 888–99, 2006.
44. Tsuji, H. and Ikada, Y., Properties and morphologies of poly(l-lactide): 1. Annealing condition effects on properties and morphologies of poly(l-lactide). *Polymer*, 36, 2709–16, 1995.
45. Lei, X.-X., Lu, H., Lu, L., Xu, H.-Q., Zhou, Y.-G., Zou, J., Improving the Thermal and Mechanical Properties of Poly(l-lactide) by Forming Nanocomposites with an *in Situ* Ring-Opening Intermediate of Poly(l-lactide) and Polyhedral Oligomeric Silsesquioxane. *Nanomaterials (Basel)*, 9, 748, 2019.
46. Zhou, Q., Gong, Y., Gao, C., Microstructure and mechanical properties of poly(L-lactide) scaffolds fabricated by gelatin particle leaching method. *J. Appl. Polym. Sci.*, 98, 1373–9, 2005.
47. Stephen, R., Ranganathaiah, C., Varghese, S., Joseph, K., Thomas, S., Gas transport through nano and micro composites of natural rubber (NR) and their blends with carboxylated styrene butadiene rubber (XSBR) latex membranes. *Polymer*, 47, 858–70, 2006.
48. Liu, Y., Liu, M., Yang, S., Luo, B., Zhou, C., Liquid Crystalline Behaviors of Chitin Nanocrystals and Their Reinforcing Effect on Natural Rubber. *ACS Sustainable Chem. Eng.*, 6, 325–36, 2018.

49. Silverstein, J.S., Casey, B.J., Natoli, M.E., Dair, B.J., Kofinas, P., Rapid Modular Synthesis and Processing of Thiol–Ene Functionalized Styrene–Butadiene Block Copolymers. *Macromolecules*, 45, 3161–7, 2012.
50. ten Brummelhuis, N., Diehl, C., Schlaad, H., Thiol–Ene Modification of 1,2-Polybutadiene Using UV Light or Sunlight. *Macromolecules*, 41, 9946–7, 2008.
51. Auvergne, R., Desroches, M., Clerc, S., Carlotti, S., Caillol, S., Boutevin, B., New biobased epoxy hardeners: Thiol-ene addition on oligobutadiene. *React. Funct. Polym.*, 72, 393–401, 2012.
52. Bertin, A. and Schlaad, H., Mild and Versatile (Bio-)Functionalization of Glass Surfaces via Thiol–Ene Photochemistry. *Chem. Mater.*, 21, 5698–700, 2009.
53. Song, Q., Chen, H., Yu, W., Ni, H., Hu, P., Zhao, K., Multi-stimuli Responsive Polyurethane Based on Quadruple Hydrogen Bonding Interactions. *Mater. Today: Proc.*, 16, 1405–9, 2019.
54. Das, D. and Veziroğlu, T.N., Hydrogen production by biological processes: A survey of literature. *Int. J. Hydrogen Energy*, 26, 13–28, 2001.
55. Li, C., Yuan, Z., Ye, L., Facile construction of enhanced multiple interfacial interactions in EPDM/zinc dimethacrylate (ZDMA) rubber composites: Highly reinforcing effect and improvement mechanism of sealing resilience. *Compos. Part A: Appl. Sci. Manuf.*, 126, 105580, 2019.
56. Wang, L., Hua, J., Wang, Z., Facile design of heat-triggered shape memory ethylene-vinyl acetate copolymer/nitrile-butadiene thermoplastic vulcanizates via zinc dimethacrylate induced interfacial compatibilization. *Polym. Test.*, 76, 481–9, 2019.
57. Wei, Z., Lu, Y., Meng, Y., Zhang, L., Improved understanding of *in-situ* polymerization of zinc dimethacrylate: The solid bulk polymerization. *Polymer*, 53, 1409–17, 2012.
58. Yuan, Z., Li, W., Li, C., Ye, L., Construction of multiple crosslinking networks in EPDM rubber: Synergistic reinforcing effect of graphene-zinc dimethacrylate on EPDM and improvement mechanism of sealing resilience. *Compos. Part A: Appl. Sci. Manuf.*, 121, 254–64, 2019.
59. Wang, Y., Xu, C., Chen, Z., Chen, Y., Improved fracture toughness of dynamically vulcanized poly(vinylidene fluoride)/silicone rubber filled zinc dimethacrylate composite. *Polym. Test.*, 39, 53–60, 2014.
60. Nie, J., Mou, W., Ding, J., Chen, Y., Bio-based epoxidized natural rubber/chitin nanocrystals composites: Self-healing and enhanced mechanical properties. *Compos. Part B: Eng.*, 172, 152–60, 201.
61. Nasseri, R., Deutschman, C.P., Han, L., Pope, M.A., Tam, K.C., Cellulose nanocrystals in smart and stimuli-responsive materials: A review. *Mater. Today Adv.*, 5, 100055, 2020.
62. Nofar, M., Sacligil, D., Carreau, P.J., Kamal, M.R., Heuzey, M.-C., Poly (lactic acid) blends: Processing, properties and applications. *Int. J. Biol. Macromol.*, 125, 307–60, 2019.

63. Cao, X., Ding, B., Yu, J., Al-Salem, S., Cellulose nanowhiskers extracted from TEMPO-oxidized jute fibers. *Carbohydr. Polym.*, 90, 1075–80, 2012.
64. Xu, C., Cui, R., Fu, L., Lin, B., Recyclable and heat-healable epoxidized natural rubber/bentonite composites. *Compos. Sci. Technol.*, 167, 421–430, 2018.
65. Wu, W., Xu, C., Zheng, Z., Lin, B., Fu, L., Strengthened, Recyclable Shape Memory Rubber Films with Rigid Filler Nano-Capillary Network. *J. Mater. Chem. A*, 7, 6901–6910, 2019.
66. Liu, X., Zhou, T., Liu, Y., Zhang, A., Yuan, C., Zhang, W., Cross-linking Process of Cis-Polybutadiene Rubber with Peroxides Studied by Two-dimensional Infrared Correlation Spectroscopy: A Detailed Tracking. *RSC Adv.*, 5, 10231–10242, 2015.
67. Trovatti, E., de Medeiros Ferreira, A., Carvalho, A.J.F., Ribeiro, S.J.L., Gandini, A., Sleeving nanocelluloses by admicellar polymerization. *J. Colloid Interface Sci.*, 408, 256–8, 2013.
68. Trovatti, E., Cunha, A.G., Carvalho, A.J.F., Gandini, A., Furan-modified natural rubber: A substrate for its reversible crosslinking and for clicking it onto nanocellulose. *Int. J. Biol. Macromol.*, 95, 762–8, 2017.
69. Brink, C., Österberg, E., Holmberg, K., Tiberg, F., Using poly(ethylene imine) to graft poly(ethylene glycol) or polysaccharide to polystyrene. *Colloids Surf.*, 66, 149–56, 1992.
70. Murphy, E.B. and Wudl, F., The world of smart healable materials. *Prog. Polym. Sci.*, 35, 223–251, 2010.
71. Lee, J.K., Wang, Y.-M., Cho, S., Wudl, F., Heeger, A.J., New approach for forming bulk-heterojunction solar cells comprising a π-conjugated polymer and C60. *Org. Electron.*, 10, 1223–7, 2009.
72. Williams, K.A., Nowak, M.J., Dormann, E., Wudl, F., Organic conductors based on diradicals: the benzobisdithiazole (BBDT) system. *Synth. Met.*, 14, 233–8, 1986.
73. Patil, A.O., Rughooputh, S.D.D.V., Wudl, F., Poly(p-phenylene vinylene): Incipient doping in conducting polymers. *Synth. Met.*, 29, 115–9, 1989.
74. Fernandes, T.F., Trovatti, E., Freire, C.S.R., Silvestre, A.J.D., Neto, C.P., Gandini, A. *et al.*, Preparation and characterization of novel biodegradable composites based on acylated cellulose fibers and poly(ethylene sebacate). *Compos. Sci. Technol.*, 71, 1908–13, 2011.
75. George, J., Varughese, K.T., Thomas, S., Dynamically vulcanised thermoplastic elastomer blends of polyethylene and nitrile rubber. *Polymer*, 41, 1507–17, 2000.
76. Gattás-Asfura, K., Weisman, E., Andreopoulos, F., Micic, M., Muller, B., Sirpal, S. *et al.*, Nitrocinnamate-Functionalized Gelatin: Synthesis and "Smart" Hydrogel Formation via Photo-Cross-Linking. *Biomacromolecules*, 6, 1503–9, 2005.
77. Le Xuan, H. and Decker, C., Photocrosslinking of acrylated natural rubber. *J. Polym. Sci. Part A: Polym. Chem.*, 31, 769–80, 1993.

78. Kloosterboer, J.G., Network formation by chain crosslinking photopolymerization and its applications in electronics, in: *Electronic Applications*, pp. 1–61, Springer Berlin Heidelberg, Berlin, Heidelberg, 1988.
79. Bagheri, A. and Jin, J., Photopolymerization in 3D Printing. *ACS Appl. Polym. Mater.*, 1, 593–611, 2019.
80. Mallakpour, S. and Zadehnazari, A., Thermoplastic Vinyl Polymers: From Macro to Nanostructure. *Polym.-Plast. Technol. Eng.*, 52, 1423–66, 2013.
81. Asmussen, S.V. and Vallo, C.I., Facile preparation of silver-based nanocomposites via thiol-methacrylate 'click' photopolymerization. *Eur. Polym. J.*, 79, 163–75, 2016.
82. Nair, D.P., Cramer, N.B., Scott, T.F., Bowman, C.N., Shandas, R., Photopolymerized thiol-ene systems as shape memory polymers. *Polymer*, 51, 4383–9, 2010.
83. Huang, Y., Yuan, J., Tang, J., Yang, J., Zeng, Z., One step synthesis of monodisperse thiol-ene clickable polymer microspheres and application on biological functionalization. *Eur. Polym. J.*, 110, 22–30, 2019.
84. Xu, K., Cantu, D.A., Fu, Y., Kim, J., Zheng, X., Hematti, P. *et al.*, Thiol-ene Michael-type formation of gelatin/poly(ethylene glycol) biomatrices for three-dimensional mesenchymal stromal/stem cell administration to cutaneous wounds. *Acta Biomater.*, 9, 8802–14, 2013.
85. Diekmann, A., Giese, U., Schaumann, I., Polycyclic aromatic hydrocarbons in consumer goods made from recycled rubber material: A review. *Chemosphere*, 220, 1163–78, 2019.
86. Dong, F., Zhao, P., Dou, R., Feng, S., Amine-functionalized POSS as crosslinkers of polysiloxane containing γ-chloropropyl groups for preparing heat-curable silicone rubber. *Mater. Chem. Phys.*, 208, 19–27, 2018.
87. Xu, C., Nie, J., Wu, W., Fu, L., Lin, B., Design of self-healable supramolecular hybrid network based on carboxylated styrene butadiene rubber and nano-chitosan. *Carbohydr. Polym.*, 205, 410–9, 2019.
88. Chen, D., Yi, S., Wu, W., Zhong, Y., Liao, J., Huang, C. *et al.*, Synthesis and characterization of novel room temperature vulcanized (RTV) silicone rubbers using Vinyl-POSS derivatives as cross linking agents. *Polymer*, 51, 3867–78, 2010.
89. Mensah, B., Gupta, K.C., Kang, G., Lee, H., Nah, C., A comparative study on vulcanization behavior of acrylonitrile–butadiene rubber reinforced with graphene oxide and reduced graphene oxide as fillers. *Polym. Test.*, 76, 127–37, 2019.
90. Das, A., Sallat, A., Böhme, F., Suckow, M., Basu, D., Wießner, S. *et al.*, Ionic Modification Turns Commercial Rubber into a Self-Healing Material. *ACS Appl. Mater. Interfaces*, 7, 20623–30, 2015.
91. Iñiguez, M.E., Conesa, J.A., Fullana, A., Recyclability of four types of plastics exposed to UV irradiation in a marine environment. *Waste Manage.*, 79, 339–45, 2018.

92. Xiang, H., Yin, J., Lin, G., Liu, X., Rong, M., Zhang, M., Photo-crosslinkable, self-healable and reprocessable rubbers. *Chem. Eng. J.*, 358, 878–90, 2019.
93. Xu, C., Cao, L., Huang, X., Lin, B., fu, L., Self-Healing Natural Rubber with Tailorable Mechanical Properties Based on Ionic Supramolecular Hybrid Network. *ACS Appl. Mater. Interfaces*, 9, 29363–29373, 2017.
94. Chino, K. and Ashiura, M., Themoreversible Cross-Linking Rubber Using Supramolecular Hydrogen-Bonding Networks. *Macromolecules*, 34, 9201–4, 2001.
95. Xiang, H.P., Qian, H.J., Lu, Z.Y., Rong, M.Z., Zhang, M.Q., Crack healing and reclaiming of vulcanized rubber by triggering the rearrangement of inherent sulfur crosslinked networks. *Green Chem.*, 17, 4315–25, 2015.
96. Zhou, J. and Sheiko, S., Reversible shape-shifting in polymeric materials. *J. Polym. Sci. Part B: Polym. Phys.*, 54, n/a-n/a, 1365–1380, 2016.
97. Sun, D., Sun, G., Zhu, X., Guarin, A., Li, B., Dai, Z. *et al.*, A comprehensive review on self-healing of asphalt materials: Mechanism, model, characterization and enhancement. *Adv. Colloid Interface Sci.*, 256, 65–93, 2018.
98. Imbernon, L. and Norvez, S., From landfilling to vitrimer chemistry in rubber life cycle. *Eur. Polym. J.*, 82, 347–76, 2016.
99. Zhai, L., Narkar, A., Ahn, K., Self-healing polymers with nanomaterials and nanostructures. *Nano Today*, 30, 100826, 2020.
100. Capelot, M., Montarnal, D., Tournilhac, F., Leibler, L., Metal-Catalyzed Transesterification for Healing and Assembling of Thermosets. *J. Am. Chem. Soc.*, 134, 7664–7, 2012.
101. Montarnal, D., Capelot, M., Tournilhac, F., Leibler, L., Silica-Like Malleable Materials from Permanent Organic Networks. *Science*, 334, 965, 2011.
102. Yue, L., Bonab, V.S., Yuan, D., Patel, A., Karimkhani, V., Manas-Zloczower, I., Vitrimerization: A Novel Concept to Reprocess and Recycle Thermoset Waste via Dynamic Chemistry. *Glob. Chall.*, 3, 1800076, 2019.
103. He, S., Tan, Y., Xiao, X., Zhu, L., Guo, Y., Li, M. *et al.*, Substituent effects on electronic character of the CN group and trans/cis isomerization in the C-substituted imine derivatives: A computational study. *J. Mol. Struct.: THEOCHEM*, 951, 7–13, 2010.
104. Suresh, C.H. and Koga, N., An isodesmic reaction based approach to aromaticity of a large spectrum of molecules. *Chem. Phys. Lett.*, 419, 550–6, 2006.
105. Krishnakumar, B., Sanka, R.V.S.P., Binder, W.H., Parthasarthy, V., Rana, S., Karak, N., Vitrimers: Associative dynamic covalent adaptive networks in thermoset polymers. *Chem. Eng. J.*, 385, 123820, 2020.
106. Krishnakumar, B., Prasanna Sanka, R.V.S., Binder, W.H., Park, C., Jung, J., Parthasarthy, V. *et al.*, Catalyst free self-healable vitrimer/graphene oxide nanocomposites. *Compos. Part B: Eng.*, 184, 107647, 2020.
107. Spiesschaert, Y., Guerre, M., Imbernon, L., Winne, J.M., Du Prez, F., Filler reinforced polydimethylsiloxane-based vitrimers. *Polymer*, 172, 239–46, 2019.

108. Liu, H., Zhang, H., Wang, H., Huang, X., Huang, G., Wu, J., Weldable, malleable and programmable epoxy vitrimers with high mechanical properties and water insensitivity. *Chem. Eng. J.*, 368, 61–70, 2019.
109. Gamardella, F., Guerrero, F., De la Flor, S., Ramis, X., Serra, A., A new class of vitrimers based on aliphatic poly(thiourethane) networks with shape memory and permanent shape reconfiguration. *Eur. Polym. J.*, 122, 109361, 2020.
110. Altuna, F.I., Hoppe, C.E., Williams, R.J.J., Epoxy vitrimers with a covalently bonded tertiary amine as catalyst of the transesterification reaction. *Eur. Polym. J.*, 113, 297–304, 2019.
111. Yang, X., Guo, L., Xu, X., Shang, S., Liu, H., A fully bio-based epoxy vitrimer: Self-healing, triple-shape memory and reprocessing triggered by dynamic covalent bond exchange. *Mater. Des.*, 186, 108248, 2020.
112. Zhang, H. and Xu, X., Improving the transesterification and electrical conductivity of vitrimers by doping with conductive polymer wrapped carbon nanotubes. *Compos. Part A: Appl. Sci. Manuf.*, 99, 15–22, 2017.
113. Qiu, M., Wu, S., Tang, Z., Guo, B., Exchangeable interfacial crosslinks towards mechanically robust elastomer/carbon nanotubes vitrimers. *Compos. Sci. Technol.*, 165, 24–30, 2018.
114. Bai, L. and Zheng, J., Robust, reprocessable and shape-memory vinylogous urethane vitrimer composites enhanced by sacrificial and self-catalysis Zn(II)–ligand bonds. *Compos. Sci. Technol.*, 190, 108062, 2020.
115. Liu, Z., Zhang, C., Shi, Z., Yin, J., Tian, M., Tailoring vinylogous urethane chemistry for the cross-linked polybutadiene: Wide freedom design, multiple recycling methods, good shape memory behavior. *Polymer*, 148, 202–10, 2018.
116. Pire, M., Norvez, S., Iliopoulos, I., Le Rossignol, B., Leibler, L., Imidazole-promoted acceleration of crosslinking in epoxidized natural rubber/dicarboxylic acid blends. *Polymer*, 52, 5243–9, 2011.
117. Pire, M., Norvez, S., Iliopoulos, I., Le Rossignol, B., Leibler, L., Epoxidized natural rubber/dicarboxylic acid self-vulcanized blends. *Polymer*, 51, 5903–9, 2010.
118. Tang, Z., Liu, Y., Guo, B., Zhang, L., Malleable, Mechanically Strong, and Adaptive Elastomers Enabled by Interfacial Exchangeable Bonds. *Macromolecules*, 50, 201, 7584–7592, 2017.
119. Cao, L., Fan, J., Huang, J., Chen, Y., A robust and stretchable cross-linked rubber network with recyclable and self-healable capabilities based on dynamic covalent bonds. *J. Mater. Chem. A*, 7, 4922–33, 2019.
120. Feng, Z., Hu, J., Zuo, H., Ning, N., Zhang, L., Yu, B. *et al.*, Photothermal-Induced Self-Healable and Reconfigurable Shape Memory Bio-Based Elastomer with Recyclable Ability. *ACS Appl. Mater. Interfaces*, 11, 1469–79, 2019.

5

Self-Healing Bacterial Cementitious Composites

R. Preetham[1], R. Hari Krishna[1*], M.N. Chandraprabha[2†] and R. Sivaramakrishna[3]

[1]*Department of Chemistry, M.S. Ramaiah Institute of Technology, Bangalore, India*
[2]*Department of Biotechnology, M.S. Ramaiah Institute of Technology, Bangalore, India*
[3]*Department of Chemistry, REVA Independent PU College, Bangalore, India*

Abstract

Concrete is one of the most commonly used construction materials around the world and plays an indispensable role to mankind. However, deterioration and formation of cracks, a natural phenomenon that occurs in these materials, results in laborious and capital-intensive maintenance and repair work. Hence, a cost effective and low maintenance technique is of importance. Self-healing of the affected cementitious materials has received prominent interest, owing to its long lasting nature and low maintenance. Among the various self-healing techniques, microbial self-healing approach is distinguished by its potential for long lasting, rapid and active crack repair, while also being environmentally friendly. Through this chapter provides comprehensive information about the basic concepts of self-healing concrete are expressed using microbes, various types of bacteria as self-healing agents and different mechanistic aspects of bacterial self-healing process in cement concretes. The factors influencing the efficiency of bacterial self-healing concretes are discussed. This chapter also explores the key reactions parameters, nutrient requirements for bacterial growth for self-healing. Potential challenges in microbial crack treatment are discussed, and recommendations for future work are expressed.

**Corresponding author*: rhk.chem@gmail.com; rhk.chem@msrit.edu
†*Corresponding author*: Chandra@msrit.edu

Inamuddin, Mohd Imran Ahamed, Rajender Boddula and Tariq Altalhi (eds.) Self-Healing Smart Materials and Allied Applications, (123–152) © 2021 Scrivener Publishing LLC

Keywords: Cement concrete, crack, bacteria, self-healing, biomineralization, calcium carbonate, bio-concrete

5.1 Introduction

Cementitious materials are one of the most popular means of producing structural elements in the construction industry. Though the structural purpose of the material is served, the service life is greatly dependant on various factors like the loading conditions [1], surrounding environment [2, 3] and hydration conditions [4]. In most cases, the failure is gradual and is initiated by minor cracks [5]. The initial cracks tend to form due to irreversible microstructural changes in the composites. The microstructures tend to change and initiate cracks when the material experiences shrinkage due to inadequate water during the hydration process [6]. The cracks also tend to form when the external force on the structures induces a plastic deformation [7]. Irrespective of the method of the initial crack formation, the breaking or failure of the structure is due to the propagation of these initial microcracks. Microcracks on concrete increases the permeability of the material and hence aids higher rates of sulfate and chloride attacks since the ions can easily penetrate into the depths of the material and induce degradations and expansion related crack propagations [8]. The cracks are also responsible for the accelerated corrosion of the steel bars in case of reinforced concrete. The formation of microcracks cannot be avoided in a structure as the occurrence is due to a range of different causes as discussed. Efforts in the directions of closing these cracks in the initial stage may protect the structure from any further damages and also retain the strength. If only there was a process that could close the cracks in the initial stage, the eventual failure could be avoided and the structure would recover and regain its strength to a great extent. Passive treatments, in which the cracks are detected and sealed using sealants that are either injected or sprayed into the cracks, have been attempted. Common sealants used are epoxy resins, acrylics, waxes, chlorinated rubbers, polyurethane etc. However, passive treatments although applicable for treating the surface cracks, they have various other limitations associated with the chemical sealants used such as poor sustainability, improper concrete-sealant bonding, lower heat and weather resistance, high susceptibility to degradation etc. Most importantly, difficulty in early detection of these micro cracks sets the limitation to measures taken to close the cracks. The detection and repair should also be a continuous process since the cracks may occur at different points of time on the structure. This makes monitoring and maintenance, a tedious and an expensive affair. The most effective way of maintaining the structural integrity of the section is by a method which allows an autonomous repair of cracks as and when it is formed. This is widely known as the self-healing method of concrete repair.

Self-healing of cementitious materials can be autogenous or autonomous depending on the crack healing mechanism involved. Autogenous self-healing relies on the intrinsic composition of cementitious materials and occurs due to the hydration of cement material and carbonation of dissolved $Ca(OH)_2$ resulting in precipitation of $CaCO_3$. The formation of $CaCO_3$ has been reported as a white residue that fills the openings. The possible mechanism of its emergence is due to the dissolved calcium hydroxide being dissipated into the crack surface. The free Ca^{2+} ions available readily interact with the atmospheric CO_2 to form crystalline $CaCO_3$. The reaction scheme is as given below [9].

$$H_2O + CO_2 \leftrightarrow H_2CO_3 \leftrightarrow H^+ + HCO^{-3} \leftrightarrow 2H^+ + CO_3^{2-} \quad (5.1)$$

$$Ca^{2+} + CO_3^{2-} \leftrightarrow CaCO_3 \text{ (pH water} > 8) \quad (5.2)$$

$$Ca^{2+} + HCO_3^- \leftrightarrow CaCO_3 + H^+ (7.5 < pH_{water} < 8) \quad (5.3)$$

The blocking of cracks may also occur due to the impurities in the water and free particles separated from the concrete matrix. The swelling of the hydrated cementitious matrix in the available space is also a possibility in closing the cracks [10]. Figure 5.1 represents all the possible processes of self-healing as discussed above.

Earlier hypothesis that self-healing is a result of further hydration of unhydrated cementitious components [11] was disproved concluding that this could be possible only in young concretes where there is a higher content of the unhydrated cements while the aged concretes are more likely to form $CaCO_3$ [12].

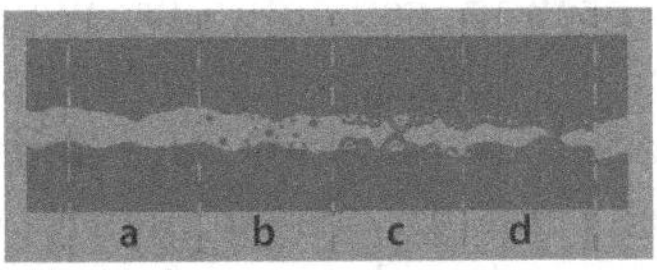

Figure 5.1 Possible mechanisms of indigenous healing of cementitious material (a–d) different stages of self-healing process.

The presence of water is crucial for the occurrence of most of the above processes. Some of the reported studies on autogenous self-healing is highlighted below. Cementitious specimen's damage due to cryogenic spillage is reported to be healed by allowing it to cure. The specimens were also mixed with steel fibers in the casting stage to minimize the cracking [13]. The healing of the concrete specimen over the period of 28 days observed over a crack is shown in Figure 5.2. Self-healing in polymer modified concrete is observed to take longer duration than that of normal concrete, however, with greater healing efficiency [14]. The healing mechanism reported by the authors is due to the hydration of the unhydrated cement material present in the matrix as discussed above. In case of recycled aggregates (RA) concrete, the crack closing ability of the material tends to reduce for high percentage of RA. However, irrespective of RA percentage in concrete, the crack closing for fine cracks is slow when compared to the larger cracks due to inadequacy of space, slow diffusion of water and CO_2 [15]. The admixtures generally used in concrete also tend to affect the healing capacity of the concrete. Improvement of the healing capacity by additions of silica fumes and ground granulated blast furnace slag (GGBS) are investigated and proved to be effective [16]. GGBS has proven to have a higher time of curing than that of Portland cements, resulting in higher availability of unreacted GGBS in the matrix, which in-turn is helpful in healing cracks in aged concrete [17]. The degree of damage is an important parameter in healing, and there exists a threshold beyond which the repair/healing marginally decreases [18].

Although advantageous, there are several limitations to autogenous self-healing process like presence of appropriate moisture and Ca^{2+} ions to trigger the healing process and for sustained healing, requirement of the cracks to be narrow enough to be completely healable. Typically, cracks of size 30–50 μm can be healed by this mechanism under optimal conditions. Proper closing of cracks cannot be expected for deeper cracks (>100 μm). However, intrinsic brittleness of most cementitious materials generally results in larger crack widths that are difficult to repair by autogenous self-healing.

High capacity self-healing cementitious materials with enhanced mixture design and incorporating suitable healing agents are some of the approaches developed to overcome the limitations of autogenous self-healing. Engineered cementitious composites (ECC), also known as the strain hardening cement-based composites, are a class of flexible concrete materials which do not immediately crack or fail under tension. The

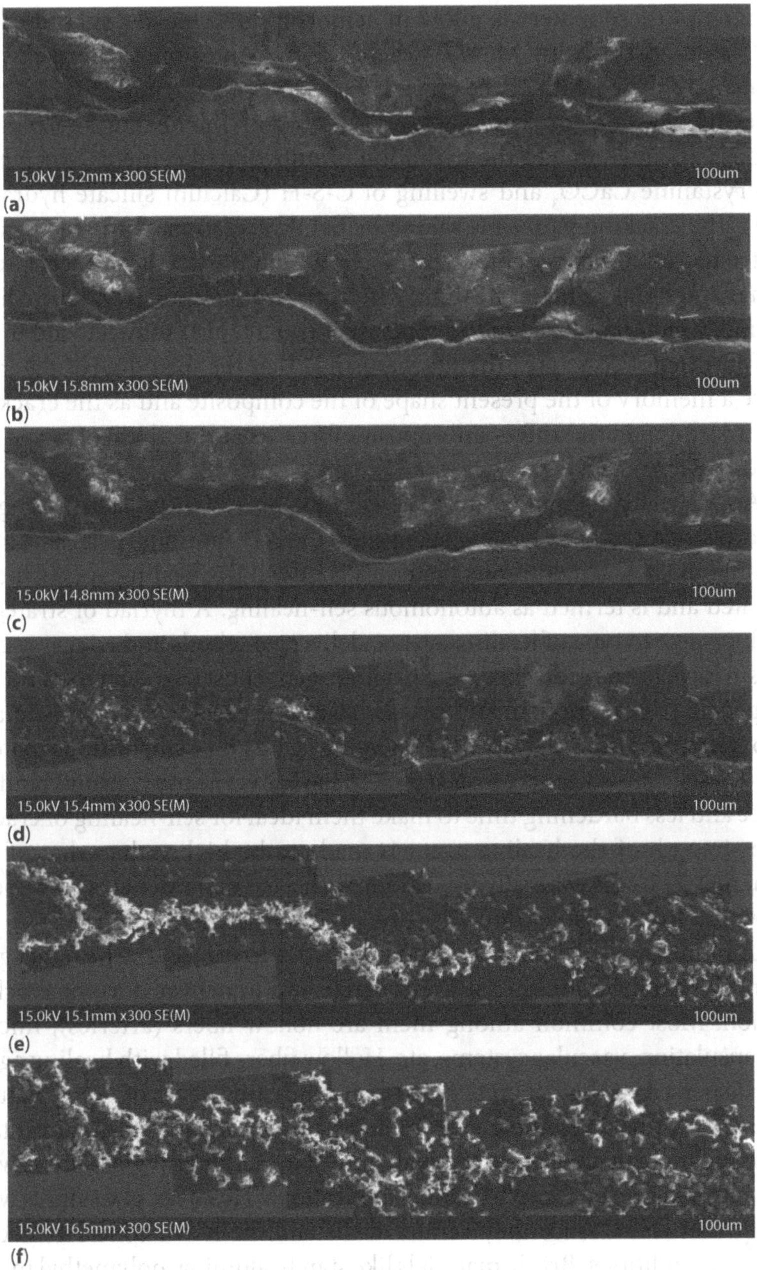

Figure 5.2 Healing of cracks over a period of 28 days (a) 0 day, (b) 1 day, (c) 3 days, (d) 7 days, (e) 14 days, (f) 28 days (Reprinted with permission from Ref. [13]. Copyright © (2019) Elsevier).

designs of these materials allow in achieving high tensile ductility with restricted crack width when it reaches a cracking point. There exists a control over the width of the crack irrespective of the length, and in most cases the width of the crack is less than 60 μm. The healing mechanism also is explained due to the hydration of unhydrated cements, formation of crystalline $CaCO_3$ and swelling of C-S-H (Calcium silicate hydrates) [19]. The cracking pattern is said to be flat propagation, with crack width restricted to 20 μm is achieved. The restriction imposed on the cracking behaviour helps in better healing, with recovery of upto 100% [20]. Shape memory materials, used in the concrete mix, have also shown to aid in the self-healing of concrete. The mechanism is that the introduced materials have a memory of the present shape of the composite and as the cracking occurs the material offers an opposite force to aid in the closing of the cracks [21].

An alternative approach to ECC is inducing enhanced self-healing by incorporating suitable healing agents into the cementitious material. Crack healing in such cementitious materials is due to the healing agents incorporated and is termed as autonomous self-healing. A myriad of strategies with respect to type of healing agents, delivery methods and trigger mechanism have been researched. The healing agent chosen in each method has a specific trigger mechanism for the healing process. Cyanoacrylate and epoxy are the two widely explored healing agents for cementitious materials. These healing agents are designed to have low viscosity, strong binding force and less hardening time to make them ideal for self-healing of cracks. The strength of the healing agent is made to be higher than that of the concrete. Epoxy is generally used along with a latent hardener, as the epoxy flows out from capsule/fibre it is cured by the hardener [22].

A number of substrates are used in practice to protect the healing agent from the harsh conditions of the concrete and to deliver it to the cracked region. Most common among them are hollow fibers (arteries), micro-encapsulation, vascular systems, etc. Hollow fibers filled with healing agent are dispersed in the concrete matrix in the casting stage of the structure. The hollow fibers resemble arteries in a natural system with an advantage of filling more healing agent in small fibre volume. Hollow glass fibers were initially considered for self-healing applications. However, glass fibres were found not viable since they cannot withstand the harsh concrete mixing process conditions. Brittle materials like starch, alumina, polymethyl methacrylate, inorganic phosphate etc. have been considered instead of glass. Alternatively, embedding steel bars during casting and removing it after curing to introduce hollow ducts in the concrete slabs have been tested for containment and release of healing agents. The trigger mechanism is such that the healing agent flows out when a crack breaks the embedded hollow fibers.

Microencapsulation as a delivery method has been demonstrated by encapsulating the active healing agent in polymeric or inorganic shell based capsules. Urea-formaldehyde, melamine-formaldehyde, polystyrene resins are commonly used as polymeric shell materials while expanded clay, silica particles or glass capsules are considered as inorganic shell materials. While polymer as shell material acts as a good sealant for the healing agent, inorganic shell materials have an advantage of strong bonding with the cementitious matrix due to their chemical similarity with the cementitious material. Triggering factors such as force, ions, pH etc. will ensure rupture of the capsule shell and release of the healing agent at the crack site.

The arteries and encapsulation methods of self-healing, although a good choice for closure of deeper cracks, have several drawbacks limiting its commercial application. The composition of the healing agents or the capsules, not being the same as that of the cementitious material, may sometimes result in expansion of crack rather than its closure. Availability of capsules that can blend easily and survive in the concrete mix is another limitation. Most importantly, success of these methods highly depends on the availability of the capsules in the cracked region. Yet when there are repetitive cracks in the same region, it is not possible to release the adhesive beyond once. An ideal crack treatment method should be able to repeatedly heal the cracks as and when they appear. Hence, there exists a need for a continuous healing system.

An alternative greener way of achieving self-healing that addresses the limitations discussed above is possible by the biotechnological method utilizing microbes. While highly innovative, it is also a natural means of self-healing of concretes. Microbe mediated self-healing relies on the biomineralization, a process of formation of mineral phases by living organisms, resulting in precipitation of calcium carbonate as a healing agent. Biomineralization is gaining momentum as an alternative and sustainable solution for self-healing of cementitious structures. It is interesting to note that, unlike most of the other methods, microbial method of self-healing does not require submerged conditions. While bacteria have been extensively researched as self-healing agents, recent literature has also demonstrated the potential use of some fungi species in self-healing of concrete. The following sections will discuss in detail the mechanism of biomineralization and its application in self-healing of cementitious structures.

5.2 Biomineralization for Self-Healing

Many species of bacteria are known to produce minerals like carbonates, sulfides, silicates, phosphates, etc., which makes them potential agents for self-healing of concrete. The mode of bacterial mineralization can be either biologically controlled mineralization (BCM), wherein the organisms have control over the mineral formation, or biologically induced mineralization (BIC), wherein the organisms have minimal or no control over the mineral formation. BCM often occurs intracellularly resulting in ordered and uniform size mineral structures. BIC is a result of cellular metabolic activity, where the mineralization may occur inside, on the cell wall or away from the cell in response to changes in mineral saturation as a result of cellular activity. The resulting minerals will have poor crystallinity, large size variations with impurities. Bacterial mineralization, unlike geological mineralization, is controlled by the organic matter secreted by the microorganisms resulting in ordered organic–inorganic composite mineral phase. Although several mineral phases are formed due to biomineralization, formation of calcium carbonate is of interest in concrete self-healing applications. Microorganisms which produce carbonates and have the ability to precipitate calcium carbonate will be the potential self-healing agents for cementitious materials. While bacteria have been extensively researched as self-healing agents, recent literature has also demonstrated the potential use of some fungi species in self-healing of concrete.

5.2.1 Bacteria as Self-Healing Agent

Bacteria, owing to their high surface to volume ratio coupled with chemically charged surface, make them the most ideal biomineralization agents among all organisms [23]. Bacteria can stay in a vegetative state and can also be designed to be in a dormant state for long time periods, making them potential agents for use in self-healing applications. Self-healing technology based on bacterial mineralization has therefore gained attention as a viable technology for concrete crack healing. Carbonate production by bacteria is governed by key factors like dissolved calcium concentration, pH, nucleation sites and dissolved inorganic carbon concentration.

Bacterial metabolic pathways can effectively alter these key factors and the cell wall provides the nucleation sites for the process of calcium carbonate precipitation.

5.2.2 Bacterial Metabolic Pathway in Self-Healing

Several autotrophic and heterotrophic bacteria like cyanobacteria, ureolytic bacteria, sulfate and nitrate reducing bacteria, methanogenic bacteria, etc. are known to induce calcium carbonate precipitation and many such species have been researched for their potential in self-healing of cementitious materials. The metabolic pathway of calcium carbonate precipitation is bacterium specific and the following paragraphs discuss few of the known mechanisms.

Broadly, the pathway for calcium carbonate precipitation can be classified as autotrophic and heterotrophic. Autotrophic pathway requires presence of CO_2 and involves microbial conversion of CO_2 to carbonate. Methanogenic and photosynthetic bacteria are known to precipitate calcium carbonate by methanogenesis and photosynthesis autotrophic pathway, respectively. While methanogenic bacteria convert CO_2 to methane with H_2 as an electron donor, photosynthesis conversion of CO_2 to methanol with H_2 or H_2S as electron donors can happen by oxygenic (cyanabacteria) or anoxygenic (purple bacteria) bacteria, respectively. Methane or methanol produced are further oxidized to bicarbonate, which in presence of Ca^{2+} ions results in calcium carbonate precipitation. The major drawback of this method is the requirement of presence of CO_2 in the environment which restricts its use only for concrete structures exposed to CO_2 and light.

By contrast, heterotrophic pathway involves precipitation of calcium carbonate by bacteria which rely on various organic compounds as energy sources. Bacteria belonging to *Bacillus*, *Arthrobacter* and *Rhodococcus* species are capable of producing various carbonates depending on the organic energy source provided. Organic compounds commonly used are calcium salts such as calcium acetate, calcium lactate, calcium citrate, etc. Organic compound used acts as a source of both energy and CO_2 making the process advantageous to autotrophic pathway. Among the several heterotrophic bacteria, bacteria producing urease and carbonic anhydrase enzymes have been explored widely for self-healing applications of cementitious materials. The mechanism of calcium carbonate precipitation in these bacteria is explained below as typical examples of heterotrophic bacteria mediated calcium carbonate precipitation.

5.2.2.1 *Urea Hydrolysis by Ureolytic Bacteria*

Ureolytic bacteria are commonly employed for crack healing of concretes. The process of metabolic activity associated mineralization in these bacteria is the basic principle of the bacterial self-healing of cracks in concrete. Ureolytic bacteria are the class of microbes that possess urease enzyme which can catalyse conversion of urea into ammonia and carbonate. The cell wall of the bacteria being negatively charged has the capability of attracting dissolved forms of the calcium ions (Ca^{2+}) to the surface. Calcium ions attracted to the cell wall interact with the carbonate to form calcium carbonates ($CaCO_3$) on the cell walls. When the concentration of the formed $CaCO_3$ exceeds the solubility limit, precipitation is observed. In a typical reaction, urease enzyme catalyzes the hydrolysis of 1 mol of the urea into 1 mol of ammonia and 1 mol of carbamic acid. Carbamic acid upon hydrolysis produces an extra mole of ammonia and carbonic acid. These products on hydrolysis form 2 mol of ammonium and 1 mol of bicarbonate, respectively. The following reactions increase the pH facilitating the precipitation of $CaCO_3$ on the cell walls of the bacteria. The reactions explained above are represented by Equations (5.4)–(5.8) [24].

$$CO(NH_2)_2 + H_2O \rightarrow NH_2COOH + NH_3 \tag{5.4}$$

$$NH_2COOH + H_2O \rightarrow NH_3 + H_2CO_3 \tag{5.5}$$

$$NH_3 + H_2O \rightarrow NH_4^+ + 2OH^- \tag{5.6}$$

$$H_2CO_3 \rightarrow HCO_3^- + H^+ \tag{5.7}$$

$$HCO_3^- + H^+ + 2NH_4^+ + 2OH^- \leftrightarrow CO_3^{2-} + 2NH_4^+ + 2H_2O \tag{5.8}$$

Reactions taking place at the interface of the cell wall where the actual precipitation takes place can be represented by Equations (5.9)–(5.10).

$$Ca^{2+} + Cell \rightarrow Cell\text{–}Ca^{2+} \tag{5.9}$$

$$Cell\text{–}Ca^{2+} + CO_3^{2-} \rightarrow Cell\text{–}CaCO_3\downarrow \tag{5.10}$$

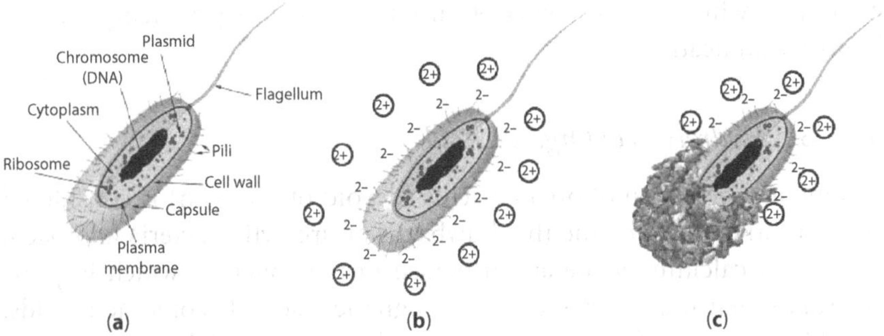

Figure 5.3 Schemetic representation of bacterium precipitating calcium carbonate (Reprinted with permission from Ref. [26] Copyright © (2016) Springer).

Precipitation of calcium carbonate due to bacterial mineralisation can be schematically represented as shown in Figure 5.3. Availability of sufficient dissolved Ca^{2+} ions is crucial for the precipitation of $CaCO_3$ on the cell walls. Calcium ions, apart from the indigenously available content in the concrete structure, can be supplemented as salts of calcium (calcium nitrate, calcium chloride, calcium lactate, etc.) along with the bacterial populations. It is interesting to note that the above discussed reactions occur even on inactive spores, biofilms and exopolymers highlighting the potential of this ureolytic bacterial pathway in commercial concrete self-healing applications [24, 25].

5.2.2.2 *Hydrolysis of CO_2 by Carbonic Anhydrase Producing Bacteria*

Carbonic anhydrase enzyme can catalyse the capture of CO_2 and its enzymatic conversion to carbonic acid as represented by Reaction (5.11).

$$CO_2 + H_2O \rightarrow H_2CO_3 \tag{5.11}$$

Carbonic acid is converted to carbonate according to Reactions (5.7) and (5.8). Calcium ions on the cell wall interact with the carbonate resulting in precipitation of calcium carbonate ($CaCO_3$) as explained by Equations (5.9) and (5.10). Although extensively explored, these methods have certain drawbacks. The major drawback of ureolytic method is the excess accumulation of ammonium ions in the concrete matrix, resulting in nitrogen oxide emissions and also risk of salt damage upon conversion to nitric acid, while less prevalence of carbonic anhydrase producing bacteria limits its application.

5.2.2.3 *Hydrolysis of Organic Acids*

Metabolic conversion of organic acids (devoid of nitrogen) is considered an alternative to overcome the drawbacks of ureolytic bacteria approach. Sources of calcium lactate are dispersed in the concrete, which in presence of oxygen and water (seeping through the cracks) become lactic acids. Aerobic oxidation of the organic acid produces CO_2, which under alkaline environment produces carbonate. Further, the dissolved Ca^{2+} ions attracted towards the cell wall interacts with carbonate resulting in precipitation of calcium carbonate as given by Equations (5.9) and (5.10). The possible reactions of breakdown for calcium lactate ($CaC_6H_{10}O_6$), calcium acetate ($CaC_4H_6O_4$) and sodium gluconate ($NaC_6H_{11}O_7$) are given by Equations (5.12)–(5.14) [27].

$$CaC_6H_{10}O_6 + 6O_2 \rightarrow CaCO_3 + 5CO_2 + 5H_2O \quad (5.12)$$

$$CaC_4H_6O_4 + 4O_2 \rightarrow CaCO_3 + 3CO_2 + 3H_2O \quad (5.13)$$

$$NaC_6H_{11}O_7 + 5.5O_2 \rightarrow 0.5Na_2CO_3 + 5.5CO_2 + 5.5H_2O. \quad (5.14)$$

Apart from the above discussed methods, calcium carbonate precipitation can also occur due to dissimilatory nitrate or sulfate reduction and ammonification of amino acids.

5.2.2.4 *Dissimilatory Nitrate Reduction*

In this method, denitrification of organic acids by denitrifying bacteria produces carbon dioxide and N_2 with increase in pH. CO_2 under alkaline conditions produces bicarbonate resulting in precipitation of calcium carbonate as given by Equations (5.9) and (5.10). Bacterial species majorly involved in this process are facultative anaerobes like *Denitrobacillus*, *Pseuodomonas*, *Thiobacillus*, *Achromobacter*, *Alcaligenes*, *Spirillum*, *Micrococcus*, etc. Although this process can be applied under anaerobic environments, it is not well developed like urease hydrolysis method.

The extent and kinetics of calcium carbonate precipitation is much higher in ureolyitc approach compared to denitrification approach. Typical reaction for denitrification is represented by Reaction (5.15).

$$\text{Organic compound} + aNO_3^- + bH^+ \rightarrow cCO_2 + dH_2O + eN_2 \quad (5.15)$$

5.2.2.5 *Dissimilatory Sulfate Reduction*

Calcium carbonate precipitation can be induced by anaerobic sulfate reduction of gypsum ($CaSO_4.H_2O$) by sulfate reducing bacteria like *Desulfovibrio* sp. Calcium carbonate precipitation occurs due to the reaction of the released calcium ions with CO_2 under alkaline environment. However, this method is not feasible since the H_2S generated during sulfate reduction can produce elemental sulfur resulting in corrosion of the concrete structures. Typical reaction for sulfate reduction is represented by Reaction (5.16).

$$6CaSO_4 + 4H_2O + 6CO_2 \rightarrow CaCO_3 + 4H_2S + 2S + 11O_2 \quad (5.16)$$

5.2.2.6 *Ammonification*

Ammonification of amino acids as a metabolic activity in certain bacteria like *Myxococcus xanthus* produces CO_2 and NH_3 (Equation (5.17)). Further hydrolysis of ammonia favours the precipitation of calcium carbonate on the cell wall of the bacteria (Equation (5.18)).

$$\text{Amino acids} + O_2 \rightarrow NH_3 + CO_2 + H_2O \quad (5.17)$$

$$NH_3 + H_2O \rightarrow NH_4 + OH^- \quad (5.18)$$

Several studies have been reported on use of various bacteria for self-healing of cementitious materials as summarised in Table 5.1. Many studies have confirmed that calcium carbonate precipitate produced by bacteria not only heals the cracks but also decreases permeability, increase the strength and durability of the concrete structure [56–62].

Table 5.1 Various microorganisms used for self-healing in cementitious concrete for crack healing.

Mechanism of precipitation	Microorganism	Nutrient	Embedment in concrete	Ref.
Ureolysis	*Bacillus sphaericus*	Urea, calcium nitrate and yeast extract	Immobilized	[1]
	Bacillus sphaericus	Urea and calcium chloride	Direct	[2]
	Bacillus sphaericus	Urea, calcium nitrate and yeast extract	Immobilized	[3]
	Bacillus sphaericus	Urea and calcium chloride	Direct	[4]
	Bacillus sphaericus	Urea, calcium nitrate and yeast extract	Immobilized	[5]
	Bacillus sphaericus	Urea, calcium chloride, calcium nitrate and yeast extract	Immobilized	[6]
	Bacillus sphaericus	Urea, calcium nitrate and yeast extract	Immobilized	[7]
	Bacillus sphaericus	Urea, calcium chloride and calcium acetate	–	[8]

(*Continued*)

Table 5.1 Various microorganisms used for self-healing in cementitious concrete for crack healing. (*Continued*)

Mechanism of precipitation	Microorganism	Nutrient	Embedment in concrete	Ref.
	Bacillus sphaericus	Urea, calcium nitrate and yeast extract	Immobilized	[9]
	S. pasteurii Pseudomonas aeruginosa	Urea and calcium chloride	Direct	[10]
	Bacillus sphaericus S. pasteurii	Urea and calcium acetate	Direct	[11]
	S. pasteurii	Urea and calcium chloride	Immobilized	[12]
	S. pasteurii	Urea, calcium nitrate and calcium chloride	–	[13]
	S. pasteurii	Urea and calcium nitrate	–	[14]
	S. pasteurii	Urea and calcium chloride	Immobilized	[15]
	S. pasteurii Bacillus cereus	Urea, nutrient broth and calcium chloride	Direct	[16]
	Bacillus amyloliquefaciens	Urea, calcium acetate yeast extract and glucose	-	[17]

(*Continued*)

Table 5.1 Various microorganisms used for self-healing in cementitious concrete for crack healing. (*Continued*)

Mechanism of precipitation	Microorganism	Nutrient	Embedment in concrete	Ref.
	Sporosarcina soli Bacillus massiliensis Arthrobacter crystallopoietes Lysinibacillus fusiformis	Urea and calcium chloride	Direct	[18]
	Bacillus mucilaginous	Saccharose, Yeast Extract, Calcium Nitrate	Immobilized	[19]
Denitrification	*Pseudomona aeruginosa and Diaphorobacter nitroreducens*	Urea, Calcium formate, Calcium nitrate, Yeast extract	Immobilized	[20]
	Diaphorobacter nitroreducens, Bacillus sphaericus	Urea, calcium formate, calcium nitrate and yeast extract	Immobilized	[21]
Silicification	Bacterium BKH1	Nutrient broth, Tryptone	Direct	[22]
	T-Bacillus subtilis	Peptone, Yeast extract	Direct	[23]
	Bacterium BKH4	Nutrient broth, Tryptone	Direct	[24]

(*Continued*)

Table 5.1 Various microorganisms used for self-healing in cementitious concrete for crack healing. (*Continued*)

Mechanism of precipitation	Microorganism	Nutrient	Embedment in concrete	Ref.
Conversion of organic acid	*Bacillus pseudofirmus*	Calcium lactate, calcium glutamate, yeast extract and peptone	Direct	[25]
	Bacillus pseudofirmus *B. cohnii*	Calcium lactate, calcium acetate, yeast extract and peptone	Direct	[26]
	B. cohnii	Calcium lactate and yeast extract	Immobilized	[27]
	Bacillus alkalinitrilicus	Calcium lactate and yeast extract	Immobilized	[28]

5.3 Strategies to Enhance the Performance of Bacterial Self-Healing

Effective and sustained bacterial self-healing of cracks is dependent on the viability and availability of the bacteria at the onset of the crack. Various strategies are adopted to enhance the durability and performance of bacteria in self-healing.

Concrete mixes being highly alkaline, bacteria that are tolerant to such high pH are the best choice. Bacteria, when introduced directly into the concrete mix, cannot withstand the harsh conditions such as high pH and nutrient/water limitation. The survival rate of bacteria directly mixed in the concrete matrix was found to be just 2 to 7% after 10 days of curing [53], due to the harsh pH and dry conditions. This limitation is overcome by using spores of bacteria, which are highly resistant dormant structures that can survive adverse environmental conditions. Under conditions viable for cell growth, the spores germinate to produce vegetative bacterial cells that can initiate the process of crack healing.

Initial studies on bioconcrete considered direct use of spores in the concrete mix supplemented with nutrients essential for the growth and metabolic activity of bacteria towards calcium carbonate precipitation. Calcium ions, additional to that available in the concrete, necessary for precipitation of calcium carbonate is provided as calcium compounds like calcium lactate, calcium glutamate and calcium acetate. However, direct use of bacteria spores in the mix of concrete is reported to have a low life span of self-healing, as the bacteria may die in 2–4 months [52]. Hence, encapsulation as a method to protect these spores from direct exposure to the harsh conditions of the concrete has been considered.

Several materials like polymers, cement additives and other minerals have been explored as a matrix for the encapsulation of spores. These encapsulations are distributed all over the matrix of the concrete, by dispersing it during the casting stage of the concrete. The cracks as formed begin to allow water into the cracked portion of the specimen. This water further tampers with the nearest encapsulations present and exposes the bacterial spores to the open cracks. A brief representation of calcite formation from the encapsulated bacteria is shown in Figure 5.4.

Diatomaceous earth, due to its desirable properties like high porosity, chemical stability and inertness, irregular morphology, has been considered as an immobilization matrix for bacterial encapsulation. Studies with diatomaceous earth encapsulated bacteria have proved effective in sealing cracks of ~0.17 mm wide [36].

Fine and coarse aggregates of expanded clay as an encapsulating material has been tested by few researchers. Successful closure of cracks of width 0.15–0.46 mm was achieved [55, 63]. Coating of the encapsulated material with a layer of geopolymer has shown to increase its performance with successful crack closure of upto 0.79 mm achieved [64]. The main drawback with the clay incorporation is the reduction in strength of concrete.

Ceramsite, a material similar to expanded clay, has been tested as an immobilization matrix by Chen *et al.* [46]. Cracks of width up to 0.5 mm were healed with subsequent increased resistance to torsion. In organic microencapsulation, the spores are encapsulated in compounds like melamine. Melamine is fairly soluble in water and hence when there exists

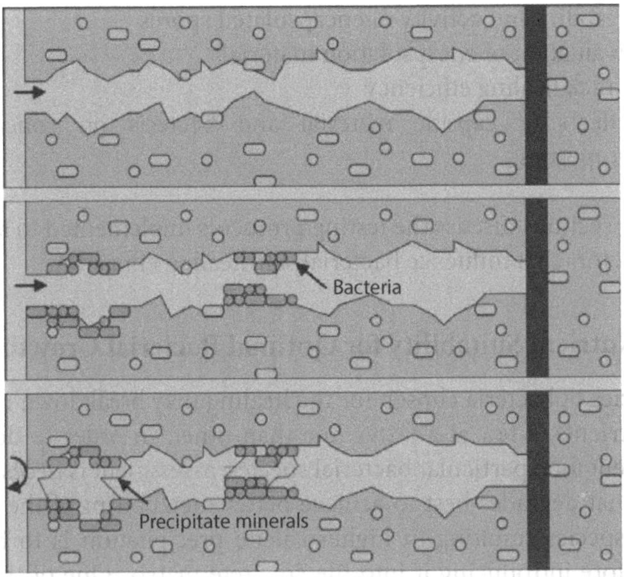

Figure 5.4 Calcite formation from the encapsulated bacteria, Adapted from Ref. [10].

a crack and water reaches the cracked portion, the material exposes the bacterial spore to the outer surface [34]. Polymers in different forms as encapsulating materials have been explored by Wang *et al.* [30, 28]. Self-healing efficiency of silica gel coated spores was higher compared to those encapsulated in polyurethane. This was attributed to higher enzymatic activity of bacterial spores in silica gel compared to polyurethane. Hydrogels have very high water absorption and retention capacity, and hence are an ideal choice for encapsulating bacterial spores. Wang *et al.* [28] have demonstrated faster healing of ~70–90% closure of crack width up to 0.5 mm using polymer hydrogels.

5.4 Evaluation of Factors Affecting Bacterial Self-Healing

Optimal efficiency of bacterial self-healing depends on several factors. Suitable testing of these factors ensures sustainable self-healing. The important evaluation parameters of a bacterial self-healing concrete is as follows,

1. Nutrient suitability for optimal bacterial growth

2. Viability and activity of encapsulated spores
3. Evaluation of encapsulation material
4. Crack healing efficiency
5. Effects of capsule material and bacteria on concrete properties.

Following sections discuss the testing protocols implemented to assess the various factors that influence bacterial self-healing efficiency.

5.4.1 Nutrient Suitability for Optimal Bacterial Growth

Each species of bacteria chosen for the healing may breakdown many different nutrients, a few at a faster rate than other. In order to decide the best nutrient for a particular bacterial species, assessment is necessary. The nutrient that degrades best to achieve better metabolism of the inherent bacterial species resulting in higher calcite precipitation is to be determined before introducing it into the concrete matrix. One of the testing protocols reported by Tziviloglou *et al.* [27] is discussed below.

Tziviloglou *et al.* [27] adopted oxygen consumption measurements to assess the best nutrient source for self-healing bacteria. The oxygen consumption measurements were performed in continuous and non-continuous modes. Alakli resistant bacteria species were considered for evaluation and three different nutrient sources suitable for growth of these bacteria were tested. Oxygen measurements under continuous mode involved tests with bacteria pre-grown on the nutrient source considered. The protocol involved suspension of cell pellets of pre-grown cultures in 0.1 M sodium sesquicarbonate/bicarbonate buffer supplemented with 0.1 M of one of the nutrient source and incubation for 1 h. Oxygen measurements were performed periodically on these incubated solutions in a simple sealed glass flask fitted with an oxygen sensor. Cell suspensions without nutrient were used as control. The choice of best nutrient was decided based on the rate of oxygen consumption of the test samples against control.

For non-continuous oxygen consumption monitoring, tests were performed in 96 well plates. Each well plate was seeded with the required volume of cell culture solution supplemented with the desired nutrient as discussed earlier. Cell suspension without nutrient was considered control. The plates were aseptically incubated at room temperature for 48 h. Unlike earlier protocol, bacteria are not pre-grown post incubation, oxygen consumption was assessed by measuring oxygen concentration in each well by an oxygen microsensor. Lower oxygen levels in wells relative to control were indicative of higher oxygen consumption.

The authors based on the different protocols use for assessing oxygen consumption concluded that bacteria pre-grown on a certain nutrient source can degrade it easily when compared to those who get exposed to the nutrient fresh.

5.4.2 Viability and Activity of Encapsulated Spores

Viability and activity of encapsulated spores will be assessed by monitoring the enzymatic activity and ability of calcium carbonate precipitation. In case of ureolytic bacteria, encapsulated spores will be dispersed in sterile medium supplemented with yeast extract and urea. Viability of the encapsulated spores will be assessed by colorimetric estimation of total ammonium nitrogen in the medium by Nesslet method to determine the amount of urea decomposed by enzymatic activity. Encapsulated spores suspended in the yeast/urea medium for 3 days are dried and subjected to thermogravimetric analysis to assess the formation of calcium carbonate precipitate on the encapsulate material. Morphology of the precipitate is analyzed by scanning electron micrographs [65].

5.4.3 Evaluation of Encapsulation Material

The properties of the encapsulation material are of importance to understand, in order to access if the material of choice serves its purpose when introduced into the concrete matrix. The prime purpose of the encapsulating material is to activate the bacterial spores when a crack appears. This is governed by the ability of the material to interact with water to ensure contact of spores with water. Hence, evaluation of encapsulation material involves test methods like leakage test, swelling test and dynamic vapour sorption test.

Leakage of the spores from the encapsulations is tested by introducing the encapsulated spores into a saline solution with repeated mixing. The extent of leakage of spores is evaluated by pour plate method on agar plates and assessing the colonies formed [65]. Direct counting of spores using a hemocytometer can also be considered.

Dynamic vapor sorption (DVS) test is carried out to understand the moisture uptake capacity of the encapsulating material. Higher moisture uptake capability ensures better spore activation. In a typical procedure, the sample is freeze dried and introduced into a DVS analyser. The vapor concentration around the sample is changed by step-wise change in the percentage humidity from 20 to 98%. The change in the sample mass was recorded for every step after equilibrium was achieved. After equilibrium is realized at 98%, desorption is carried out for the samples in a similar manner. The moisture uptake was calculated gravimetrically based on the change in mass with respect to humidity [65].

Swelling tests are performed on the encapsulation material by assessing the water absorption capability of the material. Briefly, the sample is suspended in cement filtrate or demineralized water (control) for 24 h and the water absorbed is assessed by the change in mass of the sample with respect to control sample [67].

5.4.4 Crack Healing Efficiency

The most important assessment parameter in a self-healing concrete is the crack healing efficiency. Bacterial crack healing by various mechanisms is examined by different methods. Some of the most common methods used for evaluating healing efficiency are by microscopy, water permeability, water absorption and tests on mechanical properties.

The crack filling efficiency is evaluated by microscopic techniques on the surface of concrete. The images of the initially formed cracks are captured and the area of the crack is calculated for all the cracks formed. The cracks are monitored over a period of 4 to 8 weeks and the change in the areas of the cracks is calculated from the images captured. The crack closure is represented as healing ratio or percentage of initial crack to healed crack. Higher the ratio better is the healing. Scanning electron microscopy may also be utilized to understand the morphology of the formed calcites [67]. Figure 5.5 displays the crack closure recorded by microscopy techniques over a period of 28 days.

Water absorption testes are conducted on the specimens in order to understand the voids in the specimen. The porosity along with the cracks allows water to be absorbed within the specimen. Hence, the water absorption test is done at initial and final stages of the healing to understand the healing percentage. In a typical protocol, the dry specimens are measured for initial weight and are fully submerged into water filled containers. In some cases samples are stored in a vacuum of about 25 bar in order to keep the gaps/pores free from any air or water prior to submersion. The submerged samples are left under water for about 24 to 48 h. The specimens are then patted for surface water and saturated weights are recorded. The specimens of both cracked and healed samples are tested under similar conditions. Percentage absorption is calculated to assess healing [68].

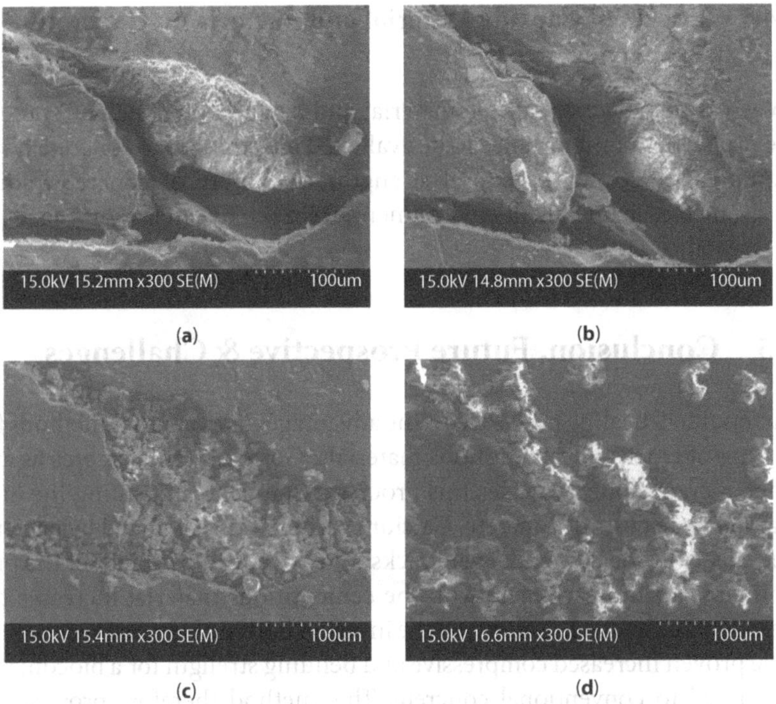

Figure 5.5 Cracks closure representation on (a) 0, (b) 3, (c) 7, and (d) 28 days of healing [67].

Lower the relative absorption achieved in the healed specimens, higher is the healing achieved.

Test of mechanical properties will distinguish if the closure of cracks achieved is sealing or healing of the cracks. Lab made concrete blocks are induced with cracks by applying high forces under Universal testing machine (UTM). The recorded measurements during cracking are stored for further use. The concrete is then allowed to heal before further testing. The healed specimens are further tested for mechanical properties by the UTM. Percentage change in the material strength is calculated with respect to the initial specimens to understand the strength regained due to the healing process. When the regained strength of the tested block is comparable to control block without cracks, it is termed as healing. Mere reduction in water absorption without any regain in strength is termed as sealing [69].

5.4.5 Effects of Capsule Material and Bacteria on Concrete Properties

To assess the effect of capsule material and bacteria on concrete, some of the properties of importance to be evaluated are strength of concrete block (compressive strength, flexural strength), durability of concrete blocks (resistance to acidic/alkali environment, freeze–thaw attack, drying, etc.), permeability, etc.

5.5 Conclusion, Future Prospective & Challenges

Bacterial self-healing has proved to be advantageous over other methods for healing of cracks in cementitious materials. Germination of spores as and when a crack appears makes this process sustainable, mitigating the high maintenance costs of concrete. Calcium carbonate precipitated by bacteria can effectively heal even deeper cracks. Chemical compatibility of the precipitated calcium carbonate with the cementitious material increases the strength the concrete. Majority of the investigations on bacterial self-healing have proven increased compressive and bending strength for a bioconcrete compared to conventional concrete. This method therefore provides an environmentally benign, sustainable, long lasting self-healing treatment of cementitious materials.

However, there are several challenges that need to be addressed to make this method commercially viable. Owing to the high cost of bacteria and nutrients, the price of bioconcrete is 2–4 times higher than the conventional concrete, limiting its large-scale applicability. Use of inexpensive nutrients such as industrial or agri waste can be explored to minimise the initial cost of making a bioconcrete. Since cracks are generally a surface phenomenon, use of bioconcrete as a coating or plaster topping instead of a complete bioconcrete structure can be an alternative solution to reduce the overall cost of the structure. Lack of standard protocols to assess the long-term viability of this method is another limitation to its commercial acceptance. Performance of bacteria may also vary geographically and environmentally which necessitates adaptation of bacteria to local conditions prior to its applicability. Although sufficient literature is available on the effective use of bacterial self-healing of cracks, investigations on the durability of this method and possibility of biocorrosion due to the impregnated bacteria has not been widely attempted. With these challenges addressed, bacterial self-healing will emerge as a durable and sustainable solution to the problems faced by the concrete industry.

References

1. Wang, J., Muhammed Basheer, P.A., Nanukuttan, S.V., Long, A.E., Bai, Y., Influence of service loading and the resulting micro-cracks on chloride resistance of concrete. *Constr. Build. Mater.*, 108, 56–66, 2016.
2. Rozière, E., Loukili, A., El Hachem, R., Grondin, F., Durability of concrete exposed to leaching and external sulphate attacks. *Cem. Concr. Res.*, 39, 1188–1198, 2009.
3. Preetham, R., Krishna, R.H., Chandraprabha, M.N., Thomas, T., Vehicular soot for improvement of chemical stability of cement composites towards acid rain and sewage like atmospheres. *Constr. Build. Mater.*, 248, 118604, 2020.
4. Han, F. and Zhang, Z., Hydration, mechanical properties and durability of high-strength concrete under different curing conditions. *J. Therm. Anal. Calorim.*, 132, 823–834, 2018.
5. Ammouche, A., Riss, J., Breysse, D., Marchand, J., Image analysis for the automated study of microcracks in concrete. *Cem. Concr. Compos.*, 23, 267–278, 2001.
6. Litorowicz, A., Identification and quantification of cracks in concrete by optical fluorescent microscopy. *Cem. Concr. Res.*, 36, 1508–1515, 2006.
7. Jeevanagoudar, Y.V., Krishna, R.H., Gowda, G., Preetham, R., Prabhakara, R., Improved mechanical properties and piezoresistive sensitivity evaluation of MWCNTs reinforced cement mortars. *Constr. Build. Mater.*, 144, 188–194, 2017.
8. Hu, S., Chen, Q., Gong, N., Effect of acid corrosion on crack propagation of concrete beams. *Sadhana*, 43, 23, 2018.
9. Edvardsen, C., Water permeability and autogenous healing of cracks in concrete. *ACI Mater. J.*, 96, 448–54, 1999.
10. Wu, M., Johannesson, B., Geiker, M., A review: Self-healing in cementitious materials and engineered cementitious composite as a self-healing material. *Constr. Build. Mater.*, 28, 571–583, 2012.
11. Neville, A.M., *Properties of Concrete*, Pearson Education Limited, London, 1995.
12. Neville, A.M., Autogenous healing—A concrete miracle. *Concr. Int.*, 24, 76–82, 2002.
13. Kim, S., Yoo, D.Y., Kim, M.J., Banthia, N., Self-healing capability of ultra-high-performance fiber-reinforced concrete after exposure to cryogenic temperature. *Cem. Concr. Compos.*, 104, 103335, 2019.
14. Elmoaty, A.E.M.A., Self-healing of polymer modified concrete. *Alexandria Eng. J.*, 50, 171–178, 2011.
15. Medjigbodo, S., Bendimerad, A.Z., Roziere, E., Loukili, A., How do recycled concrete aggregates modify the shrinkage and self-healing properties? *Cem. Concr. Compos.*, 86, 72–86, 2018.
16. Jaroenratanapirom, D. and Sahamitmongkol, R., Self-crack closing ability of mortar with different additives. *J. Met. Mater. Miner.*, 21, 9–17, 2011.

17. Park, B. and Cho, Y.C., Prediction of Self-Healing Potential of Cementitious Materials Incorporating Crystalline Admixture by Isothermal Calorimetry. *Int. J. Concr. Struct. Mater.*, 13, 36, 2019.
18. Zhong, W. and Yao, W., Influence of damage degree on self-healing of concrete. *Constr. Build. Mater.*, 22, 1137–1142, 2008.
19. Zhang, Z. and Zhang, Q., Self-healing ability of Engineered Cementitious Composites (ECC) under different exposure environments. *Constr. Build. Mater.*, 156, 142–151, 2017.
20. Yang, Y., Lepech, M.D., Yang, E.H., Li, V.C., Autogenous healing of engineered cementitious composites under wet–dry cycles. *Cem. Concr. Res.*, 39, 382–390, 2009.
21. Saiidi, M.S., Sadrossadat, M., Ayoub, C., Itani, A., Pilot study of behavior of beams reinforced with memory alloys. *J. Mater. Civil Eng. ASCE*, 6, 1543–1558, 2007.
22. Thao, T.D.P., Johnson, T.J.S., Tong, Q.S., Dai, P.S., Implementation of self-healing in concrete—Proof of concept. *IES J. Part A*, 2, 116–125, 2009.
23. Douglas, S. and Beveridge, T.J., Mineral formation by bacteria in natural microbial communities. *FEMS Microbiol. Ecol.*, 26, 79–88, 1998.
24. Tittelboom, K.V., Belie, N.D., Muynck, W.D., Verstraete, W., Use of bacteria to repair cracks in concrete. *Cem. Concr. Res.*, 40, 157–66, 2010.
25. Achal, V. and Mukherjee, A., A review of microbial precipitation for sustainable construction. *Constr. Build. Mater.*, 93, 1224–1235, 2016.
26. Seifan, M., Samani, A.K., Berenjian, A., Bioconcrete: Next generation of self-healing concrete. *Appl. Microbiol. Biotechnol.*, 100, 2591–2602, 2016.
27. Tziviloglou, E., Wiktor, V., Jonkers, H.M., Schlangen, E., Selection of Nutrient Used in Biogenic Healing Agent for Cementitious Materials. *Front. Mater.*, 4, 15, 2017.
28. Wang, J.Y., Snoeck, D., Van Vlierberghe, S., Verstraete, W., De Belie, N., Application of hydrogel encapsulated carbonate precipitating bacteria for approaching a realistic self-healing in concrete. *Constr. Build. Mater.*, 68, 110–119, 2014.
29. Achal, V., Mukerjee, A., Sudhakara Reddy, M., Biogenic treatment improves the durability and remediates the cracks of concrete structures. *Constr. Build. Mater.*, 48, 1–5, 2013.
30. Wang, J., Van Tittelboom, K., De Belie, N., Verstraete, W., Use of silica gel or polyurethane immobilized bacteria for self-healing concrete. *Constr. Build. Mater.*, 26, 532–540, 2012.
31. Achal, V. and Pan, X., Characterization of urease and carbonic anhydrase producing bacteria and their role in calcite precipitation. *Curr. Microbiol.*, 62, 894–902, 2011.
32. Wang, J.Y., Van Tittelboom, K., De Belie, N., Verstraete, W., Potential of applying bacteria to heal cracks in concrete. *2nd International Conference on Sustainable Construction Materials and Technologies*, 2010.

33. Van Tittelboom, K., De Belie, N., De Muynck, W., Verstraete, W., Use of bacteria to repair cracks in concrete. *Cem. Concr. Res.*, 40, 157–166, 2010.
34. Wang, J.Y., Soens, H., Verstraete, W., De Belie, N., Self-healing concrete by use of microencapsulated bacterial spores. *Cem. Concr. Res.*, 56, 139–152, 2014.
35. De Muynck, W., Debrouwer, D., De Belie, N., Verstraete, W., Bacterial carbonate precipitation improves the durability of cementitious materials. *Cem. Concr. Res.*, 38, 1005–1014, 2008.
36. Wang, J.Y., De Belie, N., Verstraete, W., Diatomaceous earth as a protective vehicle for bacteria applied for self-healing concrete. *J. Ind. Microbiol. Biotechnol.*, 39, 567–577, 2012.
37. Ramachandran, S.K., Ramakrishnan, V., Bang, S.S., Remediation of concrete using micro-organisms. *ACI Mater. J.*, 98, 3–9, 2001.
38. Kim, H.K., Park, S.J., Han, J.I., Lee, H.K., Microbially mediated calcium carbonate precipitation on normal and lightweight concrete. *Constr. Build. Mater.*, 38, 1073–1082, 2013.
39. Bang, S.S., Galinat, J.K., Ramakrishnan, V., Calcite precipitation induced by polyurethane-immobilized *Bacillus pasteurii*. *Enzyme Microb. Technol.*, 28, 404–409, 2001.
40. Qian, S.Z., Zhou, J., Schlangen, E., Influence of curing condition and precracking time on the self-healing behavior of engineered cementitious composites. *Cem. Concr. Compos.*, 32, 686–693, 2010.
41. Chunxiang, Q., Jianyun, W., Ruixing, W., Liang, C., Corrosion protection of cement-based building materials by surface deposition of CaCO3 by *Bacillus pasteurii*. *Mater. Sci. Eng.*, 29, 1273–1280, 2009.
42. Bang, S.S., Lippert, J.J., Yerra, U., Mulukutla, S., Ramakrishnan, V., Microbial calcite, a bio-based smart nanomaterial in concrete remediation. *Int. J. Smart Nano Mater.*, 1, 28–39, 2010.
43. Maheswaran, S., Dasuru, S.S., Rama Chandra Murthy, A., Bhuvaneshwari, B., Ramesh Kumar, V., Palani, G.S., Iyer, N.R., Krishnamoorthy, S., Sandhya, S., Strength improvement studies using new type wild strain *Bacillus cereus* on cement mortar. *Curr. Sci. India*, 106, 50–57, 2014.
44. Lee, Y.N., Calcite production by *Bacillus amyloliquefaciens* CMB01. *J. Microbiol.*, 41, 345–348, 2003.
45. Park, S.J., Park, Y.M., Chun, W.Y., Kim, W.J., Ghim, S.Y., Calcite-forming bacteria for compressive strength improvement in mortar. *J. Microbiol. Biotechnol.*, 20, 782–788, 2010.
46. Chen, H., Qian, C., Huang, H., Self-healing cementitious materials based on bacteria and nutrients immobilized respectively. *Constr. Build. Mater.*, 126, 297–303, 2016.
47. Ersan, Y., Boon, N., De Belie, N., Microbial self-healing concrete: Denitrification as an enhanced and environment-friendly approach, in: *5th International Conference on Self-Healing Materials*, Ghent, 2015.

48. Erşan, Y.Ç., Da Silva, F.B., Boon, N., Verstraete, W., De Belie, N., Screening of bacteria and concrete compatible protection materials. *Constr. Build. Mater.*, 88, 196–203, 2015.
49. Majumdar, S., Sarkar, M., Chowdhury, T., Chattopadhyay, B., and Mandal, S., Use of bacterial protein powder in commercial fly ash pozzolana cements for high performance construction materials. *Open J. Civ. Eng.*, 2, 218–228, 2012.
50. Sarkar, M., Adak, D., Tamang, A., Chattopadhyay, B., and Mandal, S., Genetically-enriched microbe-facilitated self-healing concrete—A sustainable material for a new generation of construction technology. *RSC Adv.*, 5, 105363–105371, 2015.
51. Sarkar, A., Chatterjee, A., Mandal, S., Chattopadhyay, B., An alkaliphilic bacterium BKH 4 of bakreshwar hot spring pertinent to bio-concrete technology. *J. Appl. Microbiol. Biochem.*, 6, 1742–1750, 2019.
52. Jonkers, H.M. and Schlangen, E., A two component bacteria-based selfhealing concrete. Concrete Repair, Rehabilitation and Retrofitting II. *ICCRRR*, Cape Town South Africa, 2009.
53. Jonkers, H.M., Thijssen, A., Muyzer, G., Copuroglu, O., Schlangen, E., Application of bacteria as self-healing agent for the development of sustainable concrete. *Ecol. Eng.*, 36, 230–235, 2010.
54. Sierra-Beltran, M.G., Jonkers, H.M., Schlangen, E., Characterization of sustainable bio-based mortar for concrete repair. *Constr. Build. Mater.*, 67, 344–352, 2014.
55. Wiktor, V. and Jonkers, H.M., Quantification of crack-healing in novel bacteria-based self-healing concrete. *Cem. Concr. Compos.*, 33, 763–770, 2011.
56. Han, S., Choi, E.K., Park, W., Yi, C., Chung, N., Effectiveness of expanded clay as a bacteria carrier for self-healing concrete. *Appl. Biol. Chem.*, 62, 19, 2019.
57. Magaji, A., Yakubu, M., Wakawa, Y.M., A review paper on self-healing concrete. *Int. J. Eng. Sci.*, 8, 47–54, 2019.
58. Xu, J. and Wang, X., Self-healing of concrete cracks by use of bacteria-containing low alkali cementitious material. *Constr. Build. Mater.*, 167, 1–14, 2018.
59. Xu, J., Wang, X., Zuo, J., Liu, X., Self-healing of concrete cracks by ceramsite-loaded microorganisms. *Adv. Mater. Sci. Eng.*, 2018, 1–8, 2018.
60. Tsangouri, E., A decade of research on self-healing concrete. *Sustain. Constr. Build. Mater.*, IntechOpen, London, 2019.
61. Palin, D., Wiktor, V., Jonkers, H.M., A bacteria-based self-healing cementitious composite for application in low-temperature marine environments. *Biomimetics*, 2, 13, 2017.
62. Souradeep, G., Dai, P.S., Wei, K.H., Autonomous healing in concrete by bio-based healing agents—A review. *Constr. Build. Mater.*, 146, 419–428, 2017.
63. Jonkers, H., *Bacteria-based self-healing concrete*, 56, 1–12, HERON, 2011.

64. Zhang, J., Liu, Y., Feng, T., Zhou, M., Zhao, L., Zhou, A., Li, Z., Immobilizing bacteria in expanded perlite for the crack self-healing in concrete. *Constr. Build. Mater.*, 148, 610–617, 2017.
65. Wang, J., Mignon, A., Snoeck, D., Wiktor, V., Van Vliergerghe, S., Boon, N., De Belie, N., Application of modified-alginate encapsulated carbonate producing bacteria in concrete: A promising strategy for crack self-healing. *Front. Microbiol.*, 6, 1088, 2015.
66. Snoeck, D., Van Tittelboom, K., Steuperaert, S., Dubruel, P., De Belie, N., Self-healing cementitious materials by the combination of microfibres and superabsorbent polymers. *J. Intell. Mater. Syst. Struct.*, 25, 13–24, 2014.
67. Kim, M.J., Kim, S., Yoo, D.Y., Banthia, N., Evaluation of Mechanical Property and Self-healing Capacity of Ultra-high-performance Fiber-reinforced Concrete Under the Cryogenic Condition. *J. Korean Soc. Hazard Mitig.*, 18, 231–238, 2018.
68. Nguyen, T.H., Ghorbel, E., Fares, H., Cousture, A., Bacterial self-healing of concrete and durability assessment. *Cem. Concr. Compos.*, 104, 103340, 2019.
69. Reddy, T.C.S. and Ravitheja, A., Macro mechanical properties of self-healing concrete with crystalline admixture under different environments. *Ain Shams Eng. J.*, 10, 23–32, 2019.

6

Self-Healable Solar Cells: Recent Insights and Challenges

Seyyed Alireza Hashemi[1*], Seyyed Mojtaba Mousavi[2], Sonia Bahrani[3], Seeram Ramakrishna[1], Chin Wei Lai[4†] and Wei-Hung Chiang[2]

[1]Department of Mechanical Engineering, Center for Nanofibers and Nanotechnology, National University of Singapore, Singapore, Singapore
[2]Department of Chemical Engineering, National Taiwan University of Science and Technology, Taipei, Taiwan
[3]Pharmaceutical Sciences Research Center, Shiraz University of Medical Sciences, Shiraz, Iran
[4]Nanotechnology & Catalysis Research Center, University of Malaya, Kuala Lumpur, Malaysia

Abstract

Nowadays, due to uprising concerns regarding the global warming, green energy sources such as solar cells have attracted noteworthy attention toward themselves. In this matter, researchers struggled to develop next generation of photovoltaic devices with improved power conversion efficiency as well as organic solar cells, perovskite and optimize their performance and optimize their performance; however, shorter life service of such devices compared with traditional silicon solar cells along with adverse effects of degradative environmental factors, could deteriorate their optimum performance. To address this flaw, new generation of photovoltaic devices should be coupled with self-healing performance to restore their damaged areas and boost their average life services which could thereby significantly decline upkeep/maintenance costs of solar panels. In current chapter, first the mechanism of action of self-healing platforms were discussed and hence recently developed self-healable polymeric structure, solar cells and employed strategies were summarized and reported in detail.

**Corresponding author*: s.a.hashemi0@gmail.com
†*Corresponding author*: cwlai@um.edu.my

Inamuddin, Mohd Imran Ahamed, Rajender Boddula and Tariq Altalhi (eds.) Self-Healing Smart Materials and Allied Applications, (153–180) © 2021 Scrivener Publishing LLC

Keywords: Solar cell, polymeric solar cells, self-healable, photovoltaic device

6.1 Introduction

Nowadays, improved practicality of solar cells and polymeric substrates in diverse industries raised the requirement for development of self-healable structure that can overcome degradation factors within the ecosystem and restore their damaged structure through self-healing protocol [1, 2]. In this matter, such intelligent polymeric platforms can respond to an appropriate stimulus such as electric filed, light, pH, temperature, microwave, magnetic field and ultrasound whenever it desired. Such structures can meet multi-functional applications and their protective roles were mainly categorized within three groups, viz., self-healing polymeric structures (SPS), shape memory polymeric structure (SMPS) and self-cleaning polymeric structures (SCPS) [2].

SPS provide alternatives for development of robust, sustainable, safer and long lasting polymeric platforms for diverse applications including energy, electronics, coating and transportation [3]. SMPS has a capability to return to its original/primary shape through a programmed and/or fixed approach and allow the shape memory recovery via an external stimulus, e.g., thermal energy, magnetic field, light energy and electric field [4]. What is more, SCPS boost the durability of solar cells and maintain their surface area from pollutants through self-cleaning approach [5]. Such novel platforms increase the average active service time of polymeric-based platforms, reduce the overall amount of wastes within ecosystem, decline the energy consumption and minimize the usage of raw materials as far as possible. Despite the fact that such platforms have been studied and developed for decades, however, mechanical robustness and structural stability of these platforms restricted their fundamental performance which required further analyses and investigations [2].

Herein, recent approaches in case of self-healable polymeric structure and solar cells were summarized and reported. In this regard, first mechanism of action of three main protection approaches (i.e., self-healing, shape-memory and self-cleaning) were explained in detail and thence recently developed self-healable polymeric structures and hydrogels as ideal supports for development of self-healable photovoltaic devices were reported. Next, integration of shape memory approach with self-healable polymers were examined and their potentials were highlighted. Finally, recently developed self-healable solar cells were evaluated and their pros over common structure were reported.

6.2 Functional Mechanism of Protection Approaches

6.2.1 Self-Healable Polymeric Structure

Self-healable polymeric structures exhibiting ideal capability via their self-repairing feature that boosting their life service, improving their capability and making them robust against applied mechanical loads or degradation factors [6]. Till today, mainly polymeric-based composites were developed as self-healable structures. Such novel platforms with their fantastic self-healing efficiency provide numerous advantages over traditional routes and resolve common flaws of polymer-based composites. Additionally, reinforcement of polymeric platforms with practical additives can improve their mechanical robustness, electrical conductivity and corrosion resistance which could thereby boost their overall performance upon introduction of self-healing approach [7–11]. Design and development of practical self-healing polymeric structure require several criteria which can be seen within Figure 6.1.

So far, different self-healing agents have been developed to add self-healing property to polymers which highlighted the necessities of functional materials for fabrication of such platforms. In this matter, self-healing agents were mainly incorporated within the polymeric matrix via placement in microvascular network and/or micro capsule [12, 13]. Self-healing agents using basic chemical reactions to repair the damaged surface that could include diverse compositions/components. Self-healing mechanism of such agents could occur through several approaches including (I) nonautonomous self-healing approach, in which an appropriate Stimulants such as heat of light is used to begin the cycle of self-healing via releasing suitable chemical compounds and healing the damaged polymeric structure; whereas the second approach (II) is known as autonomous self-healing approach where the self-healable chemical compound is automatically released in response to applied damage to the polymeric structure which

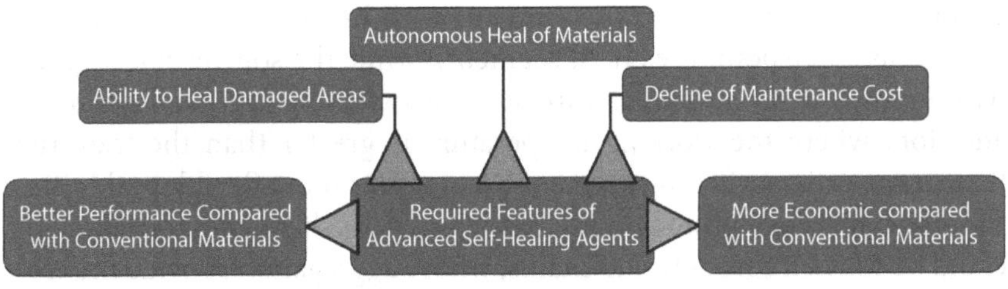

Figure 6.1 Required features of self-healing agents for development of self-healable polymeric materials [2].

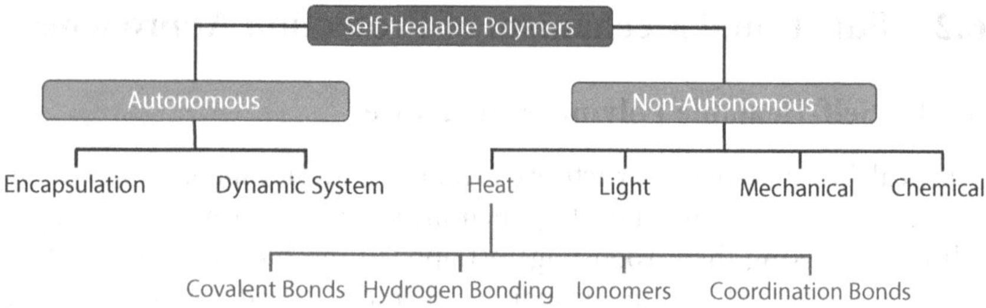

Figure 6.2 Classification of self-healable polymeric materials [2].

lead to the repair of damaged area. What is more, these two approaches covering wide range of chemical compounds as self-healing agent; a view of their mechanisms and used self-healing routes can be seen in Figure 6.2.

6.2.2 Shape Memory Polymeric Structure

The shape memory characteristic of a polymer is mainly originated from integration of specific morphology and processing conditions. A shape memory material could get its characteristic via special kinds of processing, e.g., injection molding and extruding. Thence, shape memory polymers are deformed and fixed into specific temporary 3D structures that known as the programming step of shape memory polymers. In fact, the polymeric structure restore its primary condition upon triggering an appropriate external stimulus. Such cycle (artificial shape memory programming and recovery) is able to repeat itself multiple times [2].

A SMPS includes appropriate stimuli responsive switches along with net points (Figure 6.3). Net-points turn polymeric structure into permanent shape memory materials. These points are cross-linking positions of physical intermolecular interactions and/or chemically covalent bonds between different functional groups [14]. The physical cross-linking occur within a segmented polymeric-block-copolymer. In fact, the physical cross-linking is acceded via crystallization of soft chains within the structure or vitrification [15]. In this matter, the segment with higher transition temperature act as a net-point (i.e., hard segment), while the soft/flexible segment with lower transition temperature act as a switching segment section. In addition, where the working temperature is greater than the transition temperature, the switching segment becomes far more flexible and initiate the shape memory protocol [16]. Likewise, such shape memory domains could be formed through side chains, whose aggregation provide the temporary possibility to avoid the recoiling of side chains. However, reversible covalent cross-linking between polymeric chains could be obtained

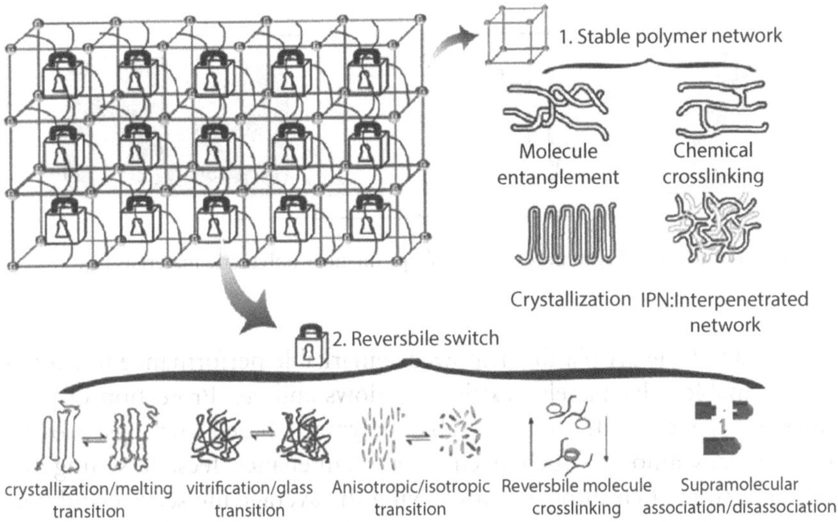

Figure 6.3 A view of net-point along with reversible switching of shape memory materials [17].

through incorporation of diverse functional groups of the segments. These functional groups are able to form reversible covalently-bonded chains via reaction with same and/or different functional groups; this reaction can also be controlled via an external stimulus [15].

Upon stretching the transition temperature of the polymeric structure, the entropy energy could be stored owing to the enlargement of shape memory chains that helps the deformation of shape memory polymeric structures [18]. Additionally, by further apply of heat to chains, these polymeric chains of SMPS will softened and oriented within a direction that could therefore help the polymeric chain to return to its primary conditions [19]. In fact, the main driving force for the shape recovery of SMPSs is the formed electric energy during the deformation of polymeric structure. Besides, different kinds of stimuli have been utilized to generate shape memory effect in various kinds of polymeric materials. In this case, the thermally founded shape memory effect is the most studied one, while other stimuli such as magnetic field, electricity, light, solvent and ultrasound have been used for generation of shape memory effect (Figure 6.4).

6.2.3 Self-Cleanable Polymeric Platforms

Another alternative for development of self-protectable photovoltaic devices is to design and fabricate self-cleanable polymeric platform.

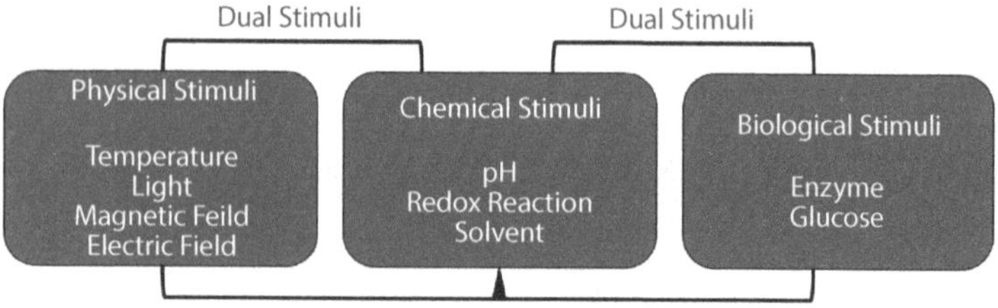

Figure 6.4 Category of stimuli responsive shape memory polymeric materials [2].

Such state of the art platform showed remarkable performance in case of self-cleanable solar panels, textiles, windows and etc. Protection of photovoltaic devices with self-cleanable polymeric-based coatings provide several pros among which decline of maintenance fees, lessening the required time of cleaning and improving the average life service and performance of solar panels can be mentioned [20, 21]. Such platforms could be categorized into two main groups, viz., hydrophobic and hydrophilic self-cleaning polymers. Both of aforementioned categories can clean their surfaces from wastes and provide ideal protection for solar panels through different pathways. The hydrophobic approach increase the contact angle of droplets with the surface, make droplets to slide and thence roll over the substrate and thereby remove pollutants from the protected surface. On the other hand, the hydrophilic approach is mainly based on the photocatalytic performance of metal oxides that degrades the dirt and remove them from their surfaces via water [22].

Superhydrophobic coating is one of the most practical approaches for protection of solar panels from degradative factors of environment and incoming dirt from surrounding areas. Different kinds of nanomaterials can be used to develop superhydrophobic self-cleanable surface, among which TiO_2, zinc oxide and nano silica are known as the most practical and economic ones [23–25]. Superhydrophobic protocol protect the surface from incoming dirt and could be governed through diverse factors. Among these factors are nanoscale ruggedness, surface strength, coating thickness and mechanical stability can be mentioned that have significant impact on the overall performance of self-cleanable surfaces [26].

Furthermore, the same as superhydrophobic surfaces, hydrophilic approach also exhibit superior potentials for development of self-cleanable substrates. This platform provide the self-cleaning possibility via photocatalytic degradation of dirt followed by sheeting of water droplets. Wettability of hydrophilic surface are so high that shows contact angle of about 0° [27]. In this matter, practical nanoparticles such as titanium

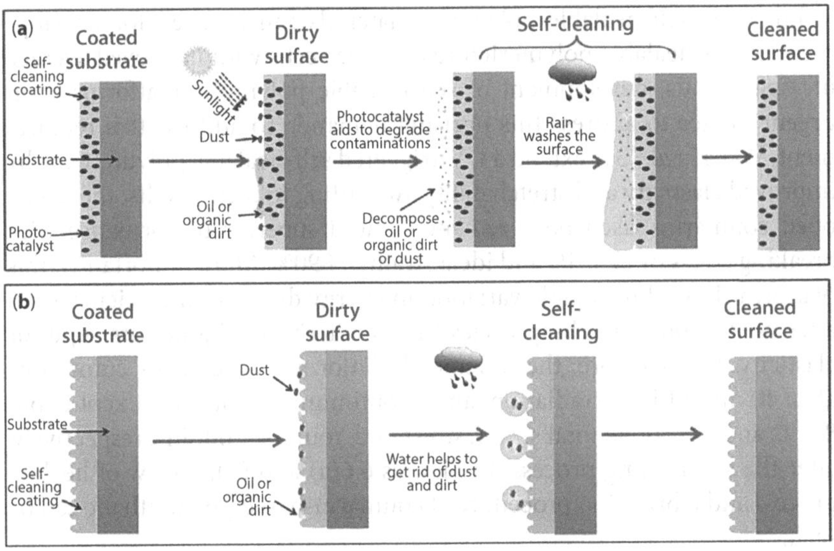

Figure 6.5 Self-cleaning approach through (a) super hydrophilic surface along with photocatalytic process and (b) superhydrophobic surface [2].

dioxide provide both photo-induced super hydrophilicity along with ideal photocatalytic performance which could thereby lead to the development of self-cleanable superhydrophilic coatings [22]. In Figure 6.5, working mechanism of self-cleanable superhydrophobic and hydrophilic surface can be seen.

6.3 Advanced Self-Healable Polymeric Materials

6.3.1 Self-Healable Polymers

Wide spread of polymeric materials in diverse industries and uprising maintenance costs of common polymeric structures, raised the demand for development of self-healable polymeric composites with extended lifetime [28]. Self-healable polymers have the ability to recover themselves through autonomous and/or non-autonomous approaches and recover some of or full performance after suffering external damages due to the applied mechanical load or harsh circumstances (e.g., radiation and high temperature) [29, 30]. Owing to the fantastic functionality of such platforms, self-healable polymers have opened new pathways for further improvement of solar cells, transistors, electrochemical sensors, electronic skin, coatings and supercapacitors [31].

Likewise, self-healable polymeric materials can be used for development of self-healable polymeric/organic solar cells with improved average life-span. Thus, development of self-healable polymeric platforms is an urgent matter to address this uprising demand. To address this requirement, Wang *et al.* [31] extend a UV-triggered self-healable polyurethane by improved elasticity and stretchability. According to their results, the developed coumarin-based polyurethane showed superior performance with breaking stress of 27 MPa and ideal strain of 890%. More importantly, this platform showed negligible variation in energy dissipation efficiency even after continuous stretching cycles that furtherly highlighted its superior elasticity. Furthermore, the developed platform recovered its cutoff area after 40 min of UV irradiation, and at optimum conditions, it kept about 84, 92 and 94% of its total strain, stress and Young's modulus, respectively, after the self-healing process. In Figures 6.6(a) and (b), a view of healing process and fabrication procedure of coumarin-based polyurethane can be seen.

In another attempt, Wong *et al.* [32] developed a coumarin embedded polyurethane via its integration with bio-based polyurethane. The technology platform could repair its applied harm by low-energy UV light with the absence of a band around 340 nm due to the 6,7-dihydroxycoumarin photo-cycload that could work six times faster after UV irradiation relative to the esculin. Lower dimerization levels of 6,7-dihydroxycoumarin (i.e., 0.006 min^{-1}) compared to esculin (0.0009 min^{-1}) will increase the self-healing efficiency of the bio-based platform created. Moreover, the cutoff areas in 6,7-dihydroxycoumarin were well-healed within the polyurethane (PUC) which showed far more better performance compared with esculin coupled polyurethane (PUE). In Figure 6.6(c) a view of healing process of developed platforms prior and after UV irradiation can be seen, where the 6,7-dihydroxycoumarin embedded polyurethane showed full recovery upon irradiation. This process occurs during the formation of cyclobutene ring between the coumarin groups of polyurethane after UV irradiation (Figure 6.6(d)).

Furthermore, SEM measurements of PUE and PUC prior to and after UV irradiation demonstrated the ideal self-healing efficiency of 6,7-dihydroxycoumarin embedded polyurethane relative to the embedded esculin (Figure 6.7(a)) in compliance with optical photos. Besides, in Figure 6.7(b) the formation mechanism of bonding orbitals between coumarin functional groups via the photochemical process can be seen. As demonstrated, coumarin dimerization cycle occurs due to the cycloaddition photochemical [2 + 2]. Two alkenes cycloading cycle is the product of contact between the largest molecular orbital occupied (HOMO) and the lowest molecular

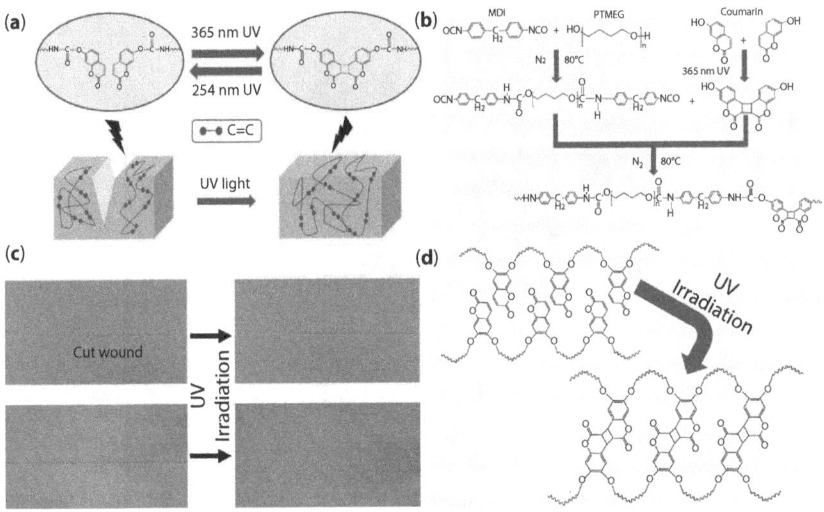

Figure 6.6 (a) Self-healing process of coumarin-based polyurethane under UV irradiation and (b) fabrication procedure of developed self-healable coumarin-based polyurethane [31]; (c) optical micrograph of cutoff areas of non-irradiated and UV irradiated PUE (upper one) and PUC (the second one) and (d) formation of cyclobutene ring between the coumarin groups of polyurethane after UV irradiation [32].

orbital unoccupied (LUMO). Until dimerizing the functional group of coumarin, the current electrons inside the orbital bonding π become excited to the orbital antibonding π* with a higher energy level through exposure to UV irradiation. Thereby, lobes with the same algebraic sign from HOMO to LUMO could overlap in order to form two sigma bonds that known as "superfacial bond" formation. On the other hand, [2 + 2] cycloaddition process of coumarin could not occur via apply of external heat. This result is due to the insufficient energy of applied heat for excitation of electrons within the π bonding orbital to π^*. Therefore, the HOMO and/or LUMO could not interact with formed sigma bonds due to the generation of antibonding orbital that could lead to electron (Figure 6.7(c)) [33, 34].

Zhang *et al.* [35] The transparent self-healing conductor based on copper nanowires (Cu NWs) and polyurethane was developed. The developed conductor is derived from core shell nanowire based on Cu NWs and self-healable polyurethane. The polyurethane part is consisted of furan rings that were crosslinked with N,N'-(4,4'-methylenediphenyl) dimaleimide through the Diels–Alder reaction to generate a healable polymeric network. The developed self-healable polymeric platform exhibited ideal sheet resistance and transparency of about 22.3 Ω sq^{-1} and 66.5% (at 590 nm), respectively. The damaged surface can recover 98% of its optimum

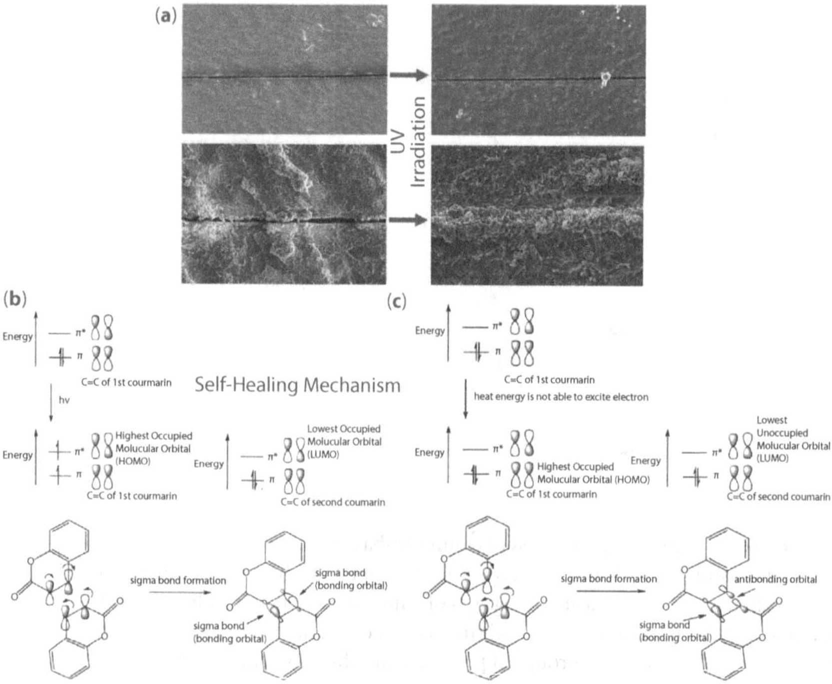

Figure 6.7 (a) SEM images of non-irradiated and UV irradiated PUE and PUC, (b) formation mechanism of bonding orbitals between coumarin groups through photochemical process [2 + 2] cycloaddition and (c) formation mechanism of bonding orbital and π* between coumarin via thermal [2 + 2] cycloaddition [32].

conductivity after being exposed to 120 °C external heat for 3 min. This process could be repeated for five times at the exact location which exhibited an ideal flexible transparent electrode for development of self-healable polymeric solar cells. In Figure 6.8(a), the self-healing performance of the developed polymeric platform after apply of external heat (120 °C) can be seen; as shown the developed self-healable conductor fully recovered itself after apply of external heat. In Figure 6.8(b), the performance of the developed transparent polymeric platform for supplying a LED light can be seen. As shown, the cut circuit recovered itself through its self-healing capability. The fabrication procedure of developed self-healable polymeric platform can be seen in Figure 6.8(c).

In another work, Wang *et al.* [36] developed a transparent self-healable polydimethylsiloxane elastomer based on the imine bond. The develop platform is derived from 1,4-diformylbenzene and polydimethylsiloxane through a simple production procedure. In this matter, the total

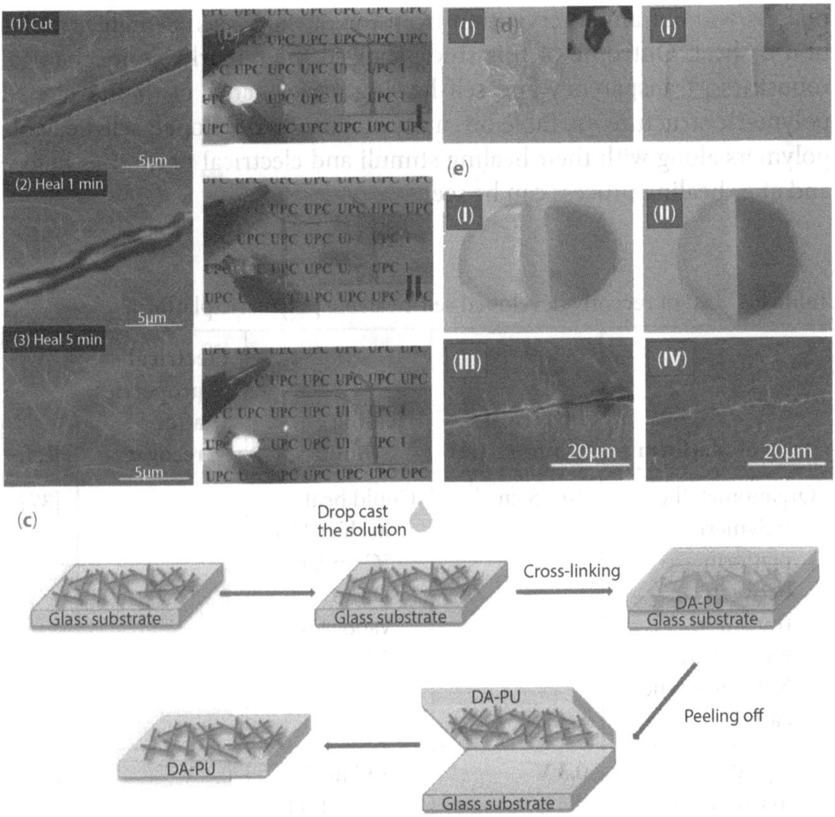

Figure 6.8 (a) SEM images of the conductor with (1) cutoff area on its surface, (2) 120 °C heating for 1 min and (3) 120 °C heating for 5 min, (b) (I) a view of composite conductor with a LED light, (II) the cut specimen and (III) restored circuit via self-healing process and (c) production procedure of self-healable nanocomposite based on the Cu NW and DA-PU [35]; (d) Photograph with photographic photographs of samples sliced into tiny pieces (I) and then modeled at room temperature for 12 h; (e) photograph of samples sliced and healed [36].

concentration of 1,4-diformylbenzene has significant effect on the gelation process of the final composition. The fabricated elastomer showed remarkable transmittance of about 80% over the visible light range which make it ideal alternative for fabrication of self-healable photovoltaic devices. The fabricated polymeric structure showed remarkable mechanical robustness and flexibility. Moreover, formation of Schiff-based linkages and presence of imine-based bonding were proved through FTIR/micro Raman spectroscopy analysis. Presence of reversible imine-based bonds can improve the touch of the damaged surfaces and lead in quick and reproducible recovery of the polymeric structure through self-healing process without any intervention from the outside, such as light irradiation or heat. Outcome of this study showed the remarkable mechanical robustness, transparency and self-healing performance of the developed polymeric structure. In Table 6.1, a list of recently developed self-healable polymers along with their healing stimuli and electrical properties before and after healing process can be seen.

Table 6.1 List of recently developed self-healable polymeric platforms.

Type of platform	Electrical properties	Healing stimuli	Electrical properties after recovery	Ref.
Organometallic polymeric platform between transition metals and N-heterocyclic carbenes	10^{-3} S cm^{-1}	Could heal itself at 150 °C under DMSO vapor for 2 h	–	[37]
Coupled epoxy matrix with Ga-In micro capsules	0.3 V	Could heal itself at RT after 160 s	0.3 V	[38]
Micro capsulation of TCNQ and TTF	–	Could cure itself at RT	–	[39]
Multilayer coated polymeric electrolyte with silver NWs	0.38 Ω sq^{-1}	Could heal itself with droplet of water after 2 min	0.42 Ω sq^{-1}	[40]
Coupled polyurethane with silver paste capsules	0.95 Ω	Could heal itself at RT after 2 weeks	1.45 Ω	[41]

(*Continued*)

Table 6.1 List of recently developed self-healable polymeric platforms.

Type of platform	Electrical properties	Healing stimuli	Electrical properties after recovery	Ref.
Reinforced hydrogen bonded polymer with Ni particles	40 S cm^{-1}	Could heal itself at RT after 3 min	35 S cm^{-1}	[42]
Embedded polymeric structure with EGaIn	3–10 Ω	Could heal itself at RT after about 10 min	1 Ω	[43]

Abbreviations: RT, room temperature and NW, nanowires.

6.3.2 Self-Healable Hydrogels

Self-healable hydrogels are among the interesting supports for development of self-healable polymeric/organic solar cells. Recently, fabrication of supermolecular gels accelerated through a non-covalent interactions, e.g., $\pi - \pi$ stacking, attraction force ,van der Waals, coordination interactions, charge transfer, hydrogen bonding, and dipole-dipole interaction have attracted considerable attention toward themselves owing to their ideal performance in matter of cell culture, drug control/release, smart materials and biology engineering [44–49]. Moreover, among this practical features of hydrogel, their self-healing performance gained more attention due to their intelligent and bionics characteristics [50]. In this matter, metallogels as self-healing hydrogels has attracted noteworthy attentions because of their fantastic electronic, redox, magnetic and optical features [51–55].

In a work by Peng *et al.* [50], they developed self-healable hydrogels and organogels via self-assembly of bis-terpyridine complex with selected metal ions. In this matter, they used a novel type of bis-terpyridine (i.e., BDT) that could form both organogel and hydrogel with superior self-healing property upon the presence of selected metal ions. Outcome of their research showed the remarkable self-healing performance of the developed gels, where the Cu^{2+}-BDT hydrogel in DMF exhibited gelation at room temperature, injectability and ideal self-healing capability. In Figure 6.9(a), self-healing capability of developed hydrogels can be seen, while their structural SEM and TEM images can be seen within Figures 6.9(b) and (c), respectively.

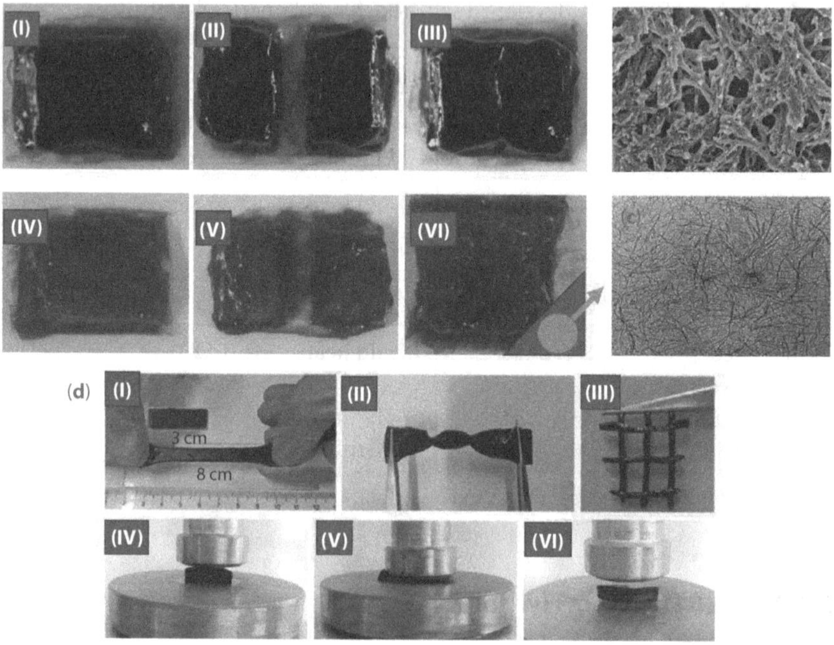

Figure 6.9 Healing process of (a) (I-III) Cu2+-BDT hydrogel (5.64 wt%) and (IV-VI) Sn^{2+}-BDT hydrogel (12.82 wt%); (b) SEM and (c) TEM images of Cu^{2+}-BDT hydrogel at 4 μm and 0.2 μm scale bars, respectively [50]; (d) mechanical properties of GO-B-PVA-KCl hydrogel that was (I) stretched, (II) twisted and (III) tailored, (IV-VI) compression performance of the developed hydrogel [56].

In another work, Peng *et al.* [56] spread a self-healable hydrogel electrolyte that was doped with graphite oxide through diol-borate ester bonding which was crosslinked with polymeric network of polyvinyl alcohol (PVA). Obtained results showed that addition of 2.3 wt% graphite oxide to the polymeric system of hydrogel exhibited an electrical conductivity of about 47.5 mS cm^{-1} along with superior ionic conductivity compared with the neat hydrogel (conductivity of 32.6 mS cm^{-1}). More importantly, the reinforced hydrogel demonstrated ideal self-healing performance that could repair the cutoff area after 5 min. Additionally, incorporated hydrogel with activated carbon showed superior different efficiency of 156 F g^{-1} at 0.3 A g^{-1} and may also recover ability after seven healing cycles with negligible deviation in the overall performance. Such platforms can be used for fabrication of self-healable flexible robots, wearable electronics and smart apparels. In Figure 6.9(d) a view of mechanical robustness of the as developed hydrogel can be seen.

Additionally, the developed hydrogel based on the diol-borate ester bonding can heal the cutoff area after elapse of 5 min and return the structure to its full performance; a view of this process can be seen in Figure 6.10(a) (I-III). Likewise, SEM images of cutoff area and healed polymeric hydrogel can be seen in Figure 6.10(a) (IV-VI). In Figure 6.10(b), the self-healing mechanism of presented platform can be seen, when the hydrogel is broken into two pieces, diol-borate ester bonds are broken together at the cutoff region interface. Once these two separated parts make contact with each other, cut diol-borate ester bonds will be prone to link with each other and form a linked structure. This recombination process of diol-borate ester bonds provide the self-healing possibility for the developed hydrogel that can also recover most of its electrical conductivity and be used for development of self-healable substrates for light-weight electronic devices. In Figure 6.10(c), the capability of hydrogel for generation of diverse shapes can be seen , while Figure 6.10(d) illustrates the potential of the fabricated hydrogel for lighting up a LED light [56, 57].

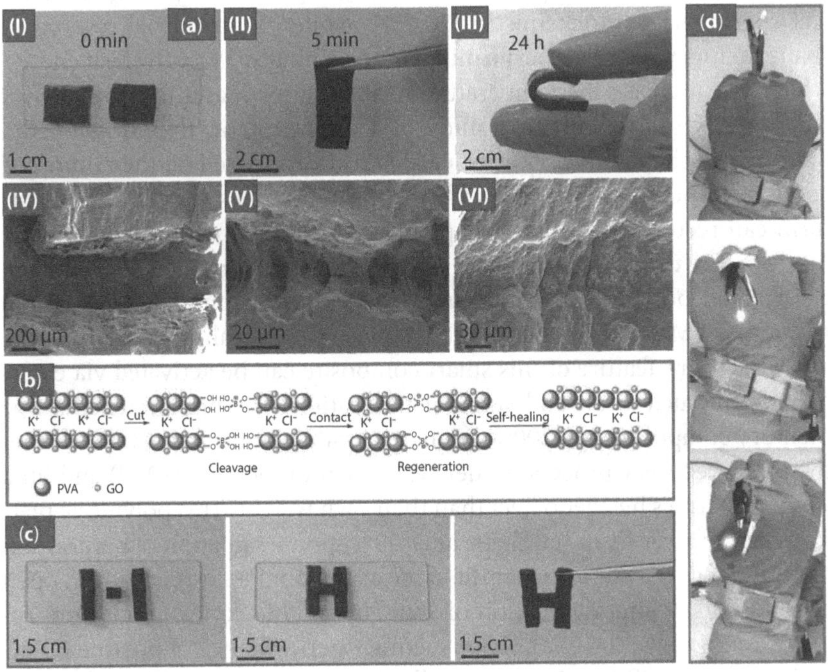

Figure 6.10 (a) (I–III) The GO-B-PVA-KCl hydrogel self-healing cycle under atmospheric circumstances, (a) (IV–VI) SEM photos of the hydrogel produced before and after reconstruction of the cutting region; (b) self-healing mechanism of the developed hydrogel, (c) tailoring the developed self-healable hydrogel into the form of H-shape and (d) lighting up a LED via the developed hydrogel [56].

6.4 Shape Memory Materials

Another alternative for development of self-healable polymeric-based solar cells is to couple them with shape memory feature. Shape memory platforms are among the smart materials that can recover their primary shape through external stimuli such as electric field [58], heat [59], magnetic field [60], light [61] and pH/chemical reactions/moisture [62, 63]. Shape memory materials were mainly consisted of an intelligent cross-linked network that can maintain its original shape though the activated elastic transition via an external stimuli; this process will lead to shape recovery [64, 65]. Moreover, light-sensitive shape memory polymeric materials are ideal candidates for advanced electronic platforms and solar cells [66]. Light-driven shape memory solar cells could be categorized within two main groups, including photothermal responsive and photochemical responsive shape memory platforms [67–69].

Photochemical responsive shape memory materials can recover their initial shape by converting the harvestable energy of sunlight into mechanical energy via photochemistry [70, 71], while photothermal responsive shape memory materials performing this task by absorbing the heat energy, which can increase the temperature of the platform and thereby recover the primary shape [68]. In a study by Cui *et al.* [66], they developed a light-driven shape memory self-healable platform based on the reinforced conductive polymeric matrix with carbon nanotubes (CNTs). This platform can recover its primary shape upon irradiation of sun light (Figure 6.11(a)). The developed shape-memory platform is consisted of reinforced chlorinated polypropylene carbonate (CPPC) with multi-walled carbon nanotubes (MWCNTs) that exhibit fantastic self-healing capability. The shape recovery feature of this smart composite can be activated via external stimuli such as IR and sunlight irradiation. The pre-stretched shape memory composite with 200% strain can produce contraction force during the shape recovery process (under IR irradiation of 2.6 ± 0.2 N) and lift a weigh 550 times heavier object than their own weight. This polymeric platform can recover its initial shape after 39 s upon irradiation of natural sun light. In fact, this platform act the same as heliophile flower that can open and close itself after irradiation of sunlight and block of sunlight, respectively. In Figures 6.11(b–e), shape memory performance of pristine CPPC and reinforced CPPC with diverse filler loadings of MWCNT can be seen.

In Figure 6.12, the shape memory performance of an artificial flower made of CPPC-MWCNT can be seen. This platform is directly responding to emitted IR irradiation and/or sunlight, close itself during blockage of beams and open its structure upon irradiation of beams. Such feature can

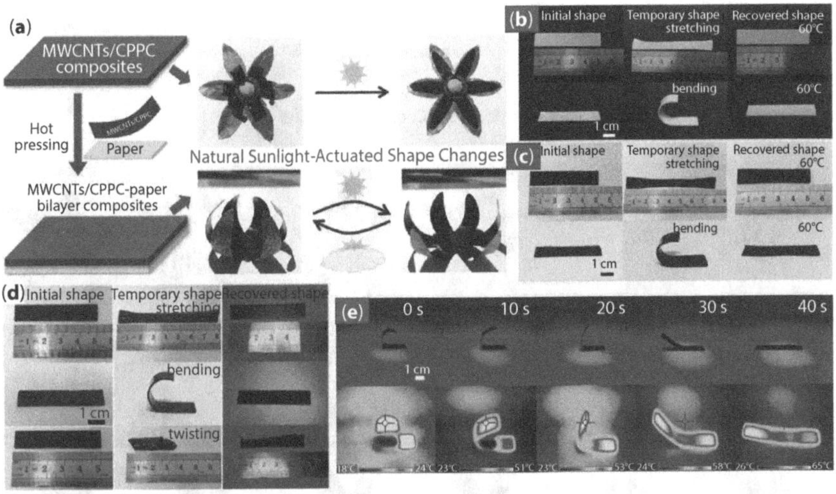

Figure 6.11 (a) Structural feature and shape memory performance of the developed platforms, (b) thermally induced shape memory performance for CPPC polymeric structure, (c) a view of thermally-induced shape memory process for the developed platform containing 2 wt% MWCNT within CPPC, (d) shape memory performance of MWCNT-CPPC composite under IR light's irradiation and (e) shape memory performance and surface temperature of the developed platform (i.e., 1 wt% MWCNT/CPPC) under IR light irradiation [66].

be used to improve the capability of photovoltaic devices and protect them from environmental factors during sunlight blockage times [66].

In addition, cyclic shape memory performance of the CPPC-MWCNT composite under IR light and sunlight irradiation can be seen in Figures 6.13(a) and (b) which exhibit rapid shape recovery response. Moreover, the fabricated composite can also heal itself upon irradiation of IR light which show simultaneous self-healing capability along with its superior shape-memory performance [66]. In Figure 6.13(c), application of this platform as a smart switch for a LED light can be seen.

6.5 Self-Healable Solar Cells

High upkeep cost of photovoltaic devices compared with common energy sources and degradative role of environmental factors raised the requirement for development of practical strategies for increasing the resistance and average life service of solar cells [72]. Among diverse types of solar cells, polymeric/organic solar cells have attracted noteworthy attention owing to their increasing power conversion efficiency (PCE) and low

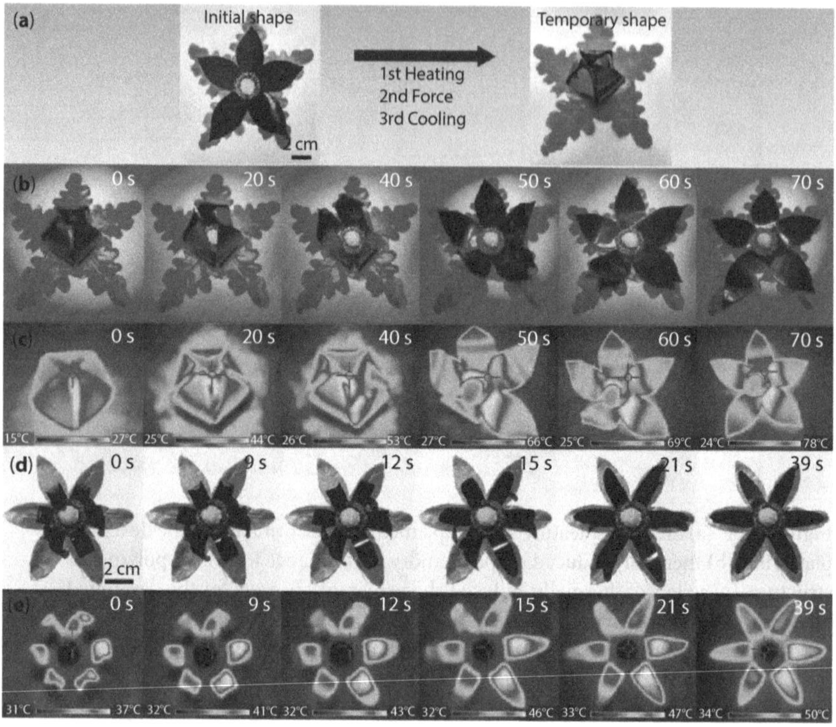

Figure 6.12 Artificial shape memory flower based on the MWCNT-CPPC composite, (b) & (c) digital and infrared images of the shape memory behavior of MWCNT-CPPC composite, (d) & (e) digital and infrared images of the shape memory MWCNT-CPPC composite from closed to open state upon sunlight irradiation [66].

fabrication cost [73–75]. Currently, polymeric solar cells are under development via diverse research groups to increase their PCE through tuning the absorption spectra [76], maintaining the 3D structure of such devices [77] and boosting their stability via encapsulation approach [78]. In this matter, development of self-healable solar cells found to be an ideal alternative to address the uprising concern regarding the life service of solar cells.

In case of encapsulation strategy, Bag *et al.* [72] developed a polymeric self-healable solar cells via its encapsulation with polyisobutylene as a printable self-healable polymeric sealant for protection of organic solar cells against degradative factors. Such sealant can be applied on the surface of any kind of polymeric solar cell by blade coating, drop casting and spin coating. In Figure 6.14, a view of 3D structure and photovoltaic performance of the device can be seen which demonstrate ideal stability within the ambient conditions.

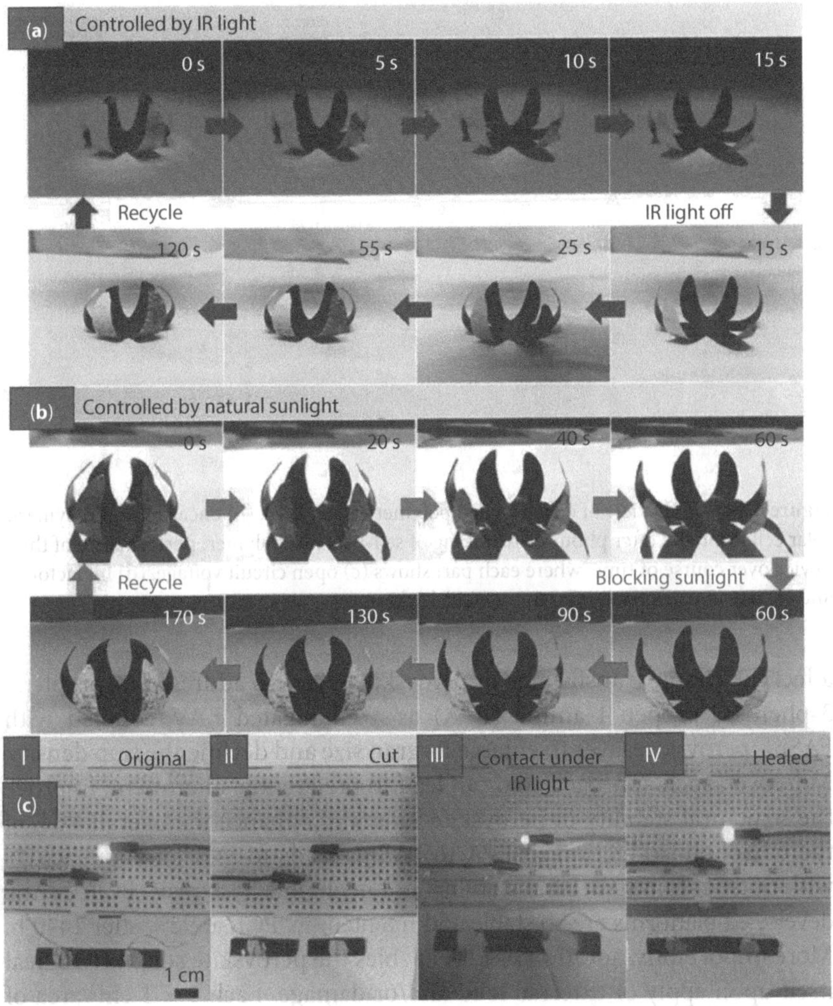

Figure 6.13 Cyclic shape memory performance of the artificial flower governed via (a) IR light and (b) sunlight irradiation; (c) fabricated circuit via the shape memory MWCNT-CPPC composite that act as an smart switch for turning LED light on or off upon IR light irradiation [66].

In another attempt, Ran *et al.* [79] developed a conjugated organic coating toward adding the self-healing performance to $FASnI_3$-based solar cells. In this matter, integration of large volume amine (LAV) with Sn-based perovskite layer shown to lead to fantastic PCE for Pb-Free perovskite photovoltaic devices. However, common types of LAVs such as butylammonium and phenylethylammonium are insulative and could adversely

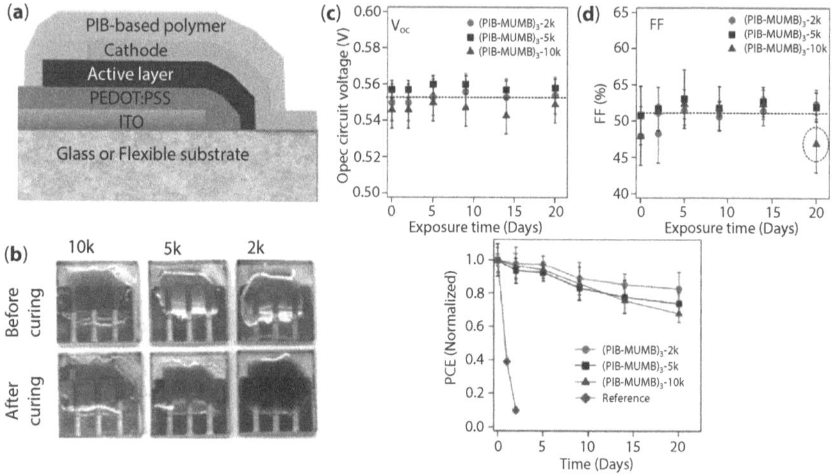

Figure 6.14 (a) 3D view of encapsulated polymeric solar cells, (b) encapsulated polymeric solar cell prior and after photo-crosslinking of self-healable polymer; performance of the device over course of time, where each part shows (c) open circuit voltage, (d) fill factor and (e) power conversion efficiency (PCE) [72].

affect the charge transfer rate of perovskite layer. To address this problem, 3-phenyl-2-propen-1-amine (PPA) as a conjugated LAV coupled with $FASnI_3$ perovskite layer to enlarge its gran size and decline the trap density. Moreover, such approach also can lead to preferential alignment, improve the structural stability of perovskite layer and enhance the overall PCE of the device. Introduction of PPA to $FASnI_3$ leads to PCE of about 9.61% with outstanding structural stability and negligible hysteresis. Besides, the developed platforms are so stable and maintain its PCE at 92% after 1440 h. More importantly, addition of PPA enables the perovskite solar cell to heal itself upon apply of external load and/or damage. Each 1×1 cm^2 area of developed Sn-based perovskite solar cell exhibit PCE of about 7.08%. In Figure 6.15(a), the chemical formula of FA and PPA along with the crystalline structure of Sn-based perovskite layer can be seen. Moreover, in Figure 6.15(b) effect of addition of diverse filler loadings of PPAI to the perovskite layer and its morphology can be seen. As shown PPAI significantly improved the grain size and structural stability of the final composition. In Figures 6.15(c) and (d), preparation of $FASnI_3$ film without and with PPAI can be seen, respectively.

Besides, cross sectional view of the developed self-healable perovskite solar cell can be seen in Figure 6.16(a). In this regard, Ag and ITO were used as cathode and anode, respectively, while C_{60}/BCP and PEDOT:PSS

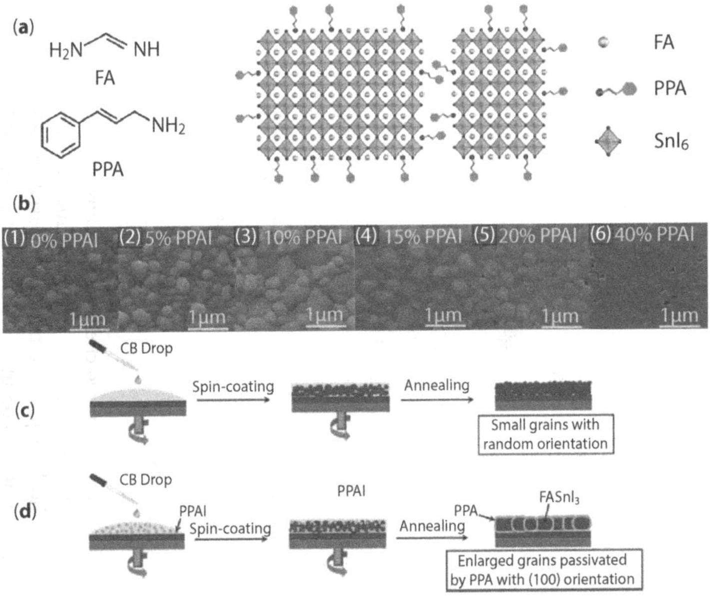

Figure 6.15 (a) Chemical formula of FA and PPA along with crystalline structure of PPAxFA1 – xSnI3, (b) SEM images of (1) 0 wt% PPAI, (2) 5 wt% PPAI, (3) 10 wt% PPAI, (4) 15 wt% PPAI, (5) 20 wt% PPAI and (6) 40 wt% PPAI with 1 μm scale bar; production procedure of FASnI3 film (c) without and (d) with PPAI [79].

selected as electron transport layer (ETL) and hole transport layer (HTL), respectively. Modified $FASnI_3$ with PPA is also selected as active perovskite layer for charge separation. In Figure 6.16(b), optimized chemical structure, HOMO and LUMO distributions of PPA and PEA molecules can be seen. In Figure 6.16(c), the possible self-healing mechanism of developed Sn-based perovskite layer can be seen. As shown, upon exposure of $FASnI_3$ perovskite layer to air, it tend to decompose under air. However, existence of PAA molecules on the surface of $FASnI_3$ prevent the decomposition of fragments (i.e., SnI_6 and FA) from moving apart or falling. Thereby, upon exposure of the enhanced perovskite film to nitrogen atmosphere, the $FASnI_3$ reconstruct itself because of the low formation energy of capped structure with PPA. In fact, even after degradation of the perovskite layer, the PPA capping layers hold all of disassembled blocks in their main space and avoid decomposition of the perovskite layer. Therefore, the $FASnI_3$ is able to reassemble itself after being exposed to N_2 atmosphere. In Figure 6.16(d) (I–VI), overall PCE and performance of the device under different circumstances can be seen in detail [79].

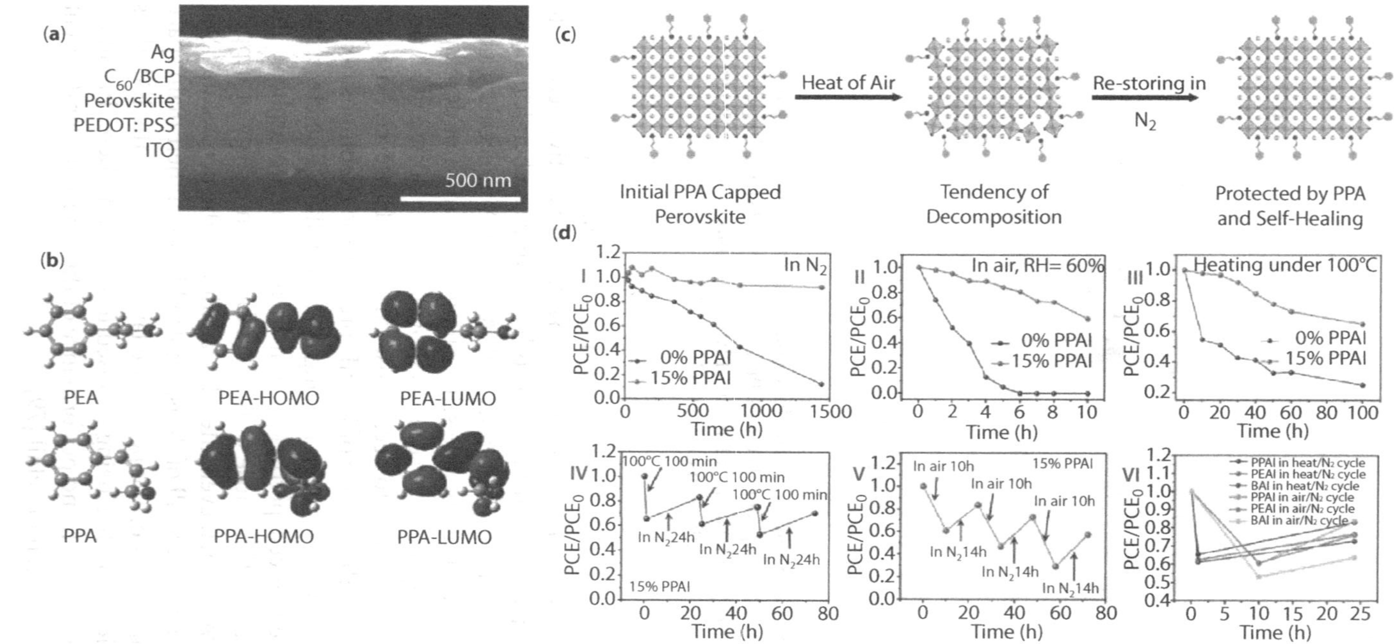

Figure 6.16 (a) Cross sectional SEM photograph of polymeric solar cell with 500 nm scale bar, (b) optimized chemical structure, HOMO and LUMO distributions of PPA and PEA molecules that were calculated via DFT at 6-31G(d) and B3LYP levels, (c) self-healing mechanism of integrated PPA with $FASnI_3$; (d) normalized power conversion efficiency of the developed self-healable photovoltaic device based on $FASnI_3$ with or without PPAI, each part show (I) performance in nitrogen atmosphere, (II) performance in ambient conditions, (III) functionality at 100 °C external heat, (IV) the self-healing performance of the device under heating-N_2 cycles, (V) the self-healing performance of the device under air-N2 cycles and (VI) PCE variation of the device under diverse testing cycles [79].

6.6 Conclusions

Ideal benefits of green energy sources such as photovoltaic devices raised the concerns about development of practical approaches to protect them from degradative environmental factors and increase their average service time. In this matter, self-healing protocol found to be an ideal alternative. In current chapter, first three main protective approaches, i.e., self-healing, shape-memory and self-cleaning, were presented and their mechanism of action were reported. Thence, recently developed self-healing and shape-memory polymeric were evaluated and reported. Finally, capability of self-healing polymeric platforms for improving photovoltaic devices were examined. Outcome of gathered data showed the superior capability of such platforms for protecting solar cells from degradative factors which can thereby minimize maintenance costs and increase the optimum life time of solar cells.

References

1. Hashemi, S.A., Ramakrishna, S., Aberle, A.G., Recent progress in flexible–wearable solar cells for self-powered electronic devices. *Energy Environ. Sci.*, 13, 685–743, 2020.
2. Thakur, S., Sustainable Nanostructural Materials for Shape Memory, Self-Healing, and Self-Cleaning Applications, in: *Dynamics of Advanced Sustainable Nanomaterials and their Related Nanocomposites at the Bio-Nano Interface*, India, pp. 235–250, Elsevier, 2019.
3. Thakur, V.K. and Kessler, M.R., Self-healing polymer nanocomposite materials: A review. *Polymer*, 69, 369–383, 2015.
4. Liu, F. and Urban, M.W., Recent advances and challenges in designing stimuli-responsive polymers. *Prog. Polym. Sci.*, 35, 3–23, 2010.
5. Kang, X., Zi, W.-W., Xu, Z.-G., Zhang, H.-L., Controlling the micro/nanostructure of self-cleaning polymer coating. *Appl. Surf. Sci.*, 253, 8830–8834, 2007.
6. Yuan, Y., Yin, T., Rong, M., Zhang, M., Self healing in polymers and polymer composites. Concepts, realization and outlook: A review. *Express Polym. Lett.*, 2, 238–250, 2008.
7. Mousavi, S.M., Hashemi, S.A., Arjmand, M., Amani, A.M., Sharif, F., Jahandideh, S., Octadecyl amine functionalized Graphene oxide towards hydrophobic chemical resistant epoxy Nanocomposites. *ChemistrySelect*, 3, 7200–7207, 2018.
8. Mousavi, S.M., Hashemi, S.A., Amani, A.M., Saed, H., Jahandideh, S., Mojoudi, F., Polyethylene terephthalate/acryl butadiene styrene copolymer incorporated with oak shell, potassium sorbate and egg shell nanoparticles

for food packaging applications: Control of bacteria growth, physical and mechanical properties. *Polym. Renewable Resour.*, 8, 177–196, 2017.
9. Mousavi, S., Arjmand, O., Hashemi, S., Banaei, N., Modification of the epoxy resin mechanical and thermal properties with silicon acrylate and montmorillonite nanoparticles. *Polym. Renewable Resour.*, 7, 101–113, 2016.
10. Mousavi, S.M., Goudarzian, N., Hashemi, S., Unsaturated polyester resins modified with cresol novolac epoxy and silica nanoparticles: processing and mechanical properties. *Int. J. Chem. Pet. Sci.*, 5, 13–26, 2016.
11. Mousavi, S., Aghili, A., Hashemi, S., Goudarzian, N., Bakhoda, Z., Baseri, S., Improved morphology and properties of nanocomposites, linear low density polyethylene, ethylene-co-vinyl acetate and nano clay particles by electron beam. *Polym. Renewable Resour.*, 7, 135–153, 2016.
12. Ghosh, B. and Urban, M.W., Self-repairing oxetane-substituted chitosan polyurethane networks. *Science*, 323, 1458–1460, 2009.
13. White, S.R., Sottos, N.R., Geubelle, P.H., Moore, Kessler, M.R., Sriram, S., Brown, E.N., Viswanathan, S., Autonomic healing of polymer composites. *Nature*, 409, 794–797, 2001.
14. Hu, J., Zhu, Y., Huang, H., Lu, J., Recent advances in shape–memory polymers: Structure, mechanism, functionality, modeling and applications. *Prog. Polym. Sci.*, 37, 1720–1763, 2012.
15. Behl, M. and Lendlein, A., Shape-memory polymers, in: *Kirk-Othmer Encyclopedia of Chemical Technology*, pp. 1–16, 2000.
16. Lendlein, A. and Kelch, S., Shape-memory polymers. *Angew. Chem. Int. Ed.*, 41, 2034–2057, 2002.
17. Meng, H. and Li, G., A review of stimuli-responsive shape memory polymer composites. *Polymer*, 54, 2199–2221, 2013.
18. Thakur, S. and Karak, N., Bio-based tough hyperbranched polyurethane–graphene oxide nanocomposites as advanced shape memory materials. *RSC Adv.*, 3, 9476–9482, 2013.
19. Kalita, H. and Karak, N., Hyperbranched polyurethane/triethanolamine functionalized multi-walled carbon nanotube nanocomposites as remote induced smart materials. *Polymer Int.*, 63, 1295–302, 2014.
20. Ganesh, V.A., Raut, H.K., Nair, A.S., Ramakrishna, S., A review on self-cleaning coatings. *J. Mater. Chem.*, 21, 16304–16322, 2011.
21. Parkin, I.P. and Palgrave, R.G., Self-cleaning coatings. *J. Mater. Chem.*, 15, 1689–1695, 2005.
22. Ragesh, P., Ganesh, V.A., Nair, S.V., Nair, A.S., A review on 'self-cleaning and multifunctional materials'. *J. Mater. Chem. A*, 2, 14773–14797, 2014.
23. Bravo, J., Zhai, L., Wu, Z., Cohen, R.E., Rubner, M.F., Transparent superhydrophobic films based on silica nanoparticles. *Langmuir*, 23, 7293–7298, 2007.
24. Lai, Y., Tang, Y., Gong, J., Gong, D., Chi, L., Lin, C., Chen, Z., Transparent superhydrophobic/superhydrophilic TiO2-based coatings for self-cleaning and anti-fogging. *J. Mater. Chem.*, 22, 7420–7426, 2012.

25. Wu, X., Zheng, L., Wu, D., Fabrication of superhydrophobic surfaces from microstructured ZnO-based surfaces via a wet-chemical route. *Langmuir*, 21, 2665–2667, 2005.
26. Das, S., Kumar, S., Samal, S.K., Mohanty, S., Nayak, S.K., A review on superhydrophobic polymer nanocoatings: Recent development and applications. *Ind. Eng. Chem. Res.*, 57, 2727–2745, 2018.
27. Long, M., Wu, D., Weimin, C., Photoinduced hydrophilic effect and its application on self-cleaning technology. *Recent Pat. Eng.*, 4, 189–199, 2010.
28. Patrick, J.F., Robb, M.J., Sottos, N.R., Moore, J.S., White, S.R., Polymers with autonomous life-cycle control. *Nature*, 540, 363–370, 2016.
29. Zhang, Y.S. and Khademhosseini, A., Advances in engineering hydrogels. *Science*, 356, eaaf3627, 2017.
30. Zhang, Q., Liu, L., Pan, C., Li, D., Review of recent achievements in self-healing conductive materials and their applications. *J. Mater. Sci.*, 53, 27–46, 2018.
31. Wang, Y., Liu, Q., Li, J., Ling, L., Zhang, G., Sun, R., Wong, C.-P., UV-triggered self-healing polyurethane with enhanced stretchability and elasticity. *Polymer*, 172, 187–195, 2019.
32. Wong, C.S., Hassan, N.I., Su'ait, M.S., Serra, M.A.P., Gonzalez, J.A.M., Granda, L.A., Badri, K.H., Photo-activated self-healing bio-based polyurethanes. *Ind. Crops Prod.*, 140, 111613, 2019.
33. Jacobs, A., *Understanding organic reaction mechanisms*, Cambridge University Press, Cambridge UK, 1997.
34. Fleming, I., *Pericyclic reactions*, Oxford University Press, USA, 2015.
35. Zhang, X., Tang, Z., Tian, D., Liu, K., Wu, W., A self-healing flexible transparent conductor made of copper nanowires and polyurethane. *Mater. Res. Bull.*, 90, 175–181, 2017.
36. Wang, P., Yang, L., Dai, B., Yang, Z., Guo, S., Gao, G., Xu, L., Sun, M., Yao, K., Zhu, J., A self-healing transparent polydimethylsiloxane elastomer based on imine bonds. *Eur. Polym. J.*, 123, 109382, 2020.
37. Williams, K.A., Boydston, A.J., Bielawski, C.W., Towards electrically conductive, self-healing materials. *J. R. Soc. Interface*, 4, 359–362, 2007.
38. Blaiszik, B.J., Kramer, S.L., Grady, M.E., McIlroy, D.A., Moore, J.S., Sottos, N.R., White, S.R., Autonomic restoration of electrical conductivity. *Adv. Mater.*, 24, 398–401, 2012.
39. Odom, S.A., Caruso, M.M., Finke, A.D., Prokup, A.M., Ritchey, J.A., Leonard, J.H., White, S.R., Sottos, N.R., Moore, J.S., Restoration of conductivity with TTF-TCNQ charge-transfer salts. *Adv. Funct. Mater.*, 20, 1721–1727, 2010.
40. Li, Y., Chen, S., Wu, M., Sun, J., Polyelectrolyte multilayers impart healability to highly electrically conductive films. *Adv. Mater.*, 24, 4578–4582, 2012.
41. Odom, S.A., Chayanupatkul, S., Blaiszik, B.J., Zhao, O., Jackson, A.C., Braun, P.V., Sottos, N.R., White, S.R., Moore, J.S., A self-healing conductive ink. *Adv. Mater.*, 24, 2578–2581, 2012.

42. Tee, B.C., Wang, C., Allen, R., Bao, Z., An electrically and mechanically self-healing composite with pressure-and flexion-sensitive properties for electronic skin applications. *Nat. Nanotechnol.*, 7, 825, 2012.
43. Palleau, E., Reece, S., Desai, S.C., Smith, M.E., Dickey, M.D., Self-healing stretchable wires for reconfigurable circuit wiring and 3D microfluidics. *Adv. Mater.*, 25, 1589–1592, 2013.
44. Babu, S.S., Praveen, V.K., Ajayaghosh, A., Functional π-gelators and their applications. *Chem. Rev.*, 114, 1973–2129, 2014.
45. Chen, J., Wang, T., Liu, M., Selective shrinkage and separation of isomeric naphthoic acids via supramolecular gelation. *Chem. Commun.*, 52, 11277–11280, 2016.
46. Ruiz-Olles, J., Slavik, P., Whitelaw, N.K., Smith, D.K., Self-Assembled Gels Formed in Deep Eutectic Solvents: Supramolecular Eutectogels with High Ionic Conductivity. *Angew. Chem.*, 131, 4217–4222, 2019.
47. Wang, Y., de Kruijff, R.M., Lovrak, M., Guo, X., Eelkema, R., van Esch, J.H., Access to Metastable Gel States Using Seeded Self-Assembly of Low-Molecular-Weight Gelators. *Angew. Chem. Int. Ed.*, 58, 3800–3803, 2019.
48. Cao, X., Ding, Q., Li, Y., Gao, A., Chang, X., Continuous multi-channel sensing of volatile acid and organic amine gases using a fluorescent self-assembly system. *J. Mater. Chem. C*, 7, 133–142, 2019.
49. Cao, X., Ding, Q., Zhao, N., Gao, A., Jing, Q., Supramolecular self-assembly system based on naphthalimide boric acid ester derivative for detection of organic amine. *Sens. Actuators B: Chem.*, 256, 711–720, 2018.
50. Peng, P., Li, Y., Song, W., Yu, X., Self-healing organogels and hydrogels constructed by self-assembled bis-terpyridine complex with selective metal ions. *Colloids Surf. A: Physicochem. Eng. Asp.*, 589, 124439, 2020.
51. Lin, Q., Zheng, F., Lu, T.-T., Liu, J., Li, H., Wei, T.-B., Yao, H., Zhang, Y.-M., A novel imidazophenazine-based metallogel act as reversible H2PO4—Sensor and rewritable fluorescent display material. *Sens. Actuators B: Chem.*, 251, 250–255, 2017.
52. Sharma, M., Sarma, P.J., Goswami, M.J., Bania, K.K., Metallogel templated synthesis and stabilization of silver-particles and its application in catalytic reduction of nitro-arene. *J. Colloid Interface Sci.*, 490, 529–541, 2017.
53. Tam, A.Y.-Y. and Yam, V.W.-W., Recent advances in metallogels. *Chem. Soc. Rev.*, 42, 1540–1567, 2013.
54. Bhattacharjee, S. and Bhattacharya, S., Pyridylenevinylene based Cu 2+-specific, injectable metallo (hydro) gel: Thixotropy and nanoscale metal-organic particles. *Chem. Commun.*, 50, 11690–11693, 2014.
55. Sasmal, S., Maji, K., Díaz, D.D., Haldar, D., Highly selective metallogel from 4-biphenylcarboxy capped diphenylalanine and FeCl3. *CrystEngComm*, 21, 4289–4297, 2019.
56. Peng, H., Lv, Y., Wei, G., Zhou, J., Gao, X., Sun, K., Ma, G., Lei, Z., A flexible and self-healing hydrogel electrolyte for smart supercapacitor. *J. Power Sources*, 431, 210–219, 2019.

57. Wang, J., Liu, F., Tao, F., Pan, Q., Rationally designed self-healing hydrogel electrolyte toward a smart and sustainable supercapacitor. *ACS Appl. Mater. Interfaces*, 9, 27745-27753, 2017.
58. Zhang, F., Xia, Y., Wang, L., Liu, L., Liu, Y., Leng, J., Conductive Shape Memory Microfiber Membranes with Core–Shell Structures and Electroactive Performance. *ACS Appl. Mater. Interfaces*, 10, 35526–35532, 2018.
59. Zhao, Q., Zou, W., Luo, Y., Xie, T., Shape memory polymer network with thermally distinct elasticity and plasticity. *Sci. Adv.*, 2, e1501297, 2016.
60. Li, W., Liu, Y., Leng, J., Programmable and shape-memorizing information carriers. *ACS Appl. Mater. Interfaces*, 9, 44792–44798, 2017.
61. Bai, J. and Shi, Z., Dynamically cross-linked elastomer hybrids with light-induced rapid and efficient self-healing ability and reprogrammable shape memory behavior. *ACS Appl. Mater. Interfaces*, 9, 27213–27222, 2017.
62. Yan, X., Wang, F., Zheng, B., Huang, F., Stimuli-responsive supramolecular polymeric materials. *Chem. Soc. Rev.*, 41, 6042–6065, 2012.
63. Huang, W., Yang, B., Zhao, Y., Ding, Z., Thermo-moisture responsive polyurethane shape-memory polymer and composites: A review. *J. Mater. Chem.*, 20, 3367–3381, 2010.
64. Liu, D., Nie, W.-C., Wen, Z.-B., Fan, C.-J., Xiao, W.-X., Li, B., Lin, X.-J., Yang, K.-K., Wang, Y.-Z., Strategy for constructing shape-memory dynamic networks through charge-transfer interactions. *ACS Macro Lett.*, 7, 705–710, 2018.
65. Du, L., Xu, Z.-Y., Fan, C.-J., Xiang, G., Yang, K.-K., Wang, Y.-Z., A fascinating metallo-supramolecular polymer network with thermal/magnetic/light-responsive shape-memory effects anchored by Fe3O4 nanoparticles. *Macromolecules*, 51, 705–715, 2018.
66. Cui, X., Chen, J., Zhu, Y., Jiang, W., Natural sunlight-actuated shape memory materials with reversible shape change and self-healing abilities based on carbon nanotubes filled conductive polymer composites. *Chem. Eng. J.*, 382, 122823, 2020.
67. Shi, Y. and Chen, Z., Function-driven design of stimuli-responsive polymer composites: Recent progress and challenges. *J. Mater. Chem. C*, 6, 11817–11834, 2018.
68. Zhang, Z.-x., Dou, J.-x., He, J.-h., Xiao, C.-x., Shen, L.-y., Yang, J.-h., Wang, Y., Zhou, Z.-w., Electrically/infrared actuated shape memory composites based on a bio-based polyester blend and graphene nanoplatelets and their excellent self-driven ability. *J. Mater. Chem. C*, 5, 4145–4158, 2017.
69. Thakur, S. and Karak, N., Multi-stimuli responsive smart elastomeric hyperbranched polyurethane/reduced graphene oxide nanocomposites. *J. Mater. Chem. A*, 2, 14867–14875, 2014.
70. Si, Q., Feng, Y., Yang, W., Fu, L., Yan, Q., Dong, L., Long, P., Feng, W., Controllable and stable deformation of a self-healing photo-responsive supramolecular assembly for an optically actuated manipulator arm. *ACS Appl. Mater. Interfaces*, 10, 29909–29917, 2018.

71. Wu, Y., Hu, J., Zhang, C., Han, J., Wang, Y., Kumar, B., A facile approach to fabricate a UV/heat dual-responsive triple shape memory polymer. *J. Mater. Chem. A*, 3, 97–100, 2015.
72. Bag, M., Banerjee, S., Faust, R., Venkataraman, D., Self-healing polymer sealant for encapsulating flexible solar cells. *Sol. Energy Mater. Sol. Cells*, 145, 418–422, 2016.
73. Chen, J.D., Cui, C., Li, Y.Q., Zhou, L., Ou, Q.D., Li, C., Li, Y., Tang, J.X., Polymer Solar Cells: Single-Junction Polymer Solar Cells Exceeding 10% Power Conversion Efficiency (Adv. Mater. 6/2015). *Adv. Mater.*, 27, 1132–1132, 2015.
74. Zhou, H., Zhang, Y., Mai, C.K., Collins, S.D., Bazan, G.C., Nguyen, T.Q., Heeger, A.J., Polymer homo-tandem solar cells with best efficiency of 11.3%. *Adv. Mater.*, 27, 1767–1773, 2015.
75. Krebs, F.C., Gevorgyan, S.A., Alstrup, J., A roll-to-roll process to flexible polymer solar cells: model studies, manufacture and operational stability studies. *J. Mater. Chem.*, 19, 5442–5451, 2009.
76. Xu, T. and Yu, L., How to design low bandgap polymers for highly efficient organic solar cells. *Mater. Today*, 17, 11–15, 2014.
77. Krebs, F.C., Polymer solar cell modules prepared using roll-to-roll methods: Knife-over-edge coating, slot-die coating and screen printing. *Sol. Energy Mater. Sol. Cells*, 93, 465–475, 2009.
78. Jørgensen, M., Norrman, K., Gevorgyan, S.A., Tromholt, T., Andreasen, B., Krebs, F.C., Stability of polymer solar cells. *Adv. Mater.*, 24, 580–612, 2012.
79. Ran, C., Gao, W., Li, J., Xi, J., Li, L., Dai, J., Yang, Y., Gao, X., Dong, H., Jiao, B., Conjugated Organic Cations Enable Efficient Self-Healing FASnI3 Solar Cells. *Joule*, 3, 3072–3087, 2019.

7

Self-Healable Core–Shell Nanofibers

Sonia Bahrani[1], Seyyed Mojtaba Mousavi[2], Seyyed Alireza Hashemi[3], Chin Wei Lai[4*] and Wei-Hung Chiang[2]

[1]*Department of Medical Nanotechnology, School of Advanced Medical Sciences and Technologies, Shiraz University of Medical Sciences, Shiraz, Iran*
[2]*Department of Chemical Engineering, National Taiwan University of Science and Technology, Taipei, Taiwan*
[3]*Department of Mechanical Engineering, Center for Nanofibers and Nanotechnology, National University of Singapore, Singapore*
[4]*Nanotechnology & Catalysis Research Centre, University of Malaya, Kuala Lumpur, Malaysia*

Abstract

Self-healing issue is the most interesting and desirable characterizes of engineered materials that is inspired from nature. The self-healing issue was developed to create engineering materials with high durability and reliability features. Encapsulation of the healing agents is considered as the more promising methods in self-healing approach. According to this, the healing agents are embedded in micro or nano capsules, vascular-based networks, and micro/nano fibers. Among these approaches, the use of core–shell nanofibers represents high promising potential to develop the self-healing nanofibers with the minimal limitations and shortcomings. The purpose of this chapter is to survey of such agents to give more insights to make progress in construction of self healable core–shell nanofiber agents and also use of them in further various areas.

Keywords: Encapsulation, core–shell nanofiber, vascular-based network, self-healing

**Corresponding author*: cwlai@um.edu.my

Inamuddin, Mohd Imran Ahamed, Rajender Boddula and Tariq Altalhi (eds.) Self-Healing Smart Materials and Allied Applications, (181–202) © 2021 Scrivener Publishing LLC

7.1 Introduction

Self-healing technologies are the interesting issues which are well-understanding from nature (such as healing processes in our bodies) and many attempts have been done to give these same characterizations toward the engineering materials [1–4]. Biological systems including humans can be self-healed in various degrees, and this behavior is retained in the vascular system [5, 6]. A mechanism of coagulation kicks following a minor injury immediately, leading to hemostasis, in which the extent of the injury is reduced. Such biological systems based on self-healing ensure humans healthy lives for longer times [7]. Similar to such mechanisms, self-healing compounds have been developed as promising candidates to generate the engineering materials with extensive service life, reduced maintenance and repair cost as well as providing durable and reliable improvement [8]. These merits can be accounted as the features that attracted the researcher's notice toward self-healing systems [9, 10]. This technology exhibits a great potential for using in advanced industries, such as using in aerospace with high assurance factor [11]. Disastrous failures could occur in general constructions and facilities, and also prostheses could subject to weariness. In this respect, the resilience of these materials makes them even more reliable to employ in acute applications such as in the structural components of aircrafts [12]. In general, the accumulation of damages caused by disastrous failures resulted in delamination or propagation of cracks, while, self-healing approach can be favored [13]. In particular, self-recovery methods can be used as alternative method as protective layer in the systems exposure to corrosion and heal the damaged part instead of using chemical toxic coatings or galvanic anode [14].

There have been numerous strategies suggested to make smart materials with the ability of self-repairing such as the releasing of healed agents, shape memory effect and the reversibility of cross-links, among which the first one has gained the most attention [9].

At initial stage, hollow-fibers and micro-capsules were utilized for self-healing systems [15], which potentially weaken the stability of the structure [16]. The critical parameters in an effective healing process is based on transfer of enough healing agents to the damaged parts and several repeated healing, in the cases of microcapsules with lack of these abilities [17]. The repeated healing concept can be achieved by embedding additional healing agents in the desired material to certify further improvements. However, in the matter of vascular network systems in which the tubes or fibers were interconnected, the repairing agents could be reserved to similar part for multiple times to heal repeatedly. In this regard, initially large-sized tubular network was suggested to be able to reserve the extra amount of healing agents [18, 19]. However, due to their influence on the decreasing mechanical stability of compounds, filling hollow glass tubulars with repairing agent was suggested that incorporated within the epoxy resins reinforced by carbon fiber and glass fibers [10, 20, 21]. This phenomenon resulted in formation of compounds which recovered its previous strength.

There are several methods to imbed self-repairing agents in various composites and materials. Fabrication of capsule-based or fiber-based healing systems have been well demonstrated for embedding self-healable agents in materials [9]. Figure 7.1 shows these processes and the related mechanism of releasing healing agents.

The encapsulation of healing materials in core–shell nanofiber systems initially was favored in 2010 [22] due to its ability to overcome problems of encapsulation strategies based on capsule-based or microchannel based systems from 2001. This current chapter aims to survey the main developments on creation of self-repairing materials through synthesizing core–shell fibers, especially nano-sized ones. Accordingly, such core–shell fibers have been utilized to embed the self-repairing agents. Releasing of repairing agents from the matrixes initiate coherency and can go through the redox or cross-linking reactions along refilling the gaps.

7.2 Self-Healing Polymers in Fabrication of Core–Shell Nanofibers

The materials with healing capability can create physical barriers and supply *in-situ* heal or repairing abilities after being exposed to damage.

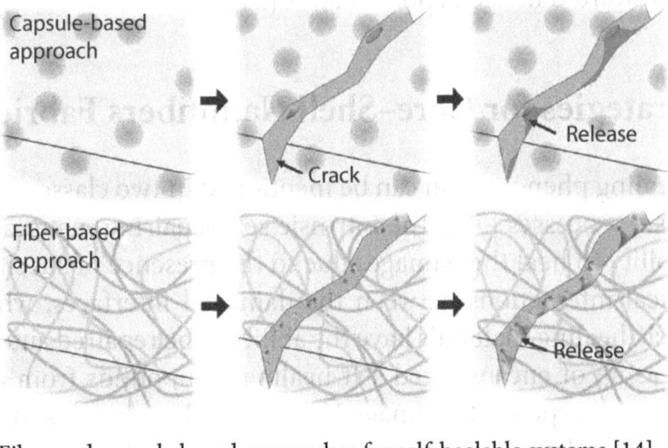

Figure 7.1 Fiber and capsule based approaches for self-healable systems [14].

Multiple materials have been recommended as healing agents candidates including polymeric hydrogels and/or supramolecular polymers, which can recover broken linkages via self-assembly processes [23]. Polydimethylsiloxane has extensively been applied as repairing agent owing to the benefits of its biocompatibility, mechanical resistance and chemical stability [24–26]. Besides, polydimethylsiloxane is flexible and transparent which can be used in electronic and optical systems, includes solar cells, touchscreens and smart tools. However, it takes time to completely cure (about 24–48 h) and after that the soft and flexible material is achieved, which is not suitable to use in fast response and stiffness material [27]. Accordingly, polydimethylsiloxan as the repairing material has been incorporated into the cores of nanofibers, in which polyetheramine and bisphenol-A diglycidyl ether of epoxy resin were separately dispersing and electrospinning in polymeric fibers. Electrospun core–shell nanofibers were fabricated using dimethylpolysiloxan as repairing capability material and polyvinylpyrrolidone as its shell. Polymer fibers were dispersed in the modified polyurethane with acrylate and then co-electrospinning method was utilized to deposit it on the steel substrates [28]. The yield of such method was confirmed by incorporating the healing agents in nano-sized hollow fibers.

Polymethylmethacrylate was also favored for encapsulating healing agents due to its chemical stability and well mechanical features. Accordingly, polymethylmethacrylate is employed to encapsulate healing agents in carbon fiber-epoxy composite through simple, economical and cost effective co-axial electrospinning approach. The mechanical robustness and excellent compatibility of polymethylmethacrylate nanofibers with the epoxy resin matrix enhances the interfacial stiffness effect while the low strain allows to heal to be initiated via nanofiber rupture with high sensitivity toward damaged epoxy matrix [29].

7.3 Strategies for Core–Shell Nanofibers Fabrication

The self-healing phenomenon can be mentioned in two classes of extrinsic and intrinsic processes [30]. The intrinsic self-repairing materials possess high capability to heal the damages due to the presence of specific reversible chemical interactions between the damaged interfaces, whereas the extrinsic self-healing materials have the ability to be repaired automatically by the releasing of encapsulated self-healing compounds from an encapsulated polymer exposed to damage [3]. Due to the advances in designing new materials and measuring process, the extrinsic self-healing technique

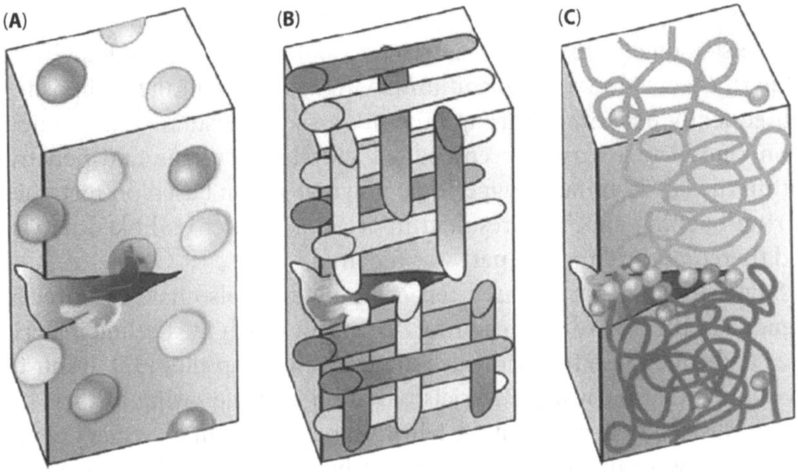

Figure 7.2 The methods for the construction of self-repairing materials. Capsulation-based approach (A), vascular-based process (B), intrinsic procedure (C) [34].

gained more attention. As example, at initial studies, encapsulated dicyclopentadiene has been employed as the self-healing agent which is released alongside the Grubbs' catalyst by crack propagating leads to the activation of a polymerization process and the gap is healed and refilled by the prepared polymer [31–33]. In the wide range of cases, the treatment materials are embedded into engineering designed materials via formation capsulation or providing a network of vascular system. Figure 7.2 represents the approaches for improving self-repairing potency in the polymeric-based compounds.

7.3.1 Capsule-Based Self-Healing

Since self-repairing method based on encapsulation of healing materials was suggested by White *et al.* [35], many improvements have been achieved over the years, in respect of self-repairing abilities of materials in basis of encapsulation and fabrication of microcapsules by incorporating various healing materials [31].

As an example, *in-situ* polymerization approach was employed using formaldehyde-urea to create microsized-capsules with 220 μm of outer diameter size and thereafter their self-healing properties were evaluated [33]. It is worth noting that *in-situ* formaldehyde-urea polymerization technique that was applied by Dietrich research group [36], enables scientifics to manage the size of microcapsules by changing the stimulation rate. In this process, liquid dicyclopentadiene embedded microcapsules were distributed in the epoxy matrice, in contact with embedded Grubbs' catalyst. By creation a crack and initiate to propagate in the composite, it subsequently penetrated through the microsized-capsules. That phenomenon caused to release the embedded dicyclopentadiene and penetrate within the crack. In fact, triggering a polymerization based on metathesis of ring-opening approach resulted in creation of a cross-linked network and therefore the damaged part was repaired.

Silyl ester and siloxan-based repairing materials also have been recommended as the core of self-repairing systems in which the healing materials were embedded into the formaldehyde-urea microcapsules [37]. It must be mentioned that repairing materials based on siloxane owing to their high chemical stability throughout the fabrication process can be actively applied in fabrication of nano-sized scale of self-healing hollow fibrous materials. Accordingly, self-repairing systems based on two types micro-capsules were suggested [37]. The first one was 60 μm formaldehyde-urea microcapsules which embedded healing agents based on siloxane, whereas the next one was polyurethane microcapsules in 90 micrometer size which incorporated a catalyst named di-methyl di-neodecanoate tin. After being embedded in an epoxy resin, the self-healing composite was formed by microcapsules. In similar work [38], silica/polyurea hybridized microcapsules in purpose of depositing of repairing materials in basis of isocyanate was fabricated. The schematic of this process was illustrated in Figure 7.3.

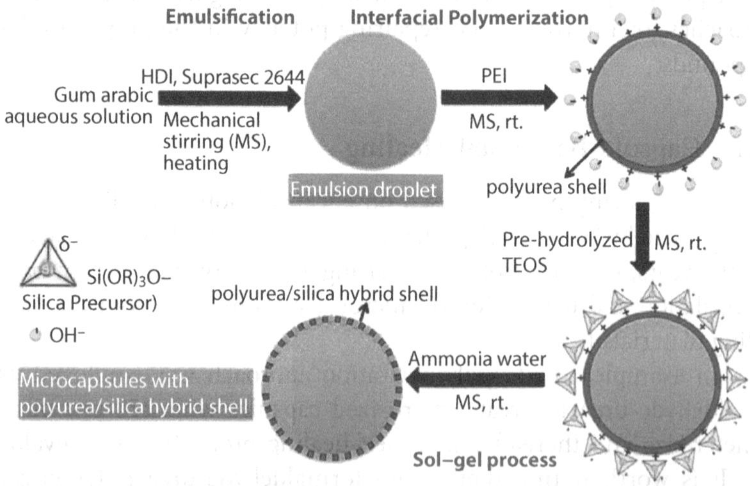

Figure 7.3 View of fabrication procedure of silica/polyurea hybridized self-repairing microcapsules [38].

The methods based on encapsulation process of the healing agents adopted with high fabrication costs, the weak linkage between the capsules and their non-uniformly dispersion in the adjacent matrice. Besides, the longitudinal size of most of the capsules was in the area of 10 μn, that is heavily structured to convey and inconsistent with scalable typical material, particularly for transport usage in tanks, aircraft and automobiles. Besides these drawbacks, self-healing systems based on encapsulation possess the higher expense of the incorporation methods, the non-uniformly dispersion of healing materials in the matrices, and lack of available repairing agents for subsequent of recovery process which was consumed once [39].

7.3.2 Vascular-Based Self-Healing

Self-healing system based on creation of vascular network has gained dramatic attention regarding design of self-repairing materials. Initially, Dry suggested a self-repairing approach based on hollow tubes in 1996 [19]. Accordingly, polymeric matrix incorporated in thin-pipette and then, the two components of an adhesive based on cyanoacrylate injected to two different pipettes. Following creating cracks by bending or impact in the polymeric matrix, self-healed materials released from the pipettes and the two joined components cross-linked, thus avoiding both formation and progressing of cracks in the damaged matrix [40]. Fiber-based self-healing incorporation was similar to healing agent embedding hollow fibers. In fact, such vascular self-repairing systems was the resemblance of the hollow tubes to vessels of blood. Based on type of healing agents and fabrication method, the diameter-size of 1-D hollow tubes in vast range of hundreds of micrometers to hundreds of nanometers in structures. Most researches on self-repairing hollow microtubes were explained in initial stages of studies, while those studies on hollow nanofibers have appeared in current years. In addition, the efficiency of the mechanical features due to hollow nano fibers-based self-healing materials and their corrosion protection capability were studied.

The vascular healing systems can be employed in reinforcing fiber polymer materials in the auto or aerospace industries. In such use, it is useful to decrease lateral scale of vascular fibers in order to reduce their interfering toward the structure of compound. In that regard, solution blowing and electrospinning can be utilized to create core–shell networks in micro or nano fibers containing healing agents. Both of the processes have extensively been used in the recent decade. To study the ability of repetitious self-repairing in the epoxy matrice, triaxial electrospinning was performed using dicyclopentadiene and its respected catalyst associated with epoxy resin and its curing agent such as amine-based material all as repairing agents [41]. These researches supported the efficacy of nano-fibrous tubes repeated process of healing.

7.4 Methods of Fabrication of Self-Healing Core–Shell Nanofibers

On account of low cost associated with acceptable efficiency, electrospinning is considered as the feasible and viable technique to produce continuous nanofibers with uniform structures [42]. There are two electrospinning methods can be utilized to create core–shell nanofibers networks packed with repairing materials, namely coaxial [43] and emulsion electrospinning [44–47].

7.4.1 Co-Electrospinning

It is worth noting that the electrospinning strategy is recognized as the simplest and most low-cost way to prepare of the nanofibers, which approves a inherent material construction for a large area of usage involving filtration applications [48, 49], engineering tissue [50], and in the advances and improving the mechanical characteristics of the composite [51, 52]. Various factors direct the electrospinning approach and influence on the characteristics and qualification of the nanofibers as well as their properties including porosity and diameter [53]. These factors are categorized as parameters of solution and materials (such as conductivity, viscosity and dielectric constant), processing factors (including the rate of solution feed, the used voltage and the distance from tip to needle), and also the environmental conditions such as humidity and temperature [54]. It is relatively low-cost and simple fabrication method to improve self-repairing materials. Nonetheless, this approach was proper only for interlayers and thin-scale of samples.

Self-healing core–shell-structured nanofibers were fabricated using electrospinning method and the beads were formed on nanofibers [28]. Two coaxial nozzle systems were employed for co-electrospinning, which was illustrated in Figure 7.4.

In one work, the core of fibers was filled by either hardener or resin of healing agent systems based on siloxan, while the solution of polyvinylpyrrolidone was utilized as the shell of system. Then, the co-electrospin nanofiber was incorporated to polymeric matrice. The outer diameters such fabricated core–shell nanofibers and beads were hundreds of nano- and

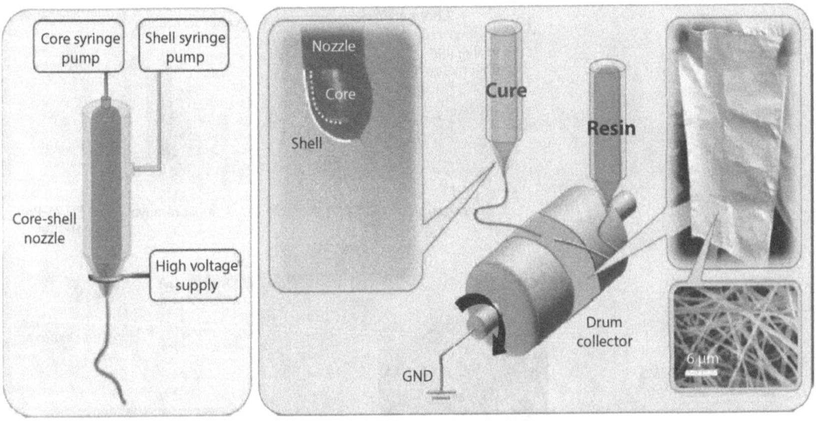

Figure 7.4 View of coaxial electrospun method [43].

micro-meters, respectively. Core–shell nanofibers containing healing agents were constructed and treated catalysts were individually introduced to the cores of the various nanofibers via coaxial electrospinning of the shell and core compounds. Once, the core–shell fibers were cut, the curing materials and healing agents were triggered, and along with polymerization create a shielding layer [27].

Electrospun core–shell nanofibers containing two-components polydimethyl siloxan (the monomer of resin and curing agent) as repairing materials were directly capsulated into the 3-D sponge-shape bulky sample matrices of uncured polydimethyl siloxan and dispersed within there (Figures 7.5(a–b)) [55]. The cross-sectional SEM and optical microscopic images of the samples were illustrated in Figures 7.5(c–e). The sponges displayed a 100% efficiency and recovery after encountering damage in the shearing assays.

In one work, amine and monomer of epoxy were encapsulated in core–nanofibers shell via co-axial electrospinning, in which the influence of different repairing agent materials on the capability of repairing was studied [56, 57]. Their results demonstrated any adverse effect of obtained nanofibers encapsulated in the epoxy matrix on the flexural features of the epoxy resin, whereas the presence of encapsulated nanofibers improved dramatically the resistance of epoxy toward delamination and proved the self-repairing behavior of the proposed composite.

Co-axial electrospinning method was utilized to create two kinds of core-fibers shell using polyacrylonitrile (PAN) in the shell; the first one using monomer of resin as repairing material in core and the latter the cure agent in its core [43]. Both of nanofibers were incorporated into two

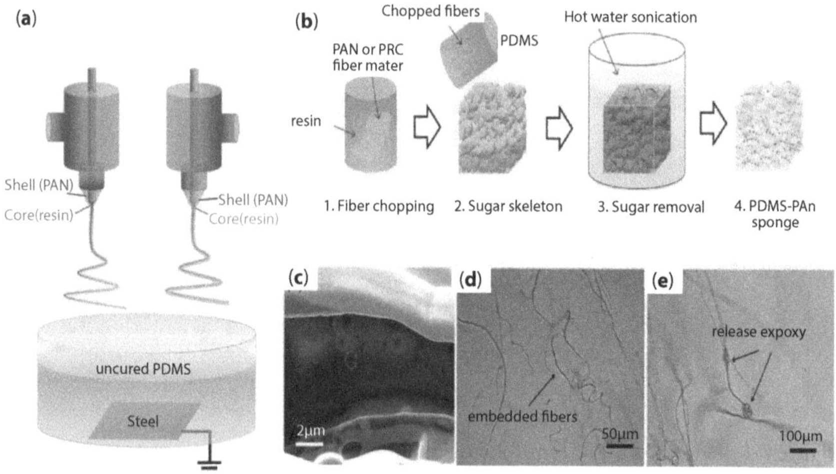

Figure 7.5 (a) Dual coaxial electrospinng core–shell fibers incorporated into uncured polydimethylsiloxane, (b) Fabrication process of self-repairing sponge, (c) The inner SEM image of 3-D compound. Several cutting edges of nanofibers containing self-repairing materials are defined by red circles, (d) The optical microscopic image of undamaged fibers, (e) and the released epoxy in a 3-D sample after exposing mechanical damaging [55].

transparent matrices, either polymer SU-8 or polydimethyl siloxan. The SEM and TEM images of fibrous materials were shown in Figures 7.6(a–b). The transmittance of fabricated nanofibers (400–800 nm of diameters of fibers) was nearly 90% (Figures 7.3(c) and (d)). The ability of healing of such compounds were assayed by its protection against corrosive.

7.4.2 Emulsion Electrospinning

Emulsion electrospinning has been applied for encapsulation the healing agents in hollow nanofibers in which the emulsified compound in the solution of polymer creates the core of the obtained hollow nanofibers, while the other compound dispersed in suspension of polymer creates the shell of prepared core–shell nanofibers [45]. Even in the need of emulsification of the core material in the polymer solution, core–shell nanofibers were formed from a one nozzle. This approach is more appropriate in comparison with co-electrospinning method in which co-axial nozzles were applied to issue two solutions simultaneously. Self-repairing abilities of silicone-based [43, 45] and epoxy-based [44] coatings were surveyed. Figure 7.7 demonstrated schematically the encapsulation of repairing material using emulsion electrospinning and improving coating based on epoxy.

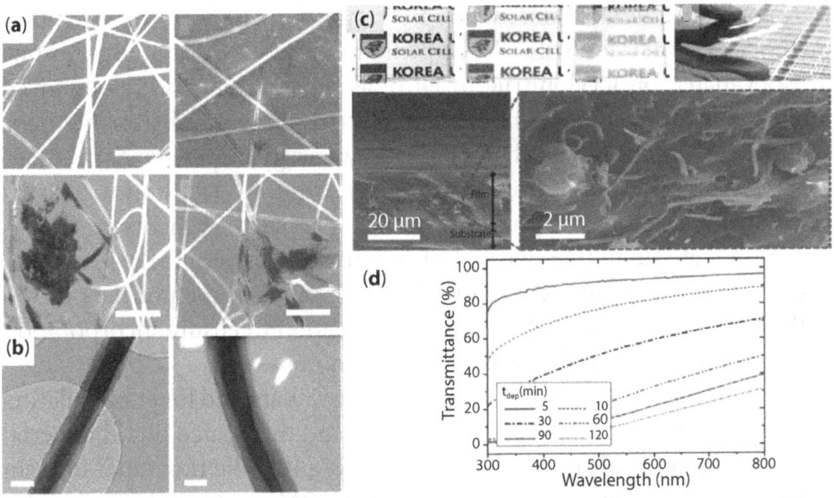

Figure 7.6 (a) SEM of resin-PAN (left), Cure-PAN (right) core–shell nanoscale fibers and the ruptured nanofibers of Resin-PAN (left, bottom) and Cure-PAN (right, bottom). (b) TEM of resin-PAN (left) and Cure-PAN (right) core–shell nanoscale fibers. (c) Transmittance of nanofiber film on glass versus wavelength at various deposition time. (d) SEM images of cross sectional of film at tdep = 120 min [43].

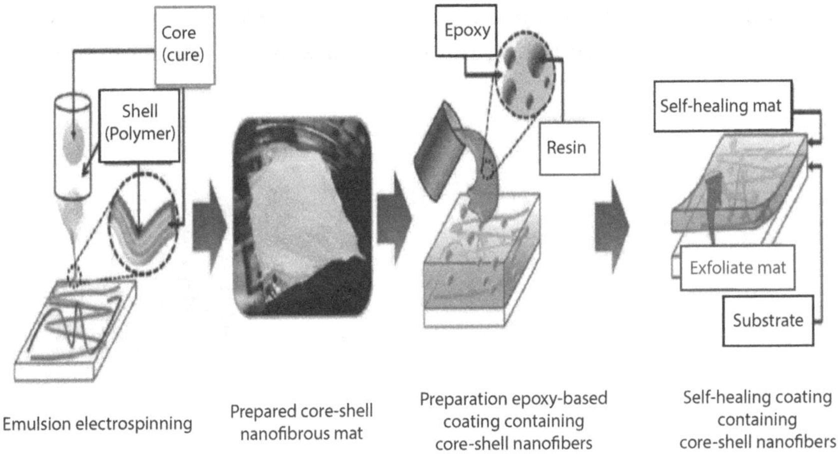

Figure 7.7 Fabrication of core–shell nanofiber using emulsion electrospinning approach to create a self-repairing coating based on epoxy [44].

Hollow nanofibers from 450 to 1,000 nm diameter sizes was fabricated using emulsion-solution blowing, co-electrospinning and emulsion electrospinning [58, 59]. In co-electrospinning, a co-axial setup was used in which dispersed polyacrylonitrile in DMF and DCPD mixed with DMF were applied as the shell and as the core, respectively [60]. For the two other methods, polyacrylonitrile/dicyclopentadiene in DMF and isophorone di-isocyanate/polyacrylonitrile in DMF as the emulsions were employed to create uniform solution-blown and electrospun core–shell nanofibers. The core of materials was characterized by energy dispersive X-ray analysis (EDXS), optical microscopy as well as scanning electron microscopy (SEM). Furthermore, they used a diffusion process to incorporate self-repairing material in free space of carbon nanotubes with the size of inner diameter of nanometer. Regarding this, carbon nanotube powders mixed with DCPD solution and dispersed by sonication to ensure incorporation of the repairing material into the carbon nanotubes.

Dual-emulsion electrospinning technique was applied for encapsulating simultaneously the resin based on siloxane and the treated liquid material as repairing materials in the nanofibers cores [46]. The monomer of resin in hexane and the emulsified curing liquid in DMF/polyacrylonitrile solution simultaneously electrospun of dual distinct nozzles to get core–shell self-repairing nanofibers including either curing liquid/resin in the core of fibers and formation of polyacrylonitrile as the shell of the nanofibers (Figure 7.8). The nanofibers had 444 nm outer diameter. the core liquid was affirmed with differential thermos gravimetry (DTG) and TGA.

A different approach was employed to construct self-repairing core–shell nanofibers incorporating the composites. Shell (polyacrylonitrile)/

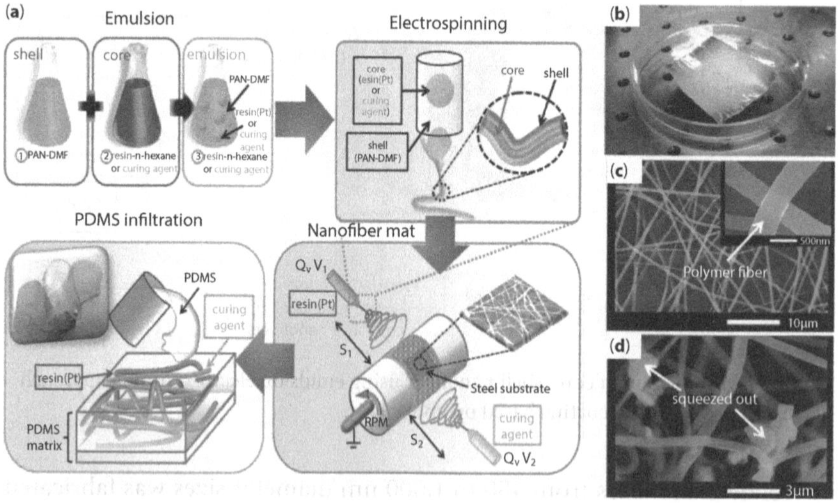

Figure 7.8 Representation of fabrication nanofibers healing materials, (b) Macroscopic nanofiber mat, (c) Magnified nanofibers within mat, (d) Squeezed out of core materials [46].

core (curing liquid) nanofibers electrospin on a metallic substrate employing a polyacrylonitrile/curing liquid emulsion [44, 47]. Then, the epoxy/resin emulsion poured on the nanofibers deposited on the metal substrate and the capsule resin microdroplets were created within the epoxy matrix. Therefore, a heterogeneous matrice consist of core/shell nanofibers and the microdroplets of capsule were created (Figure 7.9a). The optimized ratio of curing liquid to resin was achieved to be 1:10. The morphology of core–shell nanofibers were represented in Figures 7.9(b–f). The corrosion protection using the self-repairing composites was analyzed by electrochemically measurements.

Though the emulsion electrospinning is the simple method to construct core–shell nanofibers, significant drawbacks of that process are the slow formation of core, disordered formation of healing agents in the emulsion and the difficulties in selection materials. To certify that the process of core creation is powerful and to avoid beads creation onto the fiber, the influence of the sizes of internal core to external shell ratio were optimized to 0.4. this ratio leads to well core–shell structure, which characterized by SEM, TGA, and DTG. The transmission-electron microscopy (TEM) also was utilized to identify the core-shell composition of the self-repairing nanofibers. Construction of self-repairing compounds via electrospinning

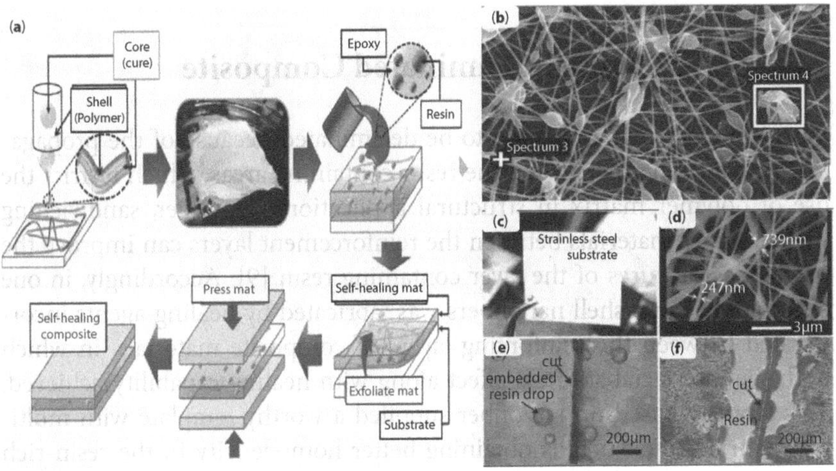

Figure 7.9 (a) View of creation of self-repairing materials embedding the core–shell nanofibers and microdroplets of capsule, (b) SEM images of nanofibers containing cure material in core, (c) Macroscopy image, (d) SEM image of mat of core nanofiber–shell containing encapsulated cure in core, (e) Optical image from a film cut of solidified epoxy containing encapsulated resin droplets, (f) Released resin [47].

approach within 5 min was demonstrated greatly lucid with the transparency of 93% and remarkably anticorrosion efficiency within corrosion trials.

7.4.3 Solution-Blown

Solution blowing is a process to spin nanofibers from solutions of polymer via high flow velocity of gas to form fiber [61, 62]. Healing agents embedded in self-healable polymers including polyethylene terephthalate and polydimethylsiloxane polymers (PDMS) form encapsulated core/shell nanofibers [27, 63]. In this research, the repairing materials was incorporated into fibers thorough the approach of solution blow. Such procedure as a flexible technique for changing the size in preparation process of core-shell fibers and the fibers in the nano/micro ranges can be achieved. The fatigue and static analysis were utilized to illustrate the self-repairing potency of such systems.

Epoxy monomer and hardener was employed and embedded individually in the cores of micro- and nanofibers that were arrayed in healing agent system via solution blowing of the polymer known as polyvinylidene difluoride polymer using compressed air. Once activated, epoxy starts to harden (about 5 min) and cured completely within an hour [27].

7.5 Self-Healing in Laminated Composite

Fiber–polymer materials tend to be delaminated because of the propagation of nano/micro cracks in the resin-containing areas, which restrict the use of polymer matrix in structural applications. However, sandwiching the nanofiber materials between the reinforcement layers can improve the mechanical features of the layer containing resin [9]. Accordingly, in one example the core–shell nanofibers was fabricated by healing agents incorporated between the reinforcing layers of composite materials, in which both the interfacial stiffness effect along with healing capability achieved. This strategy based on nanofiber supplied a worthy template with multiple main benefits such as obtaining better homogeneity in the resin-rich parts, by passing the reduction of mechanical properties due to embedding of large-sizes fibers, and increasing significantly in the volume ratio of the available healing agent to treat the creation nano/micro gaps in the resin-rich part to enhance the recovery cycles. In fact, incorporation of nanofibers among reinforcing layers results in improved mechanical features associated with the fact that incorporating healing agents in nanofibers is both simple and economical [64]. In addition, the shell wall material has a well chemical compatibility with the matrix and a low strain threshold.

Reinforcing effects of the nanofibers were traced upon embedding of nanofiberous materials among the reinforcing layer of polymer matrice composites [65–67]. It also has been revealed that core–shell nanotubes including repairing agents has a healing effect in mechanical reinforcement [68]. Using such nanofibers for self-repairing fiber reinforced polymer composite was primarily investigated by Ray *et al.* in 2012 [60]. This research group encapsulate isophorone diisocyanate and dicyclopentadiene in the nanofibers using emulsion-electrospinning procedure and the polyacrylonitrile shell of such fibers. Accordingly, the repairing properties of the core–shell nanofiber materials were evaluated by incorporating the electrospun fibrous materials among the fourth and third layers of common epoxy/carbon compound. Following cracking the compound, both resin and curing agent able to release within 24 h. The layer of the nanoscale fibers which rich in resin was detected using the microscopic procedure to certify releasing the repairing compound.

Regarding, in the similar work, encapsulation of dicyclopentadiene in polyacrylonitrile nanofibers was reported, in which the nanofiberous material were located among the layers of carbon in the composite and the Grubbs' catalyst was distributed in the matrice of epoxy [69]. To assess the self-healing performance, three-point bending experiment was performed. The results demonstrated nearly 100% recovery of rigidity after damage and respective healing process. The repairing action of nanofibers in interlayers was also confirmed by SEM.

The hollow nanofibers of polyacrylonitrile using epoxy resin and also curing agent of amine between five inter-layers of epoxy/carbon composites was indicated [17]. The mechanism of the self-repairing process and respective FeSEM and TEM characterization of total repairing approaches in the reinforcing composite involving rupture of nanofiber, trigger and solidification of repairing agents was schematically illustrated in Figure 7.10. They assessed self-healing performance and mechanical properties of the compound. In the latter work, epoxy and its curing material herein amine-based ones were embedded in the nanofiber using the shell of poly (styrene-acrylonitrile) [70]. They realized that poly (styrene-acrylonitrile) resulted in greater yield and more extent of encapsulated material in comparison to polyacrylonitrile, even so, the more value of repairing agents did not increase the repairing efficiency. The shell of Poly (methyl methacrylate) was employed to embed the curing agent of amine and epoxy resin with high performance of self-healing [29]. They illustrated that reinforced core–nanofiber shell, the diameter size of nanofiber has the notable

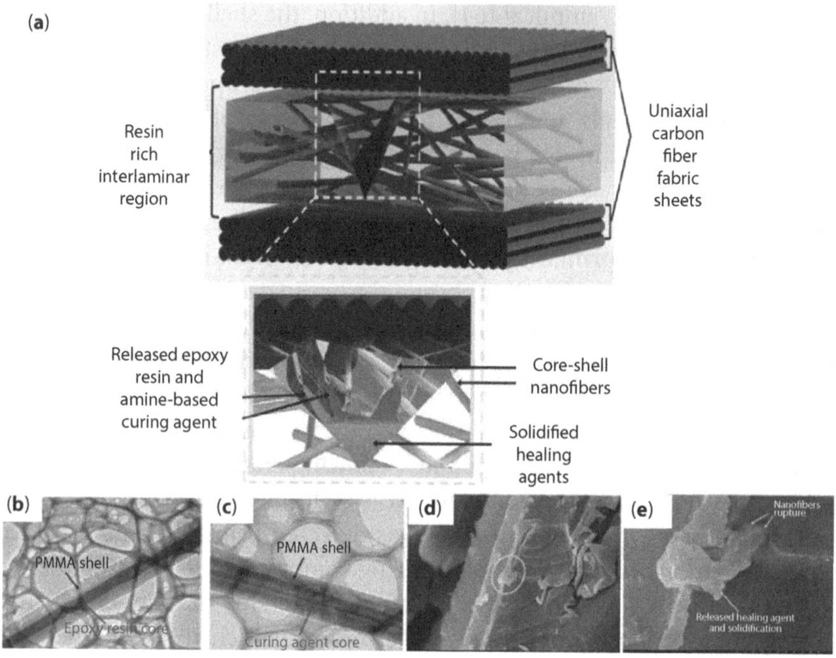

Figure 7.10 (a) Core–nanofiber shell with self-healing capability into the laminated structure. (b, c) TEM of epoxy and amin based cure agent-PMMA embedded nanofibers. (d, e) FeSEM of releasing repairing agent from cut-open nanofibers in epoxy/carbon composite [29].

influence toward enhancing the mechanical features of the fiber and self-repairing capability.

Self-repairing epoxy/glass fiber composite was developed by depositing core-nanofibers shell embedding healing agents amongst the layers of glass fibers [71]. They employed resin of epoxy as a repairing agent and polymethyl meta acrylate as its outer shell in core–shell materials. However, utilization of inner shell prolonged the healing lifetime performance. The results confirmed enhancing self-healing ability and mechanical characterizes via embedding tri-axial nanofiberous materials within the epoxy/glass fibers composites.

7.6 Beneficial Self-Repairing Systems on Basis of Core–Shell Nanofibers

The inversely relationship between composite mechanical strength and the fibrous diameter size was investigated [72]. Large size of hollow tubes significantly has a side effect on the mechanical characteristics of the compound strength, which was well demonstrated [64]. Since 2010, extensive attempts have been done for encapsulation of repairing materials within small-sized of hollow tubes, pioneered by Park *et al.* [28], to overcome the difficulties along with the use of microcapsules and hollow microfibers. Due to the weight limitations associated to the basic composite materials, it is usually preferred to use thin layer of nanofibers instead of using small micrometer capsules. Likewise, electrospun nanotubes were used for micro-scaled tube replacement.

The method based on embedded healing agents in core–shell nanofiber has more benefits compared to the problems along with the capsules [73]. The important one is the scale size of fibers in the limit of a few hundred nanometers [35, 74, 75]. Also, the inter-connected and entangled fibers enable the effective transportation of repairing materials within the formed three-dimensional network utilization of the fibers [69, 71]. As well, such formed fibers have a lighter weight compared to capsules. Meanwhile, the fibers can be a part of more transparent materials.

In fact, encapsulation of repairing materials in nanofibers were significantly gainful for the impressive self-repairing compared to the self-repairing in basis of capsulation [76]. As a matter of fact, the interconnected nanofibers form a linked network with the capability of providing repairing materials to release in the damaged and depletion parts and subsequently facilitate its healing. A self-healing nanofiber with thin layer illustrated effective healing ability and recovery the mechanical features after tensile, delamination and bending damages.

7.7 Conclusion

Fabrication of core–shell fibers with nano-metric diameter size and incorporating healing agents into core-fibers shell via different methods including coaxial electrospinning, emulsion electrospinning and also solution blown techniques provide multiple advantages in comparison to encapsulation methods. In addition, owing to the cost-effectiveness of such approaches based on core–shell nanofibers, they can be used in the construction of self-healing materials.

References

1. Burattini, S. *et al.*, A self-repairing, supramolecular polymer system: healability as a consequence of donor-acceptor pi-pi stacking interactions. *Chem. Commun. (Camb.)*, 44, 6717–9, 2009.

2. Jones, A.R. *et al.*, Self-healing thermoplastic-toughened epoxy. *Polymer*, 74, 254–261, 2015.
3. Döhler, D. *et al.*, Principles of Self-Healing Polymers. Macromol. *Rapid Commun.*, 34, 203–220, 2013.
4. Liu, D. *et al.*, Core–shell molecularly imprinted polymer nanoparticles with assistant recognition polymer chains for effective recognition and enrichment of natural low-abundance protein. *Acta Biomater.*, 10, 2, 769–775, 2014.
5. Doan, T.Q. *et al.*, Characterization of core-shell microstructure and self-healing performance of electrospun fiber coatings. *Polymer*, 107, 263–272, 2016.
6. Sedghi, R. and Shaabani, A., Electrospun biocompatible core/shell polymer-free core structure nanofibers with superior antimicrobial potency against multi drug resistance organisms. *Polymer*, 101, 151–157, 2016.
7. Trask, R.S., Williams, G.J., Bond, I.P., Bioinspired self-healing of advanced composite structures using hollow glass fibres. *J. R. Soc. Interface*, 4, 13, 363–71, 2007.
8. Wool, R.P., Self-healing materials: A review. *Soft Matter*, 4, 3, 400–418, 2008.
9. Esmaeely Neisiany, R. *et al.*, Core–shell nanofibers for developing self-healing materials: Recent progress and future directions. *Mater. Des. Process. Commun.*, 1–7, 2019.
10. Trask, R.S. and Bond, I.P., Biomimetic self-healing of advanced composite structures using hollow glass fibres. *Smart Mater. Struct.*, 15, 3, 704–710, 2006.
11. Liu, J. *et al.*, A novel polychloromethylstyrene coated superparamagnetic surface molecularly imprinted core–shell nanoparticle for bisphenol A. *J. Mater. Chem.*, 21, 25, 9232–9238, 2011.
12. Chen, Y. and Guan, Z., Self-assembly of core–shell nanoparticles for self-healing materials. *Polym. Chem.*, 4, 18, 4885–4889, 2013.
13. Guerreiro, A.R. *et al.*, Selection of imprinted nanoparticles by affinity chromatography. *Biosens. Bioelectron.*, 24, 8, 2740–2743, 2009.
14. An, S. *et al.*, A review on corrosion-protective extrinsic self-healing: Comparison of microcapsule-based systems and those based on core-shell vascular networks. *Chem. Eng. J.*, 344, 206–220, 2018.
15. Ghosh, S.K. Self-Healing Materials: Fundamentals, Design Strategies, and Applications. Self-Healing Materials, pp. 1-28, 2008.
16. Zaidi, S.A., Molecular imprinted polymers as drug delivery vehicles. *Drug Deliv.*, 23, 7, 2262–2271, 2016.
17. Neisiany, R.E. *et al.*, Encapsulation of epoxy and amine curing agent in PAN nanofibers by coaxial electrospinning for self-healing purposes. *RSC Adv.*, 6, 74, 70056–70063, 2016.
18. Motuku, M., Vaidya, U.K., Janowski, G.M., Parametric studies on self-repairing approaches for resin infused composites subjected to low velocity impact. *Smart Mater. Struct.*, 8, 5, 623–638, 1999.

19. Dry, C., Procedures developed for self-repair of polymer matrix composite materials. *Compos. Struct.*, 35, 3, 263–269, 1996.
20. Trask, R.S., Williams, G.J., Bond, I.P., Bioinspired self-healing of advanced composite structures using hollow glass fibres. *J. R. Soc. Interface*, 4, 13, 363–371, 2007.
21. Williams, G., Trask, R., Bond, I., A self-healing carbon fibre reinforced polymer for aerospace applications. *Compos. Part A: Appl. Sci. Manuf.*, 38, 6, 1525–1532, 2007.
22. Park, J.-H. and Braun, P.V., Coaxial Electrospinning of Self-Healing Coatings. *Adv. Mater.*, 22, 4, 496–499, 2010.
23. Loh, X.J. *et al.*, Controlled drug release from biodegradable thermoresponsive physical hydrogel nanofibers. *J. Control. Release*, 143, 2, 175–82, 2010.
24. Keller, M.W., White, S.R., Sottos, N.R., A Self-Healing Poly(Dimethyl Siloxane) Elastomer. *Adv. Funct. Mater.*, 17, 14, 2399–2404, 2007.
25. Yuan, Y.C. *et al.*, Self-Healing Polymeric Materials Using Epoxy/Mercaptan as the Healant. *Macromolecules*, 41, 14, 5197–5202, 2008.
26. Cho, S.H. *et al.*, Polydimethylsiloxane-Based Self-Healing Materials. *Adv. Mater.*, 18, 8, 997–1000, 2006.
27. Lee, M.W., Yoon, S.S., Yarin, A.L., Solution-Blown Core-Shell Self-Healing Nano- and Microfibers. *ACS Appl. Mater. Interfaces*, 8, 7, 4955–62, 2016.
28. Park, J.H. and Braun, P.V., Coaxial electrospinning of self-healing coatings. *Adv. Mater.*, 22, 4, 496–9, 2010.
29. Esmaeely Neisiany, R. *et al.*, Facile strategy toward fabrication of highly responsive self-healing carbon/epoxy composites via incorporation of healing agents encapsulated in poly(methylmethacrylate) nanofiber shell. *J. Ind. Eng. Chem.*, 59, 456–466, 2018.
30. Fischer, E., Einfluss der Configuration auf die Wirkung der Enzyme. *Ber. Dtsch. Chem. Ges.*, 27, 3, 2985–2993, 1894.
31. Ullah, H. *et al.*, The Potential of Microencapsulated Self-healing Materials for Microcracks Recovery in Self-healing Composite Systems: A Review. *Polym. Rev.*, 56, 3, 429–485, 2016.
32. Samadzadeh, M. *et al.*, Tung oil: An autonomous repairing agent for self-healing epoxy coatings. *Prog. Org. Coat.*, 70, 4, 383–387, 2011.
33. Brown, E.N. *et al.*, *In situ* poly(urea-formaldehyde) microencapsulation of dicyclopentadiene. *J. Microencapsul.*, 20, 6, 719–30, 2003.
34. Döhler, D. *et al.* Self-Healing Polymers: From Principles to Applications, Design of Self-Healing Materials, Principles of Self-Healing Polymers, 5-60, 2013. *<10.1002@9783527670185.ch1.pdf>*.
35. White, S.R. *et al.*, Autonomic healing of polymer composites. *Nature*, 409, 6822, 794–797, 2001.
36. Beluomini, M.A. *et al.*, Electrochemical sensors based on molecularly imprinted polymer on nanostructured carbon materials: A review. *J. Electroanal. Chem.*, 840, 343–366, 2019.

37. Cho, S.H., White, S.R., Braun, P.V., Self-Healing Polymer Coatings. *Adv. Mater.*, 21, 6, 645–649, 2009.
38. Wu, G. *et al.*, Robust microcapsules with polyurea/silica hybrid shell for one-part self-healing anticorrosion coatings. *J. Mater. Chem. A*, 2, 30, 11614–11620, 2014.
39. Blaiszik, B.J., Sottos, N.R., White, S.R., Nanocapsules for self-healing materials. *Compos. Sci. Technol.*, 68, 3–4, 978–986, 2008.
40. Golker, K. *et al.*, Influence of Composition and Morphology on Template Recognition in Molecularly Imprinted Polymers. *Macromolecules*, 46, 4, 1408–1414, 2013.
41. Monfared Zanjani, J.S. *et al.*, Repeated self-healing of nano and micro scale cracks in epoxy based composites by tri-axial electrospun fibers including different healing agents. *RSC Adv.*, 5, 89, 73133–73145, 2015.
42. Reneker, D.H. and Yarin, A.L., Electrospinning jets and polymer nanofibers. *Polymer*, 49, 10, 2387–2425, 2008.
43. An, S. *et al.*, Highly flexible transparent self-healing composite based on electrospun core–shell nanofibers produced by coaxial electrospinning for anti-corrosion and electrical insulation. *Nanoscale*, 7, 42, 17778–85, 2015.
44. Lee, M.W. *et al.*, Hybrid self-healing matrix using core-shell nanofibers and capsuleless microdroplets. *ACS Appl. Mater. Interfaces*, 6, 13, 10461–8, 2014.
45. Kang, Y. *et al.*, Electrospun nanofibers with a core–shell structure of silicon nanoparticles and carbon nanotubes in carbon for use as lithium-ion battery anodes, *J. Mater. Chem. A*, 2, 15094–15101, 2014.
46. Lee, M.W. *et al.*, Self-healing transparent core–shell nanofiber coatings for anti-corrosive protection. *J. Mater. Chem. A*, 2, 19, 7045–7053, 2014.
47. Lee, M.W. *et al.*, Hybrid Self-Healing Matrix Using Core–Shell Nanofibers and Capsuleless Microdroplets. *ACS Appl. Mater. Interfaces*, 6, 13, 10461–10468, 2014.
48. Savage, N. and Diallo, M.S., Nanomaterials and Water Purification: Opportunities and Challenges. *J. Nanopart. Res.*, 7, 4–5, 331–342, 2005.
49. Nabeela Nasreen, S.A.A. *et al.*, Nanomaterials: Solutions to Water-Concomitant Challenges. *Membranes*, 9, 3, 40, 2019.
50. Enayati, M.S. *et al.*, Development of electrospun poly (vinyl alcohol)-based bionanocomposite scaffolds for bone tissue engineering. *J. Biomed. Mater. Res. A*, 106, 4, 1111–1120, 2018.
51. Enayati, M.S. *et al.*, Effect of nanofiller incorporation on thermomechanical and toughness of poly (vinyl alcohol)-based electrospun nanofibrous bionanocomposites. *Theor. Appl. Fract. Mech.*, 99, 44–50, 2019.
52. Gopiraman, M. *et al.*, Structural and mechanical properties of cellulose acetate/graphene hybrid nanofibers: Spectroscopic investigations. *Express Polym. Lett.*, 7, 6, 554–563, 2013.
53. Huang, Z.-X. *et al.*, The technique of electrospinning for manufacturing core–shell nanofibers. *Mater. Manuf. Processes*, 33, 2, 202–219, 2017.

54. Haider, A. *et al.*, PLGA/nHA hybrid nanofiber scaffold as a nanocargo carrier of insulin for accelerating bone tissue regeneration, *BioMed Res. Int.*, 9, 314-326, 2014.
55. Lee, M.W. *et al.*, Self-healing three-dimensional bulk materials based on core-shell nanofibers. *Chem. Eng. J.*, 334, 1093–1100, 2018.
56. Cuvellier, A. *et al.*, Coaxial electrospinning of epoxy and amine monomers in a pullulan shell for self-healing nanovascular systems. *Polym. Test.*, 69, 146–156, 2018.
57. Cuvellier, A. *et al.*, Selection of healing agents for a vascular self-healing application. *Polym. Test.*, 62, 302–310, 2017.
58. Nesterova, T., Dam-Johansen, K., Kiil, S., Synthesis of durable microcapsules for self-healing anticorrosive coatings: A comparison of selected methods. *Prog. Org. Coat.*, 70, 4, 342–352, 2011.
59. Patrick, J.F., Sottos, N.R., White, S.R., Microvascular based self-healing polymeric foam. *Polymer*, 53, 19, 4231–4240, 2012.
60. Sinha-Ray, S. *et al.*, Encapsulation of self-healing materials by coelectrospinning, emulsion electrospinning, solution blowing and intercalation. *J. Mater. Chem.*, 22, 18, 9138–9146, 2012.
61. Zhuang, X. *et al.*, Solution blowing of submicron-scale cellulose fibers. *Carbohydr. Polym.*, 90, 2, 982–987, 2012.
62. Behrens, A.M. *et al.*, *In Situ* Deposition of PLGA Nanofibers via Solution Blow Spinning. *ACS Macro Lett.*, 3, 3, 249–254, 2014.
63. Lee, M.W. *et al.*, Fatigue of Self-Healing Nanofiber-based Composites: Static Test and Subcritical Crack Propagation. *ACS Appl. Mater. Interfaces*, 8, 28, 18462–70, 2016.
64. Kousourakis, A. and Mouritz, A.P., The effect of self-healing hollow fibres on the mechanical properties of polymer composites. *Smart Mater. Struct.*, 19, 8, 1–9, 2010.
65. Razavi, S.M.J. *et al.*, Effect of neat and reinforced polyacrylonitrile nanofibers incorporation on interlaminar fracture toughness of carbon/epoxy composite. *Theor. Appl. Mech. Lett.*, 8, 2, 126–131, 2018.
66. Razavi, S.M.J. *et al.*, Fracture assessment of polyacrylonitrile nanofiber-reinforced epoxy adhesive. *Theor. Appl. Fract. Mech.*, 97, 448–453, 2018.
67. Vijay Kumar, V. *et al.*, Electrospun nanofiber interleaving in fiber reinforced composites—Recent trends. *Mater. Des. Process. Commun.*, 1, 1, 1–8, 2019.
68. Naeimirad, M. *et al.*, Recent advances in core/shell bicomponent fibers and nanofibers: A review. *J. Appl. Polym. Sci.*, 135, 21, 46265–46288, 2018.
69. Wu, X.-F. *et al.*, Electrospinning core-shell nanofibers for interfacial toughening and self-healing of carbon-fiber/epoxy composites. *J. Appl. Polym. Sci.*, 129, 3, 1383–1393, 2013.
70. Neisiany, R.E. *et al.*, Towards the development of self-healing carbon/epoxy composites with improved potential provided by efficient encapsulation of healing agents in core-shell nanofibers. *Polym. Test.*, 62, 79–87, 2017.

71. Sinha-Ray, S. *et al.*, Encapsulation of self-healing materials by coelectrospinning, emulsion electrospinning, solution blowing and intercalation. *J. Mater. Chem.*, 22, 18, 9138–9146, 2012.
72. Kousourakis, A. and Mouritz, A.P., The effect of self-healing hollow fibres on the mechanical properties of polymer composites. *Smart Mater. Struct.*, 19, 8, 085021, 2010.
73. Neisiany, R.E. *et al.*, Self-healing and interfacially toughened carbon fibre-epoxy composites based on electrospun core-shell nanofibres. *J. Appl. Polym. Sci.*, 134, 31, 44956–44966, 2017.
74. Therriault, D. *et al.*, Fugitive Inks for Direct-Write Assembly of Three-Dimensional Microvascular Networks. *Adv. Mater.*, 17, 4, 395–399, 2005.
75. Hansen, C.J. *et al.*, Self-Healing Materials with Interpenetrating Microvascular Networks. *Adv. Mater.*, 21, 41, 4143–4147, 2009.
76. Chao, M.-R., Hu, C.-W., Chen, J.-L., Glass substrates crosslinked with tetracycline-imprinted polymeric silicate and CdTe quantum dots as fluorescent sensors. *Anal. Chim. Acta*, 925, 61–69, 2016.

8

Intrinsic Self-Healing Materials

Angelita Cristiane Saul* and João Henrique Zimnoch dos Santos

Universidade Federal do Rio Grande do Sul, Porto Alegre, Brazil

Abstract
Self-healing materials are substances that possess the capability to repair themselves after damages. It can or cannot depend on an external stimulus, for example, that can be increase temperature, humidity or UV light. It can also develop this ability due to the addition of an external constituent in the material formulation that will be required in case of a damage or it can have an intrinsic ability for repairing, tailor made specifically for the requirements of the application. Starting with polymers based on reversible bonds, going through chain recombination properties or supramolecular interactions, a large variety of different intrinsic self-healing materials have been developed in the last decades. This review briefly summarizes some of the most significant researches regarding this class of materials, providing support to novel studies and developments.

***Keywords*:** Self-healing, intrinsic self-healing, inverse reactions, chain recombination, covalent bonds, supramolecular bonds

8.1 Introduction

Normally, polymer reactions are irreversible, however during the past decades, many studies were developed seeking for solving this behavior that hinder polymers to modify their backbone. Self-healing materials have been continuously studied due to its application in many different fields, such as electronics, transportation and medicine. They possess the capability of repairing themselves after damages, which is a striking property that can extend functional lifetime, reliability and energy efficiency of the products and, thus, lower the cost [1, 2]. According to

**Corresponding author*: ita.saul@gmail.com

Inamuddin, Mohd Imran Ahamed, Rajender Boddula and Tariq Altalhi (eds.) Self-Healing Smart Materials and Allied Applications, (203–236) © 2021 Scrivener Publishing LLC

Samadzadeh and co-workers [3], studies show that 2 to 5% of GBP of developed countries are used in maintenance of industrial machinery and marine devices, due to their prompt to corrosion. Aiming maintenance cost reduction, for example, many studies were held in the field of self-healing materials in order to reduce oxidation in metal surfaces by healing of cracks on the protective coating avoiding the contact of polyelectrolytes, responsible for oxidation, to the metal surface. Efforts in the field of self-healing materials were conducted on intrinsic and extrinsic self-healing areas [4–6].

An extrinsic self-healing material stands for a system whose polymer matrix cannot heals itself without an external agent. The most common application of the external agent is by encapsulating it in an inert capsule/shell and then adding it to the system. The core of the capsule will only be released in case a crack or a change in the inertness of the capsule happen [7–30].

According to Zhang *et al.*, intrinsic self-healing polymers are based on specific behavior of polymeric matrices that enables crack healing under stimulus (normally heating) [5]. Autonomic healing with no manual intervention is rarely available. The authors have initially classified the processes in two modes: (i) physical interactions, most of the time heating leading to polymer surface rearrangement, and (ii) chemical interactions, based on the reversibility of polymer dynamic covalent and/or noncovalent bonds, such as Diels–Alder bonds, for example. A few years later a third mode was included to this list: (iii) supramolecular interactions [31], which are based

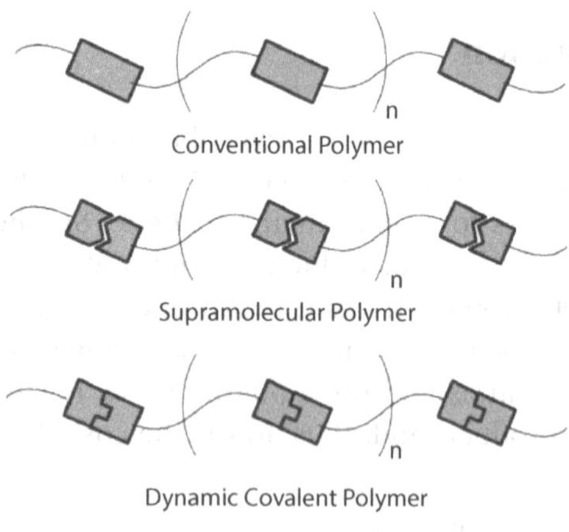

Figure 8.1 Schematic representations of conventional polymer, supramolecular polymer and dynamic covalent polymer connected by covalent bonds, noncovalent bonds, and dynamic covalent bonds, respectively [35].

on secondary noncovalent interactions, such as hydrogen bonds or π–π stacking interactions. All this kind of polymers, once exposed to external stimuli, such as heating, will respond to it by reorganizing their chain and restoring their initial properties in almost totality.

A translation of how polymers react according to each mode of healing was demonstrated by Otsuka and collaborators and it is depicted in Figure 8.1.

8.2 Inverse Reactions and Chain Recombination

The recombination of chains has been reported as one among all self-healing methods studied until today. It takes advantage of weaknesses of the material, such as low resistance to hydrolysis or thermal degradation, and turns it into its favor [32–34].

Takeda *et al.* [33] reported the production of a self-healing system based on polycarbonate (PC). PC suffers severe damages in its mechanical strength due to thermal degradation and hydrolysis [32]. In their study they have worked with different amounts of Na_2CO_3 and followed the structure of deterioration and recovery reactions presented in Scheme 8.1.

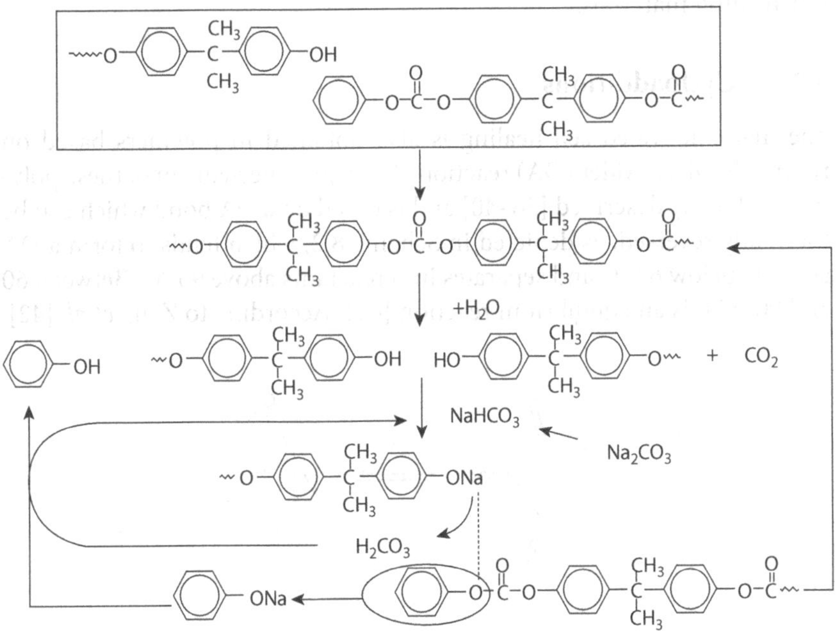

Scheme 8.1 Deterioration and recovery reactions of PC [33].

Samples were prepared with amounts of Na_2CO_3 ranging from 0.1 to 10,000 ppm and treated at 120 °C in a steam chamber and subsequently at 130 °C under nitrogen atmosphere. The molecular weight recovery weight on concentration of Na_2CO_3 was monitored by size exclusion chromatography (SEC) and it was noted that the M_w was stable after 20 days when the concentration of sodium carbonate was 100 to 1,000. At 0.1 ppm and 10,000 ppm M_w drops significantly. Sample with 100 ppm of sodium carbonate was transparent even after 30 days. The best result of tensile strength recovery was 98%, indicating that a simple change in the PC composition could result in an innovative and very useful material.

8.3 Reversible (Covalent) Bonds

While most of the polymers are structurally irreversible due to their strong and fixed covalent bonds, there is a sort of reorganizable polymers whose bonds are a result of monomer chains reversibly linked by covalent and/or noncovalent bonds, so their properties can be changed and optimized even after polymerization [35]. There are an enormous number of studies in this area and some will be presented further on to exemplify the many different polymer with reversible bonds used as self-healing materials.

8.3.1 Cycloadditions

Thermally activated self-healing is also obtained in polymers based on reversible Diels–Alder (DA) reaction. The repair mechanism of these polymers is largely described [36–40] and is based on a DA bond which can be thermally reversed, as depicted in Scheme 8.2, which tends to form a DA adducts below 60 °C and separates into reactants above 90 °C. Between 60 and 90 °C it is an equilibrium reaction [41]. According to Zang *et al.* [42],

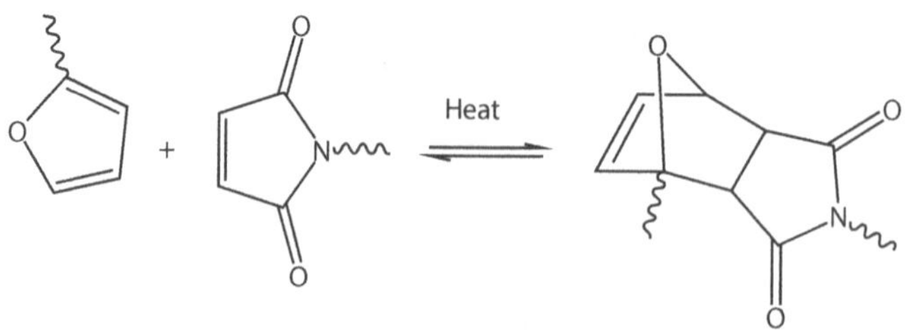

Scheme 8.2 Mechanism of polymerization and repair for Diels-Alder cross linked polymers [50].

it is preferable that all systems are let for enough time at 60 °C to reform all DA adducts. The healing time and temperature dramatically influence the interfacial shear strength of healed specimens.

Several compounds can be used in the production of a reversible polymer, such as anthracene-maleimide [43, 44], cyclopentadiene [45], fulvene derived [46, 47], however, the most important systems are furan–maleimine pairs reacted via Diels–Alder (DA) reaction [48]. The bonds originated in the DA reaction are weaker than any other in the polymer chain, so they are preferably broken under loading. After heating, and subsequent cooling, the broken ends recombine to reconstruct the chain [49]. The weak bonds mentioned act as a directional region for fracture, protecting the whole chain in order to be broken in the self-healable region. These polymers have a tensile strength comparable to other thermoset polymers used in structural applications [50].

Kötteritzsch *et al.* synthesized terpolymers with functional moieties that could reversibly crosslink by DA reaction [51]. In this work, the authors synthesized methacrylate-based linear self-healing polymers containing the two corresponding moieties needed for DA reaction in the side chain of these polymers. No crosslinker or second polymer was required to achieve self-healing properties; however, protective groups have been introduced to the furan and maleimide units to prevent crosslinking whilst polymerization. The desired polymer is illustrated by Scheme 8.3. The resulting protected polymer retro DA react upon heating, so the maleimide is

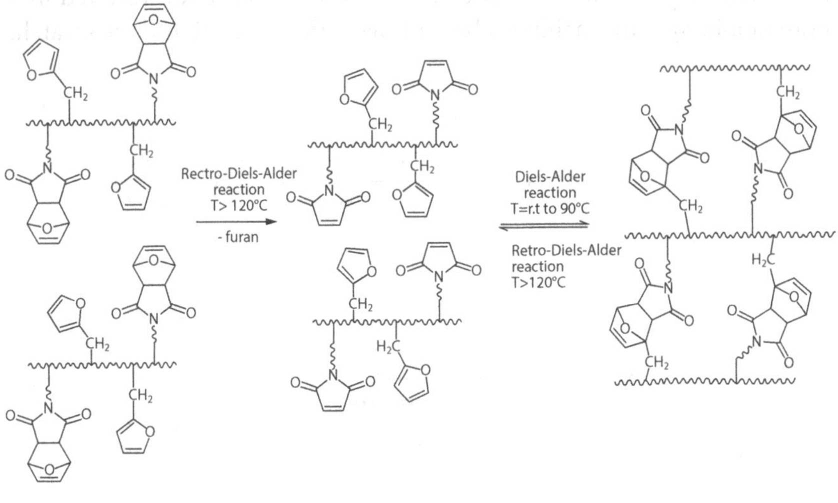

Scheme 8.3 Schematic representation of the self-healing concept via reversible DA reactions [51].

released of its protection by furan, enabling the crosslinking after cooling down. This structure can again be broken via retro DA and repeat this cycle multiple times.

The synthesis of maleimide methacrylate (MIMA) monomer used in this work is depicted in Scheme 8.4. Initially DA reaction takes place at 90% yield reaction. Thereafter, the resulting compound is reacted with ethanolamine, with a yield of 66%, and then reacted with methacryloyl chloride, obtaining a yield of 86%.

Finally, to obtain the reversible crosslinking polymer, the previously prepared MIMA was reacted with furfuryl methacrylate (FMA) and alkyl methacrylate (AMA) at 90 °C (Scheme 8.5). Ethyl 2-bromoisobutyrate (EBiB) was used as initiator, CuBr as the catalyst, N,N,N′,N′′,N′′-pentamethyldiethylenetriamine (PMDETA) or 1,1,4,7,10,10-hexamethyltriethylene-tetramine (HMTETA) as the ligand and reacted at 70 °C.

Kötteritzsch *et al.* tried to understand the importance of the co-monomer composition. For this purpose, they used MMA, BMA and LMA. Films of polymers produced with concentration of co-monomer varying from 10 to 67% were prepared and a scratch was performed on the surface of each film in order to evaluate their self-healing properties. The scratches were monitored by SEM and AFM techniques during annealing in different conditions. It was observed that polymers with the lower the Tg of the co-monomer or the higher the temperature and time of annealing to which the polymer is being submitted, the more fluid is the polymer, enabling a better flow to fill the gaps of the scratches. In addition, it was noted that polymers with lower content of functionalization resulted in a more mobile system, enabling a better flow of the material into the scratch.

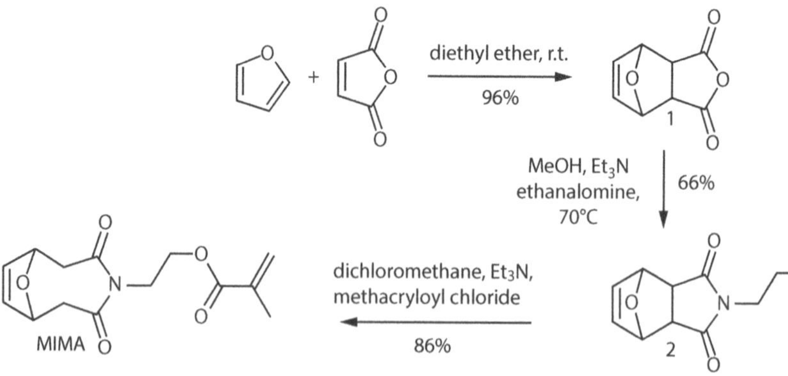

Scheme 8.4 Schematic representation of the synthesis of the maleimide methacrylate (MIMA) [51].

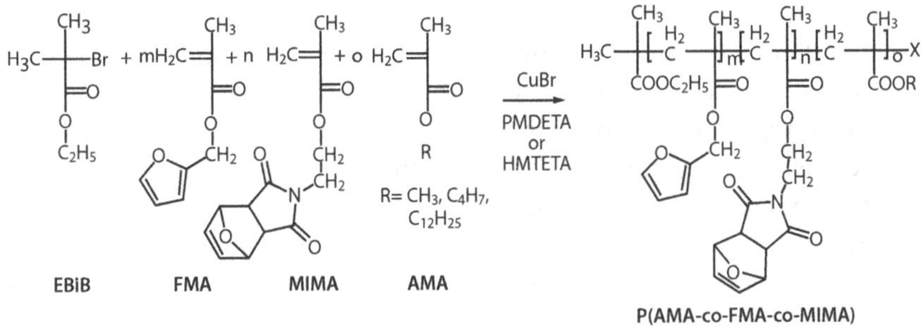

Scheme 8.5 Schematic representation of the copolymerization of AMA, FMA, and MIMA by ATRP polymerization [51].

Peterson *et al.* [41] and Zang *et al.* [42] have proposed the introduction of self-healing properties of glass fiber reinforced epoxy composites via thermal reversible DA adduct in the interphase of layers. However, the second study describes a more difficult development since carbon fiber, which was the matter of this work, has an extremely inert surface and a relatively small diameter when compared with glass fiber, matter of the first mentioned study. Due to the inertness of carbon fiber, Zang's group studied its surface treatment, aiming to add more polar functionalities to it. The fibers were oxidized with concentrated HNO_3 at 115 °C, for different amount of times. They were then treated with tetraethylenepentamine (TEPA) to graft amino groups which are required for the further reaction with a saturated solution of bismaleimide (BIM). All the steps and conditions of reactions are described in the Scheme 8.6.

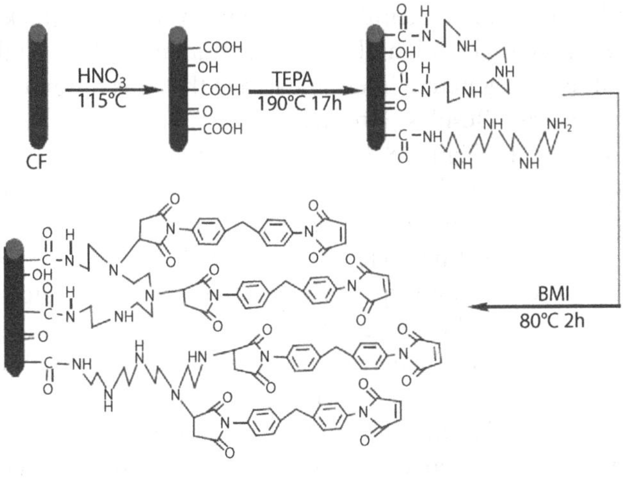

Scheme 8.6 Steps for grafting of maleimide groups on carbon fiber surface [42].

It was reported that the self-healing efficiency in the interfaces of the treated fibers with the epoxy matrix was strongly dependent on the oxidation time, since corrosion of the carbon fiber by exposure to nitric acid leads to a larger surface area. Fibers that were oxidized during 60 min presented a self-healing efficiency of 82%, 3 times better than untreated fibers or fibers oxidized for only 30 min. The efficiency was measured in terms of tensile force necessary to displace the fiber inside the epoxy matrix. Loss of healing efficiency was also measured and resulted in 16% loss in both 2nd and 3rd healings.

An external stimulus, such as temperature, is frequently necessary for the healing process. Most of the literature suggests oven heating of the polymers. However, in large scale, it cannot be possible for many different reasons. An alternative was described by Park *et al.* [49] in which the group studied the healing ability via electrical resistive heating. In this study, a graphite/epoxy substrate was used and coated with self-healing composite which matrix was based on a dicyclopentadiene adduct.

After the analysis to identify what was the current needed to reach the healing temperature, DSC curves were compared. It was seen that the composite cured at 150 °C for 20 h and 5 h show an endothermic peak around 100 and 60 °C, meanwhile the one cured at 120 °C for 3 h have an endothermic peak around 40 °C. This behavior indicates that the electrical resistive heating is adequate to the healing process and that the higher the degree of crosslinking of the polymer (longer cure time and temperatures), the higher the glass transition temperature becomes.

In 2010, the same group proposed a different approach for electrical heating [50]. A composite produced from carbon fibers embedded with a polymer based on bis-maleimide tetrafuran. A 0.55 mm thick composite panel was prepared, with a fiber volume fraction of 0.46, and exposed to a flexural test, to generate internal damage that was observed by X-ray micro-tomography. Besides internal fracture, a small fiber breakage occurred, as seen in Figure 8.2. The sample was heated during 15 min through application of electrical current and sample temperature, measured with an infrared thermometer, was between 100 and 110 °C during this time. A second heating cycle was performed at the same temperature, for 1 h. After the heating step, the sample recovered its original shape, demonstrating that the healing can be obtained without any other manual intervention. A load-displacement test identified that, in both cases, the healing recovered only 60 to 65% of composite properties.

A second run of trials had the samples bended with a 0.5 mm thick aluminum foil covering the load head to avoid fiber breakage. Therefore, the size of damage was relatively small (Figure 8.3). Healing was performed

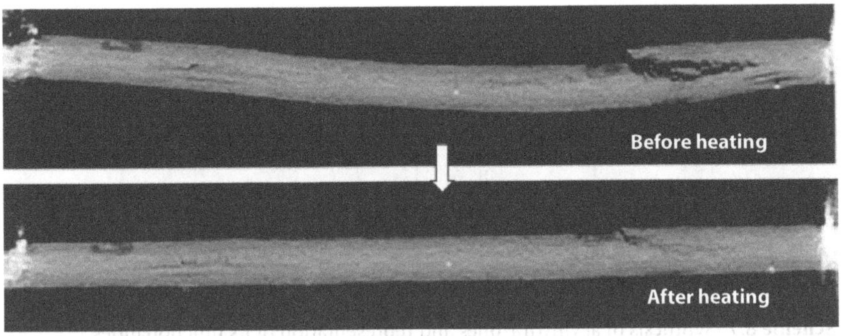

Figure 8.2 Section views of first trial sample using X-ray tomography [50].

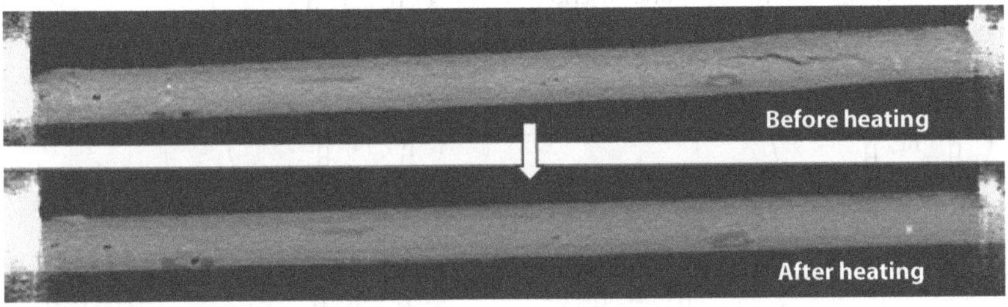

Figure 8.3 Section views of second trial sample using X-ray tomography [50].

with electrical current being applied during 1, 2 and 3 h to evaluate the healing times. The recoveries measured were 86.4, 88.4 and 97.8%, leading us to the conclusion that the longer the healing time, the higher the healing efficiency.

8.3.2 Reversible Acylhydrazones

Acylhydrazones play an important role in the field of self-healing materials. They are obtained through hydrazide and an aldehyde condensation. Systems based on acylhydrazones have an intrinsic reversibility through the imino group (Scheme 8.7). They are capable of reversibly exchange their components through the additions of aldehyde or hydrazine (Scheme 8.8). This reversibility is achieved under mild conditions, simple as heat or a reduction in the environment pH, resulting in acid catalysis [52, 53]. Furthermore, amine groups provide additional hydrogen bonding, reinforcing the self-healed structure [48].

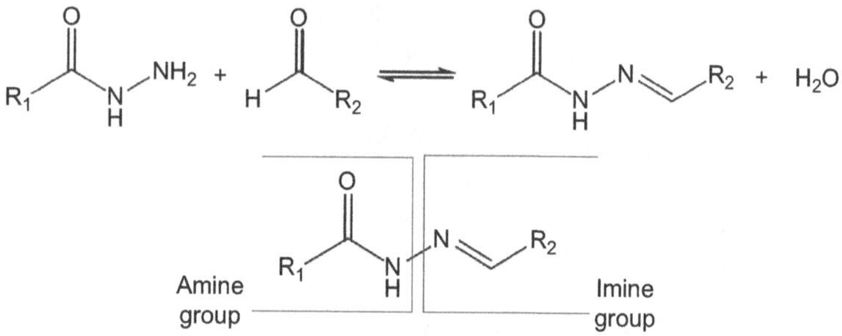

Scheme 8.7 Synthesis of acylhydrazones and representation of its functionalities.

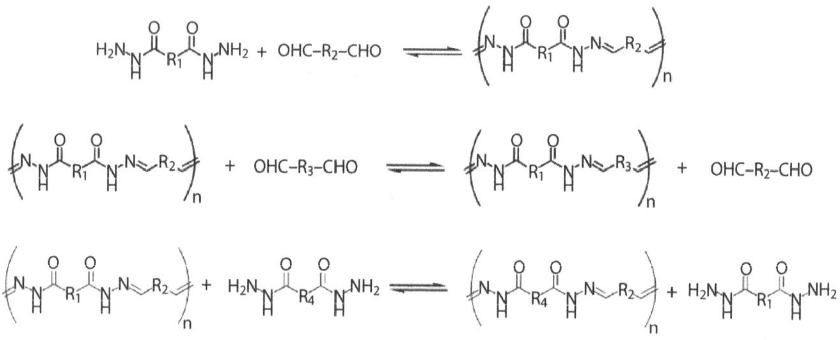

Scheme 8.8 Modification of acylhydrazones by dialdehyde exchange and dihydrazine exchange in a polyacylhydrazone [53].

Based on the acylhydrazones studies of Lehn *et al.* [52–54], one of the first self-healing systems based on acylhydrazones was published by Deng *et al.* [55] in 2010. In this work they report the construction of a novel reversible polymer gel, based on the dynamic covalent chemistry that acylhydrazones have to offer. The two aldehyde end groups of poly(ethylene oxide) (PEO) were led to react with the aldehyde groups of tris[(4-formyl phenoxy)methyl]ethane, generating a network with acylhydrazones bonds as cross-links.

The sol–gel phase transition ability was verified by its response to acidity change and it demonstrated in Figure 8.4. An initial gel was prepared in solvent DMF and its pH adjusted to ca. pH 5. After gel aging overnight, a small amount of concentrated HCl was added to the gel surface. It took 1.5 h for a complete conversion into a sol. Final pH dropped to 3–4. Then, an equivalent amount of triethylamine was added, and the sol returned to a gel state in 20 s.

Figure 8.4 Sol-gel phase transition of the covalent polymer gel in response to alteration of aidity [55].

Self-healing ability of the gel was observed when two gel specimens, prepared with the same polymer gel, each tinted in a different color, were cut in half and then a half of each specimen were brought into contact for 7 h, without any outside intervention and at ambient temperature. After this time, the junction line of the two halves is clear (Figure 8.5), however the gels have merged together as a whole. The healed specimen didn't show any signal of a junction line and maintained its integrity even after vigorous squeeze by tweezers.

Lehn and coworkers [56] have also reported the obtention of a doubly active self-healing polymer network by using of reversible covalent bonds and dynamic non-covalent hydrogen bonding interactions at the same time. This double dynamicity can be achieved in the polymer by polycondensation reaction between a siloxane-based dialdehyde (bis-tosylated polydimethylsiloxane) and carbohydrazide, through the formation of

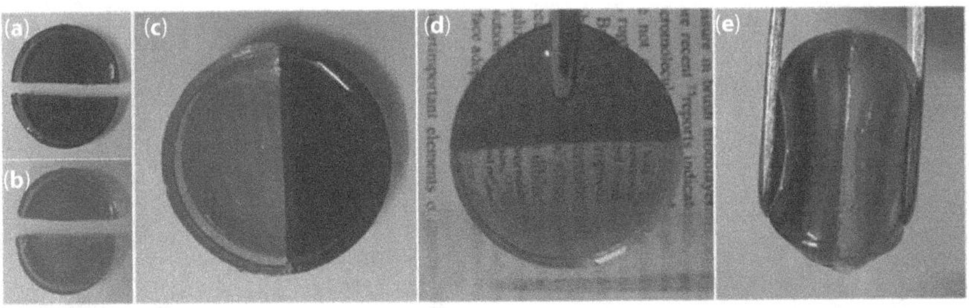

Figure 8.5 Sol–gel phase transition of the covalent polymer gel in response to alteration of acidity. (a) cracked gel black specimen; (b) cracked gel pink specimen; (c) merged bicolor gel; (d) merged bicolor gel lifted by tweezers; (e) merged bicolor gel squeezed by tweezers [55].

bis-iminourea type subunits, that may be considered as bis-acylhydrazone as seen in Scheme 8.9. To evaluate the polymer promptness to self-healing, the resulting polymeric material was diluted in a chloroform solution and a small amount of an acid catalyst was added. Evaporation of the solvent resulted in a soft and stretchable film. This film, when cut in two pieces and brought together after a few minutes, resulted in a completed healed film. The healing occurred at room temperature and with application of gentle initially pressure. 90% of the original strain was retained by the healed polymer. In addition, it was observed that the rupture stress (at maximum elongation) of the self-healed polymer was 0.5 MPa, different from the original polymer that had permanent deformation beyond 115%, indicating a healing efficiency of, approximately, 75%.

An interesting approach for stable structure for a self-healable acylhydrazone polymer was proposed by Yu *et al.* [57] Working in the field of cartilage tissue engineering and dealing with the particular requirements of the physiological environment, the group developed a polymer combining DA reaction and hydrazine bond (Scheme 8.10). In this case, DA is responsible for the maintenance of polymer structural integrity (Scheme 8.11) and mechanical strength while the dynamic covalent acylhydrazone bond is responsible for the self-healing property.

Sol–gel phase transition was tested just like performed by Deng *et al.* (2010) and here previously described. It was possible to observe the sol-gel transition in single crosslinked polymers (SCP) as pH was decreased or increased. The same behavior was not seen in double crosslinked polymers (DCP), since DA bonds did not cleave under pH influence maintaining the original structure.

The stability of the DA bonds increases the mechanical resistance of polymer and was confirmed via mechanical test and the obtained compressive

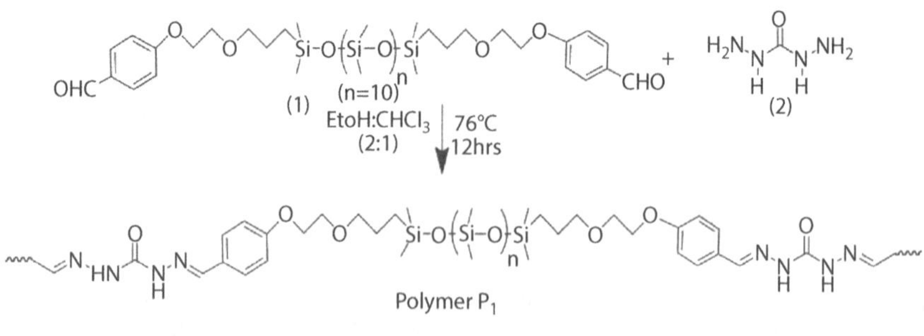

Scheme 8.9 Synthesis of the dynamic polymer (P1) by polycondensation of the monomeric components (1) bis-aldehyde and (2) carbohydrazide [56].

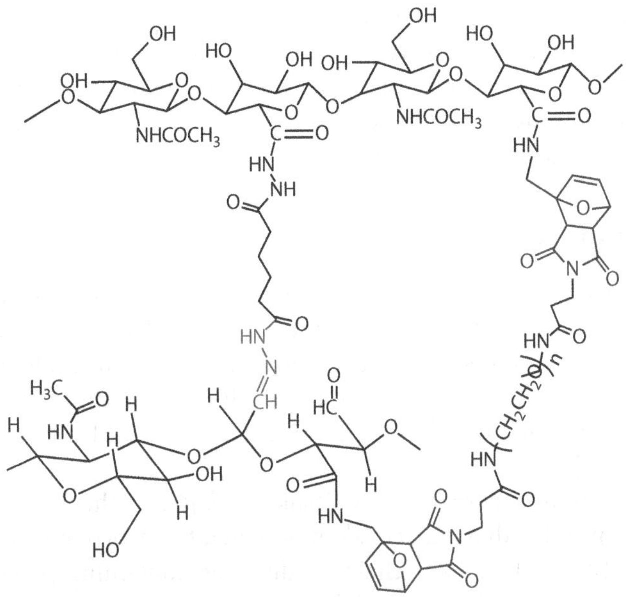

Scheme 8.10 Demonstration of chemical structure of the double crosslinked system [57].

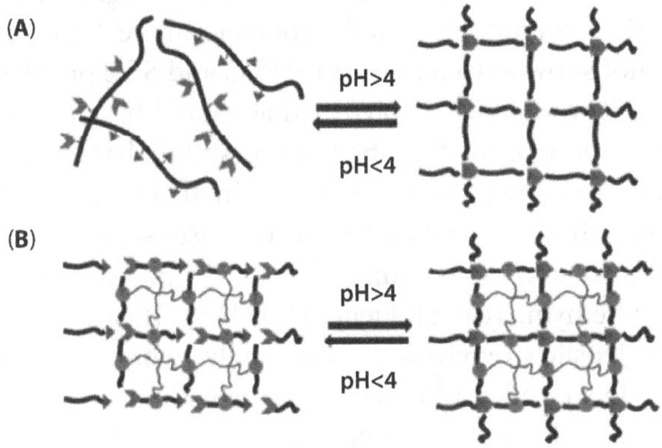

Scheme 8.11 Crosslinking diagram of a (A) single crosslinked network and (B) a double crosslinked network [57].

stress–strain curves in which it was seen that original and healed polymers did not show significantly different curves. However, static compressive moduli of DCP were much higher than those of SCP.

The self-healing property was confirmed by cutting polymer gel specimens in half and putting the halves together again for 3 h without outside intervention. After this time the healed SCP specimens could stand its own gravity while DCP healed specimens could additionally bear some pulling force. Furthermore, an additional reaction of aldehyde groups and amines of local cartilage occurred, increasing the stability and acceptance of DCP in the physiological environment.

8.3.3 Disulfides

Dynamic disulfide groups can also be inserted into polymers in order to obtain a self-healable material. Many studies were already reported in this field, mainly covering specific reversible reactions as disulfide exchange/metathesis [58–61], first described in 2010 [62], thiol–disulfide exchange [63, 64], radical mediated disulfide fragmentation [65] and reduction–oxidation [66].

Disulfide exchange was used by Otsuka *et al.* [62] in the development of a photo-responsive dynamic covalent polymer, by incorporating disulfide functionality into the main chain. A disulfide-containing polyester was irradiated for 60 min. It was possible to see via HPLC chromatograms that a 3rd component was present after irradiation. It was also possible to see, via SEC, that the molecular weight distribution broadened without dislocating the peak. The confirmation that these changes were initiated by irradiation is the comparison with the control sample, that was kept in the dark and did not show any changes on HPLC and SEC profiles.

Gao *et al.* [67] worked on a polymer that should have a better mechanical stability than previously described in polysulfide-based material, whose tensile strength was lower than 0.8 Mpa). In their work they combined the higher strength of polyurethanes, in this case isophorone diisocyanate (IPDI), and the exchangeable disulfide bonds of a polysulfide oligomer, 2,5-diamino-3,6-dimethylmercaptotoluene (DMTDA). Their aim was to increase initial polymer tensile strength and, additionally, to have a healing efficiency close to 100%. Even under a moderate thermal stimulus, polysulfide-based polyurethane offered good self-healing ability mainly because of the disulfide metathesis in the soft segments. The healing efficiency was greater than 90% when healing process was held for 24 h at 75 °C and 4 h at 100 °C, which was at least three times better than similar polysulfide-based materials previously reported. Self-healing effect is demonstrated in Figure 8.6.

Asua *et al.* [68] evaluated the self-healing performance of water based dispersion of polyurethane-ureas (PUU) containing aromatic disulfides in comparison with similar systems containing C–C bonds (aromatic rigid bonds) instead of the disulfide. Linear and cross-link urea units were tested. Films cast from these different dispersions were prepared

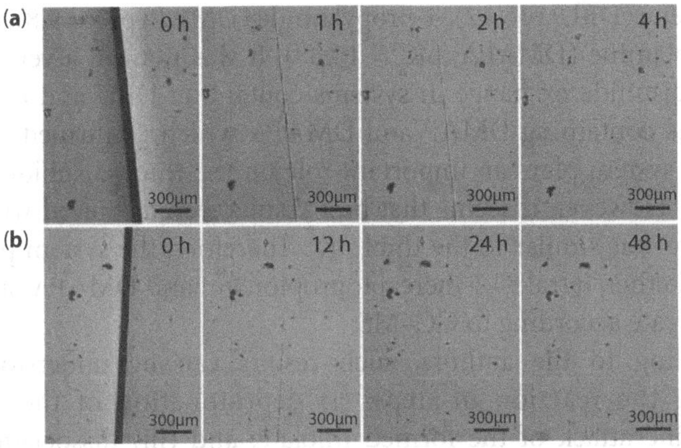

Figure 8.6 Optical microscope images after different healing conditions: (a) 4 h at 100 °C and, (b) 48 h at 75 °C [67].

and scratched in order to evaluate the material self-healing efficiency. After 2 h at 60 °C a complete gap closure of linear urea-disulfide system was observed while linear urea-aromatic rigid bonds were still visible. Due to the mobility of the chain, cross-linked urea systems took some longer time for complete gap closure but, again, system with disulfide bonds was the one which achieved complete healing after 20 days. Film casts were also tested by means of thermal, mechanical and rheological analysis. According to tensile tests, the presence of disulfide bonds in the linear system decreases the tensile strength of the PUU. Nevertheless, cross-linked systems do not show this behavior, indicating that a combination of S–S bonds and cross-linked ureas should be the best system in test.

Different from disulfide–disulfide reactions, Pepels *et al.* [69] have worked on a disulfide–thiol system. It is known that disulfide–thiol reaction generates new disulfides and thiols, which can continuously react to provide the cure of the material.

$$R_1\text{-S-S-}R_1 + R_2\text{-SH} \longrightarrow R_1\text{-S-S-}R_2 + R_1\text{-SH} \qquad (8.1)$$

The authors evaluated the influence of chemical structure and catalysts in the development of a disulfide bond-based self-healing thermoset system. Different grades of polysulfide and mercaptoalkanoates were evaluated and formation of final product was accompanied by GC-MS. The authors also tested different tertiary amines that were used to provide the basicity necessary to the reaction, such as 1,8-diazabicyclo[5.4.0] undec-7-ene (DBU; pK_a 12), 4-propylamide (DMAP; $pK_a = 9.8$), dimethylaminoprolamine (DMAPA; $pK_a = 10.38$). It was noticed a very high rate of thiol–disulfide exchange in systems containing DBU and much lower in systems containing DMAP and DMAPA which confirmed that basicity of the system plays an important role on the thiol–disulfide exchange reactions. However, it seems that best results were achieved when pH of the system was similar to the thiol pK_a. Therefore, the system polysulfide diglycidyl ether, tetrakis(2-mercaptopropionate) and DMAPA offered best exchange rate according to GC–MS.

According to the authors, such results can be understood when analyzing the reaction in steps: (i) deprotonation of the thiol, (ii) nucleophilic attack of the formed thiolate, and (iii) dissociation of the intermediate product in one of the two possible thiolates [69]. Low rate of nucleophilic attack and high deprotonation rate (when system has a high number of thiolates), confer a better stability of the thiolate. This stability determines which intermediate product will be the leaving group. A self-healing efficiency of 99% was confirmed via tensile tests, after letting two cut halves into contact for 24 h. However, because thiols are not stable to oxidation, this efficiency is decreased with time, letting space for new studies on this topic.

8.3.4 Alkoxyamines (Radicals)

Before the extensive use of alkoxyamines in nitroxide-mediated radical polymerization (NMP), the process was initially described by Hawker *et al.* According to them, these procedures, many of them based on nitroxide free radicals, such as 2,2,6,6-tetramethylpiperidinyloxy (TEMPO) can result in materials, with controlled weights and narrow polydispersity [70]. The explanation for this controlled radical polymerization may be due to the reversible termination of the growing polymeric radical by the stable nitroxide free radical and is well explained in Scheme 8.12. In their work, Hawker and collaborators describe their efforts to understand potential diffusion of nitroxide free radicals in different polymerization steps. Using a specially designed HPLC technique, they proved that the migration of the nitroxide radical during polymerization is simple and merely depends on the use of different unimolecular initiators, enabling a precise control of molecular structure.

The study of graft polymer with thermo reversible bonds was held by Higaki *et al.* [71]. The authors concluded that dynamic synthesis of graft polymer can be achieved by reacting side chain and chain end alkoxyamine units. Atom transfer radical polymerization (ATPR) technique was

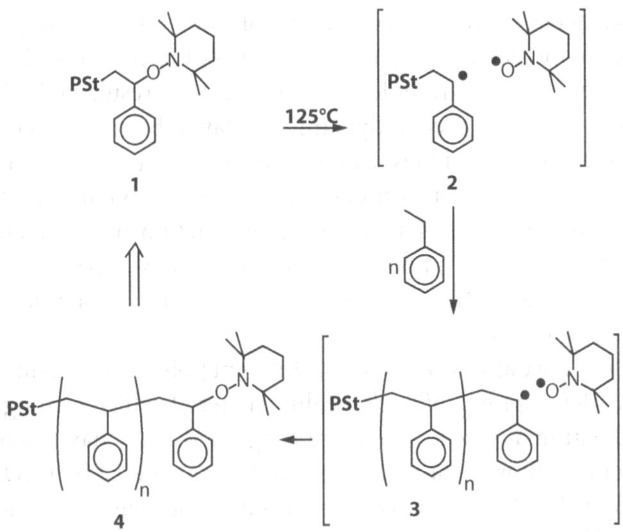

Scheme 8.12 Inactive species (1) is fragmented into two radical species: A nitroxide free radical and a polymeric radical (2). This polymeric radical can undergo chain extension with another monomer, however, maintaining its radical nature (3) to later recombine with nitroxide radical to generate a new inactive material (4) [70].

explored to obtain a methyl methacrylate (MMA) polymer with alkoxyamine units in the side chain. Then a TEMPO-based alkoxyamine-terminated polystyrene was prepared through conventional NMP technique. Both polymers were brought together at 100 °C under argon atmosphere and chain M_n was accompanied by size exclusion chromatography (SEC). After 24 h, M_n was constant indicating that the system equilibrium was reached. The reversibility of the obtained polymer was evaluated by dissolving the polymer in anisole and heating the solution at 100 °C. The equilibrium system was reached after 30 h. Resulting molecular weight was significantly close to the molecular weight of the MMA derived polymer, proving that the material can self-heal in response to heating.

In a later paper [72], Higaki and co-workers described the development of a reaction model for the thermodynamic covalent cross-linking system. In their studies they explored the reaction comparing the GPC profiles of polymers obtained in different reaction times and concentration. It was noticed that when crosslinking reaction is held at 0.5 wt%, the equilibrium of reaction, noticed when M_n became constant, was reached in 24 h reaction. However, same GPC profile was noted after only 6 h reaction when working at 2.0% concentration condition. This happened due to the different behavior of intermolecular and intramolecular cross-linking reactions.

In high concentrations conditions, intermolecular cross-linking is favored, increasing the molecular weight with few cross-linked points. Meanwhile, at low concentrations, intramolecular cross-linking results in highly cross-linked polymers with thermodynamically favorable molecular weights. Since alkoxyamines moieties can be easily incorporated into various monomers, some variations on covalent polymers containing alkoxyamines were described by this same group as reaction of alkoxyamine diol and adipoyl chloride [35], poly(4-vinylpyridine) [73], acrylate monomers [74], obtaining many different polymers that could be potentially used in self-healing materials.

Zhang *et al.* have also worked with a different polymer base and described a successful development of a self-healing material, which healing could be conducted within a wide temperature range [75]. They have incorporated intermolecular alkoxamines moieties in polystyrene (PS). Electron spin spectroscopy (ESR) spectroscopy was used for studying free radical formation at different temperatures. It was seen a first signal at about 125 °C, indicating the start of the alkoxyamine moiety cleavage around this temperature. The higher temperature that can be used is limited to PS degradation, however a more stable reversibility was achieved below 200 °C. The measured healing efficiency of cracks, generated by compression, was 26.4% for PS sample with styrene/crosslinker ratio of 5/1 and 32.6% for ratio 10/1, under healing temperature of 130 °C. According to the authors, this behavior happens because the higher the cross-linking density of the polymer the more the range of motion of the radicals is reduced. Healing efficiency of 70.1% was reached when temperature was raised for 145 °C, due to the increase in the mobility of the radicals.

Due to the successful achievement with stiff matrices, self-healing polyurethane (PU) was also obtained by the same authors through the synthesis of an alkoxyamine diol which hydroxy group participates in the production of PU [76]. It was seen that all the studied PU showed better reversibility than PS systems, previously described. This was assigned to the Tg of the soft segments of PU, which is generally lower than at room temperature, which in turn allows better mobility of the radicals. Healing efficiency, measured by tensile test, up to 70% was obtained with good healing efficiencies after successive rupture-healing tests (Figure 8.7).

Photochemical stimulus is an alternative to thermally triggered repair systems [6, 77] and this alternative was also explored by Telitel *et al.* [78]. The group described for the first time the obtention of self-healing materials under irradiation at ambient temperatures. The object of study was a polymer based on 4-arm acrylic oligomer, with the chain ends modified with azide functionalities, cross-linked with alkoxyamines, functionalized

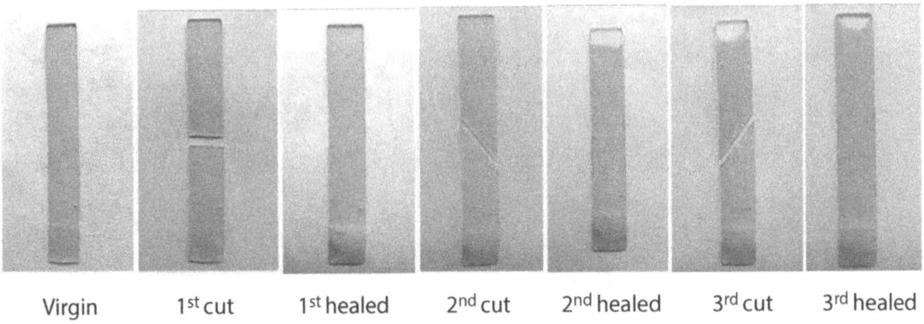

Figure 8.7 Healing of specimens in 3 healing cycles [76].

with two alkyne groups. Those materials were cross-linked forming an organized covalent network.

Photodissociation of this network was performed using a lamp with irradiance of 100 mW/cm in a range of wavelength of 250–300 nm. The formation of nitroxide radicals under UV irradiation was monitored by ESR and was confirmed, suggesting that once cleaved, the radicals would condensate again, repairing films prepared with the polymer. Self-healing ability was checked by preparing 60 μm layers over silicon wafers and performing 600 nm depth scratches on the surface of the specimens. Samples were submitted to UV irradiation and periodically scanned by atomic force microscopy (AFM). The total efficiency of cleavage restoration was achieved after 24 h of irradiation.

Alternatively of alkoxyamines, a radical self-healing system of diarylbibenzofuranone-based polymer was successfully developed by Imato and coworkers [79]. A system based on a PU pre-polymer and diarylbenzofuranone (DABBF) (Scheme 8.13) was used as a dynamic covalent bond. DABBF is a dimer of arylbenzofuranone, largely used as antioxidant,

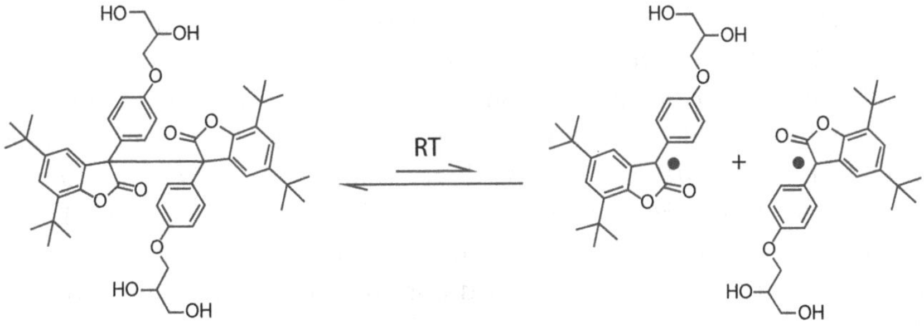

Scheme 8.13 Chemical structures of DABBF—noncleaved and cleaved forms [79].

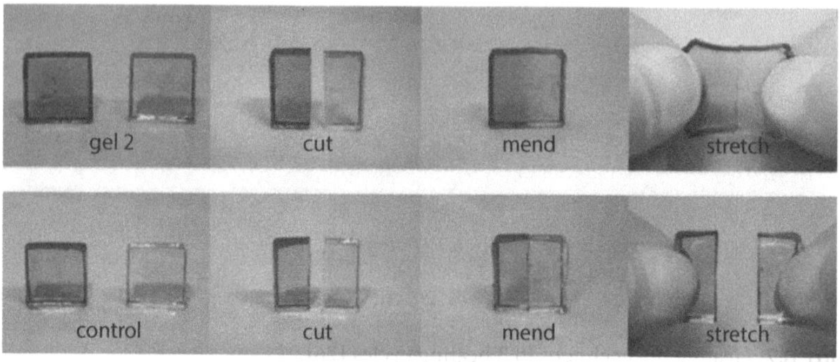

Figure 8.8 Self-healing behavior of cross-linked polymer gels (one piece is tinted with indigo for better visibility of cut session) [79].

known by its efficiency at room temperature, so the system was also tested in this condition.

The prepared specimens of polymer/gel embedded in dimethylformamide (DMF) were cut in half with a razor blade. The fresh surfaces were wet with DMF to avoid hydrogen bonding of urethane groups, and then fresh surface of pieces of two different specimens were brought together and kept at room temperature for 24 h. Self-healing confirmed after this period. Manual stretching did not cause any damage to the samples, as seen in Figure 8.8 so they had a tensile test performed at them, confirming a healing efficiency of 98%. A control sample, without DABBF, was prepared for comparison. This control sample did not show any sign of healing. The authors mentioned that the self-healing property was seen even after 5 days of being separated.

8.3.5 Transesterification

Self-healing based on transesterification through reaction of esters of vinylphenylboronic acid (VPBE) with 3,6-dioxa-1,8-octanedithiol (DODT) and pentaerythritol tetrakis(3-mercaptopropionate (PTMP) and a photoinitiator was performed by Cash *et al.* to produce a boronic ester network, as illustrated in Scheme 8.14 [80]. Polymers with different ratios of DODT and PTMP were prepared. The samples were cured for 30 min under irradiation of 365 nm. After this period, samples were cut in half and let to heal in contact to each other, with auxiliary drops of water providing enough moisture so surface-exposed boronic ester groups could convert to a disassociated state, allowing a readily bond formation with the diols groups (Scheme 8.15). In later evaluations, it was concluded that healing

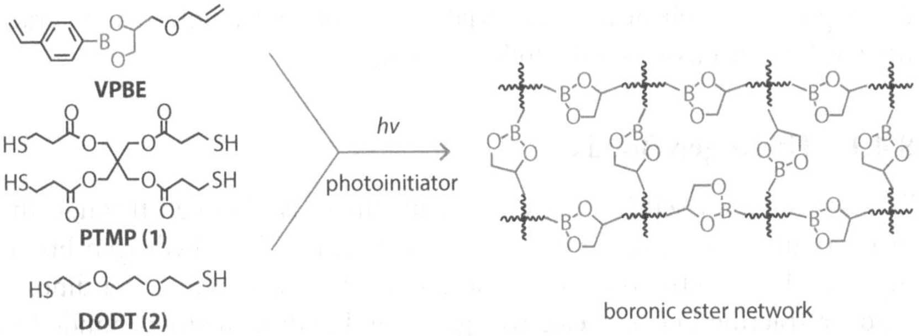

Scheme 8.14 Synthesis of boronic ester network materials via photoinitiated thiol-ene curing [80].

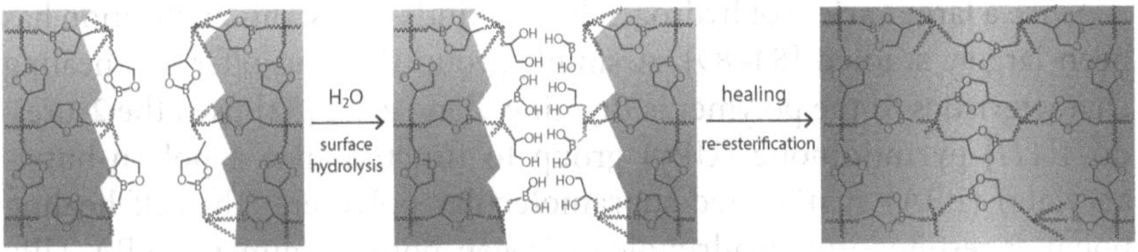

Scheme 8.15 Mechanism of healing proposed by Cash *et al.* [80].

also occur in high humidity environments (>85%), however a minor addition of water makes the reaction more efficient.

The authors emphasized that the dried samples were not tacky, so self-healing should not be considered self-adhesion. After 3.5 days the original joint had disappeared and after 4 days, healed samples were manually stretched to twice its length and no damage was noticed. When evaluating the healing efficiency in a quantitative way, 75:25 DODT:PTMP samples showed excellent performance, even after 3 healing recovering almost 100% of initial tensile strength and elongation at break.

8.4 Supramolecular Interactions

Supramolecular polymers are materials whose dynamic behavior is consequence secondary non-covalent interactions within the polymeric structures, such as hydrogen bonds, π–π interactions, ionic- and metal–ligands interactions [81, 82]. These supramolecular bonds can be strengthened through the change of the binding moieties and can even reach similar strength of dynamic covalent bonds [81]. Some of the studies held on the development of self-healing materials based on supramolecular interactions will be discussed on the following pages.

8.4.1 Hydrogen Bonds

This widely used self-healing class of polymers is obtained through the incorporation of strong and reversible non-covalent hydrogen-bonding moieties into the polymer structure. Hydrogen bonding is a dipole–dipole attraction between electronegative and hydrogen atoms bonded to another electronegative atom [2]. In an event of a fracture, the active ends of hydrogen-bonds network are exposed because the strength of those bonds is lower than the covalent bonds in the molecule. Those ends can be recombined by simply bringing them together [83].

Since a large variety of hydrogen bonds kinds exists, this application has been largely studied [84–89]. A simple group that can offer self-healing characteristics to the polymer is thymine. Zeng *et al.* [82] used the 2-ureido-4[1*H*]-pyrimidinone (UPy) group to functionalize a poly(n-butyl-acrylate) (PBA) synthesized supramolecular polymers. The self-healing polymer resulted in a quadrupole hydrogen-bonding dimer of PBA–UPy (Scheme 8.16). Self-healing ability was confirmed when two pieced of fractured PBA–UPy films were brought together, without an external load and recovered 40, 81 and 100% of their self-adhesion strength in 10, 3 and 50 h, respectively. It was also concluded that the adhesion properties were determined by the surface density of chains that can interpenetrate across the contacting interface. The PBA-UPy showed excellent self-healing ability even after the surfaces were separated and brought together into contact multiple times, which, according to the authors, is attributed to the reversibility, stability, long lifetime and strength of UPy–UPy bonds.

The interaction of hydrogen-bonding van der Waals force healing ability was proposed and studied by Chen *et al.* [90]. The group has evaluated the

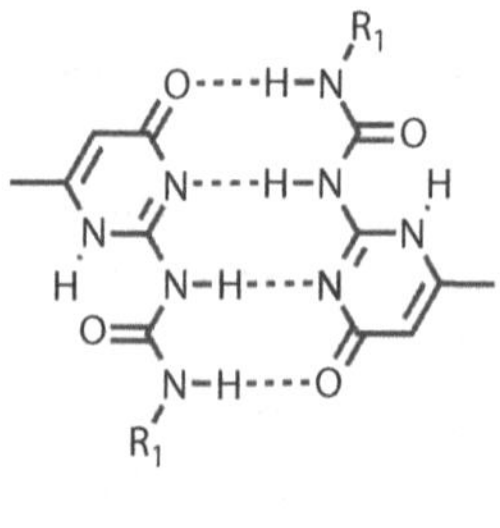

Scheme 8.16 Hydrogen bonds between two PUy groups [82].

self-healing efficiency of a poly(propylene-carbonate)–polyurethane(urea) elastomer (PPC–PU) at different temperatures. Samples with different amounts of PPC were produced and it was seen that a 100% self-healing efficiency could be reached at 50 °C, in samples with a larger amount of carbonyl groups on the poly(propylene-carbonate) segment. This is an indication that self-healing process of PU takes advantages of van der Waals interaction. Also, the mobility of the amorphous chain of PPC contributed to reduce the rigidity of the PU chain. This introduced strong van der Waals forces, promoting better interaction between molecular chains thus activating the dynamic exchange of "hydrogen-bonding pairs".

8.4.2 π–π Interaction

Despite of the fact of being weaker than hydrogen-bonds, π–π stacking interaction is one of the most important supramolecular interactions [81]. A large number of studies are found in the application of chain-folding structures (Scheme 8.17), as the more stable supramolecular systems in self-healing material, indicating that this is a promising field of study [91–95].

Hart *et al.* [91] have described π–π stacking interactions in chain-folding complexes. A perylene-terminated polyether (π-electron rich) blended with a chain-folding diimide (π-electron deficient) was developed giving place to a novel supramolecular healing system. To quantify healing efficiencies a test strip film was prepared and then damaged by cutting. The damaged film was heated up to 75 °C, temperature in which the healing started. Then the film was heated to 100 °C and to 125 °C, when the cut had almost completely healed itself. Tensile strength of the healed specimens was the same as the original samples, indicating a recovery of 100% of its efficiency. This behavior was repeated even after 3 cycles of damage/healing.

A on a tweezer-type π–π stacking interaction-based system was described by Burattini *et al.* [94]. This system has an identical polyamide

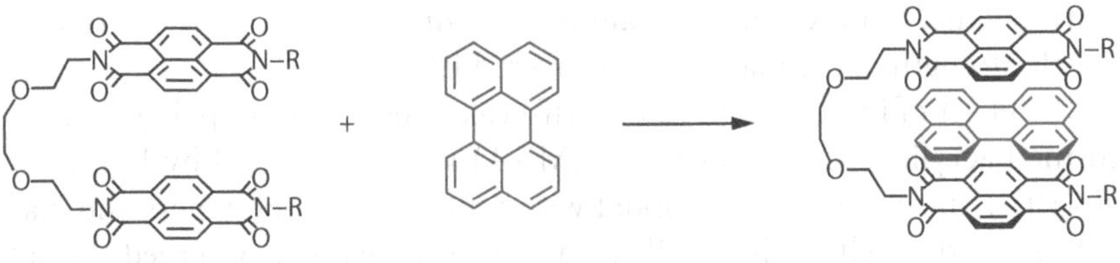

Scheme 8.17 Complexation of perylene by complementary π-π stacking with the chain-folding bis-diimide preceptor [91].

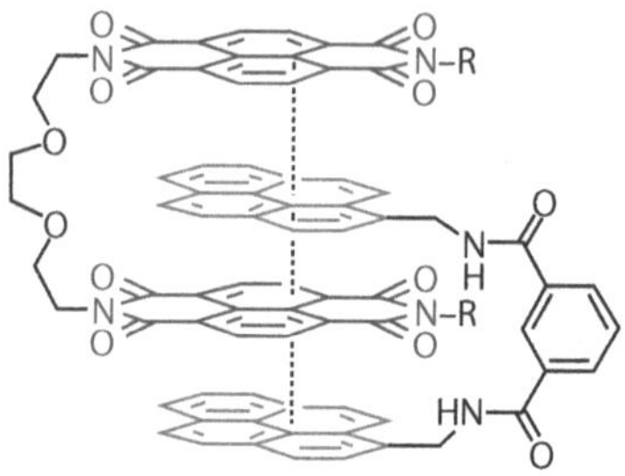

Scheme 8.18 Model structure of chain-folding bis-pyrenyl group and chain-folding polyimide [94].

backbone to that described before however, however it has tweezer-type bis-pyrenyl end groups (Scheme 8.18), strongly enhancing binding constants, capable to optimize polymer mechanical resistance. The later system offered a much better elongation at break and tensile strength than systems based on simple perenyl group. The edges of the polymer fracture samples were put together and heated at 140 °C for different periods of time to evaluate their prompt to self-healing. The original tensile strength was 100% recovered and achieved after approximately 160 min of healing time. The high healing temperature requirement is still constantly reported and opens space for many different approaches on π–π stacking structure developments in the future.

8.4.3 Ionomers (Ballistic Stimulus)

Ionomers are polymers which have up to 20 mol% of ionic species in its structure and which properties are a result of ionic interactions within the polymer [96, 97]. These polymers present ability to self-heal themselves upon ballistic impact. The impacted polymer largely elongates itself before the failure and then returns to its original position promoting closure of cavity created [98]. It was suggested that the energy brought by the projectile is transferred to the ionomer and, by reaching its melting point, rearrangement of the semi crystalline structure starts and part of original tensile strength of the material is recovered [97, 99, 100].

The effect of ionic content in the self-healing properties of poly(ethylene-co-methacrylic acid) copolymers (EMAA) was investigated by Kalista *et al.* [101]. The authors have worked with four EMAA copolymers and Na^+ ionomers, that self-healing ability have been previously observed. Initial mechanical properties, such as dynamic mechanical analysis (DMA), tensile strength and elongational viscosity were determined and ballistic

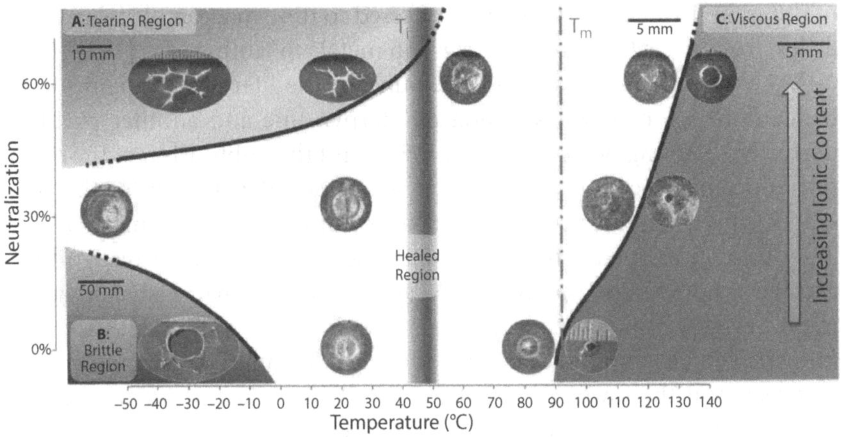

Figure 8.9 Self-healing "phase diagram" showing ballistic puncture-reversal as a function of temperature and ionic content [101].

studies were performed in a range of temperature from −50 to 140 °C, so the effect of ionic content and temperature on self-healing processed could be detailed explored. As a result of this work, an accurate healing profile was provided. All findings are summarized on Figure 8.9.

By examining self-healing ability in the conditions already mentioned, it was noticed that as temperatures decreased, less ionic material did not present enough strength to maintain a rigid framework for elastic recovery and allow the contact of the surfaces again. However, samples with higher ionic contents, due to ionic clusters, polymer mobility were strongly reduced, inhibiting any elastic energy transfer. In both cases, temperature is needed to give more mobility to the structure, bringing it to a molten state. Samples with moderated ionic content did not present any restriction to the healing process. According to the authors, the loss of the healing ability in high temperatures happens due to a transition from an elastic to dominant viscous response, that also affect the mobility of the semi crystalline segments.

According to Varley *et al.*, the level of stretching enables the polymer to seal the cavity as it returns to its original position. It is also important to have a balance of elastic strength to rebound and a polymer molten enough to melt and chemically diffuse across interfacial boundaries [96].

8.4.4 Metallopolymers

Metallopolymers are compounds with metal centers that can combine the properties of both polymers and metal–ligand interactions. A diverse range of different metal centers can be harnessed to tune macromolecular properties, from transition- and main-group metals to lanthanides [102–104].

Many ligand–metal systems are found in the literature, ranging from benzothiazoles, carbazoles, imidazole, terpyridine and another pyridine derivatives, among others [48, 103, 105]. Until the publication of the study of Bode *et al.* [106], metal complexes were not utilized often for the generation of self-healing properties, in particular in polymers films/coatings. They have developed a model system based on crosslinked terpyridine-acrylates copolymers (methyl methacrylate, butyl methacrylate and lauryl methacrylate) through the addition of Fe(II) sulfate and investigated the influence of the comonomer in the self-healing ability of the materials. It was seen that the lower the T_g of the comonomer, the better the healing behavior. Terpyridine-lauryl methacrylate copolymer (T_g = −65 °C) was capable to heal a crack 250 μm long and 35 μm wide, while terpyridine-butyl methacrylate copolymer (T_g = −2 °C) healed only a crack 180 μm × 10 μm and terpyridine-methyl methacrylate (T_g = 74 °C) did not offer any healing ability.

Sandmann *et al.* [104] have investigated by light microscopy, at different temperatures, the influence of several divalent metal salts in a triazole-pyridine-based self-healing polymer coating in the self-healing efficiency of the metallopolymer films. It was possible to note that, by crosslinking the copolymer with the metal salts, as depicted in Scheme 8.19, the material properties changed from thermoplastic feature polymers to polymers with elastomeric properties. Independent of the used metal salt, T_g was increased from −85 to −40 °C, just because of the additional crosslinks generated by the metals. Because of the strong complex bond of Fe(II) and Co(II) tiazole–pyridine complexes, only salts based on these ions were evaluated with regard to the self-healing ability. It was verified that healing temperature is dependent on the salt. Healing temperatures were 50

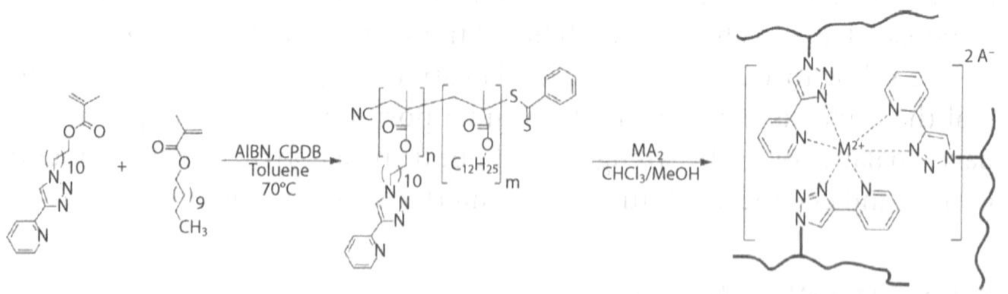

Scheme 8.19 Schematic representation of the synthesis of triazole-pyridine-based copolymer and the subsequent crosslinking with metal salt into copolymer networks [102].

to 100 °C for $FeCl_2$, 75 °C for $FeBr_2$, 75 to 100 °C for $Fe(OAc)_2$, $Fe(OTf)_2$, $CoCl_2$ and $CoBr_2$, and 100 °C for $Co(BF_4)_2$. Self-healing time was dependent on the length of the scratch and varied between 5.5 and 26.5 h.

8.5 Conclusion

In the past decade, impressive achievements in the field of self-healing materials have already been reported and new opportunities and special requirements of novel materials open a broad field for new researches and developments. Nevertheless, there are still many future challenges that need to be overcome such as a much better understanding of all self-healing mechanisms, reduced dependence of external stimuli (light, temperature, humidity, etc.) or the development of systems more affordable to enable commercial exploration of systems developed.

Through this study we attempted to accomplish the clarification of the intrinsic self-healing method in the repairing of materials that can be used in industrial, medical or pharmaceutical applications. Each self-healing method can give rise to desirable characteristics that strongly influence on the final use of material. A deeper understanding of the behavior of materials chemical base is required. However knowledge could only be gained by analyzing analytical and experimental results of previous works and by connecting them, confirming that still there is much more to be explored and understood.

In conclusion, the development of self-healing materials is becoming every day more promising and may be the key for expand materials lifetime, bringing quality and cost benefits on industrial applications and also may be the key for improvements in quality of life of many people that are dependent on medical solutions that currently require permanent monitoring and frequent repairs.

References

1. Zhang, H., Xia, H., Zhao, Y., Poly(vinyl alcohol) Hydrogel Can Autonomously Self-Heal. *ACS Macro Lett.*, 1, 1233–1236, 2012.
2. Thangavel, G., Lee, P.S., Tan, M.W.M., Advances in self-healing supramolecular soft materials and nanocomposites. *Nano Convergence*, 6, 29, 1–18, 2019.
3. Samadzadeh, M., A review on self-healing coatings based on micro/nanocapsules. *Prog. Org. Coat.*, 68, 159–164, 2010.

4. Abdolah Zadeh, M. and van der Zwaag, S., Routes to extrinsic and intrinsic self-healing corrosion protective sol-gel coatings: A review, in: *Self-Healing Materials*, pp. 1–18, 2013.
5. Zhang, M.Q., Yuan, Y., Yin, T., Rong, M., Self healing in polymers and polymer composites. Concepts, realization and outlook: A review. *eXPRESS Polym. Lett.*, 2, 4, 238–250, 2008.
6. Zhang, M.Q. and Rong, M.Z., Design and synthesis of self-healing polymers. *Sci. China*, 55, 5, 648–676, 2012.
7. Challener, C., The Inteligence Behind Smart Coatings. *JCT Coatingstech*, 3, 50–55, 2006.
8. Cho, S.H., White, S.R., Braun, P.V., Self-Healing Polymer Coatings. *Adv. Mater.*, 21, 6, 645–649, 2009.
9. Kumar, A., Stephenson, L., Murray, J., Self-healing coatings form steel. *Prog. Org. Coat.*, 55, 244–253, 2006.
10. Hughes, A.E., Cole, I.S., Muster, T.H., Varley, R.J., Designing green, self-healing coatings form metal protection. *NPG Asia Mater.*, 2, 4, 143–151, 2010.
11. Brown, E.N., Kessler, M.R., Sottos, N.R., White, S.R., *In situ* poly(urea-formaldehyde) microencapsulation of dicyclopentadiene. *J. Microencapsulation*, 20, 719–730, 2003.
12. Suryanarayana, C., Chowdoji Rao, K., Kumar, D., Preparation and characterization of microcapsules. *Prog. Org. Coat.*, 63, 72–78, 2008.
13. White, S.R., Sottos, N.R., Geubelle, P.H., Moore, J.S., Kessler, M.R., Sriram, S.R., Brown, E.N., Viswanathan, S., Autonomic healing of polymer composites. *Nature*, 409, 794–797, 2001.
14. Baghdachi, J., Smart Coatings, in: *ACS Symposium Series*, Wachington DC, 2009.
15. Nistor, C.L., Donescu, D., Perichaud, A., Ballout, W., Ghiurea, M., Microencapsulation of an acrylate monomer in silica particles. *J. Sol-Gel Sci. Technol.*, 57, 164–171, 2001.
16. Di Credico, B., Griffini, G., Levi, M., Turri, S., Microencapsulation of a UV-responsible photocromic dye by means of novel UV-screening polyurea-based shells for smart coating applications. *Appl. Mater. Interfaces*, 5, 6628–6634, 2013.
17. Kendig, M., Hon, M., Warren, L., Smart corrosion inhibiting coatings. *Prog. Org. Coat.*, 47, 183–189, 2003.
18. Puri, R.G., Meshram, P.D., Sirsam, R.S., Recent developments in smart coatings for corrosion protection. *Paintindia*, 5, 55–61, May 2012.
19. Cirimina, R., Sciortino, M., Alonzo, G., Schrijver, A.d., Pagliaro, M., From molecules to systems: Sol–gel microencapsulation in silica-based material. *Chem. Rev.*, 111, 765–789, 2011.
20. Yadav, S.K., Khilar, K.C., Suresh, A.K., Release rates from semi-crystalline polymer microcapsules formed by interfacial polycondensation. *J. Membr. Sci.*, 125, 213–218, 1997.

21. Park, S.-J., Shin, Y.-S., Lee, J.-R., Preparation and Characterization of Microcapsules Containing Lemon Oil. *J. Colloid Interface Sci.*, 241, 502–508, 2001.
22. Brown, E.N., Sottos, N.R., White, S.R., Fracture Testing of a Self-Healing Polymer Composite. *Exp. Mech.*, 42, 4, 372–378, 2002.
23. Landfester, K. and Weiss, C.K., Encapsulation by miniemulsion polymerization, in: *Modern Techniques for Nano- and Microreactors/-reactions. Advances in Polymer Science,* Caruso F. (ed.), vol 229. Springer, Berlin, Heidelberg, 2010.
24. Mahdavian, A.R., Sarrafi, Y., Shabankareh, M., Nanocomposite particles with core–shell morphology III: Preparation and characterization of nano Al2O3-poly(styrene-methyl methacrylate) particles via miniemulsion polymerization. *Polym. Bull.*, 63, 329–340, 2009.
25. Blaiszik, B.J., Sottos, N.R., White, S.R., Nanocapsules for self-healing materials. *Compos. Sci. Technol.*, 68, 978–986, 2008.
26. Brady, R.F., Jr, Twenty-first century materials: Coatings that interact with their enviroment. *ACS Symp. Ser.*, Chapter 1, 957, 3–11, 2007.
27. Esteves, A.C.C., Barros-Tommons, A., Trindade, T., Nanocompósitos de Matriz Polimérica: Estratégias de Síntese de Materiais Híbridos. *Quím. Nova,* 27, 5, 798–806, 2004.
28. Keller, M.W., White, S.R., Sottos, N.R., A Self-healing Poly(dimehyil siloxane) Elastomer. *Adv. Funct. Mater.*, 17, 2399–2404, 2007.
29. Ciriminna, R., Sciortino, M., Alonzo, G., Schrijver, A. de, Pagliaro, M. From Molecules to Systems: Sol–Gel Microencapsulation in Silica-Based Materials. *Chem. Rev.*, 111, 2, 765–789, 2011.
30. Tiarks, F., Landfester, K., Antonietti, M., Silica Nanoparticles as Surfactants and Fillers for Latexes Made by Miniemulsion Polymerization. *Langmuir 20,* 17, 5775–5780, 2001.
31. Zhang, Q.M. and Rong, M.Z., Basics of self-healing: state of the art, in: *Self-healing polymers and polymer composites,* pp. 1-82, John Willey & Sons Inc., New Jersey, 2011.
32. Takedaa, K., Tanahashia, M., Unno, H., Self-repairing mechanism of plastics. *Sci. Technol. Adv. Mater.*, 4, 435–444, 2003.
33. Takeda, K., Unno, H., Zhang, M., Polymer Reaction in Polycarbonate with Na2CO3. *J. Appl. Polym. Sci.*, 93, 920–926, 2004.
34. Imaizumi, K., Ohba, T., Ikeda, Y., Takeda, K., Self-repairing Mechanism of Polymer Composite. *Mater. Sci. Res. Int.*, 7, 4, 249–253, 2001.
35. Otsuka, H., Aotani, K., Higaki, Y., Amamoto, Y., Takahara, A., Thermal Reorganization and Molecular Weight Control of Dynamic Covalent Polymers Containing Alkoxyamines in Their Main Chains. *Macromolecules,* 40, 1429–1434, 2007.
36. Cordier, P., Tournilhac, F., Soulié-Ziakovic, C., Leibler, L., Self-healing and thermoreversible rubber from supramolecular assembly. *Nature,* 451, 977–980, 2008.
37. Garcia, S.J., Effect of polymer architecture on the intrinsic self-healing character of polymers. *Eur. Polym. J.*, 53, 118–125, 2014.

38. Kersey, F.R., Loveless, D.M., Craig, S.L., A hybrid polymer gel with controlled rates of cross-link rupture and self-repair. *J. R. Soc. Interface*, 4, 373–380, 2007.
39. Zhang, Y., Broekhuis, A.A., Picchioni, F., Thermally Self-Healing Polymeric Materials: The Next Step to Recycling Thermoset Polymers. *Macromolecules*, 42, 1906–1912, 2009.
40. Zhang, W., Duchet, J., Gérard, J., Self-healable interaces based on thermo-reversible Diels–Alder reactions in carbon fiber reinforced composites. *J. Colloid Interface Sci.*, 430, 61–68, 2014.
41. Peterson, A.M., Jensen, R.E., Palmese, G.R., Thermoreversible and remendable glass–polymer interface for fiber-reinforced composites. *Compos. Sci. Technol.*, 71, 5, 586–592, 2011.
42. Zhang, W., Duchet, J., Gérard, J., Self-healable interfaces based on thermo-reversible Diels–Alder reactions in carbon fiber reinforced composites. *J. Colloid Interface Sci.*, 430, 61–68, 2014.
43. Yoshie, N., Saito, S., Oya, N., A thermally-stable self-mending polymer networked by Diels–Alder cycloaddition. *Polymer*, 52, 6074–6079, 2011.
44. Syrett, J.A., Mantovani, G., Barton, W.R.S.B., Price, D., Haddleton, D.M., Self-healing polymers prepared via living radical polymerization. *Polym. Chem.*, 1, 102–106, 2010.
45. Park, J.S., Takahashi, K., Guo, Z., Wang, Y., Bolanos, E., Hamann-Schaffner, C., Murphy, E., Wudl, F., Hahn, H.T., Towards Development of a Self-Healing Composite using a Mendable Polymer and Resistive Heating. *J. Compos. Mater.*, 42, 26, 2869–2881, 2008.
46. Wei, Z., Yang, J.H., Du, X.J., Xu, F., Zrinyi, M., Osada, Y., Li, F., Chen, Y.M., Dextran-Based Self-Healing Hydrogels Formed by Reversible Diels–Alder Reaction under Physiological Conditions. *Macromol. Rapid Commun.*, 34, 1464–1470, 2013.
47. Roy, N., Bruchmannband, B., Lehn, J.-M., Dynamers: dynamic polymers asself-healing materials, *Chem. Soc. Rev.*, 44, 3786–3807, 2015.
48. Hager, M.D., van der Zwaag, S., Schubert, U.S., *Self-Healing Materials*, Advances in Polymer Science, 273, Springer, Heidelberg, Germany, 2016.
49. Park, J.S., Takahashi, K., Guo, Z., Wang, Y., Bolanos, E., Hamann-Schaffner, C., Murphy, E., Wudl, F., and Hahn, H.T., Towards development of a self-healing composite ausing a mendable polymer and resistive heating. *J. Compos. Mater.*, 42, 26, 2869–2881, 2008.
50. Park, J.S., Darlington, T., Starr, A.F., Takahashi, K., Riendeau, J., Hahn, H.T., Multiple Healing effect of thermally activated self-healing composites based on Diels–Alder reaction. *Compos. Sci. Technol.*, 70, 2154–2159, 2010.
51. Kötteritzsch, J., Stumpf, S., Hoeppener, S., Vitz, J., Hager, M.D., Schubert, U.S., One-Component Intrinsic Self-Healing Coatings Based on Reversible Crosslinking by Diels–Alder Cycloadditions. *Macromol. Chem. Phys.*, 214, 1636–1649, 2013.

52. Skene, W.G. and Lehn, J.-M.P., Dynamers: Polyacylhydrazone reversible covalent polymers, component exchange, and constitutional diversity. *PNAS*, 101, 22, 8270–8275, 2004.
53. Lehn, J.-M., Dynamers: Dynamic molecular and supramolecular polymers. *Prog. Polym. Sci.*, 30, 814–831, 2005.
54. Lehn, J.-M., From supramolecular chemistry towards constitutional dynamic chemistry and adaptive chemistry. *Chem. Soc. Rev.*, 36, 151–160, 2007.
55. Deng, G., Tang, C., Li, F., Jiang, H., Chen, Y., Covalent Cross-Linked Polymer Gels with Reversible Sol–Gel Transition and Self-Healing Properties. *Macromolecules*, 43, 1191–1194, 2010.
56. Roy, N., Buhlerb, E., Lehn, J.-M., Double dynamic self-healing polymers: Supramolecular and covalent dynamic polymers based on the bis-iminocarbohydrazide motif. *Polym. Int.*, 63, 1400–1405, 2014.
57. Yu, F., Cao, X., Du, J., Wang, G., Chen, X., Integrity, Self-healing and Tissue-adhesive Property Formed by Combining Diels-Alder Click Reaction and Acylhydrazone Bond. *ACS Appl. Mater. Interfaces*, 7, 43, 24023–24031, 2015.
58. Lei, Z.Q., Xiang, H.P., Yuan, Y.J., Rong, M.Z., Zhang, M.Q., Room-Temperature Self-Healable and Remoldable Cross-linked Polymer Based on the Dynamic Exchange of Disulfide Bonds. *Chem. Mater.*, 26, 6, 2038–2046, 2014.
59. Rekondo, A., Martin, R., Ruiz de Luzuriaga, A., Cabañero, G., Catalyst-free room-temperature self-healing elastomers based on aromatic disulfide metathesis. *Mater. Horiz.*, 1, 237–240, 2013.
60. Abdolahzadeh, M., Esteves, C.C., van der Zwaag., S., Garcia, S.J., Healable Dual Organic–Inorganic Crosslinked Sol–Gel Based Polymers: Crosslinking Density and Tetrasulfide Content Effect. *Polym. Chem.*, 52, 14, 1953–1961, 2014.
61. Martin, R., Rekondo, A., Ruiz de Luzuriaga, A., Cabañero, G., The processability of a poly(urea-urethane) elastomer reversibly crosslinked with aromatic disulfide bridges. *J. Mater. Chem. A*, 2, 5710–5715, 2014.
62. Otsuka, H., Nagamo, S., Kobashi, Y., Maeda, T., Takahara, A., A dynamic covalent polymer driven by disulfitde metathesis under photoirradiation. *Chem. Commun.*, 46, 1150–1152, 2010.
63. Gyarmati, B., Némethy, Á., Szilágyi, A., Reversible disulphide formation in polymer networks: A versatile functional group from synthesis to applications. *Eur. Polym. J.*, 49, 1268–1286, 2013.
64. Zhou, L., Chen, M., Zha, X., Rapid degradation of disulfide-based thermosets through thiol-disulfide exchange reaction. *Polymer*, 120, 1–8, 2017.
65. Fairbanks, B.D., Singh, S.P., Brwman, C.N., Anseth, K.S., Photodegradable, Photoadaptable Hydrogels via Radical-Mediated Disulfide Fragmentation Reaction. *Macromolecules*, 44, 2444–2450, 2011.
66. Bapat, A.P., Ray, J.G., Savin, D.A., Sumerlin, B.S., Redox-Responsive Dynamic-Covalent Assemblies: Stars and Miktoarm Stars. *Macromoleculer*, 46, 2188–2198, 2013.

67. Gao, W., Quan, Y., Bie, M., Zhu, J., Zhang, W., Self-healing, reprocessing and sealing abilities of polysulfide-based polyurethane. *Polymer*, 151, 27–33, 2018.
68. Asua, J.M., Nevejansa, S., Ballarda, N., Fernández, M., Reck, B., García, S.J., The challenges of obtaining mechanical strength in self-healing polymers containing dynamic covalent bonds. *Polymer*, 179, 121670, 2019, https://doi.org/10.1016/j.polymer.2019.121670.
69. Pepels, M., Klumperman, B., Filot, I., Goossens, H., Self-healing based on disulfide-thiol exchange reactions. *Polym. Chem.*, 4, 4955–4965, 2013.
70. Hawker, C.J., George, B.G., Dao, J., Radical Crossover in Nitroxide Mediated "Living" Free Radical Polymerizations. *J. Am. Chem. Soc.*, 118, 11467–11471, 1996.
71. HIgaki, Y., Otsuka, H., Takahara, A., Dynamic Formation of Graft Polymers via Radical Crossover Reaction of Alkoxyamines. *Macromolecules*, 37, 1696–1701, 2004.
72. Higaki, Y., Otsuka, H., Takahara, A., A Thermodinamic Polymer Cross-Linking System Based on Radically Exchangeable Covalent Bonds. *Macromolecules*, 39, 2121–2125, 2006.
73. Sato, T., Amamoto, Y., Ohishi, T., HIgaki, Y., Takahara, A., Otsuka, H., Radical crossover reactions of a dynamic covalent polymer brush forreversible hydrophilicity control. *Polymer*, 55, 4586–4592, 2014.
74. Su, J., Amamoto, Y., Sato, T., Ohishi, T., Higaki, Y., Takahara, A., Reversible cross-linking reactions of alkoxyamine-appended polymers under bulk conditions for transition between flow and rubber-like states. *Polymer*, 55, 1474–1480, 2014.
75. Zhang, M.Q., Yuan, C., Rong, M.Z., Zhang, Z.P., Yuan, Y.C., Self-healing of Polymers via Synchronous Covalent Bond Fission/Radical Recombination. *Chem. Mater.*, 23, 5076–5081, 2011.
76. Zhang, M.Q., Yuan, C., Rong, M.Z., Self-healing polyurethane elastomer with thermaly reversible alkoxyamines as crosslinkages. *Polymer*, 55, 1782–1791, 2014.
77. Ghosh, B. and Urban, M.W., Self-repairing Oxetane-Substituted Chitosan Polyurethane Networks. *Science*, 323, 5920, 1458–1460, 13 March 2009.
78. Telitel, S., Amamoto, Y., Poly, J., Soppera, O., Lalevéea, J., Matyjaszewski, K., Introduction of Self-Healing Properties into Covalent Polymer Networks via the Photodissociation of Alkoxyamine Junctions. *Polym. Chem.*, 2013, https://pubs.rsc.org/en/content/articlelanding/2014/py/c3py01162c.
79. Imato, K., Takahara, A., Otsuka, H., Nishihara, M., Kanehara, T., Amamoto, Y., Self-Healing of Chemical Gels Cross-Linked by Diarylbibenzofuranone-Based Trigger-Free Dynamic Covalent Bonds at Room Temperature. *Angew. Chem. Int.*, 51, 1138–1143, 2012.
80. Cash, J.J., Kubo, T., Bapat, A.P., Sumerlin, B.S., Room-temperture self-healing polymers based on dynamic-covalent boronic esters. *Macromolecules*, 48, 2098–2106, 2015.

81. Enke M., Döhler D., Bode S., Binder W.H., Hager M.D., Schubert U.S. Intrinsic Self-Healing Polymers Based on Supramolecular Interactions: State of the Art and Future Directions. In: *Self-healing Materials. Advances in Polymer Science*, Hager M., van der Zwaag S., Schubert U. (eds.), vol 273. Springer, Cham, 2015. https://doi.org/10.1007/12_2015_345
82. Zeng, H., Faghihnejad, A., Feldman, K.E., Yu, J., Tirrell, M.V., Israelachvili, J.N., Hawker, C.J., Kramer, E.J., Adhesion and Surface Interactions of a Self-Healing Polymer with Multiple Hydrogen-Bonding Groups. *Adv. Funct. Mater.*, 24, 16, 2322–2333, 2013.
83. Mynar, J. and Aida, T., The gift of healing. *Nature*, 451, 895–896, 21 February 2008.
84. Zhang, A., Yang, L., Lin, Y., Yan, L., Lu, H., Wang, L., Self-healing Supramolecular Elastomers Based on the Multi-Hydrogen Bonding of Low-Molecular Polydimethylsiloxanes: Synthesis and Characterization. *J. Appl. Polym. Sci.*, 129, 5, 2435–2442, 2013.
85. Chino, K. and Ashiura, M., Thermoreversible Cross-LInking Rubber Using Supramolecular Hydrogen-Bonding Networks. *Macromolecules*, 34, 9201–9204, 2001.
86. Cordier, P., Tournilhac, F., Soulié-Ziakovic, C., Leibler, L., Self-healing and thermoreversible rubber from supramolecular assembly. *Nature*, 451, 977–980, 21 February 2008.
87. Montarnal, D., Cordier, P., SOulié-Ziakovic, C., Tournilhac, F., Leibler, L., Synthesis of Self-Healing Supramolecular Rubbers from Fatty Acid Derivatives, Diethylene Triamine and Urea. *J. Polym. Sci.: Part A: Pol. Chem.*, 46, 7925–7936, 2008.
88. Montarnal, D., Tournilhac, F., Hidalgo, M., Couturier, J.-L., Leibler, L., Versatile One-Pot Synthesis of Supramolecular Plastics and Self-Healing Rubbers. *J. Am. Chem. Soc.*, 131, 7966–7967, 2009.
89. Cui., J. and del Campo, A., Multivalent H-bonds for self-healing hydrogels. *Chem. Commun.*, 48, 9302–9304, 2012.
90. Chen, J., Li, F., Luo, Y., Shi, Y., Ma, X., Zhang, M., Boukhvalov, D., Luo, Z., A self-healing elastomer based on an intrinsic non-covalent cross-linking mechanism. *J. Mater. Chem. A*, vol, 15207–15214, 2019.
91. Hart, L.R., Nguyen, N.A., Harries, J.L., Mackay, M.E., Colquhoun, H.M., Hayes, W., Perylene as an electron-rich moiety in healable, complementary π–π stacked, supramolecular polymer systems. *Polymer*, 69, 293–300, 2015.
92. Burattini, S., Greenland, B.W., Merino, D.H., Weng, W., Seppala, J., Colquhoun, H.M., Hayes, W., Mackay, M.E., Hamley, I.W., Rowan, S.J., A healable Supramolecular Polymer Blend Based on Aromatic π–π Stacking and Hydrogen-Bonding Interactions. *J. Am. Chem. Soc.*, 132, 12051–12058, 2010.
93. Burattini, S., Colquhoun, H.M., Greenland, B.W., Hayes, W., A novel self-healing supramolecular polymer system. *Faraday Discuss.*, 143, 251–264, 2009.

94. Burattini, S., Greenland, B.W., Hayes, W., Mackay, M.E., Rowan, S.J., Colquhoun, H.M., A Supramolecular Polymer Based on Tweezer-Typer π–π Stacking Interactions: Molecular Design for Healability and Enhanced Toughness. *Chem. Mater.*, 23, 6–8, 2011.
95. Burattini, S., Colquhoun, H.M., Fox, J.D., Friedmann, D., Greenland, B.W., Harris, P.J., Hayes, W., Mackay, M.E., Rowan, S.J., A self-repairing, supramolecular polymer system: Healability as a consequence of donor–acceptor π–π stacking interactions. *Chem. Commun.*, 44, 6717–6719, 2009, https://pubs.rsc.org/en/content/articlelanding/2009/cc/b910648k.
96. Varley, R., Ionomers as SelfHealing Polymers, in: *Self Healing Materials. An Alternative Approach to 20 Centuries of Material Science*, pp. 95–114, Springer, Heidelberg, Germany, 2007.
97. Kalista, S.J., Jr, Ward, T.C., Oyetunji, Z., Self-Healing of Poly(Ethylene-co-Methacrylic Acid) Copolymers Following Projectile Puncture. *Mech. Adv. Mater. Struct.*, 14, 391–397, 2007.
98. Varley, R.J. and van der Zwaag, S., Development of a quasi-static test method to investigate the origin of self-healing in ionomers under ballistic conditions. *Polym. Test.*, 27, 11–19, 2008.
99. Kalista, S.J., Jr and Ward, T.C., Thermal characteristics of the self-healing response in poly(ethylene-co-methacrylic acid) copolymers. *J. R. Soc. Interface*, 4, 405–411, 2007.
100. Varley, R.J. and van der Zwaag, S., Towards an understanding of thermally activated self-healing of an ionomer system during ballistic penetration. *Acta Mater.*, 56, 5737–5750, 2008.
101. Kalista, S.J., Jr, Pflug, J.R., Varley, R.J., Effect of ionic content on ballistic self-healing in EMAA copolymers and ionomers. *Polym. Chem.*, 4, 18, 4910–4926, 2013.
102. Sandmann, B., Happ, B., Kupfer, S., Schacher, F.H., Hager, M.D., Schubert, U.S., The Self-Healing Potential of Triazole-Pyridine-Based Metallopolymers. *Macromol. Rapid Commun.*, 36, 604–609, 2015.
103. Whittell, G.R., Hager, M.D., Schubert, U.S., Manners, I., Functional soft materials from metallopolymers and metallosupramolecular polymers. *Nat. Mater.*, 10, 176–188, 21 February 2011.
104. Sandmann, B., Bode, S., Hager, M.D., Schubert, U.S., Metallopolymers as an Emerging CLass of Self-Healing Materials, in: *Self-Healing polymers: From principles to applications*, pp. 239–258, Springer-Verlag, Berlin Heidelberg, 2013.
105. Bentz, K.C. and Cohen, S.M., Supramolecular Metallopolymers: From Linear Materials to Infinite Networks. *Angew. Chem. Int. Ed.*, 57, 2–12, 2018.
106. Bode, S., Zedler, L., Schacher, F.H., Dietzek, B., Schmitt, M., Popp, J., Hager, M.D., Schubert, U.S., Self-Healing Polymers COatings Based on Crosslinked Metallosupramolecular Copolymers. *Adv. Mater.*, 25, 1634–1638, 2013.
107. Taylor, D.L. and Panhuis, M.i.h., Self-Healing Hydrogels. *Adv. Mater.*, 28, 9060–9093, 2016.

9

Self-Healable Catalysis

Bilge Coşkuner Filiz

Science and Technology Application and Research Center, Yildiz Technical University, Istanbul, Turkey

Abstract

Self-healable catalysis could be of property of heterogeneous or molecular materials based on various mechanisms. Self-healing catalysts have been already investigated in recent years, yet the studies were rarely reported. The investigations are generally focused on oxygen evolution catalysts. The self-healable catalysts could be improved by using proper metal, controlling potential or pH and usage of additive in catalysts structure is depending on the material. Nature inspired property for these artificial materials enable significant advantages for users. These smart materials could work under harsh conditions. By recovering their catalytically active sites by themselves, the high activity is provided with long lifespan. As a result maintenance costs of processes are decreased for long time applications.

Keywords: Self-healable, catalysis, recovery, oxygen evolution, hydrogen generation

9.1 Introduction

Self-healing materials are a kind of smart materials that could repair damage caused by mechanical usage or extreme environment, leading to maintain their structure or function in long-term use [1]. The design and development of new types of smart materials have gained importance day by day due to sophisticated requirements of many industries such as lighter, stronger, long-shelf life products. Self-healing property enables safety, long lifespan, low maintenance cost and limits the fatigue damage [2].

Email: coskuner@yildiz.edu.tr; bilgecoskuner@gmail.com

Inamuddin, Mohd Imran Ahamed, Rajender Boddula and Tariq Altalhi (eds.) Self-Healing Smart Materials and Allied Applications, (237–246) © 2021 Scrivener Publishing LLC

In recent years, self-healing catalysts have gained much attention for continuous specific process at steady-state [3]. Self-healing catalysts mean that catalysts have to maintain their catalytic activity and stability even though at high turnover numbers without any modification. The formation of products or by-products must not decrease the catalytic activity. The catalyst can repair itself without the addition of any material such as organic ligand, phase-transfer cations etc. into reaction medium during full turnover conditions. The development of self-healing catalysts is also one of the goals of the green chemistry [4]. The catalysts that have this feature have longer catalyst lifespan so that decreases the amount of annual catalysts requirement.

Monometallic noble metals such as Pt, Ru, Pd, Au, monometallic non-noble metals such as Co, Cu, Ni, bimetallic or multi-metallic catalysts in the form of metallic, oxide, alloy or composite have been used in catalytic process to catalyze the reactions to get higher yield and shorten production time. Besides their significant advantages, their costs are higher compared to the process inputs. Noble metals catalysts are higher price than the non-noble catalysts, which, together with these non-noble metals are easily oxidized under atmospheric conditions. So, the regeneration, storage and transfer of catalysts before and after production process are generally mass. Moreover, the activity of the catalysts decreases with time. Reusability, recyclability and regeneration properties of catalysts are the most important factors for selection right process. They have to keep their reactivity high under harsh conditions and subsequent batch of processes. The reactivity also decreases by the accumulation of product and by-products on the surface of catalysts. Sometimes they poison the catalytic sites and this lead to decrease their quantity. At this point, self-healing property of catalysts is an important property that not only enables decreasing the cost of catalytic process but also provides high reactivity for long term usages. The industrial catalysts generally have higher recyclability and stability properties and they keep their high reactivity up to limited cycles then they are required to be regenerated by some procedures such as heating up to determine temperature to burn the accumulated by-products on the surface of catalysts. This procedure extends the lifetime of the catalysts up to years. Sometimes these procedures are not enough to gain reactivity of catalyst, so they are required to be replaced with new materials which increase the process costs of products. At that point, self-healable property of catalyst increases the lifetime of catalysts and decreases cost as well. The self-healing property means that catalyst could turn back its initial internal and external properties by repairing itself during process [5].

9.2 Self-Healable Catalysis Applications

Self-healing catalysts have been already investigated, yet the studies were rarely reported. Self-healable catalysts have been gaining more attention in recent years for steady-state oxygen and hydrogen evolution reactions. These applications are basics of future energy systems, and these systems are now on research. Novel and green approaches are required to investigate and advance the future energy systems.

Production of hydrogen and water splitting are important topics due to being promising candidates among renewable energy options. Several techniques have been used and developed for the efficient and user-friendly mobile applications. In this part of the chapter, application of self-healable catalysts and their mechanism for energy field are presented.

9.2.1 Oxygen Evolution Catalysts

Water splitting is one of the trendiest researches in the energy field. Water is an attractive medium for storing solar energy and water splitting provides production of hydrogen in presence of catalysts. Self-healing property of the catalyst is a desired property for the electrocatalysts. These catalysts generally work under harsh alkali conditions which effect their performance and shorten the life [5, 6].

Electrocatalysts are one of the most promising technologies that provide large-scale production of oxygen with low cost. The formation of by-products such as hydroxide forms of metals on anode and hydride on cathode lead to discharge and consequently degradation of electrodes [7]. One of the most curial evaluation factors of catalysts is turnover number. This is defined as the number of catalytic cycles for catalysts to get in a deactivated form. Deactivated catalysts, used in finite number of cycles, are repaired by chemical activation and pre-catalysts are obtained to be used in reactions. On the other hand, self-healing mechanism provided infinite catalytic cycles and has an equilibrium between pre-catalysts and catalysts [5].

Metal oxides show the highest activity in the oxygen evolution reactions (OER) and they dissolved during the process due to corrosion. The catalytic cycle consists of four oxygen (proton) releasing step containing 4 electron and proton exchange under highly oxidative medium [5, 8]. In high pH medium, OH^- ions neutralize the release protons. In neutral medium, OH^- ion concentration changes between 10^{-5} and 10^{-9} M so the strongest base is oxide. On acidic mediums, metal oxides easily react with H^+ ions and effect from corrosion. Especially, some noble metals, earth-abundant metals are easily dissolved in acidic medium [8].

Self-healing property related with materials, enables autonomous healing from the damage due to reductive dissolution of material itself. Self-healing is an interesting strategy to increase durables of catalysts and prolong catalytic lifetime. Sometimes, compounds of catalyst could be so expensive, dangerous or impossible to change frequently. Nanomaterials also have a short catalytic life time due to their easy degradation and structural damages. There are only limited papers that have reported about known self-healing ability [9]. Together with this, non-noble metals such as Co, Mn and Ni balance their ions in the presence of phosphate (P) and borate (B) anions and formed active phase during OER cycle [5]. Molecular mechanism is based on O_2/H_2O cycles at cobalt centers that includes oxidation states of cobalt such as Co^{+2}, Co^{+3} and Co^{+4} during catalysis. High spin Co^{+2} ion is unstable. Higher oxidation states (Co^{+3} and Co^{+4}) of cobalt have low spin and are inert in oxygen–atom ligand. Metal ions form in medium by dissolution of metal oxide during cycle based on ligand substitution rates. During the catalytic cycle, cobalt electro-catalysts are unstable [8].

Water splitting catalysts generally produced from non-noble metals easily corrode in neutral medium. Self-healing of catalysts inhibits this corrosion and allows to use neutral/natural water. The chemical steps of kinetic model are taken into consideration for designing self-healing water-splitting catalysts. The applied potential, that are below the turnover, and solution pH provides controlling the repairment of catalytic material. During water splitting catalysts are cycled between catalytically active (catalyst) and precatalytic resting (pre-catalyst) states. In the electro oxidation process, catalysts converted into inactive material which are required to regenerate via chemical process [5]. Self-repair mechanism relies on the *in-situ* formation of active sites. For instance, in water splitting mechanism of O_2/H_2O cycle, cobalt centers change among Co^{2+}, Co^{3+} and Co^{4+} oxo oxidation states. The dynamic equilibrium between oxidation states during the catalytic cycles provides self-repair of electrodes [10].

Self-healing enables working in the wide range of pH such as close to neutral pH values so the requirement to external chemical for adjusting pH is not necessary any more [11]. Limited amount of researchers have studied this issue. Kanan and Nocera improved self-repairing Co-oxide catalysts for >pH 5 mediums, then they reported Ni based system. Zharieve and Dau published their work on manganese-based catalysts. Studies are generally on neutral-alkali pH values. On the other hand, electrocatalysts on anode shows instability in acidic mediums. Stahl *et al.* reported Nocera-type Co-oxide catalysts for acidic fluoride-containing mediums (pH 3.7). They stated that lower pH values than 3.5 led to formation of homogeneous catalysts in the electrode and they could not re-form catalysts on electrode [3, 8, 10, 12–14].

The deposition of metal hydroxide such as Ni [6], Co [7] etc. type metals on the surface of electrodes enables self-healing properties with high turnover frequency. In the preparation and adhering steps of these materials on to the electrodes surface, several polymers and minerals were used. Liang *et al.* preferred sodium tartrate and polyvinyl pyrrolidone (PVP) to get self-repairable $Ni(OH)_2$ films on electrode. These materials inhibit precipitation due to their coordinative and protective properties. The $Ni(OH)_2$ electrodes work well in high alkali medium to get hydrogen electrocatalysis toward glycerol oxidation with a high transformation efficiency and good stability while self-repairing [6]. Kuroda *et al.* modified brucite-type cobalt hydroxide with tripodal ligand tris(hydroxymethyl)aminomethane to get hybrid cobalt nanosheet (Co-ns). An organic modification of metal hydroxide provided highly stable self-repairing and dispersed catalyst layer on a nickel anode under cycled potential with high activity. Also this layer protected the nickel anode from corrosion in alkaline medium. The tripodal ligands selectively and strongly connected to basal planes hydroxides. Also these ligands employed various functionality to the electrode [7].

Self-healing properties of catalysts come from their electro-chemical reactions. The electrostatic interactions and charges of the electrode materials do not strongly affect self-healing. This property is supplied to catalyst by acting as oxygen evolution reaction catalytic proper and protective layer. Besides, the catalytic layer is precipitated partially during electrolysis cycles, application of constant current in the process repair the catalyst. Also, nano size and homogeneously dispersion of particles were reported to be the most important parameters of self-healing property. These features are provided by surface modification thanks to tripodal ligands. The use of a tripodal ligand is quite effective to provide these functions to the cobalt hydroxide nanosheets [7]. Metal ions are required to re-oxidize from an alkaline electrolyte. The change in current results with pH change of electrolyte [15]. Generally, self-healing property of catalysts is obtained by reversing catalyst corrosion at open circuit upon re-application of an anodic potential [16]. Several researchers reported self-healing pH range for their catalysts. Actually, the critical pH value of self-healing property depends on cell geometry of OER system and buffer concentration [5].

Surendranath *et al.* investigated OER catalysts prepared based on steady-state electrodeposition from Co^{2+} solutions in weakly basic electrodes. The Co (III) hydroxo based clusters were reported self-healable at higher pH values than 6 due to lower pH values were reported as corrosive for that material. The self-healing property of Co based catalysts reported that Co film formation and repairing property was provided by potential. Potential is a driving force for redepositing of Co ions in solution that were leached from film during cycles [17].

Lutterman and coworkers prepared ^{57}Co isotope containing Co–Pi OEC for monitoring the catalysts self-repairing mechanism. Firstly, unstable Co^{2+} ions slowly liberated from catalysts when there was no applied voltage. Then, 1.3 V applied and. re-oxidation of liberated Co^{2+} ions to repair catalyst occurred. Usage of phosphate in catalytic system enhanced cobalt release and the direct proportion of dissolution with applied voltage was reported. The existence of phosphate enable self-healing mechanism by rapid re-deposition of cobalt on the catalysts film [8].

Huynh *et al.* reported functionally stable manganese oxide (MnO_x) OER catalysts over wide range of pH (2.5–12). They presented the self-healing process based on reoxidation of MnO_x during OER in presence of chemical oxidant. It was reported that deposition of MnO_x was faster than O_2 evolution on MnO_x under same current density [18]. Najafpour *et al.* used permanganate ions to oxidize water to evolve O_2 and combine with Mn (II) ions, formed in solution, to form MnO_2 solids. Their proposed mechanism water oxidation and self-healing mechanism in water oxidation by Mn oxides in presence of Ce(IV) ions. Ce(IV) ions oxidized Mn(II) ions arrange acid–base chemistry and coordinated more acidic water than free water [9].

Gong *et al.* present self-healing tri-metallic PdAuCu nanobranch catalysts for oxygen reduction reactions. They used AuCu as a stable support and Cu counterparts were also responsible from the self-repairing of catalysts based on *in situ* formation of different oxidation states under high potential treatments up to 2,500 cycles. By application of high potential during starting and closing procedures of fuel cells, catalysts repairs itself. As a result of interaction between Cu and Pd, a synergistic effect resulted. This and high potential treatment were not only promoting the activity of catalysts but also repaired the catalysts by enhancing the mass activity [19].

Silicon based catalysts generally remain their activity in low and moderate pH values. Basic oxides such as first and second group elements do not have stability and easily corrode in acidic conditions. Besides this, first raw transition metal oxides showed stability in low pH values due to higher d-electron counts. It was also used as passivate on the other metals' surface. Due to most of metal oxides' unstableness under oxygen evolution reaction, self-healing property of catalysts gain importance for long term usage of fuel cells. Self-healing property of catalysts was determined based on the oxidation states of materials. Most of metal oxides are unstable in acidic medium and also they are dissolved during OER reaction due to O_2

evolution. The electro-configuration of the metals changed the acid–base equilibrium of the solution [9, 18].

9.2.2 Specific Catalysis Applications of Self-Healing Property

There are limited specific catalytic applications of self-healing material that have been reported.

Kinik *et al.* reported self-healing catalysts for the hydrolytic dehydrogenation of ammonia borane. They synthesized Cu based catalysts that was called SION-X, with the chemical formula of $Cu_2^{II}[(BO)(OH)_2](OH)_3$, which was the synthetic form of the mineral Jacquesdietrichite and, following *in situ* reduction. They indicated that the Cu^{II} ions in SION-X are reduced to Cu^0 nanoparticles during hydrolysis reaction, and then, following exposure to air, they were oxidized re-forming SION-X, and proved the self-healing property by using XRD analysis after each hydrolysis reaction. Also, the hydrogen generation activity of catalyst unchanged over many cycles. Without any external heat application or procedure, under ambient conditions catalyst gain its initial form and save its catalytic activity without any afford [20].

Another application of self-repairing biological material to based system to obtain oxygen is microbial photoelectrochemical solar cell. Photosynthetic organisms exhibit self-healable properties and this provides low cost durable large-scale solar power generation systems. The photosynthetic proteins extracted from organisms and their self-healing properties transfer to the power systems. Microbial fuel cells used glucose and oxygen as reactants, as photosynthetic microbial ones use carbon dioxide and water. As a result of consumption, long-term electricity is generated from sunlight without external input [21].

Kim *et al.* developed spontaneously repaired molecular catalysts that are inspired from self-repair mechanism of PSII in plants. A bipyridine-embedded UiO-type metal–organic framework was prepared for photocatalytic hydrogen evolution. 2,2'-bipyridine-5,5'-dicarboxylate as a ligand is a self-healing part of molecular catalyst. The total self-healing cycle of catalyst depends on the ratio of ligand and Pt sites and lasts 38 h [22].

Sodium alginate coated chloroperoxidase@$Ca_3(PO_4)_2$ hybrid nanocatalyst were reported to have self-healing properties in low pH medium. The coating the hybrid with sodium alginate gained the self-healing properties to the composite. Sodium alginate, polysaccharide, was used for chelating metal ions with ionic bonds between polyvalent cations and ionic G-G-rich series through polysaccharide backbone. Sodium alginate entrapping maintained the structure and function of enzymes. Also it crosslinked the metal ions that escaped from the composite structure in low pH mediums [1].

9.3 Conclusion

Self-healing could be defined for both heterogeneous and molecular catalysis that are inspired from nature. Self-healing property of oxygen evolution catalysts has been most investigated field for this topic. Using less potential than the catalytic cycle is required for self-healing. pH value of solution enables controlling potential. The self-healing of OER catalysts could be summarized: i) constant current electrolysis for the deposition of the catalyst layer, ii) presence of a highly dispersed catalyst in an alkaline electrolyte, iii) catalyst deposition ability of the anodic reaction, and iv) usage of proper additive into metal structure. Self-healing catalysis, as a rarely investigated and reported topic, has to properly gain more attention by researchers for several processes. This property enables high life-span for catalysts and in long term decreases the process investments. Besides, there is only limited research on this property, and the researcher is going to be focusing on this property in near future.

References

1. Liu, Y., Li, X., Yuan, Q., Liang, H., Self-repairing metal–organic hybrid complexes for reinforcing immobilized chloroperoxidase reusability. *Chem. Commun.*, 53, 3216–3219, https://doi.org/10.1039/C6CC10319G, 2017.
2. Mariconda, A., Longo, P., Agovino, A., Guadagno, L., Sorrentino, A., Raimondo, M., Synthesis of ruthenium catalysts functionalized graphene oxide for self-healing applications. *Polymer (Guildf)*, 69, 330–342, https://doi.org/10.1016/j.polymer.2015.04.048, 2015.
3. Bloor, L.G., Molina, P.I., Symes, M.D., Cronin, L., Low pH Electrolytic Water Splitting Using Earth-Abundant Metastable Catalysts That Self-Assemble *in Situ*. *J. Am. Chem. Soc.*, 136, 3304–3311, https://doi.org/10.1021/ja5003197, 2014.
4. Hill, C.L., Stable, Self-Assembling, Equilibrating Catalysts for Green Chemistry. *Angew. Chem. Int. Ed.*, 43, 4, 402–404, https://doi.org/10.1002/anie.200301701, 2004.
5. Costentin, C. and Nocera, D.G., Self-healing catalysis in water. *Proc. Natl. Acad. Sci. USA-Phys. Sci.*, 114, 13380–13384, https://doi.org/10.1073/pnas.1711836114, 2017.
6. Xiao, M., Miao, Y., Li, W., Yang, Y., Liang, X., Onsite Deposition of Self-Repairing Biomimetic Nanostructured Ni Catalysts with Improved

Electrocatalysis toward Glycerol Oxidation for H 2 Production. *Electrochim. Acta*, 178, 209–216, https://doi.org/10.1016/j.electacta.2015.07.185, 2015.
7. Kuroda, Y., Nishimoto, T., Mitsushima, S., Self-repairing hybrid nanosheet anode catalysts for alkaline water electrolysis connected with fluctuating renewable energy. *Electrochim. Acta*, 323, 134812, https://doi.org/10.1016/j.electacta.2019.134812, 2019.
8. Lutterman, D.A., Surendranath, Y., Nocera, D.G., A Self-Healing Oxygen-Evolving Catalyst. *J. Am. Chem. Soc.*, 131, 3838–3839, 2009.
9. Najafpour, M.M., Kompany-Zareh, M., Zahraei, A., Sedigh, D.J., Jaccard, H., Maryam, K., Britt David, R., Casey, W., Mechanism, decomposition pathway and new evidences for self-healing of manganese oxides as efficient water oxidizing catalysts: New insights. *Dalton Trans.*, 42, 14603–14611, https://doi.org/10.1039/C3DT51406D, 2013.
10. Kanan, M.W. and Nocera, D.G., *In Situ* Formation of an Water Containing Phosphate and Co2+. *Science (80-.)*, 321, 1072–1076, 2008.
11. Bediako, D.K., Costentin, C., Jones, E.C., Nocera, D.G., Proton–Electron Transport and Transfer in Electrocatalytic Films. Application to a Cobalt-Based O2 Evolution Catalyst. *J. Am. Chem. Soc.*, 135, 10492–10502, https://doi.org/10.1021/ja403656w |, 2013.
12. Zaharieva, I., Chernev, P., Risch, M., Klingan, K., Kohlhoff, M., Fischer, A., Dau, J., Electrosynthesis, functional, and structural characterization of a water-oxidizing manganese oxide. *Energy Environ. Sci.*, 5, 7081–7089, 2012, https://doi.org/10.1039/C2EE21191B, 2010.
13. Gerken, J.B., Landis, E.C., Hamers, R.J., Stahl, S.S., Fluoride-Modulated Cobalt Catalysts for Electrochemical Oxidation of Water under Non-Alkaline Conditions. *ChemSusChem*, 3, 1176–1179, https://doi.org/10.1002/cssc.201000161, 2010.
14. Nickel-borate oxygen-evolving catalyst that functions under benign conditions. *Proc. Natl. Acad. Sci. U.S.A.*, 107, 10337–10341, https://doi.org/10.1073/pnas.1001859107, 2010.
15. Galµn-Mascarós, J.R., Water Oxidation at Electrodes Modified with Earth-Abundant Transition-Metal Catalysts. *Chem. Electro. Chem.*, 1, 1–15, https://doi.org/10.1002/celc. 201402268, 2014.
16. Esswein, A.J., Surendranath, Y., Reece, Y., Nocera, D.G., Highly active cobalt phosphate and borate based oxygen evolving catalysts operating in neutral and natural waters. *Energy Environ. Sci.*, 2, 499–504, https://doi.org/10.1039/c0ee00518e, 2011.
17. Surendranath, Y., Lutterman, D.A., Liu, Y., Nocera, D.G., Nucleation, Growth, and Repair of a Cobalt-Based Oxygen Evolving Catalyst. *J. Am. Chem. Soc.*, 134, 6326–6336, https://doi.org/10.1021/ja3000084 |, 2012.
18. Huynh, M., Bediako, D.K., Nocera, D.G., A Functionally Stable Manganese Oxide Oxygen Evolution Catalyst in Acid. *J. Am. Chem. Soc.*, 136, 6002–6010, https://doi.org/10.1021/ja413147e, 2014.

19. Gong, H., Cao, X., Li, F., Gong, Y., Gu, L., PdAuCu Nanobranch as Self-Repairing Electrocatalyst for Oxygen Reduction Reaction. *ChemSusChem*, 10, 1469–1474, https://doi.org/10.1002/cssc.201700008, 2017.
20. Kinik, F.P., Nguyen, T.N., Oveisi, E., Valizadeh, B., Ebrahim, F.M., Gładysiak, A., Stylianou, K.C., Discovery of a self-healing catalyst for the hydrolytic dehydrogenation of ammonia borane. *J. Mater. Chem. A*, 7, 23830–23837, 2019.
21. Malik, S., Drott, E., Grisdela, P., Lee, J., Lee, C., Lowy, D.A., Tender, L.M., A self-assembling self-repairing microbial photoelectrochemical solar cell. *Energy Environ. Sci.*, 2, 292–298, https://doi.org/10.1039/b816417g, 2009.
22. Kim, D., Whang, D.R., Park, S.Y., Kim, D., Whang, D.R., Park, S.Y., Self-Healing of Molecular Catalyst and Photosensitizer on Metal–Organic Framework: Robust Molecular System for Photocatalytic H Evolution from Water Self-Healing of Molecular Catalyst and Photosensitizer on Metal-Organic Framework : Robust Molecular Syst. *J. Am. Chem. Soc.*, 138, 8698–8701, https://doi.org/10.1021/jacs.6b04552, 2016.

10

Self-Healing Materials in Corrosion Protection

Eiman Alibakhshi[1,2*], Bahram Ramezanzadeh[1†] and Mohammad Mahdavian[1‡]

[1]*Surface Coating and Corrosion Department, Institute for Color Science and Technology, Tehran, Iran*
[2]*Atlas Protecting Coating, Tehran, Iran*

Abstract

During the past decade, the application of polymeric coatings with intrinsic self-healing corrosion inhibition potency regarding metal substrates becomes a vital aspect for academic researches and industries. This chapter gives the recent advances and concise description regarding the potential applications of different kinds of micro/nano-carriers of healing agents for the development of smart/intrinsic self-healing anti-corrosion polymeric coatings. The self-healing materials/fillers classification, various techniques of the inhibitors loading in the carriers/containers, and their release/corrosion-inhibition mechanisms are introduced and discussed. The nano-containers are divided into five main categories with self-healing activity which are: ion-exchange based materials (i.e., clay, bentonite, montmorillonite, feldspar, layer double hydroxide (LDH), zeolite, halloysite, and kaolin), porous-structure and metal oxide materials, conductive polymers, fibril materials and lamellar-structure materials (i.e., graphene-based materials). The latest achievements regarding the development of the self-healing coatings based on these five groups of nano-containers to corrosion control of the metallic substrates are reviewed. Moreover, healing using polymerizable agents and micro/nano-capsules is proposed.

Keywords: Self-healing, micro/nano-container, corrosion-inhibition, polymerizable agents, inhibitive pigments, coating, micro/nano-capsules

**Corresponding author*: alibakhshi-ei@icrc.ac.ir
†*Corresponding author*: ramezanzadeh-bh@icrc.ac.ir
‡*Corresponding author*: mahdavian-m@icrc.ac.ir

Inamuddin, Mohd Imran Ahamed, Rajender Boddula and Tariq Altalhi (eds.) Self-Healing Smart Materials and Allied Applications, (247–296) © 2021 Scrivener Publishing LLC

10.1 Introduction

Corrosion of metal is an inevitable phenomenon, but it can be controlled by several different methods. From the past until now, one of the most promising and cost-effective ways for corrosion mitigation of engineering surfaces is the use of surface protective coatings [1, 2]. The allure of such systems demonstrates in the potency to provide a compact barrier against water and harsh/aggressive species due to their low free volume. Notwithstanding, the coating defects are unavoidable over time. This means that the corrosive species can reach the metallic substrate through chemical and/or physical defects in the polymer coating structure. The main physical factors affecting coating degradation are thermal cycles, scratches, impacts, and fatigue. Moisture, oxygen, alkali/acid aggressions, and UV radiation are also chemical factors [3, 4]. The formation of the defect in the coating may have a consequential effect on the coating structure, and hence physical, mechanical, and barrier function will be affected. Likewise, the surfaces exhibit significant growth of corrosion development, and the coatings cannot prevent it. In such cases, the presence of effective materials like corrosion inhibitors, corrosive inhibitive pigments, and micro/nano-containers are essential to control the corrosion in the defect sites and give effective long-term protection [5].

The new millennium has looked-on a metamorphosis of corrosion protection criteria, where safekeeping of metallic surfaces bestowed by prevalent protective coatings is being augmented by the impregnation of essential active species. Therefore, multi-functional coating systems that show significantly improved mechanical characteristics and possess protecting properties attributions and, along with satisfactory active inhibition, are of fundamental importance in goal-oriented designed coating systems. This is where the introduction of self-healing materials in corrosion protection becomes essential and effective. At present, the development of self-healing materials has attracted tremendous consideration in various fields, both from the fundamental and technological points of view. Conferring the Scopus data (Figure 10.1), there have been 974 papers on self-healing materials in corrosion protection announced from 1984 to end of March 2020.

The aim of this chapter book is to address new concepts and state of the art in the application of self-healing materials to the problems of metal corrosion. A potent emphasis is on the new materials that can express stimuli-responsive features and, therefore, a tremendous protection performance under demanding circumstances.

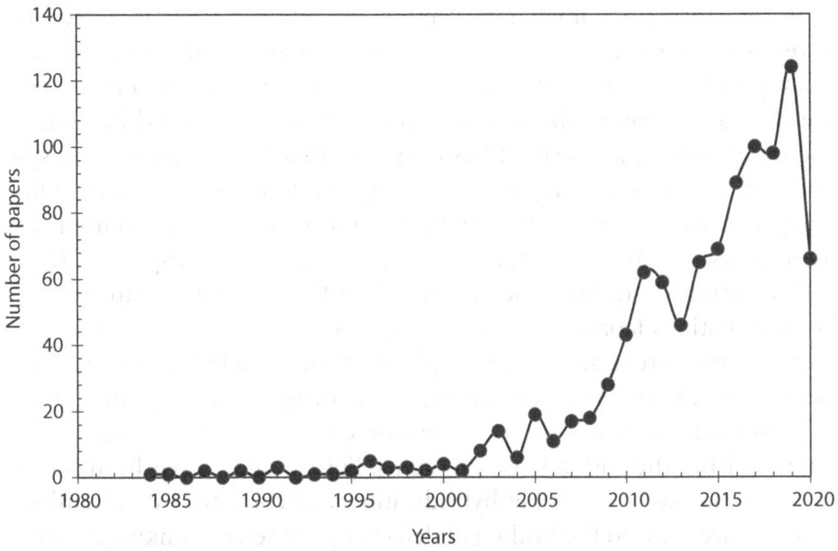

Figure 10.1 Number of publications on self-healing materials in corrosion protection announced from 1984 to the end of March 2020 conferring the Scopus data.

10.2 Self-Healing Definition

Many people may have seen Arnold's Terminator movie. The idea of creating self-healing/self-repairing/self-recovery materials has been a prominent fantasy ever since the Terminator franchise mythologized the concept in the 1980s. But, this idea has been extended in wide applications as time elapsed, and researchers are still working on developing materials that can independently heal/repair/recover themselves.

The definition of self-healing is the self-recovery of the preliminary characteristics of the material after deterioration because of the external triggering [6]. Self-healing materials have been classified as either independent (interference-free) or non-independent (foreign stimuli requirement) [7–9]. The independent (autonomic) mechanism has been frequently elaborated by impregnation of corrosion inhibitors or polymerizable healing factors into the polymeric matrix (coating). Also, the non-independent (non-autonomic) triggers include pH, water, temperature, mechanical damage, UV radiation, laser beam, and corrosive ions [10, 11].

Signs or signals may happen when a defect/damage in the coating occurs. Application of monitoring systems and sensors as a response activator (through human intervention or automatically) is a recently developed technique that has attracted attention in the field of self-healing materials for the detection of the signs or signals. In this strategy, the changes in colors, fluorescence, and transmittance could be used as coding of signals. However, it has been reported that applying these systems can lead to increasing the cost compared to the independent (autonomic) smart coating systems. Therefore, it is recommended to use them in precious, sensitive, or dangerous places that cannot be inspected [12].

The common strategies demonstrated in the literature regarding self-healing coatings fabrication is the encapsulation of healing materials, the presentation of reversible intrinsic physical/chemical hybridizations such as ionomers, Diels–Alder chemistry, and hydrogen bonding, and miscellaneous technologies [9, 13]. The healing capability of these strategies is different from the anti-corrosive systems. For instance, it has been shown that the thermally reversible hybridization detachment at a determined temperature leads to the healing of defects by the re-crosslinking polymer chains [14]. Accordingly, it seems crucial to control the external stimuli to grant the essential activation energy for bond reformation/breakage. This topic has been discussed extensively in most references [3, 6, 8] and will not be addressed in this chapter. In this section, the focus is on the self-healing materials for corrosion protection.

The classical healing concept is fundamentally based on the full recovery of the functionalities of the damaged coating [6, 15]. This phenomenon ultimately results in a significant enhancement of the systems' lifetime. However, in the case of the corrosion protection issue, partial recovery is sufficient [16]. In this regard, they are responsive materials that actively protect the metallic structures under the external environment. The employment of functional/active species is an efficient strategy for tailoring the structure and properties of the self-healing materials for corrosion protection. The several strategies that will be discussed in the following sections are as follows:

1. An inhibition of the corroded regions thanks to the presence of corrosion inhibitive pigments/inhibitors
2. The imprisonment and physical release of the inhibitor
3. Healing using polymerizable agents.

10.3 Inhibition of the Corroded Regions Thanks to the Presence of Corrosion Inhibitive Pigments/Inhibitors

The employment of corrosion inhibitive pigments/inhibitors is a swiftly progressing field in corrosion mitigation of the metallic structures [17–22]. For decades, the use of chromate compounds has gotten extensive attention due to its highly effective corrosion protection for various metals. However, due to the toxic characteristics of the hexavalent chromium, the utilization of this compound has been strongly prohibited for industrial applications [5, 23]. Regulations imposed by governmental agencies led to a vigorous probe to figure out the surrogate pigments with low perniciousness and superior performance. Zinc phosphate (ZP) pigment as the first chemical compound recommended for the replacement of the zinc chromate dominates the market [24–29]. Shao *et al.* evaluated the effect of ZP on the self-healing behavior of epoxy coating applied on steel in NaCl solution using electrochemical methods [30]. The inhibition effect of ZP pigment on the epoxy protective ability has been proved by the EIS (electrochemical impedance spectroscopy) and EN (electrochemical noise). As it is apparent from the Figure 10.2, the results of scanning electrochemical microscopy (SECM) depicted that the defected area was healed by the development of a protective layer. Huang *et al.* have surveyed the self-healing potent of ZP, which was loaded into a plasma electrolytic oxidation (PEO) coating on aluminum [31]. Outcomes have demonstrated astonishingly remarkable protection (over 11,000 h salt spray test) through the migration of ZP components (i.e., phosphate/zinc ions) from the surface to pores. This leads to the healing of the corrosion paths and longevity increments. The self-healing process of the PEO coating, including ZP during the salt spray test, has been schematically given in Figure 10.3.

Toxicity, cost, and performance are essential factors for industrial product development and commercial marketing of the corrosive inhibitive pigments [23]. Despite low toxicity, it has been shown that ZP has less corrosion inhibitive action than zinc chromate. This is because of its lower solubility in the harsh environment. Thus, some physical and/or chemical adjustments have been proposed to enrich the ZP pigment inhibitive

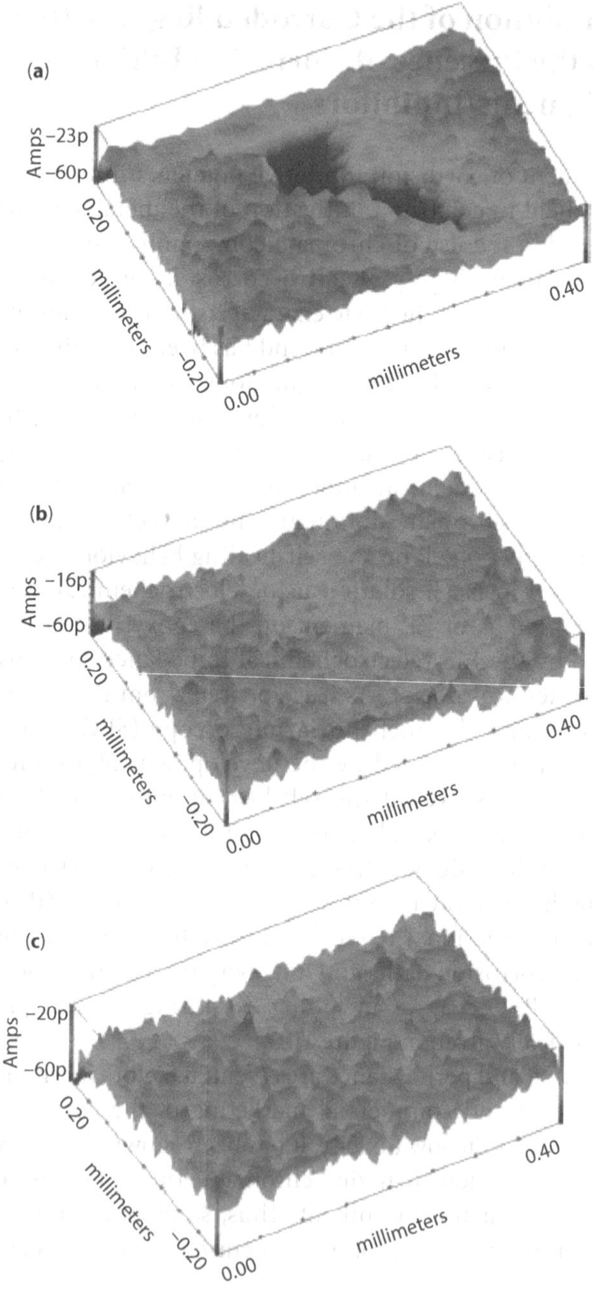

Figure 10.2 SECM images of the epoxy coating, including zinc phosphate immersed in a 3.5% NaCl solution. (a) the initial stage, (b) 36 h, and (c) 58 h after making an artificial pore (adapted from Ref. [30]).

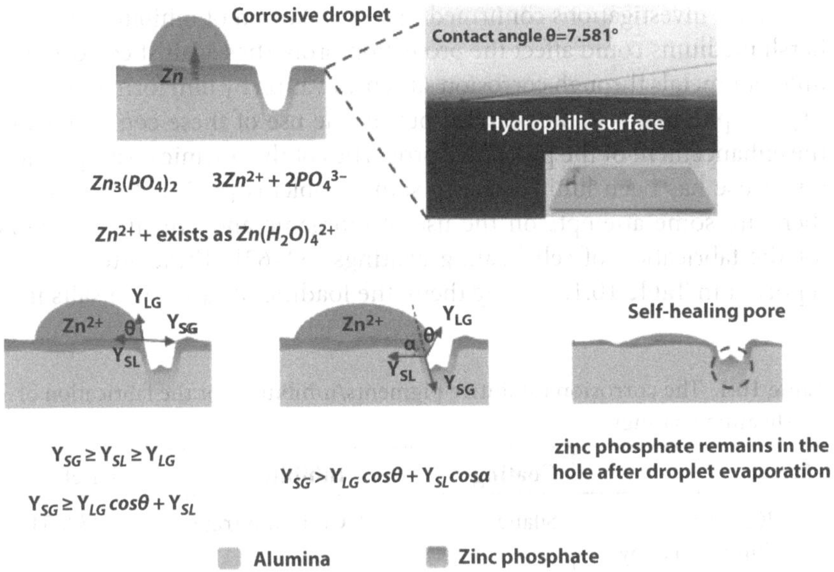

Figure 10.3 Schematic demonstration of the self-healing process of the PEO coating including ZP during the salt spray test (adapted from Ref. [31]).

performance in recent years. Zinc potassium phosphate, zinc sodium phosphate, zinc molybdenum phosphate, zinc aluminum phosphate, zinc lithium phosphate, zinc polyphosphates are classic instances of traditional zinc phosphate pigment that have been obtained through modification of the anodic, cathodic or both parts of the pigment. It has been claimed that the modified ZPs provide better corrosion inhibition activity than ZP. The inhibitive mechanism of ZPs depends on the partial solubility in water and releasing solubilized inhibitive species, protecting the metal corrosion via passive layer formation [32–41]. The outcomes of our group researches have demonstrated that the development of such pigments could effectively enhance the epoxy composite protection function by releasing the Zn^{2+} and PO_4^{3-} species [32, 34, 42].

It is observed in the literature that the composite containing nano-sized pigments has attracted more attention to developing high-performance protective coatings thanks to their greater surface area compared with the micro-sized pigments [43]. In other words, incorporation of the nano-sized pigments such as ZnO, Fe_2O_3, SiO_2, and Al_2O_3 into the organic coatings led to significant coating corrosion-protection ability enhancement resulted by the effective role of nanoparticles in the barrier performance promotion via covering the free volumes and voids and increasing the diffusion length [44–46].

Several investigations confirmed that the addition of inhibitors into the harsh mediums could affect the protection properties against corrosion of different metals through corrosion sites passivation by film formation [47–51]. Despite many studies carried out on the use of these compounds in the enhancement of the protective properties of the organic coatings, their direct use has been limited due to some problems [5, 52]. Nevertheless, there are some attempts on the use of inorganic and organic inhibitors for the fabrication of self-healing coatings [53–62]. These attempts are reported in Table 10.1. Among them, the loading of rare-earth salts into

Table 10.1 The corrosion inhibitive pigments/inhibitors for the fabrication of self-healing coatings.

Metal	Coating	Inhibitor	Ref
AA2024-T3 Aluminum alloy	Silane	Cerium nitrate	[53, 54]
Steel	Silane	Praseodymium nitrate	[55]
Galvanized steel	Silane	Cerium nitrate	[56]
AA2024-T3 Aluminum alloy	Polyurethane	Lithium oxalate	[57]
AA2024-T3 Aluminum alloy	Polyurethane	Lithium carbonate	[58]
AA2024-T3 Aluminum alloy	Epoxy	Cerium cinnamate	[59]
2017 Aluminum alloy	Silane	Chloranil	[60]
5083 Aluminum alloy	Vinylester	TiO_2	[61]
7075 Aluminum alloy	Titania	Benzotriazole (BTA)	[62]
Steel	Polydimethyl-siloxane	BTA	[70]
MA8 Magnesium alloy	PEO	8-hydroxyquinoline (8-HQ)	[71, 72]
AA2024-T3 Aluminum alloy	Silane	8-HQ	[73]

(*Continued*)

Table 10.1 The corrosion inhibitive pigments/inhibitors for the fabrication of self-healing coatings. (*Continued*)

Metal	Coating	Inhibitor	Ref
Steel	Silane	Praseodymium nitrate/ Benzimidazole (BIA)	[63]
Magnesium alloy	First layer: micro-arc oxidation Top layer: water-based epoxy-polyamide	Na_3PO_4, 2-Mercaptobenzo-thiazole (MBT)	[64]
Steel	Epoxy	ZP/Benzotriazole (BTA)	[65]
Steel	ZrO_2 ceramic-based coating	CeO_2/BTA	[66]
Steel	Epoxy ester	Zinc cations/ *Cichoriumc intybus* L leaf extract	[74]
Steel	Epoxy	Neodymium (III) nitrate hexahydrate/BIA	[67]
Steel	Epoxy-polyamide	Zinc acetate/BTA	[68]
Steel	Epoxy-polyamide	Zinc acetate/*Urtica dioica*	[69]

sol-gel silane coating has been extensively considered respecting their outstanding amended electrochemical characteristics [53–56]. More recently, the self-healing performance of a polyurethane coating on AA2024-T3 by the aid of lithium salts is also described [57, 58]. Such behavior is attributed to the lithium species leaching on the scratched area and also the creation of a compact protective layer.

Another promising candidate for healing the surface of the polymeric matrix through introducing functionalities are different classes of the inhibitors/pigments composed of varying anionic and cationic structures

[63–69]. The suggested mechanism for their self-healing performance can be elaborated based on the partial solubility of the hybrid inhibitive species in the harsh medium and diffusion toward the defect area or the liquid/metal interface where it builds a protective layer. These hybrid organic/inorganic compounds are also tabulated in Table 10.1.

The works from the Ramezanzadeh group denote a breakthrough in active corrosion protection. They have conducted a series of examinations to investigate the self-healing capability of the mixture of inorganic-praseodymium nitrate/organic-benzimidazole inhibitors embedded into the sol–gel based silane coating on the mild steel [63]. The electrochemical methods and surface analysis results showed the synergistic healing behavior of this sample by the development of a protective film on the damaged zones. The current distribution outcomes from SVET exhibited that both cathodic and anodic inhibitions exist at the damaged zones with a dominant role of the cathodic reactions retardation. Ralkhal *et al.* [67] have prepared a novel hybrid pigment with neodymium and benzimidazole (BIA) for achieving an impressive active/barrier protective coating. It was denoted that the scratches can be healed using a homogeneously distributed protective layer. Despite all the above reviews, it seems that such inhibitors still had the potential to show more improved properties. Thus, other methods are discussed, which will be discussed in the next section.

10.4 The Imprisonment and Physical Release of the Inhibitor

It was demonstrated that amenability between the inhibitor and the coating matrix, as well as film formability, play key roles in self-healing behavior. In this regard, the unsought reactions must be restricted to continuously increases the corrosion inhibitive performance over time. Likewise, the inhibitor/pigment aggregation causes the conduits to form, therefore, accelerates water penetration. To sum up, it can be said from the above-mentioned aspects that the controlled-release of inhibitor influences the self-healing characteristics of the coating. Right now, the loading of active functional agents in the host nano/micro container and capsules are effective controlled-release strategies for increased corrosion protection, especially the self-healing prevalence. In these strategies, the reactive and mobile inhibitive species are stored in a suitable setting and released when necessary (coating defects). The latest studies in this regard are presented in this section.

In the last two decades, there was a gigantic growing activity in the development of containers in self-healing materials for corrosion protection. A wide range of compounds like ion-exchange based materials, porous-structure materials, metal oxide nanoparticles, conductive polymers, fibril, nanotubes, and lamellar materials have been used in different coatings as the container for loading of various active functional agents. In this section, the most important nano/micro containers used for active corrosion protection and their working mechanisms are pursued.

10.4.1 Ion-Exchange-Based Materials

As one of the important milestones in the progress of self-healing materials development for corrosion protection, ion-exchangeable materials have been very outstanding. In these systems, the presence of harsh ions gives sufficient driving force to the ion-exchange process to displace and smart release of inhibitors in proper locations. In recent years, the systems that rely on ion-exchange triggered inhibitor release include coatings based on clay, bentonite, montmorillonite, feldspar, layer double hydroxide (LDH), zeolite, halloysite, and kaolin. They have been categorized into two groups: cation exchange and anion exchange materials. The ion-exchange capacity of a container is highly dependent on the layers' spacing and charge density of its surface (negative or positive charge). It is well accepted that there is a close relationship dependence between the inhibitor release profile and the intrinsic ion-exchange capacity of containers [7]. Accordingly, it seems crucial to control the inhibitor release to improve the self-healing properties of the coatings in the long-term. This means that the release of the inhibitor must be balanced. It has been revealed that the fast release rapidly drains the inhibitor from the container, which leads to an inverse alteration in the protection and expedites corrosion. It also seems pretty probable that the slow release of the inhibitor has not been able to provide proper concentration inside the damaged zones and simultaneously may not result in self-healing behavior. It was also demonstrated that the solubility of the inorganic scaffold plays a crucial role in self-healing performance. The final point to note about the containers is the size and concentration. It has been reported that applying massive containers can lead to a decrease in the coating barrier ability accompanied by increasing the defects and strains inside the coating. The effect of these parameters on the active corrosion protection was investigated by Borisova *et al.* [75].

It is interesting to note that the natural and synthetic clays (such as pristine or montmorillonites, bentonite, kaolin, feldspar, zeolite, halloysite) are indexed in the self-healing evaluation of different coatings because of

Table 10.2 The most important cation exchanger containers with related inhibitors for the fabrication of self-healing coatings.

Metal	Coating	Container	Inhibitor	Ref
Steel	Epoxy	Pristine clay	Indole-3 butyric acid	[76]
Steel	Epoxy	Montmorillonite (MMT)	Aminotrimethyl-phosphonic acid	[77]
Steel	Epoxy ester	MMT	Zinc/BI	[78]
Steel	Silane	MMT	*Ocimum basilium* leaves	[79]
Galvanized steel	Polyvinyl butyral (PVB)	Bentonite	Cerium	[80]
Galvanized steel	PVB	Bentonite	Zinc	[81]
Steel	Epoxy ester	Kaolin	Zinc	[82]
Steel	Epoxy ester	Feldspar	Zinc	[83]
Steel	Epoxy ester	Zeolite	Zinc	[84]
Steel	Epoxy ester	Zeolite	Zinc/2-Mercaptobenzimi-dazole (MBI)	[85]
Mg-Li alloy	Epoxy	Zeolite	Cerium	[86]
AA2024-T3 Aluminum alloy	Silane	Zeolite	Cerium	[87]
AA2024 Aluminum alloy	Silane	Zeolite	Cerium	[88]
AA2024 Aluminum alloy	Silane	Zeolite	Cerium/ Diethyldithio-carbamate	[89]
AM50 Magnesium alloy	PEO	Halloysite	BTA	[90]

(Continued)

Table 10.2 The most important cation exchanger containers with related inhibitors for the fabrication of self-healing coatings. (*Continued*)

Metal	Coating	Container	Inhibitor	Ref
AA2024 Aluminum alloy	Silane	Halloysite	2-mercaptobenzo-thiazole (MBT)	[91]
Copper alloy	Acrylic latex and Polyurethane	Halloysite	BTA	[92]
Copper alloy	Acrylic latex and Polyurethane	Halloysite	MBI	[92]
Copper alloy	Acrylic latex and Polyurethane	Halloysite	MBT	[92]

their wealthy intercalation nature. All of them are cation-exchanger and used for the controlled-release of various corrosion inhibitors. The most important of these nano/micro containers with related inhibitors applied for self-healing coating fabrication are reported in Table 10.2.

In one of these studies, the self-healing behavior was seen by localized electrochemical impedance spectroscopy (LEIS) test when the indole-3 butyric acid-modified clay was added to an epoxy coating [76]. This compound improves the self-healing function thanks to high inhibitor release when the local pH increases during the corrosion process in a neutral environment. It has been reported that the inhibitor impregnation into the clay particles changes the interlayer distance, which leads to a significant alteration on the protection properties.

Modification of the Na-MMT (sodium montmorillonite clay) in order to increase its interlayer distance and release content for self-healing coating application was investigated by Ghazi *et al.* [78]. The significant enhancement in the interlayer distance is obtained after loading of benzimidazole (BIA^+)/zinc (Zn^{2+}) mixture into the Na-MMT structure. This leads to a better cation exchange (inhibitors) with Na^+ as a harmful ion in NaCl solution. The healing process was studied by EIS and surface analyses. The outcomes evidenced a continuous growth of the impedance and resistance with time because of the construction of inhibitive species over the scratched zones (Figure 10.4).

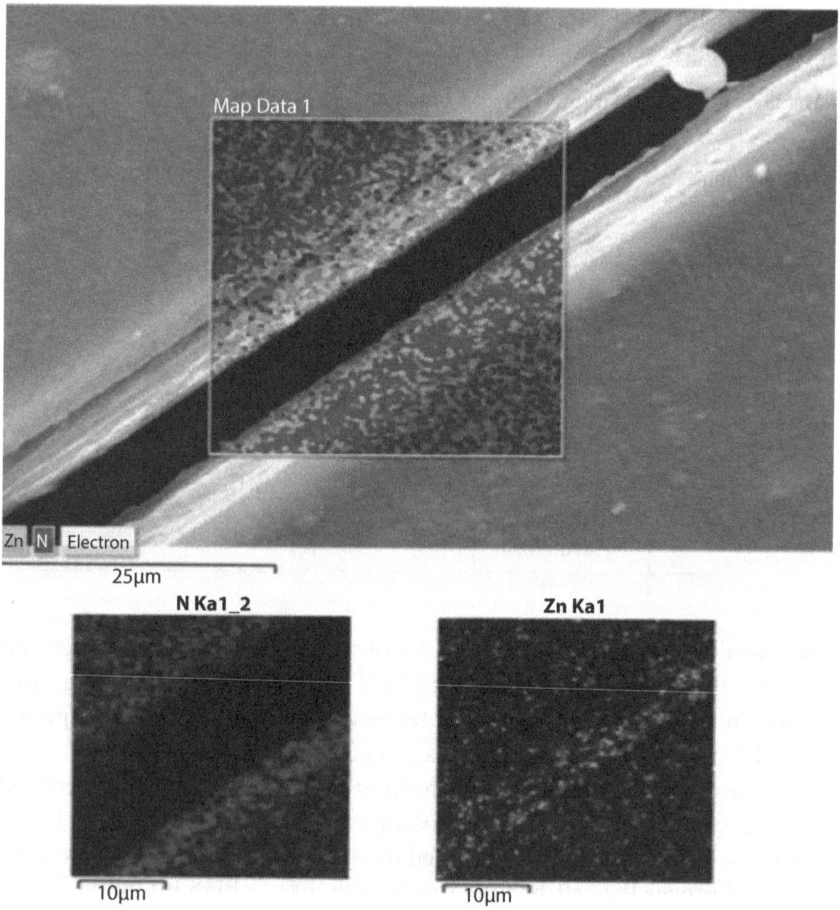

Figure 10.4 FE-SEM–EDX elemental map of mild steel coated with MMT-Zn-BIA immersed in 3.5% NaCl after 5 h (adapted from Ref. [78]).

The role of bentonite, feldspar, and kaolin as the container for self-healing coatings has been rarely reported, though early outcomes have been hopeful [80–83]. Unlike the three mentioned compounds, zeolite and halloysite structures have been extensively studied in the literature to achieve active corrosion protection [84–92].

Zeolites (aluminosilicates with a 3D porous configuration) have been found as a promising container for active species release in a controllable way thanks to the great surface area and outstanding ion-exchange potency. Rassouli *et al.* [85] have probed prolonging protection of an epoxy ester coating containing NaX zeolite nano-container embedded with zinc and 2-mercaptobenzimidazole. The prolonging protection is mainly due to the creation of the resistive layer caused by the adsorption/desorption of inorganic/organic complex inside the scratched area. Some studies have investigated the presence of cerium ions within the zeolite structure, and its role in the self-healing performance of different coatings (such as silane and epoxy) on various metal surfaces protection [86–89]. It was proved that the cerium loaded zeolite effectively protected the surface through adsorption of ionic agents during corrosion and release of cerium. The self-healing behavior of a silane coating impregnated with zeolite, including a mixture of cerium/diethyldithiocarbamate on aluminum alloy, was also stated by Ferrer *et al.* [89]. Highlights in these studies are the high loading of zeolite into the coating, which makes them difficult to use in practice. To overcome this limitation, halloysite has been suggested. Halloysite refers to a class of clay nanotubes containing two-layered aluminosilicate structure with an empty lumen with 15 nm diameter and about 1,000 nm length [92]. They are widely used as containers owing to their outstanding properties, such as great dispersibility, high interlayer adhesion, environmentally friendly nature, and cost-effective [90–92]. Sun *et al.* [90] have examined the benzotriazole (BTA) embedded halloysite nanotube for self-healing evaluation of PEO coating on magnesium alloy. The healing behavior was obtained by film formation in the corrosion zones. The growth of film and pH are proven to have a direct correlation, and the protection (film formation) increases by increasing the pH in the pores of coating. The surface of the readied coatings after the electrochemical experiments is shown in Figure 10.5. The surface morphology of the coating, including BTA loaded halloysite (Figure 10.5c), shows no significant changes without any evidence of corrosion pits. Whereas, the surface of the coating with (Figure 10.5b) and without (Figure 10.5a) halloysite shows large pits, indicating pitting corrosion.

Halloysite nanotube container is also reported as an effective container for MBI, BTA, and MBT inhibitors for self-healing coating fabrication on copper alloy [92]. The self-healing behavior follows in the order of MBI > MBT > BTA. The enhanced self-healing character is related to the higher inhibitor lumen embedding and controllable release in damaged zones. The release profile results denoted that the MBT and BTA release happens uncontrollably, while the MBI release was balanced (not fast, not slow).

So far, we have been talking about cation exchangeable compounds. However, materials with anion exchange capability are also highly capable of self-healing due to the presence of chlorine and hydroxide ions in corrosive environments. The most important nano-containers with anion exchangeability are the layered double hydroxides (LDHs), which will be discussed in detail in this section.

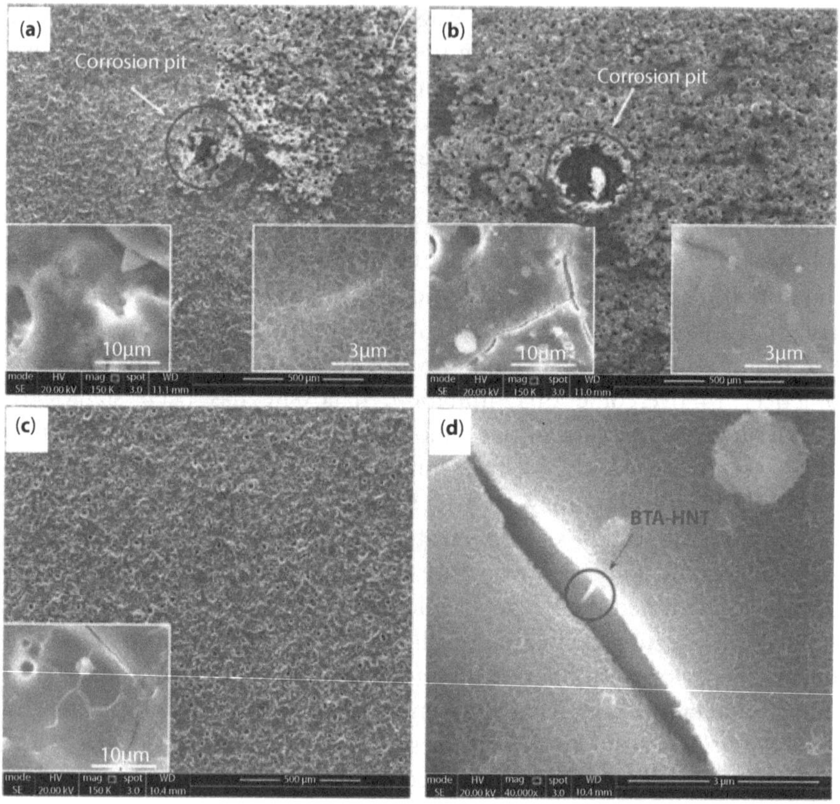

Figure 10.5 Surface morphologies of PEO coatings following electrochemical corrosion tests (a) PEO, (b) PEO with halloysite, (c) PEO with halloysite/BTA and (d) enlarged view of the figure (c) (adapted from Ref. [90]).

The general formula for LDH, known as hydrotalcite or anionic clay, is $\left[M^{2+}_{1-x}M^{3+}_{x}(OH)_2\right]^{x+} A^{n-}_{x/2} . mH_2O$, where M^{2+} and M^{3+} are the metal cations, and A^{n-} is an anion. LDHs exist in nature but are also cheap and straightforward to prepare, making them an exciting option for many applications. The LDHs are usually synthesized into two routes: first is direct preparation that leads to most facile production because of the one-step synthesis. The second route is based on an anion exchange mechanism to intercalate inhibitive anions in the LDH galleries. The major disadvantage of LDHs is that they are not selective to particular anions. One application of LDHs, which has achieved worthy attention in recent years, is their ability to be applied as a container in the field of corrosion protection. LDH structure is ideal for self-healing performance as it delivers the inhibitive species necessary for long-time protection and harsh ions away from the damaged area. LDHs are able to capture harsh anions from a corrosive environment by an anion-exchange mechanism. A wide range of LDHs has been used for active corrosion protection. This section aims to provide a sample of the available literature that contains relevant or required background information necessary to understand its self-healing behavior.

LDH was identified in the 1967s, but the corrosion protection application started in the first of the 2000s. As one of the first works, Buchheit *et al.* have studied corrosion protection behavior of epoxy coating, including Zn–Al-decavanadate hydrotalcite [93]. Simultaneously, Williams and McMurray formulated PVB coating with hydrotalcite (HT), including chromate, nitrate, and carbonate to corrosion control of AA2024-T3 surface [94]. In the next year, they have loaded benzotriazole, ethyl xanthate, and oxalate anions into hydrotalcite structure with the same purpose [95]. The results denoted the ability of these compounds to protect metals from corrosion and emphasized the need for further investigation to find their mechanism. For this reason, much investigation has been done in the following years to understand the mechanism of hydrotalcite behavior within the coating. The most important LDH containers applied for self-healing coatings fabrication are briefed in Table 10.3.

Zheludkevich *et al.* are among the forerunners, who applied the LDHs as nano-containers of inhibitors in corrosion-driven self-healing behavior investigations [96, 97]. They loaded various inorganic (NO_3^-, H_xPO_4, VO_x) and organic (MBT) inhibitors into the LDH structure and dispersed them in the bilayer systems, including sol–gel (first layer) and epoxy coating (top layer).

The vanadate-containing system in the first layer and MBT in the top layer denoted the best inhibiting performance. The longevity of the active corrosion inhibition performance is correlated to a controllable release of vanadate in the long term and fast release of MBT at abridged times. As another result, phosphate (HxPO_4) is the least efficient inhibitor for the protection of AA2024 [96]. Contrary to this study, Alibakhshi and colleagues have proposed the positive role of phosphate $\left(PO_4^{3-}\right)$ loaded Zn–Al–LDH for application in active protective coatings on mild steel [98, 99]. The prepared $Zn{-}Al{-}PO_4^{3-}$ by anion exchange route has been dispersed in a silane coating, and then an epoxy layer without $Zn{-}Al{-}PO_4^{3-}$ was applied on it. The superior active corrosion ability of the coating systems, including $Zn{-}Al{-}PO_4^{3-}$ after 600 and 1000 h exposure to salt spray cabinet, is illustrated in Figure 10.6. The released phosphate from LDH galley can be reacted with mild steel surface, forming iron phosphate on the surface or damaged area. In other words, the PO_4^{3-} is an anodic inhibitor, which highly tends to be adsorbed on the anodic sites and form insoluble iron

Table 10.3 The most important LDH container applied for self-healing coating fabrication.

Metal	Coating	LDH type	Interlayer anion	Ref
AA2024 Aluminum alloy	Silane and epoxy	Zn–Al	Vanadate, phosphate, nitrate, and MBT	[96]
AA2024 Aluminum alloy	Epoxy	Mg–Al and Zn–Al	Divanadate	[97]
Steel	Silane and epoxy	Zn–Al	Nitrate and phosphate	[98, 99]
Steel	Silane and epoxy	Mg–Al	Nitrate and phosphate	[100]
Steel	Silane	Zn–Al	Nitrate, molybdate, and phosphate	[101]
Hot-dip galvanized (HDG) steel	Polyurethane	Zn–Al	Molybdate	[102]
Steel	Epoxy	Zn–Al	MBT	[103]
Magnesium alloy	Epoxy	Mg–Al	Tungstate	[104]
AA2024 Aluminum alloy	Epoxy	Zn–Al	Aniline, and benzene derivatives	[105]
Steel	Zinc-rich epoxy	Zn–Al	MBT	[106]
AA2024 Aluminum alloy	Silane	Zn–Al	Cerium	[107]
Steel	Silane	Zn–Al	Molybdate and SiO_2	[108]

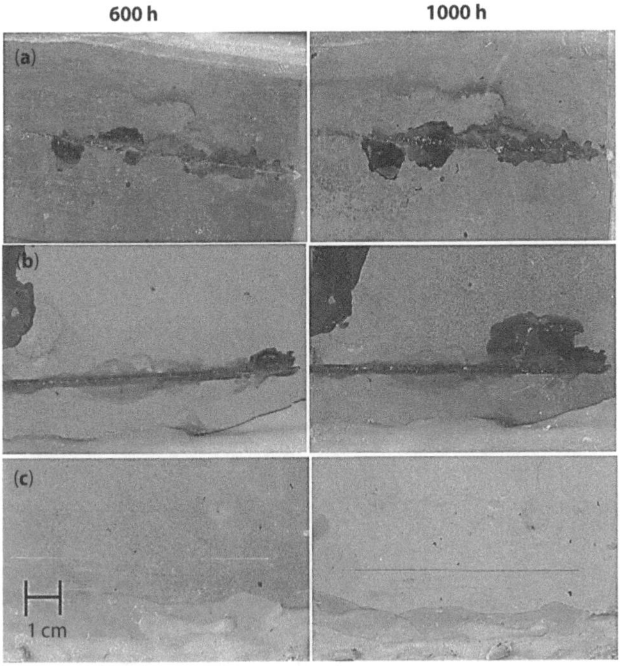

Figure 10.6 Results of salt spray test for the coating systems (a) without LDH, (b) with Zn–Al– NO_3^- LDH, and (c) with Zn–Al–PO_4^{3-} LDH after 600 and 1,000 h (adapted from Ref. [99]).

phosphate compounds. While this condition did not occur for aluminum [96]. It was also revealed that the synergistic effect of phosphate with cation release from LDH lattice (slight solubility) is an important mechanism of active inhibition rather than on-demand release reported in the literature.

This group has also assed the active protection performance of the silane coating, including Zn–Al–LDH as a container of various inhibitors (nitrate, molybdate, and phosphate) with electrochemical methods [101]. The electrochemical current noise (ECN) signal results for active corrosion protection detection of the scratched samples (Blank: bare metal, SC: silane coating, SCN: silane coating including Zn–Al–NO_3^- LDH, SCM: silane coating including Zn–Al–MoO_4^{2-} LDH, and SCP: silane coating including Zn–Al–PO_4^{3-} LDH) after 4 h immersion in a 3.5% NaCl solution is demonstrated in Figure 10.7. The ECN results showed fewer components at low frequency in the presence of SCP and SCM, and the high-frequency components were only observed for these samples. These are the sign of localized corrosion occurrence at the scratched zones, reflecting the active corrosion protection by Zn–Al–PO_4^{3-} and Zn–Al–MoO_4^{2-} LDHs. The

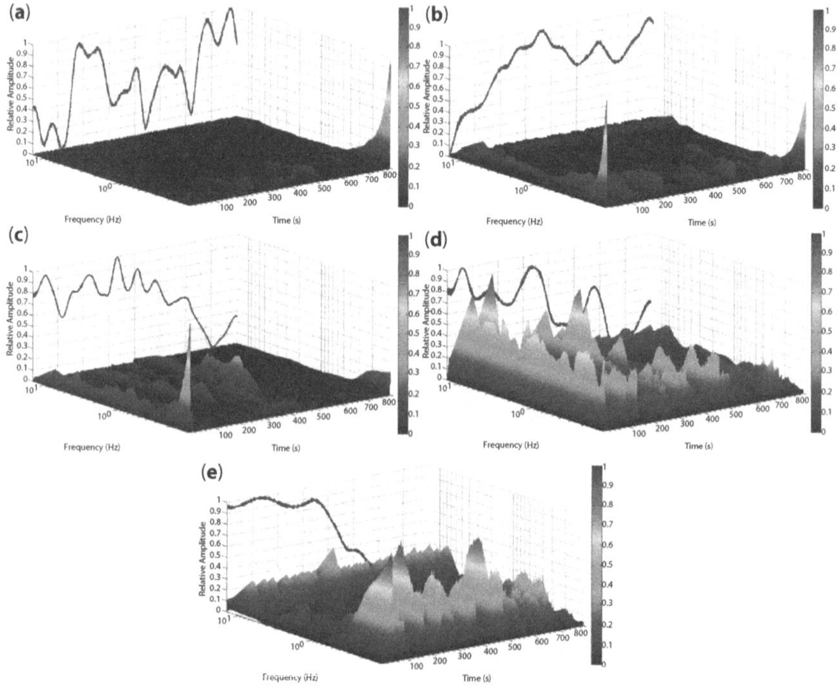

Figure 10.7 UWT spectra of the ECN signals obtained for blank (a), SC (b), SCN (c), SCM (d), and SCP (e) samples after 4 h immersion in 3.5 wt.% NaCl solution (adapted from Ref. [101]).

lowest total absolute energy of detail crystals was observed in the presence of Zn–Al–PO_4^{3-}, indicating its better protection compared to the other samples. This is related to a higher release of active species from LDH.

In an interesting research, Shkirskiy *et al.* [102] have discussed the controlling inhibitive concept of molybdate loaded LDH into polyurethane coating using environment composition. They demonstrated that the carbonate anions present in the harsh solution lead to the fast release of molybdate.

Comparing Zn–Al and Mg–Al systems, although the release of Mg–Al inhibitors is faster, Zn–Al protective behavior is better [97].

Almost all studies in this field have shown that the inhibitor release through anion exchange and partial dissolution of the LDH scaffold can lead to improved self-healing behavior. However, the physical entrapment of aggressive ions from a corrosive environment into the structure is another application of LDHs [109–111]. The ion-entrapment effect can be applied to the coating with and without scratches (intact coating) and even without inhibitors. This mechanism was comprehensively discussed for the first time by Tedim *et al.* [110]. Zn–Al–NO_3^--LDH has been prepared via the direct route and embedded into an automotive polyurethane coating. The results demonstrated that the synthesized LDH could effectively reduce the coating permeability to chlorides, thereby hence the coatings protection increase.

More recently, Alibakhshi *et al.* [109] have conceptualized the role of the LDH synthesis pH on the interlayer spacing and chloride entrapment from the harsh medium. A schematic representation for the LDHs protection mechanism with different pHs is depicted in Figure 10.8. The outcomes have repeated the beneficial inhibition effects of the LDH because of its higher chlorine entrapment from the solution and proper inhibitor release. It is also well revealed that there is a close dependence between the interlayer spacing and the corrosion inhibition ability. The LDH with a higher interlayer spacing (prepared at pH = 9.5) denoted the best performance because of its higher capacity for chloride entrapment.

Another compound that can be used in self-healing coating through the anion-exchange mechanism is layered hydroxide salts (LHS). There has not been much study on its protection behavior, but it acts exactly like LDH [112]. Wang *et al.* [112] have evaluated active corrosion protection of the dehydrated zinc hydroxide sulfate in fluorocarbon coatings. They showed

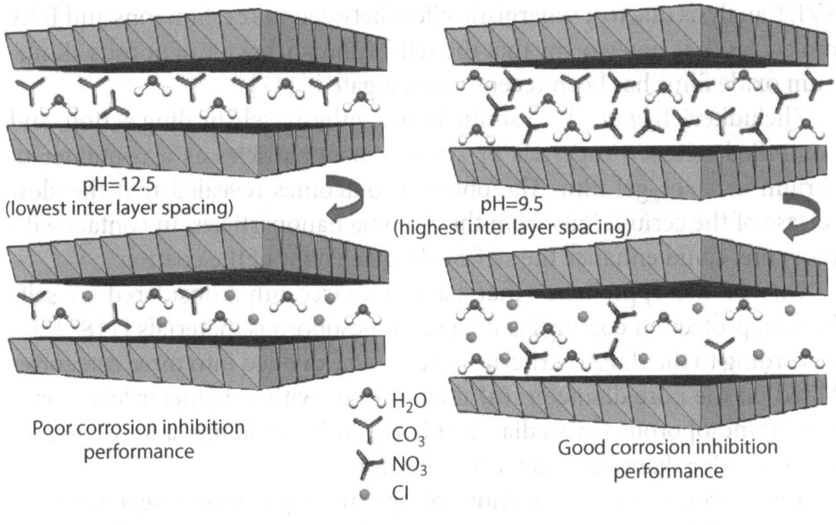

Figure 10.8 A schematic presentation for the LDHs protection mechanism with different pHs (adapted from Ref. [109]).

that the compound could act as ion traps by phase transition behavior and strengthen the active protective properties. The great advantage of LHSs compared to LDHs is the easier synthetic method to obtain the first, without the need for a complicated apparatus to maintain an inert gas flow and constant pH during the entire process of synthesis, and also the use of a single metal in the material structure. The structure of LHSs is a modification of the brucite-like structure by substitution of part of the hydroxide groups by suitable anions or water molecules.

10.4.2 Porous-Structure and Metal Oxide Materials

Porous-structure and metal oxide materials as the container can be very interesting in self-healing applications as they halt the corrosion process by their ability in long-lasting controlled-release. TiO_2 [113, 114], ZrO_2 [115], CeO_2 [116], and SiO_2 [116] are some important examples of such metal oxide materials with porous-structure that have been extensively applied for self-healing coating fabrication in recent years.

Lamaka *et al.* [113] have investigated the active corrosion protective performance of a bilayer system, including an inhibitor-embedded TiO_2 porous film and a hybrid silane film. In this study, the TiO_2 porous film acted as a container for a mixture of cerium nitrate/BTA and was able to limit the negative effects of the inhibitor on the barrier properties of the hybrid coating. The self-healing behavior of the system confirmed by SVET analysis due to a synergistic effect between the cerium ions and BTA molecules. It is noteworthy that the self-healing behavior of porous aluminum oxide films has been recently investigated [117].

Zheludkevich *et al.* [115] studied the synthesis, self-healing action, and controlled-release properties of zirconia nanoparticles as a container of cerium in a sol–gel film. The obtained outcomes revealed that the slow release of the cerium ions from the zirconia nanoparticles, in contact with moisture, could enhance the self-healing behavior of the coating.

One of the approaches that have been recently considered in self-healing protection coatings is the use of mesoporous materials [118–120]. According to the IUPAC criteria, pores can be divided into three categories based on the pore diameter: (a) microporous with a diameter less than 2 nm (b) mesoporous with a diameter between 2 and 50 nm, and (c) macroporous with a diameter greater than 50 nm.

The active corrosion protection ability of polypyrrole coating in the presence of molybdate loaded mesoporous silica has been denoted by Saremi and Yeganeh [118]. Borisova *et al.* [119] have proposed a self-healing coating by the employment of monodisperse mesoporous silica nanoparticles.

The synthesized mesoporous silica has shown to have an enhancing impact on the embedding inhibitor (BTA) capacity (slightly two times higher) since this material demonstrates a higher surface area and great pore volume compared to the previously reported ones. The BTA loaded silica without additional polyelectrolyte shell significantly amended the self-healing behavior of the SiO_x/ZrO_x sol–gel coating applied on aluminum, proved by SVET analysis. This could be attributed to the BTA molecules release in a controlled way induced by the corrosion process. Interestingly, some researchers believe that the zirconium particles in the SiO_x/ZrO_x hybrid system can act as a container of inhibitors [121].

More recently, a new porous structure material based on Zeolitic imidazole framework-8 (ZIF-8) and hollow mesoporous silica nanoparticle have been introduced as a container of BTA in epoxy coating with excellent self-healing ability [122]. The results provided a perspective on the creation of a multi-purpose nano-container for use in self-healing coatings.

10.4.3 Conductive Polymers

Regarding the favorable advantages of conductive polymers like environmental nature and facile synthesis, there are a number of studies to introduce the corrosion inhibitive behavior of this type of material for enhancement of anti-corrosive performance of various coatings. Various types of conductive polymers are considered as filler/pigment to be impregnated into coatings which polyaniline, polythiophenen, and polypyrrole can be referenced [123–127]. In the coating matrix, they undertake the metal surface passivation at pinhole regions because of their redox behavior. By shifting the corrosion potential to passive areas, the anodic protection can also be achieved. However, detection of the protection mechanism is obscured for this material, and many mechanisms can be provided [128]. In recent years, the conductive polymers were popularized as a container to the self-healing coating development. From the viewpoint of the self-healing applications, conductive polymers in coatings are expected to act as a container for the release of their doped inhibitors in response to external triggering during their reduction process [128]. In this regard, various kinds of negatively charged inhibitors such as polyoxometalates, dithiocarbamate, mercaptothiadiazole, transition metal ions [217], and ammonium salt have been doped in the conductive polymers for self-healing coating fabrication [3, 129–131]. The possible mechanism for these compounds can be elaborated based on the construction of a metal oxide passive film through the smart release of doped anions for the healing of the small defects. This behavior is exactly the same as that of chromate compounds [128].

2,5-dimercapto1,3,4-thiadiazole dipotassium salt (DMcT) doped polyaniline significantly improved the self-healing properties of the epoxy coatings applied on various aluminum alloys experimented for 2,000 h in harsh solution [130]. Corrosion was controlled thanks to the protecting ability of the DMcT leaking from polyaniline in the coating defects.

Karpakam and colleagues [132] have reported the enhanced self-healing performance of the coating, including molybdate modified polyaniline. Molybdate and phosphate anions accomplish the same behavior as in the case of polypyrrole coating [128]. These anions (when released) can react with metal, which reduces the corrosion process and heal the coating defects. In another work, the self-healing action of the PVB coating incorporated with TiO_2/polyaniline has been studied [133]. The excellent self-healing action of this coating was correlated to dual active/barrier function by polyaniline and TiO_2 nanoparticles.

Arefinia *et al.* [134] have synthesized dodecylbenzene sulfonic acid (DBSA) doped polyaniline and its self-healing behavior evaluated in an epoxy ester coating on carbon steel. The self-healing action proved in this paper, thanks to the reaction of surface metal cations with released DBSA from polyaniline, led to the creation of protecting layer on coating defects.

10.4.4 Fibril Materials

Fibril is a promising material for application as a container of corrosion inhibitors for self-healing coating applications. The fibril network structure is able to store an extensive number of inhibitive species and release them at the damaged sites. The focus of the studies in the field of self-healing coatings, especially container subject, has been mostly on cellulose fiber structures and polyaniline fibers [135–139]. However, making containers using electrospun coaxial fiber and 3D network film is also provided [140–142]. Acceptance of the first two systems (cellulose or polyaniline fibers) is in the controlled-release of inhibitor in the long-term, while the last two systems are not able to trigger or controlled-release features of the corrosion inhibitor and therefore their use is limited [15].

Cellulose fibers are classified as biodegradable polymers that are abundant in nature, eco-friendly, and have network structures. That's why Yabuki's team, as a pioneer, has done some research on the role of these compounds as a container of corrosion inhibitors in self-healing coating fabrication [15]. They denoted that the active corrosion protection of the epoxy coating can be significantly enhanced in the presence of doped cellulose nanofibers with calcium nitrate [135]. This confirmed by higher polarization resistance of the scratched coating, including this mixture compared to the coating, including only a calcium nitrate. The best self-healing function was to mix 0.5% cellulose and 2% calcium nitrate in the coating. In conclusion, the nanofiber networks act as a pathway for the controlled-release of inhibitors at the scratched zones. In another study, a direct incremental correlation between the inhibitor (oleic acid) release from the nanofibers and the pH value is discovered by this group [137]. The most significant increasing rate for polarization resistance versus the immersion time has been recorded at pH 11.4. This is mainly due to the creation of the resistive layer caused by the adsorption of the oleic acid molecules at the scratched zones. This phenomenon ultimately results in a significant enhancement of the coatings' self-healing ability.

Polyaniline can be present in various chemical structures like the Emeraldine salt with various morphologies. One of these morphologies is in the form of fibers, which is used in the active corrosion protection fields as a container for the release of inhibitors. Pirhady *et al.* [136] fabricated a coating based on cerium nitrate embedded polyaniline nanofibers for self-healing evaluation. Self-healing behavior was studied using EIS and SVET on the scratched coating, demonstrating the tendency of polyaniline fibers to form a container that controls the inhibitor release. The self-healing behavior of this system was due to the construction of pH-sensitive cerium–polyaniline complex at the scratched sites.

It should be noted that other fibers, such as hollow and carbon fibers, have been used to self-healing polymer coatings but have not been employed for corrosion protection [143–145].

10.4.5 Lamellar-Structure Materials

In the previous sections, we talked about three-layered compositions (clay, LDH, and LDS), and in this section, our focus is on the graphene-based materials for self-healing applications.

In recent years, graphene and graphene oxide (GO), as lamellar structure materials, have been studied in the formulation of coatings to elevate their mechanical properties and electrochemical characteristics due to the special properties such as low permeability and high specific surface area [146–148]. Despite all of these remarkable perks, the application of GO as an adequate additive in coatings for corrosion protection is still subject to debate, and some researchers believe that it (unfunctionalized GO) can hasten metal oxidation in harsh mediums in the long run [149].

Notwithstanding significant progress in the corrosion protection of GO/coating composites, dispersion of GO in coatings (thanks to its high surface area) has remained an important challenge for researchers outlining a big demand for new functionalized GO compounds [150, 151]. Thus, the surface functionalization of GO by the aid of covalent and/or non-covalent mechanisms is commonly practiced in the corrosion protection field. The epoxide groups on the basal plane and carboxylic at the edges of the GO surface are credible for its covalent functionalization. Also, the non-covalent functionalization of organic molecules and ions is performed by ion exchange, electrostatic forces, hydrogen bonding, π–π, and cation–π interactions [151].

It has been proved that the barrier properties of the coatings are improved through good dispersion of functionalized GO. Functionalization has been done by various compounds like polyisocyanate [152], polydopamine [153], polyaniline [147, 154, 155], SiO_2 [156], Al_2O_3 [157], Fluor [158], phenylenediamine [159, 160], aminopropyl triethoxy silane [161, 162], 3-(triethoxysilyl)propyl isocyanate [163], poly(2-butylaniline) [164], polyvinylpyrrolidone [146], and so on. It is noteworthy that all of the above studies have emphasized the barrier function of graphene-based compounds, and none of them have shown active corrosion protection and self-healing behavior.

In recent years, graphene-based materials have also been used in adsorbent and water treatment applications [165–167]. Due to the conjugated system of GO surface beside the multiple functionalities on the surface and boundary of sheets, their maximum adsorption index is much higher than that of typical adsorbents [151]. Accordingly, graphene-based materials seem to be able to be successfully used in harsh environments as a container of inhibitors. However, the functionalization of the GO surface is still necessary to gain sufficient dispersion in the coating and also to enhance the adsorption capacity. Recently, GO was applied as a solid container with a controllable release rate of inhibitors to be used in coatings for achieving anti-corrosion coatings with excellently improved dual-barrier/self-healing properties. The most important of these investigations are reported in Table 10.4. Unlike the containers of the previous sections, all the compounds presented in this table are loaded into coating in very small amounts (less than tenths of a percent) and have a prominent self-healing effect. The enhancement in the prepared coatings anti-corrosion ability is accredited to the controlled release capability of inhibitors at the coating-metal interface, resulting in the anodic/cathodic reactions control and protective film generation at defect region. The physical interactions are of remarkable importance for the adequate desorption of inhibitors, so the functionalization of GO nanosheets was also resulted in the increase of the smart inhibitors release in the corrosion simulated condition. It seems that

Table 10.4 The most important Go container applied for self-healing coating fabrication.

Metal	Coating	Functionalization agent	Inhibitor	Ref
Steel	Epoxy polyamide	Aminopropyl triethoxy silane	BIA	[151]
Steel	PVB	Polyvinyl alcohol (PVA)	cerium hydroxide	[168]
Steel	Epoxy polyamide	Polyaniline	BTA	[169]
Steel	Epoxy	Zinc phosphate	BIA	[170]
Steel	Epoxy	–	Cerium nitrate/BIA	[171]
Steel	Epoxy	Polystyrene	8-HQ	[172]
Steel	Epoxy ester	Tamarindus indiaca	Zinc	[173]
Steel	Epoxy	Cyclodextrin	BTA	[174]
Steel	Silane	–	Praseodymium nitrate	[175]
Steel	Epoxy	–	*Urtica dioica*	[176]
Steel	Epoxy polyamine	Histamine	Zinc nitrate	[177]
Steel	Epoxy	–	Cerium nitrate	[178]
Steel	Epoxy polyamide	Polydopamine	Zinc nitrate	[179]
Steel	Epoxy polyamide	Polypyrrol	Zinc nitrate	[180]
Steel	Epoxy polyamide	–	Zinc nitrate	[181]
N80	Epoxy	–	Aspartic acid intercalated LDH	[182]
Steel	Alkyd	Dopamine	Polyaniline	[183]
Steel	Epoxy polyamide	–	Zeolitic imidazolate framework-8 (ZIF-8)	[184]

the role of graphene oxide as a container of corrosion inhibitors requires more work to enter commercial scale.

10.4.6 Other Containers

In the previous sub-sections, the most important nano-containers used in the construction of self-repairing coatings were introduced. However, due to the breadth of this discussion, some of the nano-containers may not have been introduced above including cation-exchange resins [185], hydroxyapatite [186], layered titanate matrix [187], cyclodextrin [188], calcium carbonate [189, 190], cerium phosphate [191, 192], and cerium molybdate [193].

It was mentioned earlier that the number of corrosion inhibitors capable of being applied in self-healing coatings is relatively limited. One of the most dynamic research domains proposed to overcome this problem is the reservation of inhibitor species into the nano/microcapsules. It is well accepted that the nano/microcapsules are fine particles containing a core and a shell. They can be prepared according to various routes: sol–gel, interfacial polymerization, extrusion, in-situ polymerization, and coacervation [7].

The topic of encapsulation in self-healing ability is used for composites or bulk polymers with mechanical damage as well as coatings, although there are differences in these applications. These differences are due to the size of the capsules, and the use of large-sized capsules in bulk polymers can cause problems in the organic coating [194]. Polymeric capsules have been extensively considered for self-healing coating fabrication. The mechanical capsules disruption during a defect formation causes a provisional activation of the water displacement mechanism as the most evident trigger. One other important methods using polymers for encapsulation in the self-healing coating fabrication is the pH-sensitive capsules [195].

It should be noted that most of the topics in this section overlap with the previous section (nano-containers) and the next section, and therefore in this division, only the method of polyelectrolyte-based shells is mentioned.

One route proposed to gain smart storage/release action is the encapsulation of active functional agents in the polyelectrolyte capsules or layers. In this route, two layers of polyelectrolyte with contrary charges as shells are formed on the particles' surfaces as the core, and the active species are loaded inside the layers via LbL (layer-be-layer) method. The trigger mechanism here is a change in pH, and due to its change with the onset of the corrosion process, the release of active species takes place [88]. There are many examples in this regard, some of which are mentioned in this section [196–200]. Andreeva *et al.* [201] have designed a self-healing coating based on pH-sensitive polyelectrolyte and 8-HQ. The localized electrochemical test proved the long-term protection of this coating against corrosion. Plawecka *et al.* [200] proposed a technique to encapsulate the MBT in the polyelectrolyte shell for the enhancement of epoxy coating protection on a different substrate. The healing behavior was obtained using inhibition of tiny defects on the surface, proved by SVET and LEIS.

It is noteworthy that a mixture of nanocontainers or layer-by-layer method is also used recently, which causes the presence and release of more inhibitors and thus improves the self-healing behavior of the coating [198, 202, 203]. All of the introduced nano-containers have advantages and disadvantages, and efforts are underway to improve these combinations. The future of self-healing coatings seems to depend on the use of nano-containers that are multifunctional.

Regarding the nano-containers discussions, it can be said that most of them are more efficient in the closest layer to the surface like hybrid sol–gel coating or pretreatments layer, although some, such as graphene oxide, have been used in the upper layers. Also, the self-healing evaluation of studies in this section has been done using various electrochemical techniques such as SVET, scanning Kelvin probe (SKP), SECM, LEIS, EN, and EIS as well as surface analysis. Another important point is the economic debate over their use, and researchers are trying to come up with combinations that are economically viable so that they can be used in the industry. With respect to all the views and opinions, research in the field of nano-containers will continue to move towards the goals of the industry.

10.5 Healing Using Polymerizable Agents

Polymerizable healing agents can release to the defect zone, and they can form a solid polymeric film to heal the defect. One-component and two-component polymerizable healing agents have been used in smart coatings' formulation. Linseed oil is a well-known one-component healing agent that dries after exposure to the atmosphere [204–207]. Linseed oil is an eco-friendly and cheap healing agent. However, the presence of weak hydrolyzable ester bonds, between glycerol and the fatty acids, in its chemical structure, makes it unappropriated for high-performance corrosion protective coatings application. Tung oil is another example of green one-component healing agents [208, 209], which dries faster and provides more crosslinks compared to the linseed oil thanks to its higher unsaturated bonds [210]. However, it suffers from weak hydrolyzable ester bonds the same as linseed oil. Epoxy ester resins are another example of air drying healing agents. They are synthesized from the reaction of an epoxy resin, and a fatty acid originated from a drying oil [211]. Due to the presence of epoxy in the chemical structure, the epoxy ester is more resistant to the harsh environment compared to the drying oils [212]. Isocyanates are among the high-performance healing agents. Isophorone diisocyanate (IPDI) is a frequently used isocyanate based healing agent to prepare self-healing coatings [213, 214]. Isocyanates can readily react with moisture to form polyurea at defect sites. Organosilanes can also be used as healing agents as they can form a siloxane network after leaching from the container in the defect area [215, 216].

Although one-component healing agents take advantage of easy application in the coatings, it should be mentioned that polymerization of one-component film formers can also take place in the containers during service time of the coatings. Therefore, the self-healing performance will be decreased during the aging of the coating as there would be fewer monomers available to form a polymeric film in the defect area. Two-component healing agents have been developed to overcome the aging weakness of one-component healing agents and to provide higher protective performance. The components have to be separately kept in the coating before their functioning in the defect area. Different approaches have been utilized to maintain the component of the healing agents unreacted before their release in the defect area. They can be loaded separately in the same or different types of containers to prepare dual-core containers. Epoxy and amine compounds, e.g., tetraethylene pentamine (TEPA), and diethylene triamine (DETA) and have been frequently used as two-component healing agents [217–219].

Among different types of containers, the organic ones including poly(urea-formaldehyde) microcapsules (PUF-MCs), poly(urea-formaldehyde-melamine) microcapsules (PUFM-MCs), poly(methyl methacrylate) microcapsules (PMMA-MCs), self-assembled graphene oxide microcapsules (SAGO-MCs), and polyurea/polyurethane microcapsules (PU-MCs) have been frequently used for encapsulation of the healing agents. A list of containers used for encapsulation of one- and two-component healings agents is provided in Table 10.5.

10.6 Conclusion and Outlook

For years, the polymeric-based coating systems with log-time autonomous self-healing anti-corrosion characteristics have been utilized for the

Table 10.5 The self-healing coatings based on one- and two-component healing agents.

Metal	Coating	Container#1	Healing agent#1	Container #2	Healing agent#2	Ref
Steel	Epoxy	PUF-MCs	Linseed Oil	–	–	[204, 205]
Steel	Epoxy	PMMA-MCs	Linseed Oil	–	–	[206]
Galvanized Steel	Waterborne Polyurethane	SAGO-MCs	Linseed Oil	–	–	[207]
Steel	Epoxy	PUF-MCs	Tung Oil	–	–	[208]
Steel	Polyurethane	PU-MCs	Tung Oil	–	–	[209]
Steel	Epoxy	PUFM-MCs	Epoxy Ester	–	–	[211]
Steel	Epoxy	PUF-MCs	Epoxy Ester	–	–	[212]
Steel	Epoxy	PU-MCs	IPDI	–	–	[213]
Steel	Epoxy	Silica- PU-MCs	IPDI	–	–	[214]
Steel	Epoxy	PUF-MCs	Organosilanes	–	–	[215]
Steel	Epoxy	PUF-MCs	Organosilanes	–	–	[216]
Steel	Epoxy	PU-MCs	Epoxy	PU-MCs	TEPA	[217]
Steel	Epoxy	TiO_2 Nanotubes	Epoxy	SBA-15	DETA	[218]
Steel	Epoxy	Carbon Hallow Spheres	Epoxy	Carbon Hallow Spheres	Polyamine	[219]

reduction of the metals' corrosion mitigation and maintenance costs reduction. In this chapter, a large number of methods, including the conventional usage of the inhibiting fillers/additives and advance nanocontainers-based coatings with high self-healing potency, are reviewed, and their self-healing concepts have been discussed and compared in details. An overview of the numerous approaches utilized for the construction of self-healing anticorrosion coatings revealed that by incorporation of various nano-containers into the coating formulation, the coating self-healing corrosion protection ability could be adequately improved. It has been shown in this review that not only the type/concentration of the nano-container but also the other parameters, i.e., the inhibitor loading/release capacity, the type of inhibitor (organic or inorganic), the release mechanism (sensitivity of the nano-container to the various parameters, i.e., pH, temperature, mechanical triggering, etc.) and the polymer matrix mechanical characteristics are important issues that should be taken in consideration when developing the self-healing coatings. The advantages and disadvantages of the loading of the active functional agents in the host nano/micro container for application in the polymeric systems for achieving long-lasting self-healing action have been discussed in detail in this chapter.

The high cost and short-term self-healing action are two important issues that have strongly limited the large commercial scale applications of the self-healing materials in anti-corrosion coatings development. So, as a future remark for application/development of self-healing coatings/materials, combining different nano-containers with the concept of achieving multilevel self-healing features in one coating system simultaneously, would be a promising strategy for the development of the anti-corrosion systems with higher self-healing durability. Besides, the development of the advanced nano-containers based on the metal organic-based frameworks (MOFs) and carbon-based materials (i.e., graphene oxide) will be more considered in future studies. One another important issue for the future application of the self-healing materials in corrosion protective coatings is the economic aspect, leading to more consideration of the new cost-effective technologies of development of the nano-containers, which are suitable for commercial applications.

References

1. Forsgenr, A., *Corrosion Control Through Organic Coatings*, CRC Press, Taylor and Francis, United States, 2017.

2. Ahmad, Z., *Principles of corrosion engineering and corrosion control*, Butterworth-Heinemann, Oxford, 2017.
3. Hughes, A., Mol, J.M.C., Zheludkevich, M.L., *Active protective coatings: New-generation coatings for metals*, Springer, Netherlands, 2016.
4. Wang, D. and Bierwagen, G.P., Sol–gel coatings on metals for corrosion protection. *Prog. Org. Coat.*, 64, 327, 2009.
5. Montemor, M.F., Functional and smart coatings for corrosion protection: A review of recent advances. *Surf. Coat. Technol.*, 258, 17, 2014.
6. Montemor, M.F., Hybrid nanocontainer-based smart self-healing composite coatings for the protection of metallic assets, in: *Smart Compos. Coatings Membr*, pp. 183–210, Elsevier Ltd, Woodhead, Cambridge, 2015.
7. *Industrial applications for intelligent polymers and coatings*, Springer, New York, 2016.
8. Eduok, U., Ohaeri, E., Szpunar, J., Self-healing composite coatings with protective and anticorrosion potentials: classification by healing mechanism, in: *Self-Healing Compos. Mater*, pp. 123–162, Elsevier Inc, Woodhead, Cambridge, 2020.
9. Abdolahzadeh, M., Van Der Zwaag, S., Garcia, S.J., Self-healing corrosion-protective sol–gel Coatings based on extrinsic and intrinsic healing approaches. *Adv. Polym. Sci. Polym. Sci.*, 273, 185, 2016.
10. Amendola, V. and Meneghetti, M., Self-healing at the nanoscale. *Nanoscale*, 1, 74, 2009.
11. Fedrizzi, L., *Self-healing properties of new surface treatments*, European Federation of Corrosion, Taylor and Francis, United States, 2010.
12. Galvao, T.L.P., Wilhelm, M., Gomes, J.R.B., Tedim, J., Emerging trends in smart nanocontainers for corrosion applications, in: *Smart Nanocontainers*, pp. 385–398, Elsevier Inc, Netherlands, 2020.
13. Lutz, A., Mol, J.M.C., De Graeve, I., Terryn, H., Smart corrosion protection by multi-action self-healing polymeric coatings, in: *Smart Compos. Coatings Membr*, pp. 157–182, Elsevier Ltd, Woodhead, Cambridge, 2020.
14. Zhang, F., Ju, P., Pan, M., Zhang, D., Huang, Y., Li, G., Li, X., Self-healing mechanisms in smart protective coatings: A review. *Corros. Sci.*, 144, 74, 2018.
15. Yabuki, A. and Fathona, I.W., Self-healing corrosion protective coatings in transportation industries, in: *Adv. Smart Coatings Thin Film. Futur. Ind. Biomed. Eng. Appl.*, pp. 99–134, Elsevier Inc, Netherlands, 2020.
16. González-García, Y., Mol, J.M.C., Muselle, T., De Graeve, I., Van Assche, G., Scheltjens, G., Van Mele, B., Terryn, H., A combined mechanical, microscopic and local electrochemical evaluation of self-healing properties of shape-memory polyurethane coatings. *Electrochim. Acta*, 56, 9619, 2011.
17. Palimi, M.J., Alibakhshi, E., Ramezanzadeh, B., Bahlakeh, G., Mahdavian, M., Screening the anti-corrosion effect of a hybrid pigment based on zinc acetyl acetonate on the corrosion protection performance of an epoxy-ester polymeric coating. *J. Taiwan Inst. Chem. Eng.*, 82, 261–272, 2018.

18. Amirudin, A., Barreau, C., Hellouin, R., Thierry, D., Evaluation of anti-corrosive pigments by pigment extract studies, atmospheric exposure and electrochemical impedance spectroscopy. *Prog. Org. Coat.*, 25, 339, 1995.
19. Palimi, M.J., Alibakhshi, E., Bahlakeh, G., Ramezanzadeh, B., Mahdavian, M., Electrochemical investigations of the corrosion protection properties of an epoxy-ester coating filled with cerium acetyl acetonate anticorrosive pigment. *J. Electrochem. Soc.*, 164, C709–C716, 2017.
20. Alibakhshi, E., Ramezanzadeh, M., Haddadi, S.A., Bahlakeh, G., Ramezanzadeh, B., Mahdavian, M., Persian Liquorice extract as a highly efficient sustainable corrosion inhibitor for mild steel in sodium chloride solution. *J. Cleaner Prod.*, 210, 660, 2019.
21. Haddadi, S.A., Alibakhshi, E., Bahlakeh, G., Ramezanzadeh, B., Mahdavian, M., A detailed atomic level computational and electrochemical exploration of the *Juglans regia* green fruit shell extract as a sustainable and highly efficient green corrosion inhibitor for mild steel in 3.5 wt% NaCl solution. *J. Mol. Liq.*, 284, 682, 2019.
22. Alibakhshi, E., Ramezanzadeh, M., Bahlakeh, G., Ramezanzadeh, B., Mahdavian, M., Motamedi, M., *Glycyrrhiza glabra* leaves extract as a green corrosion inhibitor for mild steel in 1 M hydrochloric acid solution: Experimental, molecular dynamics, Monte Carlo and quantum mechanics study. *J. Mol. Liq.*, 255, 185–189, 2018.
23. Sinko, J., Challenges of chromate inhibitor pigments replacement in organic coatings. *Prog. Org. Coat.*, 42, 267, 2001.
24. Deya, C., Blustein, G., del Amo, B., Romagnoli, R., Evaluation of eco-friendly anticorrosive pigments for paints in service conditions. *Prog. Org. Coat.*, 69, 1, 2010.
25. Wei, D., Ma, S., Zhou, Z., Huang, Z., Yuan, A., Electrochemical behavior of potassium zinc phosphates extract and coating. *Appl. Mech. Mater.*, 154, 64, 2012.
26. Gimeno, M.J., Chamorro, S., March, R., Oró, E., Pérez, P., Gracenea, J., Suay, J., Anticorrosive properties enhancement by means of phosphate pigments in an epoxy 2k coating. Assessment by NSS and ACET. *Prog. Org. Coat.*, 77, 2024, 2014.
27. Naderi, R. and Attar, M.M., The role of zinc aluminum phosphate anticorrosive pigment in Protective Performance and cathodic disbondment of epoxy coating. *Corros. Sci.*, 52, 1291, 2010.
28. Hao, Y., Liu, F., Han, E.-H., Anjum, S., Xu, G., The mechanism of inhibition by zinc phosphate in an epoxy coating. *Corros. Sci.*, 69, 77, 2013.
29. El, O.D.A., El, H.L., Hlimi, G.Z., Jodeh, H.E.O.H.S., The Role of Zinc Phosphate Pigment in the Anticorrosion Properties of Bisphenol A Diglycidyl Ether-Polyaminoamide Coating for Aluminum Alloy AA2024-T3. *J. Bio-Tribo-Corrosion*, 5, 7, 2019.

30. Shao, Y., Jia, C., Meng, G., Zhang, T., Wang, F., The role of a zinc phosphate pigment in the corrosion of scratched epoxy-coated steel. *Corros. Sci.*, 51, 371, 2009.
31. Huang, Q., Liu, L., Wu, Z., Ji, S., Wu, H., Chen, P., Corrosion-resistant plasma electrolytic oxidation coating modified by Zinc phosphate and self-healing mechanism in the salt-spray environment. *Surf. Coat. Technol.*, 384, 125321, 2020.
32. Alibakhshi, E., Ghasemi, E., Mahdavian, M., Corrosion inhibition by lithium zinc phosphate pigment. *Corros. Sci.*, 77, 222–229, 2013.
33. Alibakhshi, E., Ghasemi, E., Mahdavian, M., Evaluation of corrosion inhibition performance of synthesized potassium zinc phosphate pigments in two different conditions during immersion time. *J. Color Sci. Technol.*, 9, 113, 2015.
34. Alibakhshi, E., Ghasemi, E., Mahdavian, M., Sodium zinc phosphate as a corrosion inhibitive pigment. *Prog. Org. Coat.*, 77, 1155, 2014.
35. Askari, F., Ghasemi, E., Ramezanzadeh, B., Mahdavian, M., Synthesis and characterization of the fourth generation of zinc phosphate pigment in the presence of benzotriazole. *Dyes Pigm.*, 124, 18–26, 2016.
36. Alibakhshi, E., Naeimi, A., Ramezanzadeh, M., Ramezanzadeh, B., Mahdavian, M., A facile synthesis method of an effective anti-corrosion nanopigment based on zinc polyphosphate through microwaves assisted combustion method; comparing the influence of nanopigment and conventional zinc phosphate on the anti-corrosion properties of an e. *J. Alloys Compd.*, 762, 730, 2018.
37. Alibakhshi, E., Ghasemi, E., Mahdavian, M., The influence of surface modification of lithium zinc phosphate pigment on corrosion inhibition of mild steel and adhesion strength of epoxy coating. *J. Sol–Gel Sci. Technol.*, 72, 359–368, 2014.
38. Alibakhshi, E., Ghasemi, M., Mahdavian, E., A comparison study on corrosion behavior of zinc phosphate and potassium zinc phosphate anticorrosive pigments. *J. Prog. Color Color. Coat.*, 5, 91, 2012.
39. Blustein, G., Romagnoli, R., Amo, B., Deya, M.C., The influence of the anion type on the anticorrosive behaviour of inorganic phosphates. *Surf. Coat. Technol.*, 150, 133, 2002.
40. del Amo, B., Blustein, G., Pérez, M., García, M., Deyá, M., Stupak, M., Romagnoli, R., A multipurpose compound for protective coatings. *Colloids Surf. A Physicochem. Eng. Asp.*, 324, 58, 2008.
41. Alibakhshi, E., Naeimi, A., Ramezanzadeh, M., Ramezanzadeh, B., Mahdavian, M., A facile synthesis method of an effective anti-corrosion nanopigment based on zinc polyphosphate through microwaves assisted combustion method; comparing the influence of nanopigment and conventional zinc phosphate on the anti-corrosion properties of an e. *J. Alloys Compd.*, 762, 730–744, 2018.

42. Alibakhshi, E., Ghasemi, E., Mahdavian, M., Optimization of potassium zinc phosphate anticorrosion pigment by Taguchi experimental design. *Prog. Org. Coat.*, 76, 224–230, 2013.
43. Yu, Y., Yeh, J., Liou, S., Chen, C., Liaw, D., Lu, H., Preparation and properties of polyimide–clay nanocomposite materials for anticorrosion application. *J. Appl. Polym. Sci.*, 92, 3573, 2004.
44. Behzadnasab, M., Mirabedini, S.M., Esfandeh, M., Corrosion protection of steel by epoxy nanocomposite coatings containing various combinations of clay and nanoparticulate zirconia. *Corros. Sci.*, 75, 134, 2013.
45. Haddadi, S.A., Ramazani, S.A.A., Mahdavian, M., Taheri, P., Mol, J.M.C., Fabrication and characterization of graphene-based carbon hollow spheres for encapsulation of organic corrosion inhibitors. *Chem. Eng. J.*, 352, 909, 2018.
46. Shi, X., Nguyen, T.A., Suo, Z., Liu, Y., Avci, R., Effect of nanoparticles on the anticorrosion and mechanical properties of epoxy coating. *Surf. Coat. Technol.*, 204, 237, 2009.
47. Ghanbari, A., Attar, M.M., Mahdavian, M., Corrosion inhibition performance of three imidazole derivatives on mild steel in 1 M phosphoric acid. *Mater. Chem. Phys.*, 124, 2010.
48. Bentrah, H., Rahali, Y., Chala, A., Gum Arabic as an eco-friendly inhibitor for API 5L X42 pipeline steel in HCl medium. *Corros. Sci.*, 82, 426, 2014.
49. Saha, S.K., Murmu, M., Murmu, N.C., Banerjee, P., Evaluating electronic structure of quinazolinone and pyrimidinone molecules for its corrosion inhibition effectiveness on target specific mild steel in the acidic medium: A combined DFT and MD simulation study. *J. Mol. Liq.*, 224, 629, 2016.
50. Qiang, Y., Zhang, S., Guo, L., Xu, S., Feng, L., Obot, I.B., Chen, S., Sodium dodecyl benzene sulfonate as a sustainable inhibitor for zinc corrosion in 26% NH4Cl solution. *J. Cleaner Prod.*, 152, 17, 2017.
51. Mahdavian, M., Tehrani-Bagha, A.R., Alibakhshi, E., Ashhari, S., Palimi, M.J., Farashi, S., Javadian, S., Ektefa, F., Corrosion of mild steel in hydrochloric acid solution in the presence of two cationic gemini surfactants with and without hydroxyl substituted spacers. *Corros. Sci.*, 137, 62–75, 2018.
52. Mahdavian, M. and Ashhari, S., Mercapto functional azole compounds as organic corrosion inhibitors in a polyester-melamine coating. *Prog. Org. Coat.*, 68, 259–264, 2010.
53. Palanivel, V., Huang, Y., Van Ooij, W.J., Effects of addition of corrosion inhibitors to silane films on the performance of AA2024-T3 in a 0.5 M NaCl solution. *Prog. Org. Coat.*, 53, 153, 2005.
54. Kasten, L.S., Grant, J.T., Grebasch, N., Voevodin, N., Arnold, F.E., An XPS study of cerium dopants in sol–gel coatings for aluminum 2024-T3. *Surf. Coat. Technol.*, 140, 11, 2001.
55. Samiee, R., Ramezanzadeh, B., Alibakhshi, E., Corrosion inhibition performance and healing ability of a hybrid silane coating in the presence of praseodymium (III) cations. *J. Electrochem. Soc.*, 165, 777, 2018.

56. Trabelsi, W., Cecilio, P., Ferreira, M.G.S., Montemor, M.F., Electrochemical assessment of the self-healing properties of Ce-doped silane solutions for the pre-treatment of galvanised steel substrates. *Prog. Org. Coat.*, 54, 276, 2005.
57. Visser, P., Liu, Y., Zhou, X., Hashimoto, T., Mol, A.J.M.C., Terryn, H.A., The corrosion protection of AA2024-T3 aluminium alloy by leaching of lithium-containing salts from organic coatings. *Faraday Discuss.*, 180, 511, 2015.
58. Liu, Y., Soc, J.E., Liu, Y., Visser, P., Zhou, X., Lyon, S.B., Hashimoto, T., Curioni, M., Gholinia, A., Thompson, G.E., Smyth, G., Gibbon, S.R., Graham, D., Mol, J.M.C., Terryn, H., Protective film formation on AA2024-T3 aluminum alloy by leaching of lithium carbonate from an organic coating. *J. Electrochem. Soc.*, 163, C45, 2016.
59. Shi, H., Han, E.H., Lamaka, S.V., Zheludkevich, M.L., Liu, F., Ferreira, M.G.S., Cerium cinnamate as an environmentally benign inhibitor pigment for epoxy coatings on AA 2024-T3. *Prog. Org. Coat.*, 77, 765, 2014.
60. Quinet, M., Neveu, B., Moutarlier, V., Audebert, P., Ricq, L., Corrosion protection of sol–gel coatings doped with an organic corrosion inhibitor: Chloranil. *Prog. Org. Coat.*, 58, 46, 2007.
61. Yabuki, A., Urushihara, W., Kinugasa, J., Sugano, K., Self-healing properties of TiO2 particle-polymer composite coatings for protection of aluminum alloys against corrosion in seawater. *Mater. Corros.*, 62, 907, 2011.
62. Shanaghi, A. and Kadkhodaie, M., Investigation of high concentration of benzotriazole on corrosion behaviour of titania-benzotriazole hybrid nano-structured coating applied on Al 7075 by the sol–gel method. *Corros. Eng. Sci. Technol.*, 52, 332, 2017.
63. Samiee, R., Ramezanzadeh, B., Mahdavian, M., Alibakhshi, E., Assessment of the smart self-healing corrosion protection properties of a water-base hybrid organo-silane film combined with non-toxic organic/inorganic environmentally friendly corrosion inhibitors on mild steel. *J. Cleaner Prod.*, 220, 340, 2019.
64. Liu, D., Han, H., Song, Y., Shan, D., Enhancing the self-healing property by adding the synergetic corrosion inhibitors of Na3PO4 and 2-mercaptobenzothiazole into the coating of Mg alloy. *Electrochim. Acta*, 323, 134796, 2019.
65. Miao, M., Yuan, X., Wang, X., Lu, Y., Liu, J., X-y, Y., X-g, W., J-k, L., One step self-heating synthesis and their excellent anticorrosion performance of zinc phosphate/benzotriazole composite pigments. *Dyes*, 141, 74, 2017.
66. Dezfuli, S.M. and Sabzi, M., A Study on the Effect of Presence of CeO2 and Benzotriazole on Activation of Self-Healing Mechanism in ZrO2 Ceramic-Based Coating. *Int. J. Appl. Ceram. Technol.*, 15, 1248, 2018.
67. Ralkhal, S., Ramezanzadeh, B., Shahrabi, T., Studying dual active/barrier and self-healing reinforcing effects of the Neodymium (III)-Benzimidazole hybrid complex in the epoxy coating/mild steel system. *J. Alloys Compd.*, 790, 141, 2019.
68. Ramezanzadeh, B., Ghasemi, E., Askari, F., Mahdavian, M., Synthesis and characterization of a new generation of inhibitive pigment based on zinc

acetate/benzotriazole: Solution phase and coating phase studies. *Dyes Pigm.*, 122, 331–345, 2015.

69. Salehi, E., Naderi, R., Ramezanzadeh, B., Synthesis and characterization of an effective organic/inorganic hybrid green corrosion inhibitive complex based on zinc acetate/*Urtica dioica. Appl. Surf. Sci.*, 396, 1499, 2017.
70. Qian, H., Xu, D., Du, C., Zhang, D., Li, X., Huang, L., Deng, L., Tu, Y., Mol, J.M.C., Terryn, H., Dual-action smart coating with a self-healing superhydrophobic surface and anti-corrosion properties. *J. Mater. Chem. A*, 5, 2355, 2017.
71. Gnedenkov, A.S., Sinebryukhov, S.L., Mashtalyar, D.V., Inhibitor-containing composite coatings on Mg alloys: Corrosion mechanism and self-healing protection. *Solid State Phenom.*, 245, 89, 2016.
72. Gnedenkov, A.S., Sinebryukhov, S.L., Mashtalyar, D.V., Gnedenkov, S.V., Localized corrosion of the Mg alloys with inhibitor-containing coatings: SVET and SIET studies. *Corros. Sci.*, 102, 269, 2016.
73. Rani, A., Somaiah, D., Poddar, M., Scratch cell test: A simple, cost effective screening tool to evaluate self-healing in anti-corrosion coatings. *J. Mater. Eng. Perform.*, 23, 3328, 2014.
74. Sanaei, Z., Bahlakeh, G., Ramezanzadeh, B., Active corrosion protection of mild steel by an epoxy ester coating reinforced with hybrid organic/inorganic green inhibitive pigment. *J. Alloys Compd.*, 728, 1289, 2017.
75. Borisova, D., Akcakayıran, D., Schenderlein, M., Mohwald, H., Shchukin, D.G., Nanocontainer-based anticorrosive coatings: Effect of the container size on the self-healing performance. *Adv. Funct. Mater.*, 23, 3799, 2013.
76. Hang, T.T.X., Truc, T.A., Olivier, M.G., Vandermiers, C., Guérit, N., Pébre, N., Corrosion protection mechanisms of carbon steel by an epoxy resin containing indole-3 butyric acid modified clay. *Prog. Org. Coat.*, 69, 410, 2010.
77. Thi, T., Hang, X., Anh, T., Hoai, T., Ke, V., Corrosion protection of carbon steel by an epoxy resin containing organically modified clay. *Surf. Coat. Technol.*, 201, 7408, 2007.
78. Ghazi, A., Ghasemi, E., Mahdavian, M., Ramezanzadeh, B., Rostami, M., The application of benzimidazole and zinc cations intercalated sodium montmorillonite as smart ion exchange inhibiting pigments in the epoxy ester coating. *Corros. Sci.*, 94, 207–217, 2015.
79. Izadi, M., Shahrabi, T., Mohammadi, I., Ramezanzadeh, B., Synthesis of impregnated Na+-montmorillonite as an eco-friendly inhibitive carrier and its subsequent protective effect on silane coated mild steel. *Prog. Org. Coat.*, 135, 135, 2019.
80. Williams, G., McMurray, H.N., Worsley, D.A., Cerium (III) inhibition of corrosion-driven organic coating delamination studied using a scanning kelvin probe technique. *J. Electrochem. Soc.*, 149, B154, 2002.

81. Williams, G., McMurray, H.N., Loveridge, M.J., Inhibition of corrosion-driven organic coating disbondment on galvanised steel by smart release group II and Zn(II)-exchanged bentonite pigments. *Electrochim. Acta*, 55, 1740, 2010.
82. Tabatabaei, M., Davoudi, M., Ramezanzadeh, M., Ghasemi, E., Construction of a smart active/barrier anti-corrosion system based on epoxy-ester/zinc intercalated kaolin nanocontainer for steel substrate. *Constr. Build. Mater.*, 247, 118555, 2020.
83. Davoodi, M., Ghasemi, E., Ramezanzadeh, B., Mahdavian, M., Designing a zinc-encapsulated Feldspar as a unique rock-forming tectosilicate nanocontainer in the epoxy coating ; Improving the robust barrier and self-healing anti-corrosion properties. *Constr. Build. Mater.*, 243, 118215, 2020.
84. Rassouli, L., Naderi, R., Mahdavain, M., The role of micro/nano zeolites doped with zinc cations in the active protection of epoxy ester coating. *Appl. Surf. Sci.*, 423, 571, 2017.
85. Rassouli, L., Naderi, R., Mahdavian, M., Study of the active corrosion protection properties of epoxy ester coating with zeolite nanoparticles doped with organic and inorganic inhibitors. *J. Taiwan Inst. Chem. Eng.*, 85, 207–220, 2018.
86. Wang, Y., Zhu, Y., Li, C., Song, D., Zhang, T., Zheng, X., Yan, Y., Zhang, M., Wang, J., Shchukin, D.G., Smart epoxy coating containing Ce-MCM-22 zeolites for corrosion protection of Mg–Li alloy. *Appl. Surf. Sci.*, 369, 384, 2016.
87. Dias, S.A.S., Marques, A., Lamaka, S.V., Simões, A., Diamantino, T.C., Ferreira, M.G.S., The role of Ce(III)-enriched zeolites on the corrosion protection of AA2024-T3. *Electrochim. Acta*, 112, 549, 2013.
88. Dias, S.A.S., Lamaka, S.V., Nogueira, C.A., Diamantino, T.C., Ferreira, M.G.S., Sol–gel coatings modified with zeolite fillers for active corrosion protection of AA2024. *Corros. Sci.*, 62, 153, 2012.
89. Ferrer, E.L., Rollon, A.P., Mendoza, H.D., Lafont, U., Garcia, S.J., Double-doped zeolites for corrosion protection of aluminium alloys. *Microporous Mesoporous Mater.*, 188, 8, 2014.
90. Sun, M., Yerokhin, A., Bychkova, M.Y., Shtansky, D.V., Levashov, E.A., Matthews, A., Self-healing plasma electrolytic oxidation coatings doped with benzotriazole loaded halloysite nanotubes on AM50 magnesium alloy. *Corros. Sci.*, 111, 753, 2016.
91. Shchukin, D.G., Lamaka, S.V., Yasakau, K.A., Zheludkevich, M.L., Ferreira, M.G.S., Möhwald, H., Active anticorrosion coatings with halloysite nanocontainers. *J. Phys. Chem. C*, 112, 958, 2008.
92. Abdullayev, E., Abbasov, V., Tursunbayeva, A., Portnov, V., Ibrahimov, H., Mukhtarova, G., Lvov, Y., Self-healing coatings based on halloysite clay polymer composites for protection of copper alloys. *ACS Appl. Mater. Interfaces*, 5, 4464, 2013.
93. Buchheit, R.G., Guan, H., Mahajanam, S., Wong, F., Active corrosion protection and corrosion sensing in chromate-free organic coatings. *Prog. Org. Coat.*, 47, 174, 2003.

94. Williams, G. and McMurray, H.N., Anion-exchange inhibition of filiform corrosion on organic coated AA2024-T3 aluminum alloy by hydrotalcite-like pigments. *Electrochem. Solid-State Lett.*, 6, B9, 2003.
95. Williams, G. and McMurray, H.N., Inhibition of filiform corrosion on polymer coated AA2024-T3 by hydrotalcite-like pigments incorporating organic anions. *Electrochem. Solid-State Lett.*, 7, B13, 2004.
96. Tedim, J., Poznyak, S.K., Kuznetsova, A., Raps, D., Hack, T., Zheludkevich, M.L., Ferreira, M.G.S., Enhancement of active corrosion protection via combination of inhibitor-loaded nanocontainers. *ACS Appl. Mater. Interfaces*, 2, 1528, 2010.
97. Zheludkevich, M.L., Poznyak, S.K., Rodrigues, L.M., Raps, D., Hack, T., Dick, L.F., Nunes, T., Ferreira, M.G.S., Active protection coatings with layered double hydroxide nanocontainers of corrosion inhibitor. *Corros. Sci.*, 52, 602, 2010.
98. Alibakhshi, E., Ghasemi, E., Mahdavian, M., Ramezanzadeh, B., Farashi, S., Fabrication and characterization of $PO4^{3-}$ intercalated Zn-Al-layered double hydroxide nanocontainer. *J. Electrochem. Soc.*, 163, C495, 2016.
99. Alibakhshi, E., Ghasemi, E., Mahdavian, M., Ramezanzadeh, B., A comparative study on corrosion inhibitive effect of nitrate and phosphate intercalated Zn–Al-layered double hydroxides (LDHs) nanocontainers incorporated into a hybrid silane layer and their effect on cathodic delamination of epoxy topcoat. *Corros. Sci.*, 115, 159–174, 2017.
100. Alibakhshi, E., Ghasemi, E., Mahdavian, M., Ramezanzadeh, B., Farashi, S., Active corrosion protection of Mg–Al–PO43–LDH nanoparticle in silane primer coated with epoxy on mild steel. *J. Taiwan Inst. Chem. Eng.*, 75, 248, 2017.
101. Alibakhshi, E., Ghasemi, E., Mahdavian, M., Ramezanzadeh, B., Fabrication and characterization of layered double hydroxide/silane nanocomposite coatings for protection of mild steel. *J. Taiwan Inst. Chem. Eng.*, 80, 924–934, 2017.
102. Shkirskiy, V., Keil, P., Hintze-Bruening, H., Leroux, F., Vialat, P., Lefèvre, G., Ogle, K., Volovitch, P., Factors affecting $MoO4^{2-}$ inhibitor release from Zn2Al based layered double hydroxide and their implication in protecting hot dip galvanized steel by means of organic coatings. *ACS Appl. Mater. Interfaces*, 7, 25180, 2015.
103. Dong, Y., Wang, F., Zhou, Q., Protective behaviors of 2-mercaptobenzothiazole intercalated Zn-Al-layered double hydroxide coating. *J. Coat. Technol. Res.*, 11, 1, 2014.
104. Li, D., Wang, F., Yu, X., Wang, J., Liu, Q., Yang, P., He, Y., Wang, Y., Zhang, M., Anticorrosion organic coating with layered double hydroxide loaded with corrosion inhibitor of tungstate. *Prog. Org. Coat.*, 71, 302, 2011.
105. Stimpfling, T., Leroux, F., Hintze-bruening, H., Organo-modified layered double hydroxide in coating formulation to protect AA2024 from corrosion. *Colloids Surf. A Physicochem. Eng. Asp.*, 458, 147, 2014.

106. Hayatdavoudi, H. and Rahsepar, M., Smart inhibition action of layered double hydroxide nanocontainers in zinc-rich epoxy coating for active corrosion protection of carbon steel substrate. *J. Alloys Compd.*, 711, 560, 2017.
107. You, Z., Jianhua, L.I.U., Yingdong, L.I., Mei, Y.U., Xiaolin, Y.I.N., Songmei, L.I., Enhancement of active anticorrosion via Ce-doped Zn–Al layered double hydroxides embedded in sol-gel coatings on aluminum alloy. *J. Wuhan Univ. Technol.*, 32, 1199, 2017.
108. Olya, N., Ghasemi, E., Ramezanzadeh, B., Mahdavian, M., Synthesis, characterization and protective functioning of surface decorated Zn–Al layered double hydroxide with SiO2 nano-particles. *Surf. Coat. Technol.*, 387, 125512, 2020.
109. Alibakhshi, E., Ghasemi, E., Mahdavian, M., The effect of interlayer spacing on the inhibitor release capability of layered double hydroxide based nanocontainers. *J. Cleaner Prod.*, 251, 119676, 2020.
110. Tedim, J., Kuznetsova, A., Salak, A.N., Montemor, F., Snihirova, D., Pilz, M., Zheludkevich, M.L., Ferreira, M.G.S., Zn–Al layered double hydroxides as chloride nanotraps in active protective coatings. *Corros. Sci.*, 55, 1, 2012.
111. Zheludkevich, M.L., Tedim, J., Ferreira, M.G.S., Smart coatings for active corrosion protection based on multi-functional micro and nanocontainers. *Electrochim. Acta*, 82, 314, 2012.
112. Wang, L., Wu, W., Sun, W., Yang, Z., Wang, S., Liu, G., Partially dehydrated zinc hydroxide sulfate nanoplates reinforced coating for corrosion protection. *Chem. Eng. J.*, 373, 8, 2019.
113. Lamaka, S.V., Zheludkevich, M.L., Yasakau, K.A., Serra, R., Poznyak, S.K., Ferreira, M.G.S., Nanoporous titania interlayer as reservoir of corrosion inhibitors for coatings with self-healing ability. *Prog. Org. Coat.*, 58, 127, 2007.
114. Lamaka, S.V., Zheludkevich, M.L., Yasakau, K.A., Montemor, M.F., Patricia, C., Ferreira, M.G.S., TiOx self-assembled networks prepared by templating approach as nanostructured reservoirs for self-healing anticorrosion pre-treatments. *Electrochem. Commun.*, 8, 421, 2006.
115. Zheludkevich, M.L., Serra, R., Montemor, M.F., Ferreira, M.G.S., Oxide nanoparticle reservoirs for storage and prolonged release of the corrosion inhibitors. *Electrochem. Commun.*, 7, 836, 2005.
116. Zhao, Y., Zhang, Z., Shi, L., Zhang, F., Li, S., Zeng, R., Corrosion resistance of a self-healing multilayer film based on SiO2 and CeO2 nanoparticles layer-by-layer assembly on Mg alloys. *Mater. Lett.*, 237, 14, 2019.
117. Yabuki, A., Nagayama, Y., Wahyudhin, I., Porous anodic oxide film with self-healing ability for corrosion protection of aluminum. *Electrochim. Acta*, 296, 662, 2019.
118. Saremi, M. and Yeganeh, M., Application of mesoporous silica nanocontainers as smart host of corrosion inhibitor in polypyrrole coatings. *Corros. Sci.*, 86, 159, 2014.

119. Borisova, D., Mohwald, H., Shchukin, D.G., Mesoporous Silica Nanoparticles for Active Corrosion Protection. *ACS Nano*, 5, 1939, 2011.
120. Sun, S., Zhao, X., Cheng, M., Wang, Y., Li, C., Hu, S., Facile preparation of redox-responsive hollow mesoporous silica spheres for the encapsulation and controlled release of corrosion inhibitors. *Prog. Org. Coat.*, 136, 105302, 2019.
121. Zheludkevich, M.L., Serra, R., Montemor, M.F., Yasakau, K.A., Salvado, I.M.M., Ferreira, M.G.S., Nanostructured sol-gel coatings doped with cerium nitrate as pre-treatments for AA2024-T3 Corrosion protection performance. *Electrochim. Acta*, 51, 208, 2005.
122. Zhou, C., Li, Z., Li, J., Yuan, T., Chen, B., Ma, X., Jiang, D., Epoxy composite coating with excellent anticorrosion and self-healing performances based on multifunctional zeolitic imidazolate framework derived nanocontainers. *Chem. Eng. J.*, 385, 123835, 2020.
123. Gupta, G., Birbilis, N., Cook, A.B., Khanna, A.S., Polyaniline-lignosulfonate epoxy coating for corrosion protection of AA2024-T3. *Corros. Sci.*, 67, 256, 2013.
124. Jadhav, N., Jensen, M.B., Gelling, V., Tungstate and vanadate-doped polypyrrole/aluminum flake composite coatings for the corrosion protection of aluminum 2024-T3. *J. Coat. Technol. Res.*, 12, 259–275, 2015.
125. Hien, N.T.L., Garcia, B., Pailleret, A., Deslouis, C., Role of doping ions in the corrosion protection of iron by polypyrrole films. *Electrochem. Acta*, 50, 1747, 2005.
126. Ahmad, S., Ashraf, S.M., Riaz, U., Corrosion studies of polyaniline/coconut oil poly (esteramide urethane) coatings. *Polym. Adv. Technol.*, 16, 541, 2005.
127. Lei, Y., Qiu, Z., Tan, N., Du, H., Li, D., Liu, J., Liu, T., Zhang, W., Chang, X., Polyaniline/CeO2 nanocomposites as corrosion inhibitors for improving the corrosive performance of epoxy coating on carbon steel in 3.5% NaCl solution. *Prog. Org. Coat.*, 139, 105430, 2020.
128. Abu-Thabit, N. and Makhlouf, A.S., Recent advances in polyaniline (PANI)-based organic coatings for corrosion protection, in: *Handb. Smart Coatings Mater. Prot.*
129. Rohwerder, M., Intelligent corrosion protection by conducting polymers, in: *Smart Coatings II*, pp. 274–287.
130. Kendig, M. and Hon, M., Environmentally triggered release of oxygen-reduction inhibitors from inherently conducting polymers. *Corrosion*, 60, 1024, 2004.
131. Wessling, B., New insight into organic metal polyaniline morphology and structure. *Polymers (Basel)*, 2, 786, 2010.
132. Karpakam, V., Kamaraj, K., Sathiyanarayanan, S., Venkatachari, G., Ramu, S., Electrosynthesis of polyaniline–molybdate coating on steel and its corrosion protection performance. *Electrochim. Acta*, 56, 2165, 2011.

133. Radhakrishnan, S., Siju, C.R., Mahanta, D., Patil, S., Madras, G., Conducting polyaniline-nano-TiO2 composites for smart corrosion resistant coatings. *Electrochem. Acta*, 54, 1249, 2009.
134. Arefinia, R., Shojaei, A., Shariatpanahi, H., Neshati, J., The results have emphasized the positive protection effects of the EB possibly due to its good adhesive property toward the metal substrate and proper interfacial interaction with polymer matrix restricting further diffusion of corrosive ions through the. *Prog. Org. Coat.*, 75, 502, 2012.
135. Yabuki, A., Kawashima, A., Fathona, I.W., Self-healing polymer coatings with cellulose nanofibers served as pathways for the release of a corrosion inhibitor. *Corros. Sci.*, 85, 141, 2014.
136. Pirhady Tavandashti, N., Ghorbani, M., Shojaei, A., Gonzalez-Garcia, Y., Terryn, H., Mol, J.M.C., PH responsive Ce(III) loaded polyaniline nanofibers for self-healing corrosion protection of AA2024-T3. *Prog. Org. Coat.*, 99, 197, 2016.
137. Yabuki, A., Shiraiwa, T., Fathona, I.W., PH-controlled self-healing polymer coatings with cellulose nanofibers providing an effective release of corrosion inhibitor. *Corros. Sci.*, 103, 117, 2016.
138. Qiu, S., Chen, C., Cui, M., Li, W., Zhao, H., Wang, L., Corrosion protection performance of waterborne epoxy coatings containing self-doped polyaniline nanofiber. *Appl. Surf. Sci.*, 407, 213, 2017.
139. Hao, Y., Zhao, Y., Yang, X., Hu, B., Ye, S., Song, L., Li, R., Self-healing epoxy coating loaded with phytic acid doped polyaniline nanofibers impregnated with benzotriazole for Q235 carbon steel. *Corros. Sci.*, 151, 175, 2019.
140. Zhang, K., Wang, L., Liu, G., Copper(II) 8-hydroxyquinolinate 3D network film with corrosion inhibitor embedded for self-healing corrosion protection. *Corros. Sci.*, 75, 38, 2013.
141. Mitchell, T.J. and Keller, M.W., Coaxial electrospun encapsulation of epoxy for use in self-healing materials. *Polym. Int.*, 62, 860, 2013.
142. Park, J.H. and Braun, P.V., Coaxial electrospinning of self-healing coatings. *Adv. Mater.*, 22, 496, 2010.
143. Pang, J.W.C. and Bond, I.P., A hollow fibre reinforced polymer composite encompassing self-healing and enhanced damage visibility. *Compos. Sci. Technol.*, 65, 1791, 2005.
144. Williams, G., Trask, R., Bond, I., A self-healing carbon fibre reinforced polymer for aerospace applications. *Compos. Part A Appl. Sci. Manuf.*, 38, 1525, 2007.
145. Zhou, H., Wang, H., Niu, H., Gestos, A., Lin, T., Robust, self-healing superamphiphobic fabrics prepared by two-step coating of fluoro-containing polymer, fluoroalkyl silane, and modified silica nanoparticles. *Adv. Funct. Mater.*, 23, 1664, 2013.
146. Zhang, Z., Zhang, W., Li, D., Sun, Y., Wang, Z., Hou, C., Chen, L., Cao, Y., Liu, Y., Mechanical and anticorrosive properties of graphene/epoxy resin

composites coating prepared by in-situ method. *Int. J. Mol. Sci.*, 16, 2239, 2015.
147. Lin, Y.T., Don, T.M., Wong, C.J., Meng, F.C., Lin, Y.J., Lee, S.Y., Lee, C.F., Chiu, W.Y., Improvement of mechanical properties and anticorrosion performance of epoxy coatings by the introduction of polyaniline/graphene composite. *Surf. Coat. Technol.*, 374, 1128, 2019.
148. Javidparvar, A.A., Naderi, R., Ramezanzadeh, B., Bahlakeh, G., Graphene oxide as a pH-sensitive carrier for targeted delivery of eco-friendly corrosion inhibitors in chloride solution: Experimental and theroretical investigations. *J. Ind. Eng. Chem.*, 72, 196–213, 2019.
149. Schriver, M., Regan, W., Gannett, W.J., Zaniewski, A.M., Crommie, M.F., Zettl, A., Graphene as a long-term metal oxidation barrier: Worse than nothing. *ACS Nano*, 7, 5763, 2013.
150. Ding, J., ur Rahman, O., Peng, W., Dou, H., Yu, H., A novel hydroxyl epoxy phosphate monomer enhancing the anticorrosive performance of waterborne Graphene/Epoxy coatings. *Appl. Surf. Sci.*, 427, 981, 2018.
151. Kasaeian, M., Ghasemi, E., Ramezanzadeh, B., Mahdavian, M., Bahlakeh, G., Construction of a highly effective self-repair corrosion-resistant epoxy composite through impregnation of 1H-Benzimidazole corrosion inhibitor modified graphene oxide nanosheets (GO-BIM). *Corros. Sci.*, 145, 119, 2018.
152. Ramezanzadeh, B., Ghasemi, E., Mahdavian, M., Changizi, E., Mohamadzadeh Moghadam, M.H., Covalently-grafted graphene oxide nanosheets to improve barrier and corrosion protection properties of polyurethane coatings. *Carbon N. Y.*, 93, 555–573, 2015.
153. Cui, M., Ren, S., Zhao, H., Xue, Q., Wang, L., Polydopamine coated graphene oxide for anticorrosive reinforcement of water-borne epoxy coating. *Chem. Eng. J.*, 335, 255, 2018.
154. Hayatgheib, Y., Ramezanzadeh, B., Kardar, P., Mahdavian, M., A comparative study on fabrication of a highly effective corrosion protective system based on graphene oxide-polyaniline nanofibers/epoxy composite. *Corros. Sci.*, 133, 358–373, 2018.
155. Ramezanzadeh, B., Mohamadzadeh Moghadam, M.H., Shohani, N., Mahdavian, M., Effects of highly crystalline and conductive polyaniline/graphene oxide composites on the corrosion protection performance of a zinc-rich epoxy coating. *Chem. Eng. J.*, 320, 363–375, 2017.
156. Ramezanzadeh, B., Haeri, H., Ramezanzadeh, M., A facile route of making silica nanoparticles-covered graphene oxide nanohy- brids (SiO2-GO); fabrication of SiO2-GO/epoxy composite coating with superior barrier and corrosion protection performance. *Chem. Eng. J.*, 303, 511, 2016.
157. Yu, Z., Di, H., Ma, Y., Lv, L., Pan, Y., Zhang, C., He, Y., Fabrication of graphene oxide-alumina hybrids to reinforce the anti-corrosion performance of composite epoxy coatings. *Appl. Surf. Sci.*, 351, 986, 2015.

158. Yang, Z., Sun, W., Wang, L., Li, S., Zhu, T., Liu, G., Liquid-phase exfoliated fluorographene as a two dimensional coating filler for enhanced corrosion protection performance. *Corros. Sci.*, 47, 2855, 2016.
159. Ramezanzadeh, B., Niroumandrad, S., Ahmadi, A., Mahdavian, M., Mohamadzadeh Moghadam, M.H., Enhancement of barrier and corrosion protection performance of an epoxy coating through wet transfer of amino functionalized graphene oxide. *Corros. Sci.*, 103, 283–304, 2016.
160. Ramezanzadeh, B., Bahlakeh, G., Moghadam, M.H.M., Impact of size-controlled p-phenylenediamine (PPDA)-functionalized graphene oxide nanosheets on the GO-PPDA/Epoxy anti-corrosion, interfacial interactions and mechanical properties enhancement: Experimental and quantum mechanics investigations. *Chem. Eng. J.*, 335, 737, 2018.
161. Ahmadi, A., Ramezanzadeh, B., Mahdavian, M., Hybrid silane coating reinforced with silanized graphene oxide nanosheets with improved corrosion protective performance. *RSC Adv.*, 6, 54102, 2016.
162. Ramezanzadeh, B., Ahmadi, A., Mahdavian, M., Enhancement of the corrosion protection performance and cathodic delamination resistance of epoxy coating through treatment of steel substrate by a novel nanometric sol-gel based silane composite film filled with functionalized graphene oxide nanosheets. *Corros. Sci.*, 109, 182–205, 2016.
163. Parhizkar, N., Ramezanzadeh, B., Shahrabi, T., Enhancement of the corrosion protection properties of a hybrid sol–gel based silane film through impregnation of functionalized graphene oxide nanosheets. *J. Electrochem. Soc.*, 164, C1044, 2017.
164. Chen, C., Qiu, S., Cui, M., Qin, S., Yan, G., Zhao, H., Wang, L., Xue, Q., Achieving high performance corrosion and wear resistant epoxy coatings via incorporation of noncovalent functionalized graphene. *Carbon N. Y.*, 114, 356, 2017.
165. Chen, X., Zhou, S., Zhang, L., You, T., Xu, F., Adsorption of heavy metals by graphene Oxide/cellulose hydrogel prepared from NaOH/urea aqueous solution. *Materials (Basel)*, 9, 582, 2016.
166. Matsuura, K.C.K.T. and Ethylenediamine, E.D.A., Removal of heavy metals and pollutants by membrane adsorption techniques. *Appl. Water Sci.*, 8, 19, 2018.
167. Yang, S., Chen, S., Chang, Y., Cao, A., Liu, Y., Wang, H., Removal of methylene blue from aqueous solution by graphene oxide. *J. Colloid Interface Sci.*, 359, 24, 2011.
168. Lin, Y., Chen, R., Zhang, Y., Lin, Z., Liu, Q., Liu, J., Wang, Y., Gao, L., Wang, J., Sandwich-like polyvinyl alcohol (PVA) grafted graphene: A solid-inhibitors container for long term self-healing coatings. *Chem. Eng. J.*, 383, 123203, 2020.
169. Hao, Y., Zhao, Y., Li, B., Song, L., Guo, Z., Self-healing effect of graphene@ PANI loaded with benzotriazole for carbon steel. *Corros. Sci.*, 163, 1082, 2020.

170. Asaldoust, S. and Ramezanzadeh, B., Synthesis and characterization of a high-quality nanocontainer based on benzimidazole-zinc phosphate (ZP-BIM) tailored graphene oxides; a facile approach to fabricating a smart self-healing anti-corrosion system. *J. Colloid Interface Sci.*, 564, 230, 2020.
171. Javidparvar, A.A., Naderi, R., Ramezanzadeh, B., Manipulating graphene oxide nanocontainer with benzimidazole and cerium ions: Application in epoxy-based nanocomposite for active corrosion protection. *Corros. Sci.*, 165, 108379, 2020.
172. Daradmare, S., Pradhan, M., Raja, V.S., Parida, S., Encapsulating 8-hydroxyquinoline in graphene oxide-stabilized polystyrene containers and its anti-corrosion performance. *J. Mater. Sci.*, 51, 10262, 2016.
173. Akbarzadeh, S., Ramezanzadeh, M., Ramezanzadeh, B., Bahlakeh, G., A green assisted route for the fabrication of a high-efficiency self-healing anti-corrosion coating through graphene oxide nanoplatform reduction by *Tamarindus indiaca* extract. *J. Hazard. Mater.*, 390, 122147, 2020.
174. Liu, C., Zhao, H., Hou, P., Qian, B., Wang, X., Guo, C., Wang, L., Efficient Graphene/Cyclodextrin-Based Nanocontainer: Synthesis and Host-Guest Inclusion for Self-Healing Anticorrosion Application. *ACS Appl. Mater. Interfaces*, 10, 36229, 2018.
175. Samiee, R., Ramezanzade, B., Mahdavian, M., Alibakhshi, E., Bahlakeh, G., Graphene oxide nano-sheets loading with praseodymium cations: Adsorption–desorption study, quantum mechanics calculations and dual active-barrier effect for smart coatings fabrication. *J. Ind. Eng. Chem.*, 78, 143–154, 2019.
176. Nikpour, B., Ramezanzadeh, B., Bahlakeh, G., Mahdavian, M., Synthesis of graphene oxide nanosheets functionalized by green corrosion inhibitive compounds to fabricate a protective system. *Corros. Sci.*, 127, 240–259, 2017.
177. Motamedi, M., Ramezanzadeh, M., Ramezanzadeh, B., Saadatmandi, S., Enhancement of the active/passive anti-corrosion properties of epoxy coating via inclusion of histamine/zinc modified/reduced graphene oxide nanosheets. *Appl. Surf. Sci.*, 488, 77, 2019.
178. Asghar, A., Naderi, R., Ramezanzadeh, B., Epoxy-polyamide nanocomposite coating with graphene oxide as cerium nanocontainer generating effective dual active/barrier corrosion protection. *Compos. Part B*, 172, 363, 2019.
179. Habibiyan, A., Ramezanzadeh, B., Mahdavian, M., Bahlakeh, G., Kasaeian, M., Rational assembly of mussel-inspired polydopamine (PDA)-Zn (II) complex nanospheres on graphene oxide framework tailored for robust self-healing anti-corrosion coatings application. *Chem. Eng. J.*, 391, 123630, 2019.
180. Mohammadkhani, R., Ramezanzadeh, M., Saadatmandi, S., Ramezanzadeh, B., Designing a dual-functional epoxy composite system with self-healing/barrier anti-corrosion performance using graphene oxide nano-scale platforms decorated with zinc doped-conductive polypyrrole nanoparticles with great environmental stability and non-tox. *Chem. Eng. J.*, 382, 122819, 2020.

181. Kasaeian, M., Ghasemi, E., Ramezanzadeh, B., Mahdavian, M., Graphene oxide as a potential nanocarrier for Zn(II) to fabricate a dual-functional active/passive protection; sorption/desorption characteristics and electrochemical evaluation. *J. Ind. Eng. Chem.*, 73, 162, 2019.
182. Zhong, F., He, Y., Wang, P., Chen, C., Xie, P., Li, H., Chen, J., One-step hydrothermal synthesis of reduced graphene oxide/aspartic acid intercalated layered double hydroxide for enhancing barrier and self-healing properties of epoxy coating. *React. Funct. Polym.*, 145, 104380, 2019.
183. Yang, N., Yang, T., Wang, W., Chen, H., Li, W., Polydopamine modified polyaniline-graphene oxide composite for enhancement of corrosion resistance. *J. Hazard. Mater.*, 377, 142, 2019.
184. Ramezanzadeh, M., Ramezanzadeh, B., Mahdavian, M., Bahlakeh, G., Development of metal-organic framework (MOF) decorated graphene oxide nanoplatforms for anti-corrosion epoxy coatings. *Carbon N. Y.*, 161, 231, 2020.
185. Williams, G., Geary, S., McMurray, H.N., Smart release corrosion inhibitor pigments based on organic ion-exchange resins. *Corros. Sci.*, 57, 139, 2012.
186. Snihirova, D., Lamaka, S.V., Taryba, M., Salak, A.N., Kallip, S., Zheludkevich, M.L., Ferreira, M.G.S., Montemor, M.F., Hydroxyapatite microparticles as feedback-active reservoirs of corrosion inhibitors. *ACS Appl. Mater. Interfaces*, 2, 3011, 2010.
187. Tang, Y., Cao, Z., Xu, J., Cang, H., Jing, W., Titanate matrices as potential corrosion inhibitor nanocontainers. *Corros. Sci.*, 88, 487, 2014.
188. Khramov, A.N., Voevodin, N.N., Balbyshev, V.N., Donley, M.S., Hybrid organo-ceramic corrosion protection coatings with encapsulated organic corrosion inhibitors. *Thin Solid Films*, 447–448, 549, 2004.
189. Raj, R., Morozov, Y., Calado, L.M., Taryba, M.G., Kahraman, R., Shakoor, R.A., Montemor, M.F., Calcium carbonate particles loaded with triethanolamine and polyethylenimine for enhanced corrosion protection of epoxy coated steel. *Corros. Sci.*, 167, 108548, 2020.
190. Snihirova, D., Lamaka, S.V., Montemor, M.F., "SMART" protective ability of water based epoxy coatings loaded with CaCO3 microbeads impregnated with corrosion inhibitors applied on AA2024 substrates. *Electrochim. Acta*, 83, 439, 2012.
191. Morozov, Y., Calado, L.M., Shakoor, R.A., Raj, R., Kahraman, R., Taryba, M.G., Montemor, M.F., Epoxy coatings modified with a new cerium phosphate inhibitor for smart corrosion protection of steel. *Corros. Sci.*, 159, 108128, 2019.
192. Calado, L.M., Taryba, M.G., Morozov, Y., Carmezin, M.J., Montemor, M.F., Novel smart and self-healing cerium phosphate-based corrosion inhibitor for AZ31 magnesium alloy. *Corros. Sci.*, 170, 108648, 2020.
193. Kartsonakis, I.A., Balaskas, A.C., Koumoulos, E.P., Charitidis, C.A., Kordas, G.C., Incorporation of ceramic nanocontainers into epoxy coatings for

the corrosion protection of hot dip galvanized steel. *Corros. Sci.*, 57, 30, 2012.

194. Hughes, A.E., Cole, I.S., Muster, T.H., Varley, R.J., Designing green, self-healing coatings for metal protection. *NPG Asia Mater.*, 2, 143, 2010.
195. García, S.J., Fischer, H.R., Van Der Zwaag, S., A critical appraisal of the potential of self healing polymeric coatings. *Prog. Org. Coat.*, 72, 211, 2011.
196. Kopec, M., Szczepanowicz, K., Mordarski, G., Podgorna, K., Socha, R.P., Nowak, P., Warszynski, P., Hack, T., Self-healing epoxy coatings loaded with inhibitor-containing polyelectrolyte nanocapsules. *Prog. Org. Coat.*, 84, 97, 2015.
197. Sonawane, S.H., Bhanvase, B.A., Jamali, A.A., Dubey, S.K., Kale, S.S., Pinjari, D.V., Kulkarni, R.D., Gogate, P.R., Pandit, A.B., Improved active anticorrosion coatings using layer-by-layer assembled ZnO nanocontainers with benzotriazole. *Chem. Eng. J.*, 189–190, 464, 2012.
198. Izadi, M., Shahrabi, T., Ramezanzadeh, B., Synthesis and characterization of an advanced layer-by-layer assembled Fe3O4/polyaniline nanoreservoir filled with Nettle extract as a green corrosion protective system. *J. Ind. Eng. Chem.*, 57, 263, 2018.
199. Shchukin, D.G., Zheludkevich, M., Yasakau, K., Lamaka, S., Ferreira, M.G.S., Möhwald, H., Layer-by-Layer Assembled Nanocontainers for Self-Healing Corrosion Protection. *Adv. Mater.*, 18, 1672, 2006.
200. Plawecka, M., Snihirova, D., Martins, B., Szczepanowicz, K., Warszynski, P., Montemor, M.F., Self healing ability of inhibitor-containing nanocapsules loaded in epoxy coatings applied on aluminium 5083 and galvanneal substrates. *Electrochim. Acta*, 140, 282, 2014.
201. Andreeva, D.V., Fix, D., Mo, H., Shchukin, D.G., Self-Healing anticorrosion coatings based on pH-sensitive polyelectrolyte/inhibitor sandwichlike nanostructures. *Adv. Mater.*, 20, 2789, 2008.
202. Montemor, M.F., Snihirova, D.V., Taryba, M.G., Lamaka, S.V., Kartsonakis, I.A., Balaskas, A.C., Kordas, G.C., Tedim, J., Kuznetsova, A., Zheludkevich, M.L., Ferreira, M.G.S., Evaluation of self-healing ability in protective coatings modified with combinations of layered double hydroxides and cerium molibdate nanocontainers filled with corrosion inhibitors. *Electrochim. Acta*, 60, 31, 2012.
203. Hou, P., Liu, C., Zhao, H., Layer-by-Layer Self-assembled Graphene Oxide Nanocontainers for Active Anticorrosion Application. *Int. J. Electrochem. Sci.*, 14, 3055, 2019.
204. Lang, S. and Zhou, Q., Synthesis and characterization of poly(urea-formaldehyde) microcapsules containing linseed oil for self-healing coating development. *Prog. Org. Coat.*, 105, 99, 2017.
205. Wang, H. and Zhou, Q., Evaluation and failure analysis of linseed oil encapsulated self-healing anticorrosive coating. *Prog. Org. Coat.*, 118, 108, 2018.

206. Navarchian, A.H., Najafipoor, N., Ahangaran, F., Surface-modified poly(methyl methacrylate) microcapsules containing linseed oil for application in self-healing epoxy-based coatings. *Prog. Org. Coat.*, 132, 288, 2019.
207. Li, J., Feng, Q., Cui, J., Yuan, Q., Qiu, H., Gao, S., Yang, J., Self-assembled graphene oxide microcapsules in Pickering emulsions for self-healing waterborne polyurethane coatings. *Compos. Sci. Technol.*, 151, 282, 2017.
208. Weishaar, A., Carpenter, M., Loucks, R., Sakulich, A., Peterson, A.M., Evaluation of self-healing epoxy coatings for steel reinforcement. *Constr. Build. Mater.*, 191, 125, 2018.
209. Tatiya, P.D., Mahulikar, P.P., Gite, V.V., Designing of polyamidoamine-based polyurea microcapsules containing tung oil for anticorrosive coating applications. *J. Coat. Technol. Res.*, 13, 715, 2016.
210. Samadzadeh, M., Boura, S.H., Peikari, M., Ashrafi, A., Kasiriha, M., Tung oil: An autonomous repairing agent for self-healing epoxy coatings. *Prog. Org. Coat.*, 70, 383, 2011.
211. Cotting, F., Koebsch, A., Aoki, I.V., Epoxy Self-Healing Coating by Encapsulated Epoxy Ester Resin in Poly (Urea-Formaldehyde-Melamine) Microcapsules. *Front. Mater.*, 6, 314, 2019.
212. Zhang, C., Wang, H., Zhou, Q., Preparation and characterization of microcapsules based self-healing coatings containing epoxy ester as healing agent. *Prog. Org. Coat.*, 125, 403, 2018.
213. Attaei, M., Calado, L.M., Taryba, M.G., Morozov, Y., Shakoor, R.A., Kahraman, R., Marques, A.C., Montemor, M.F., Autonomous self-healing in epoxy coatings provided by high efficiency isophorone diisocyanate (IPDI) microcapsules for protection of carbon steel. *Prog. Org. Coat.*, 139, 105445, 2020.
214. Attaei, M., Vale, M., Shakoor, A., Kahraman, R., Montemor, M.F., Marques, A.C., Hybrid shell microcapsules containing isophorone diisocyanate with high thermal and chemical stability for autonomous self-healing of epoxy coatings. *J. Appl. Polym. Sci.*, 137, 10, 2020.
215. US9879142B2 - Microencapsulation of organic silanes and their use as self healing materials—Google Patents, (n.d.). https://patents.google.com/patent/US9879142B2/en (accessed April 11, 2020).
216. Kongparakul, S., Kornprasert, S., Suriya, P., Le, D., Samart, C., Chantarasiri, N., Prasassarakich, P., Guan, G., Self-healing hybrid nanocomposite anticorrosive coating from epoxy/modified nanosilica/perfluorooctyl triethoxysilane. *Prog. Org. Coat.*, 104, 173, 2017.
217. Yi, H., Deng, Y., Wang, C., Pickering emulsion-based fabrication of epoxy and amine microcapsules for dual core self-healing coating. *Compos. Sci. Technol.*, 133, 51, 2016.
218. Poornima Vijayan, P. and Al-Maadeed, M.A.S.A., TiO2 nanotubes and mesoporous silica as containers in self-healing epoxy coatings. *Sci. Rep.*, 6, 1, 2016.

219. Haddadi, S.A., Ramazani, A., Mahdavian, M., Taheri, P., Mol, J.M.C., Gonzalez-Garcia, Y., Self-healing epoxy nanocomposite coatings based on dual-encapsulation of nano-carbon hollow spheres with film-forming resin and curing agent. *Compos. Part B Eng.*, 175, 107087, 2019.

11

Self-Healable Conductive Materials

M. Ramesh[1]*, L. Rajeshkumar[2], D. Balaji[2], V. Bhuvaneswari[2] and S. Sivalingam[3]

[1]*Department of Mechanical Engineering, KIT-Kalaignarkarunanidhi Institute of Technology, Coimbatore, India*
[2]*Department of Mechanical Engineering, KPR Institute of Engineering and Technology, Coimbatore, India*
[3]*Department of Mechanical Engineering, Coimbatore Institute of Technology, Coimbatore, India*

Abstract

The conductivity is always given the prime importance of any circuit to work properly, even minor loose connection pave path to catastrophe of the complete system. The circuit failure deemed to have human intervention to get rid out this problem. The self-healable conductive materials are unique in nature such that they regain the working state of the system without human assistance. As a matter of fact, self-healing means heal by itself but in other case it is not quite complex, when assessed, this technique for circuit connections it was found very hard to establish one such connections. The connection also could be established with the involvement of human beings, as how the soldering, brazing works are being carried out. The connection is very thin layer which could provide the complete flow of electricity. The growth in the field of automation, starts doing the complex works without human interventions. Very complex connections are done by the programmed machines which requires human skills in programming. In these case, one way or other humans are involved, but healing by themselves is the better option. In this technique human brain is also involved but only at the initial phase. This kindle the interest among the researchers to develop such novel materials. So, the curiosity put in various forms to attain the conductivity over the circuits. In this chapter, the discussion is all about the various methods by which the self-healing is achieved for conductive materials. The complex conductivity is established even as a coating, so that is also touched upon a bit.

**Corresponding author*: mramesh97@gmail.com

Inamuddin, Mohd Imran Ahamed, Rajender Boddula and Tariq Altalhi (eds.) Self-Healing Smart Materials and Allied Applications, (297–320) © 2021 Scrivener Publishing LLC

Keywords: Self-healing, conductive materials, polymers, coatings, reversible materials

11.1 Introduction

The self-healing specifically for conductive material is quite an attractive factor for researchers so they started developing conductors which possess self-healing capability. This material is stated as self-healing driven electrically, which is remarkable in terms of control instantaneously along with considerable atom economy. Few researchers stated that faster in scaling for time in auto healing and self-defense for electronics are still adding value for electrochemical impetuses [1]. In general, the connection gets interrupted owing to the load when applied mechanically. This applied load affects the conductive nature of bonds present in the polymers, paving path to micro crack formation. The consequence of the micro crack, affects the conductivity which in turn increases the state of resistance in the circuit near the crack vicinity. Once the resistance shoots up, it directly contributes to raise the temperature and insufficient current magnitude to the circuit. It means that circuit required attention to restore the original state. The temperature rises near the micro-crack that develops a moveable phase in that zone, which alarm the healing process to be initiated. Here the action of healing materials triggered, the micro-crack gets filled by the healable materials. So, the original property of the electric network is completely attained by this process [2]. The healable materials can be of various forms like polymers, composites, elastomers, microcapsules, additives and coatings. These kinds of materials are embedded with conductive materials which establish the conductivity in the circuit.

Figure 11.1 reveals the suitable self-healable conductors to restore the specific damages like cracks and cuts. The primary hurdle in developing such materials is that it should satisfy the auto healing ability without comprising the properties related to electronics. The various strategies involved in developing those materials are unveiled in this chapter.

11.2 Self-Healing Materials

11.2.1 Elastomers

The self-healing capability of material to establish the conductivity to the circuit is of many varieties. One of them is elastomers; the elastomers

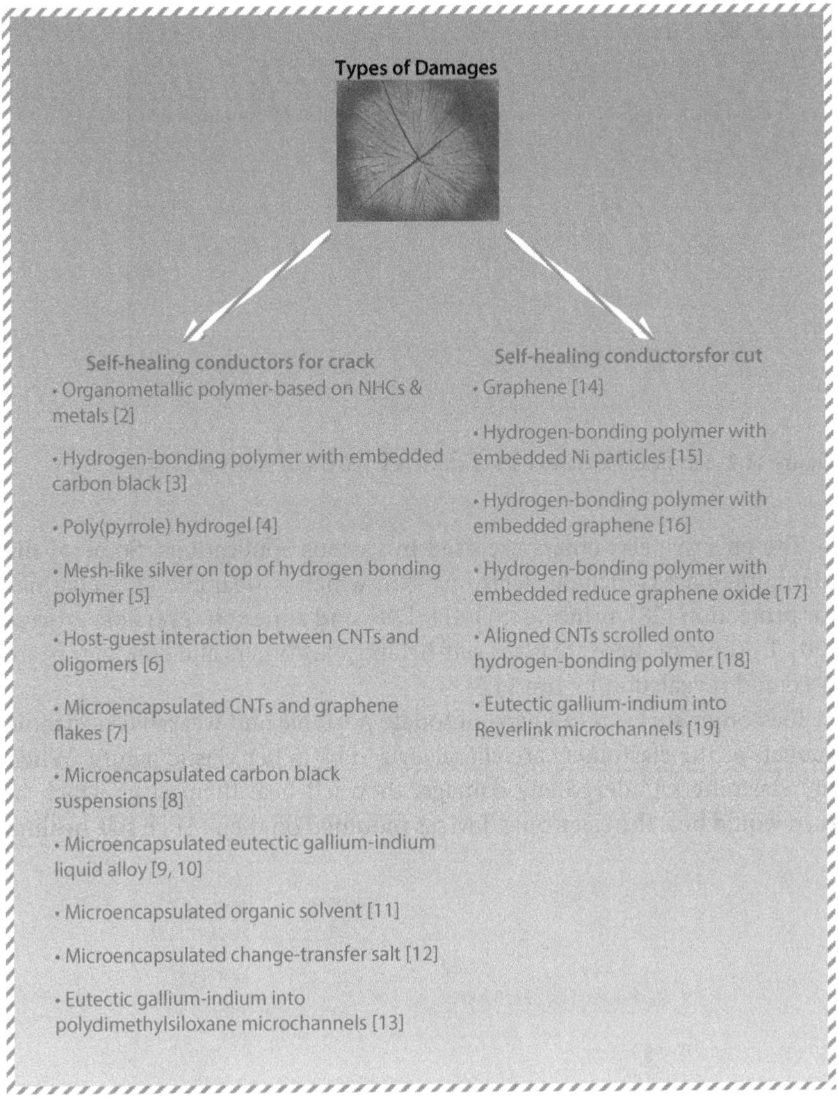

Figure 11.1 Self-healable conductors for healing cracks and cuts.

are further categorized into many variants. Few for the vision to proceed towards various healing abilities of those are polyurethane based, Polydimethyl siloxane (PDMS) based and others. As the categorization further grows long, different materials with their efficiencies have to be converged and assessed. The following consolidation reveals only 100% and near 100% efficient elastomers (Figure 11.2).

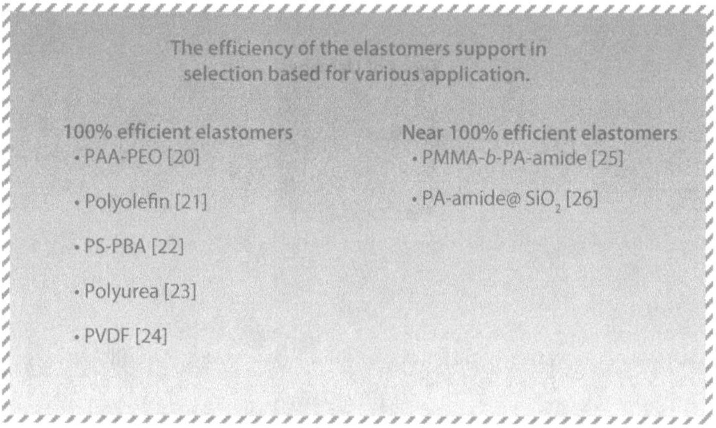

Figure 11.2 Self-healable elastomers of efficiency 100% and near 100%.

The effective elastomers are used in various applications. Some of the highlighted areas include electronic skin which is wearable [27], coatings for protection [28], printing (3D/4D) [29], and adhesion reversible process [30]. In some of these areas the self-healing elastomers and conductors are used and revealed in Figure 11.3.

Elastomer surfaces can accommodate portable and wearable electronic gadgets as the elastomers are characterized by better elastic nature. When the elastomers undergo any damage, they self-heal themselves which in turn would heal the electronic devices mounted on them. Such self-healing

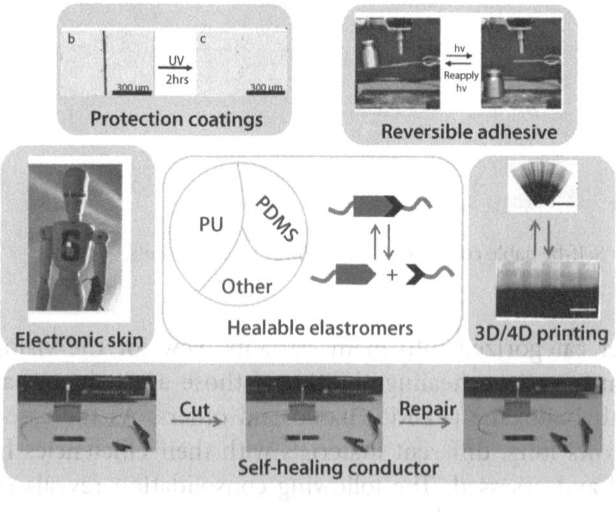

Figure 11.3 Schematic representation of various applications [30–34].

approach in coated surfaces may eradicate scratches on the surface and safeguards the base metal substrate from being open to environment. Bonds in the chain of polymers present in adhesive material may disintegrate itself and reduces the viscosity of the material steeply which is responsible for the creation of fresh and tight bonding. When the external stimulus is removed from the material, the disintegrated bonds again forms thus resulting in solid adhesion of the elastic materials onto the substrate. 3D or 4D printing of self-healing elastic materials requires these viscosity transition mechanisms for their successful fabrication. When the conductive materials are incorporated into self-healing elastic materials, they turn into self-healing conductive materials. Such materials may repair the substrates which are lying within the conductive route thus enabling unobstructed conduction. Such conductive self-healable materials find its advanced applications in electronic gadgets and skins. Electronic gadgets for wearing can be fabricated by using elastomers as the elastic substrate materials. Electronic materials can be made to be enhanced in terms of their sustainability, durability and safety by inducing the self-heal ability in elastic conductors and this has been in lime light of research in the recent days [35]. Specifically, the elastic self-healing conductors are characterized by comfort and better sensitiveness towards strain that make them to be applied in sensors [36], skins [33], super capacitors [37] and interactive interfacial materials between machine and humans [38]. Dispersing the thin film inorganic materials over the surface of self-healing elastic materials is the most popular way of manufacturing the self-healing conductive elastic chemi-resistor materials [39, 40]. Few researchers manufactured self-healing elastomer through drop-casting of molecular mutant of nano gold particle coating over di-sulfide polyurethane cross-linked substrate [41]. Such materials are experimented as self-healing materials for 30 min in room temperature and restored to its original state in 16 h. By this restoration the material possesses better temperature sensing ability along with high strain rates.

Fillers such as ionic liquids, graphene and multi-walled CNTs which are inherently conductive are also embedded over the self-healable polymer walls to manufacture the self-healable conductive materials [36, 42–45]. First self-sustained self-healing conductive elastomer was developed by embedding nickel nano particles over supramolecular organic material for electronic skins [33]. When the material was tested it had an electrical conductivity of 40 S/cm while it could heal back to its 90% initial state, when damaged, after 15 s at room temperature and its mechanical characteristics reverted back in 10 min of time. Due to inversion of resistive forces of such elastic materials upon application of external forces, they can be readily applied to measure various mechanical forces like tactile pressure and flexions as shown in Figure 11.4c. Some researchers tried to use the self-healing elastic polymer conductors made of nano CNT conductive structures along with metal–ligand elastomer bond materials [38, 46–48]. Recently a self-healing metal–ligand bonded material having positive/negative piezo-resistivity tuning was developed by supramolecular hierarchical bonding [47]. Such developed materials can be employed in human body for the detection of deep breathing, pronunciation, coughing and other miniature signals. These sensors had a good autonomy, a rapid response rate of 2 min, an efficiency of 89% even after healed for three times and an appreciable repeatability in self-healing process such that the materials exhibited enhanced detection, sensitiveness and flexibility that remained smooth beyond 10,000 bending cycles. Next generation electronic materials could possibly be developed with piezo-resistive materials coupled with biomimetic self-healing ability. Such a material was made of PUDA/CNT self-healing composites by few researchers which had an electrical conductivity of 27.3 S/m when 3% by weight of CNT was used and this material can be used in electronic sensor applications [32]. This PUDA/CNT material restored by 91% from damage to its actual conductivity within 10 s by means of irradiation of NIR light. The material before and after self-healing exhibited better reliability, durability and sensitivity in the process of sensing strain even beyond 5,000 bend-straightening cycles.

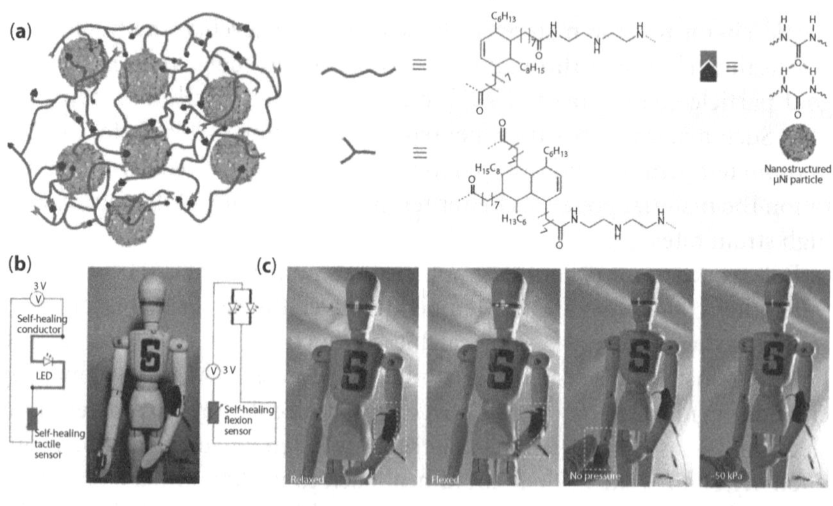

Figure 11.4 (a) Conductive self-healing material, (b) wooden mannequin embedded with tactile sensor circuit and flexion and (c) demonstration of tactile sensor circuit and flexion [33].

11.2.2 Reversible Materials

Major cause of a self-healing polymer is in conductive polymers in which the dynamic reversible bonds are initiated (Figure 11.5). Few researchers started to introduce conductive polymer network based on electrical which is the initial self-healing conductor [2]. For self-healing investigations, organometallic polymers are effectively used which is made up of N-heterocyclic carbenes (NHCs) and different metal salts (M¼Ni, Cu, Pt, Pd, Rh and Ag). To make the carbene moiety to be more secure numerous NHCs were arranged from the various compositions of R¼ alkyl, aryl [49–52]. Structured organometallic polymers were made ready for use with the featuring conductivities in the order of 10^{-3} S/cm with several amalgamation of metallic salts. To produce minute scrape which is most important to study the self-healing property, the researchers applied a sharp razor blade. Initial sharpened side appearance of the led was able to see when the thin layer of it was made warm to a temperature of 200 °C with the time period of 25 min.

Led with the solvent vapor of dimethyl sulfoxide which was made hot up to the temperature of 150 °C for 2 h recovered a conductive film which was spoiled mechanically. Around from 800 to 0 nm the penetrant of failure was decreased. Even though there was an observation regarding electronic behavior which was relatively weak. This kind of problem has to be eliminated by improving the measure of the resisting power of a specified material to the flow of an electric current which has to be improved at least to 1 S/cm for fewer requirements. Self-healing of entire carbon-based conductors provide a probable plan of action with double or triple imine polymers [53–57]. For further examination imine metathesis or transamination of reversibility of the dynamic bonds in solution creates the chances for self-healing conductors [58–60].

Due to the more electrical conductivity and thermal stability properties, graphene is an important conducing material which could be used effectively for the application of self-healing properties. Action of returning the carrier mobility in graphene conductors was investigated by Yu and coworkers [14]. Mechanical scrubbing of SiO_2 used to produce graphene in which the failures were attained by the application of continuous attack

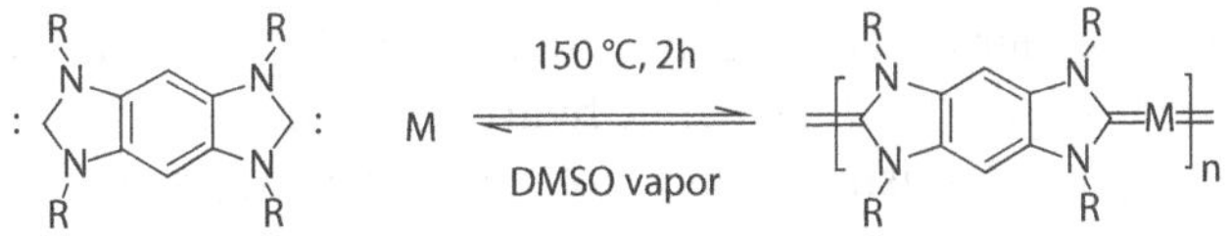

Figure 11.5 Reversible dynamic self-healing process [2].

of the shells with plasma for 5 s. There was an observation of failure in the materials which were made up of carboxyl groups, carbonyl and hydroxyl with the surface of graphene by X-ray photoelectron spectrum (XPS) analysis. By altering the microstructure of the material from 300 to 900 °C, flawed graphene self-healing capacity was improved. Raman spectroscopy and high-resolution transmission electron microscopy (HR-TEM) used to examine the self-healing property and its performance. At the end of the healing method, signals of Raman vanished fully which was produced after the plasma treatment. Self-healing of graphene is most advanced and significant for manufacturing of fabricating miniature structures of micrometer scales and smaller in the absence of external carbon medium. Some experimenters focused on the continuous examination of self-healable single-walled carbon nanotubes (CNTs) [61, 62].

11.3 Self-Healing Conductive Materials

11.3.1 Polymers

Fixing a conductive material in to a matrix is secondary thought of generating self-healing conductors. If the conductive substance added goes beyond the filtration edge since mix only convert in to conductive then it is very much challenging to understand. Conductive containers to fill them decide the self-healing property and physical ability of the material. Conductive material and matrix play a vital role to attain the self-healable composite [63] in terms of chemical interaction of substances and sufficient contact area accordingly. Further research has been focused embedding inorganic metal particle in to organic supramolecular polymer matrix [15]. Effective diffusion between the matrixes was achieved by hydrogen bonding in the supramolecular oligomers, thin nickel oxide layer on the particle surface and usage of large spherical particles which creates decrement in stage accumulation as shown in Figure 11.6. By changing the composition of nickel particles, the glass transition temperature and electrical conductivity were improved from −20 to 10 °C and 40 S/cm respectively. Conductivity of the material is disturbed by cutting the sample.

So instead of cutting, pressing the cut surfaces curing of the material was attained in order to reduce conductivity resulting in restored conductivity (up to 90% after 15 s). Dynamic hydrogen-bonding network decides the self-healing mechanism of supramolcular polymer. The major application of this type was examined which are tactile sensor and flexible sensor. There was an investigation carried out for further self-healing examination

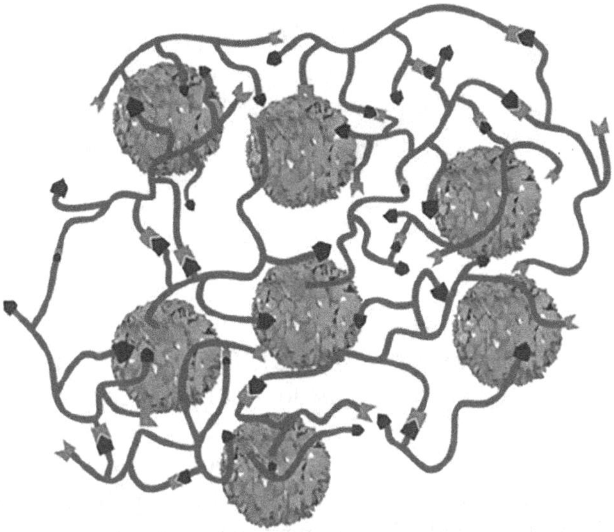

Figure 11.6 Schematic representation of polymer and nickel particles [15].

with several anodes for lithium-ion batteries [3]. Compared to graphite anodes, the performance of lithium-ion batteries were ten times better by using silicon micro particles as an anode. Even though the performance or capabilities are immediately reduced after only ten cycles which cause failure and particle separation. To avoid this kind of activity the anode should be capsuled by a self-healable material. If a carbon black nanoparticleis dissolved with self-healable polymer, it increases the conductivity by a value of 0.25 S/cm.Therefore, compared to graphite anode, silicon-covered micro particle anode gives better capability by more than six times. The life time cycles increases more than ten times with micro particle anodes due to its higher mechanical tensile and rigid collaborations between the silicon surfaces. Due to this reason,the self-healable coating manages a bulk variation lithiation/delithiation better. Inserting graphene networks with self-healable hydrogel was explained by Li and coworkers [16]. Relatively easiest method was used to manufacture a graphene-poly (N,N dimethylacrylamide) crosslinked polymer.

Required cross linked composites will be attained through free radical polymerization of graphene hydrogel with dipping in to an N,N-dimethylacrylamide solution. Next to cut the material, joining of half-blocks were done and made warm or hot to a temperature of 37 °C for 12 h. The rearrangement of polymer chains of the newly removed surfaces decides the anticipated healing mechanism among the joined surfaces. Healing of surface with sharp edge could be achieved by immediate study

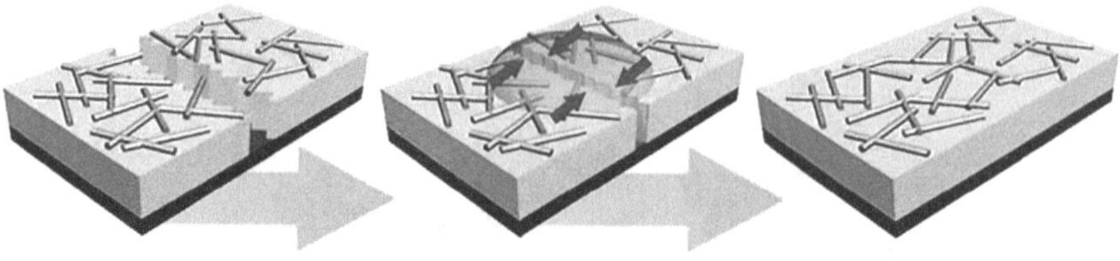

Figure 11.7 Polymer healing process [5].

of combining graphene layers. Due to this method around sixty percentage of actual conductivity was stored back. By using IR laser radiation, the other healing method could be done for 2 h by maintaining the heat less than 50 °C.

Figure 11.7 depicts that the healing ability water simulated nano silver wire-based polymer. The composition comprises of poly acrylic acid hyaluronic acid and branchedpolyethyleneimine were deposited over silicon wafer. The requirement is film with mesh structure it is achieved by the process of drop casting using the conductive layer of silver nanowire. At last, polyvinylpyrrolidone was deposited over the silver mesh structure to allowrobust adhesion between silver nanowires and polymer film by the help of hydrogen bonds. The scratches occurred in conductive film were self-healed combination of deionized water droplet.

11.3.2 Capsules

Self-healing is achieved by capsule is another renowned technique. In this technique the conductive material is placed in microencapsulated form, which when triggered mechanically, releases the healing core agent. This healing material that comes out from the capsule without any other stimulant automatically restores the electric connection over the faulty circuit. There are various types of conductive self-healable materials that are encapsulated to attain the conductors. Figure 11.8 depicts the healing process with the support of microcapsule. It comprises of known polyurea-formaldehyde and metal oxide as a passivation film. The ensuing shape of capsule was ellipsoidal. Once the damage occurs over the layer the capsules releases metal alloy oxide which closes the circuit. The current flows and the lamp lits and the healing takes place both electrically and mechanically. Some other researches successfully demonstrated the novel conductive auto-healing wire that shows complete healing when the wire is damaged [19]. The conductive channel is comprised of eutectic gallium–indium which acts as a polymer. Figure 11.9 shows the healing of wire in

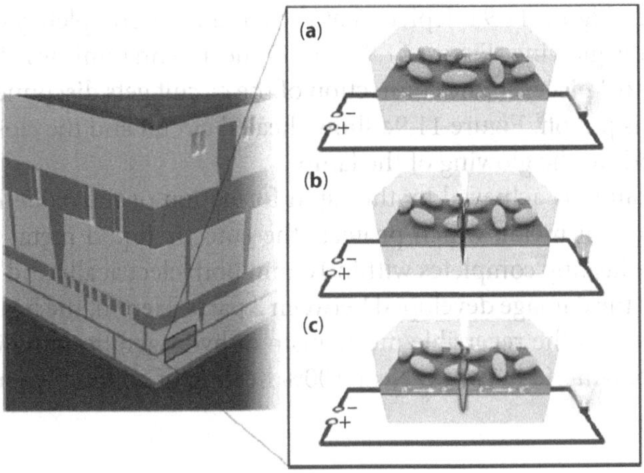

Figure 11.8 Conductive unit: (a) Multiple layer of microelectronic device, (b) interrupted and ruptured capsules with damage, and (c) Metal alloy occupied the damage [9].

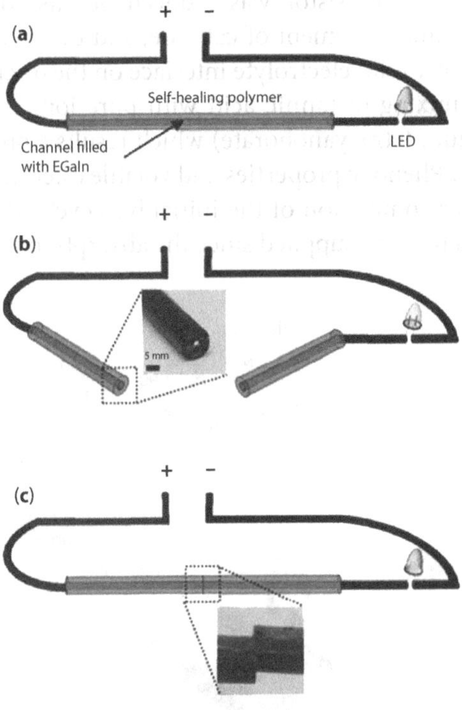

Figure 11.9 Conductive wire self-healable: (a) Originalself-healable conductive wire, (b) cut made using scissors, and (c) restoring of the 2 parts [19].

the circuit. Figure 11.9a depicts that the circuit has completely closed the lamp and is glowing. Figure 11.9b shows the antenna unit was broken by scissor into 2 pieces so the connection of the circuit gets disconnected, and the lamp is put off. Figure 11.9c shows healed circuit and the closes circuit which leads to the glowing of the lamp.

The healing is achieved by the rapid formation of the oxide layer over the surfaces of metal, which protects the eutectic liquid metal to stay in place. The healing completes within 10 min both electrically and mechanically after the damage developed between 2 parts. Herein, the healing takes place based on the reversible mechanism of hydrogen bonding inside the sheet of polymeric network. Near 100% healing of conductive material is achieved.

11.3.3 Liquids

Self-healable ionic liquid gated transistor was revealed by some researchers after a few times [64]. Figure 11.10 shows the chemical and general configurations of electric double layer transistor and the used reagents. The behavior of the transistor was reduced because of humiliation of alpha zinc oxide and decrement of cathode, and the creation of superoxides. Formation of a solid electrolyte interface on the α-ZnO surface could be achieved by mixing of tannic acid with pure ionic liquid (1-ethyl-3-methylimidazolium tetracyanoborate) which results from unstable to stable onset voltage. Phenolic properties and volatile ester decide this binding mechanism. After completion of the initial bias cycle, this layer is strong and not required to be re-supplied since the absorption is more. The above

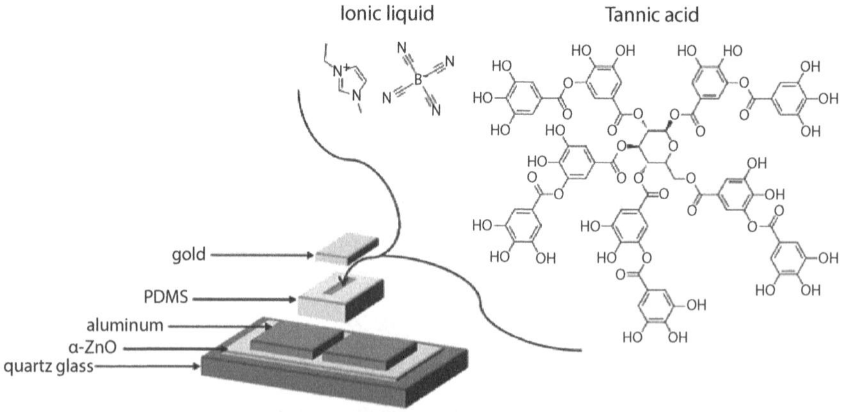

Figure 11.10 Schematic transistor with ionic liquid structure [64].

mentioned self-healable conductors suggest the chance to have future electronic gadgetswith considerable economic as well as safety enhanced properties andoutstanding lifetime.

11.3.4 Composites

The different types of materials are aiding in reconstructing the original state of the electric compounds. Herein the discussion is all about materials in combination that acts as a composite material. It is a known fact that the considerable parameter to heal specifically depends on the various media of conduction like liquids in ionic state, metal, metal alloys, polymers and carbon. The researchers started to assess various materials which possess effective healing capability with better conductivity. They found gold nanowire (AgNWs) attracted many due to its remarkable conductivity property.The nanowire made of gold has high aspect ratio which allows to have remarkable conductivity filtration network for considerable applications in auto-healing substances [65, 66]. Pei and coworkers are developed successfully an auto-healing conductive film made out of polymers. They are embedded with AgNWs over the polymer surface, as the surface possesses the natural healing network comprised of Diels–Alder (D–A) [67]. The builds this hybrid material as they used drop casting technique. The considerable parameter to taken care of is temperature, which they suggest that healing shall be carried out in the room temperature, as a higher temperature may affect the performance of the material.

The recent research reveals that the polyeletrolyte which assembled as multiple layers of film possess the ability of auto-healing with water as catalyst [68, 69]. Let us consider this example, Sun and his support workers developed the facile route to build the self-healable conductive electric film. The hybrid material comprises of AgNWs over the poly acrylic acid–hyaluronic acid (PAA–HA) and branched polyethylenimine (bPEI) as a film [69]. The functionality of the material is demonstrated in the following Figure 11.11. They used a LED blub circuit for this purpose. Initially, the circuit is completely closed so the bulb glows. The cut is made over the circuit with the help of a knife and immediately the LED bulb puts off. Thereafter, the developed composite material is applied over it, which heals the circuit and then the bulb glows. The SEM image also shows various magnifications to understand the healing process at micron level. Likewise, the healing obtained by the hybrid material to make the circuit ready to work. Other than this water stimulant and many other composite materials are developed which are consolidated and given in Figure 11.12.

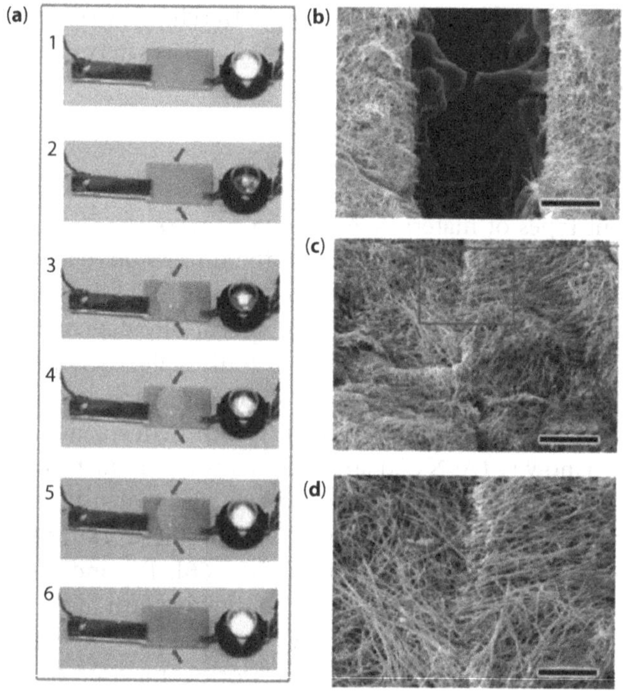

Figure 11.11 (a) AgNWs/(bPEI/PAA-HA) film circuit with LED: (a1 to a6) various state of healing; b–d the images from SEM (b) with cut and (c, d) being healed [69].

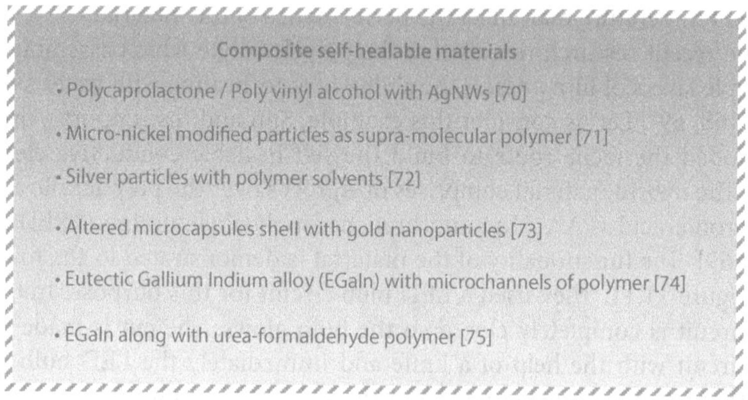

Figure 11.12 Self-healing conductive composite materials.

Researchers never stopped at this level, and in the above case, restoring cycle is limited. Wang *et al.* developed nanomotors which are capable of detecting the damages and healing it effectively and the restoring cycle count is increased by multiple times [76]. A dual layer self-healable

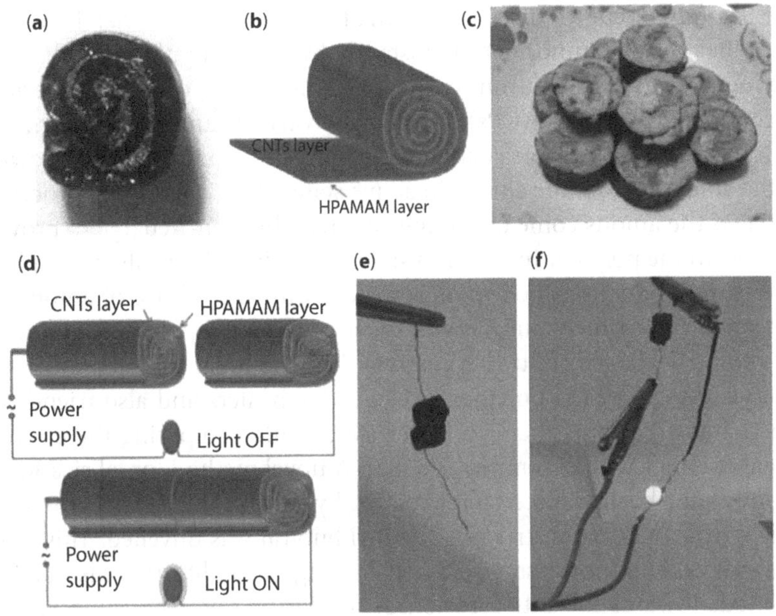

Figure 11.13 Self-healing process of conductive composite [77].

conductive polymeric material made up of HPAMAM and CNTs was developed replicating a traditional Japanese food Sushi. During the manufacturing of the material, the CNTs were laid onto the HPAMAM substrates and the rolling of them into double layered sheets was carried out. Surface resistance of CNTs were initially measured and found to be 741 ± 110 Ω/cm^2. In Figure 11.13a, the gray layers depict CNTs and thick layers depict HPAMAM materials. UV radiation renders a clear visibility of sushi structures since the HPAMAM materials have the better fluorescence under UV light. The fraction of HPAMAM materials was about 80 % and the volume fraction of CNT was maintained as low as possible at the time fabrication of self-healing materials which induced the self-healing ability of composites without any external stimulus at room temperature, when damaged. After self-healing these materials could withstand the self-weight and even the weight of clip or wire. These are illustrated in Figures 11.13d, e and f [77].

11.3.5 Coating

The techniques discussed above restore the original state of the establishing the lost connection. This approach is getting self-healed by coating.

The coating herein for discussion is made of polymer material. It is capable of healing the minor defects instantly by spreading the thin film over the metal surface to confront corrosion. This type of polymers is termed as conductive polymers (CPs). It has many variants among which include PANI and PPy as healable coating that engrossed many researchers owing to the ability to hold the anions in active state. Once the healing process is initiated the anions come out of the coating which anti-charge to provide support to the polymer in oxidized state. As a whole the healing completes by ensuing reduction process of polymers with the release of anions. To consolidate the timely unveiling of the anions by the polymers is considered to be healable conductive coatings. The highlight to note in this scenario is these types of polymers are effective healers and also friendly to the vicinity that puts them in better state while comparing the conventional coatings with chromium [78–82]. A novel method for fabrication of organic auto-healing material for coating by entrenching black carbon and polypyrrole over the matrix of polyvinyl butyral was unveiled. This novel material coating possesses good self-healing capability that also acts as blockade towards the anticorrosion. A self-healing protective mechanism for corrosion which is redox based is used to initiate the release of agents for self-healing was also devised. The novel protective coating is comprised of PANI as shell as core of self-healing material.

In addition to that, the entire unit was altered by nanoparticles of gold (AuNPs) to avoid straight contact between the CP and metal [83]. Figure 11.14 above depicts when a type of corrosion happens, the unit releases the active element to prevent the corrosion. The designed AuNPs have eye on triggered signal which should lead to the unit reduction. When the system is in the ideal state, the shell made up of PANI effectively blockades by

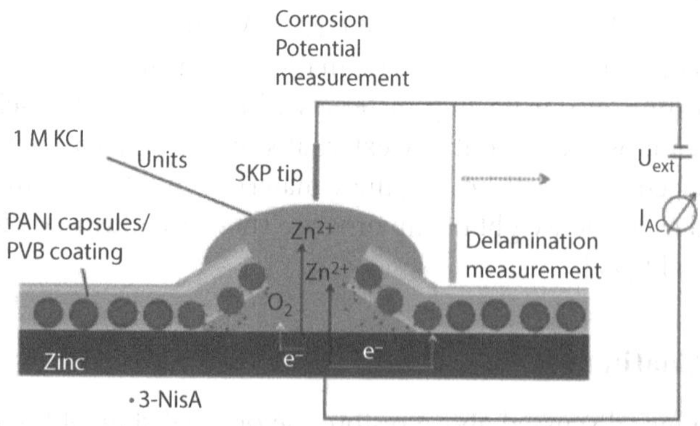

Figure 11.14 Redox-initiated self-healing coating [83].

releasing the acid 3-nitrosalicylic which is the part of core. The PANI in the shell reduces resulting to increased permeability over the shell and lets out the inhibitor prevent the corrosion.

11.4 Conclusion

Self-healing itself is an attractive area of research when comes to repair of electric and electronic circuits which is another remarkable parameter to have eye on. In this chapter, we started with auto-healing conductive materials in various forms starting with elastomers, reversible bonds, polymers, capsules, liquids, composite and coatings. The various segmentation itself depicts that it has very border scope and future for the next generation electronic gadgets. The variants also unveiled that the major accidents can be avoided by short circuit with effective faster healing techniques. In other ways, it can even save human life by avoiding shock, by way of designing the circuit after healing the heavy current which might be re-routed to earth. Other than composite, still hybrid self-healing materials could be developed to cater the needs of gadgets trend. The scope gets wider which supports in the generation of many novel materials that improve the intellectual property status of that nation. So, the major contribution in safety aspect is one of remarkable parameters to address in future. The considerable amount of publications pop up yearly and is expected to highly grow in the coming years. The future direction for the researchers then is developing novel aerogel based conductive self-healing materials.

References

1. Kemp, M., Go, Y.M., Jones, D.P., Nonequilibrium thermodynamics of thiol/disulfide redox systems: A perspective on redox systems biology. *Free Radic. Biol. Med.*, 44, 6, 921–937, 2008.
2. Williams, K.A., Boydston, A.J., Bielawski, C.W., Towards electrically conductive, self-healing materials. *J. R. Soc. Interface*, 4, 13, 359–362, 2007.
3. Wang, C., Wu, H., Chen, Z., McDowell, M.T., Cui, Y., Bao, Z.A., Self-healing chemistry enables the stable operation of silicon microparticle anodes for high-energy lithium-ion batteries. *Nat. Chem.*, 5, 12, 1043–1049, 2013.
4. Hur, J., Im, K., Kim, S.W., Kim, J., Chung, D.Y., Kim, T.H., Jo, K.H., Hahn, J.H., Bao, Z.A., Hwang, S., Park, N., Polypyrrole/agarose-based electronically conductive and reversibly restorable hydrogel. *ACS Nano*, 8, 10, 10066–10076, 2014.

5. Li, Y., Chen, S.S., Wu, M.C., Sun, J.Q., Polyelectrolyte multilayers impart healability to highly electrically conductive films. *Adv. Mater.*, 24, 33, 4578–4582, 2012.
6. Zhang, D.L., Ju, X., Li, L.H., Kang, Y., Gong, X.L., Li, B.J., Zhang, S., An efficient multiple 6. healing conductive composite via host–guest inclusion. *Chem. Commun. (Camb.)*, 51, 29, 6377–6380, 2015.
7. Odom, S.A., Tyler, T.P., Caruso, M.M., Ritchey, J.A., Schulmerich, M.V., Robinson, S.J., Bhargava, R., Sottos, N.R., White, S.R., Hersam, M.C., Moore, J.S., Autonomic restoration of electrical conductivity using polymer-stabilized carbon nanotube and graphene microcapsules. *Appl. Phys. Lett.*, 101, 4, 043106, 2012.
8. Kang, S., Jones, A.R., Moore, J.S., White, S.R., Sottos, N.R., Microencapsulated carbon black suspensions for restoration of electrical conductivity. *Adv. Funct. Mater.*, 24, 20, 2947–2956, 2014.
9. Blaiszik, B.J., Kramer, S.L.B., Grady, M.E., McIlroy, D.A., Moore, J.S., Sottos, N.R., White, S.R., Autonomic restoration of electrical conductivity. *Adv. Mater.*, 24, 3, 398–401, 2012.
10. Blaiszik, B.J., Jones, A.R., Sottos, N.R., White, S.R., Microencapsulation of gallium–indium (Ga–In) liquid metal for self-healing applications. *J. Microencapsul.*, 31, 4, 350–354, 2014.
11. Odom, S.A., Chayanupatkul, S., Blaiszik, B.J., Zhao, O., Jackson, A.C., Braun, P.V., Sottos, N.R., White, S.R., Moore, J.S., A self-healing conductive ink. *Adv. Mater.*, 24, 19, 2578–2581, 2012.
12. Odom, S.A., Caruso, M.M., Finke, A.D., Prokup, A.M., Ritchey, J.A., Leonard, J.H., White, S.R., Sottos, N.R., Moore, J.S., Restoration of conductivity with TTF-TCNQ charge-transfer salts. *Adv. Funct. Mater.*, 20, 11, 1721–1727, 2010.
13. So, J.H., Thelen, J., Qusba, A., Hayes, G.J., Lazzi, G., Dickey, M.D., Reversibly deformable andmechanically tunable fluidic antennas. *Adv. Funct. Mater.*, 19, 22, 3632–3637, 2009.
14. Chen, J.H., Shi, T.W., Cai, T.C., Xu, T., Sun, L.T., Wu, X.S., Yu, D.P., Self-healing of defected graphene. *Appl. Phys. Lett.*, 102, 10, 103107, 2013.
15. Tee, B.C.K., Wang, C., Allen, R., Bao, Z.N., An electrically and mechanically self-healing composite with pressure- and flexion-sensitive properties for electronic skin applications. *Nat. Nanotechnol.*, 7, 12, 825–832, 2012.
16. Hou, C.Y., Duan, Y.R., Zhang, Q.H., Wang, H.Z., Li, Y.G., Bio-applicable and electroactive near-infrared laser-triggered self-healing hydrogels based on graphene networks. *J. Mater. Chem.*, 22, 30, 14991–14996, 2012.
17. Peng, R.G., Yu, Y., Chen, S., Yang, Y.K., Tang, Y.H., Conductive nanocomposite hydrogels with self-healing property. *RSC Adv.*, 4, 66, 35149–35155, 2014.
18. Sun, H., You, X., Jiang, Y.S., Guan, G.Z., Fang, X., Deng, J., Chen, P.N., Luo, Y.F., Peng, H.S., Selfhealable electrically conducting wires for wearable microelectronics. *Angew. Chem. Int. Ed.*, 53, 36, 9526–9531, 2014.

19. Palleau, E., Reece, S., Desai, S.C., Smith, M.E., Dickey, M.D., Self-healing stretchable wires for reconfigurable circuit wiring and 3D microfluidics. *Adv. Mater.*, 25, 11, 1589–1592, 2013.
20. Burnworth, M., Tang, L., Kumpfer, J.R., Duncan, A.J., Beyer, F.L., Fiore, G.L., Rowan, S.J., Weder, C., Optically healable supramolecular polymers. *Nature*, 472, 7343, 334–337, 2011.
21. Cromwell, O.R., Chung, J., Guan, Z., Malleable and self-healing covalent polymer networks through tunable dynamic boronic ester bonds. *J. Am. Chem. Soc.*, 137, 20, 6492–6495, 2015.
22. Mozhdehi, D., Ayala, S., Cromwell, O.R., Guan, Z., Self-healing multiphase polymers via dynamic metal–ligand interactions. *J. Am. Chem. Soc.*, 136, 46, 16128–16131, 2014.
23. Xu, J.H., Chen, W., Wang, C., Zheng, M., Ding, C.D., Jiang, W., Tan, L.H., Fu, J.J., Extremely stretchable, self-Healable elastomers with tunable mechanical properties: Synthesis and applications. *Chem. Mater.*, 30, 17, 6026–6039, 2018.
24. Cao, Y., Morrissey, T.G., Acome, E., Allec, S., II, Wong, B.M., Keplinger, C., Wang, C., A transparent, self-healing, highly stretchable ionic conductor. *Adv. Mater.*, 29, 10, 1605099, 2017.
25. Chen, Y. and Guan, Z., Multivalent hydrogen bonding block copolymers self-assemble into strong and tough self-healing materials. *Chem. Commun.*, 50, 74, 10868–10870, 2014.
26. Williams, G.A., RyoheiIshige, C., Olivia, R., Chung, J., Takahara, A., Guan, Z., Mechanically robust and self-healable superlattice nanocomposites by self assembly of single component "sticky" polymer grafted nanoparticles. *Adv. Mater.*, 27, 26, 3934–3941, 2015.
27. Wang, S., Oh, J.Y., Xu, J., Tran, H., Bao, Z., Skin-inspired electronics: an emerging paradigm. *Acc. Chem. Res.*, 51, 5, 1033–1045, 2018.
28. Xu, J.H., Ye, S., Ding, C.D., Tan, L.H., Fu, J.J., Autonomous self-healing supramolecular elastomer reinforced and toughened by graphitic carbon nitride nanosheets tailored for smart anticorrosion coating applications. *J. Mater. Chem. A*, 6, 14, 5887–5898, 2018.
29. Zhang, Z.P., Rong, M.Z., Zhang, M.Q., Polymer engineering based on reversible covalent chemistry: A promising innovative pathway towards new materials and new functionalities. *Prog. Polym. Sci.*, 80, 39–93, 2018.
30. Michal, B.T., Spencer, E.J., Rowan, S.J., Stimuli-responsive reversible two-level adhesion from a structurally dynamic shape-memory polymer. *ACS Appl. Mater. Interfaces*, 8, 17, 11041–11049, 2016.
31. Wang, Z., Yang, Y., Burtovyy, R., Luzinov, I., Urban, M.W., UV-induced self-repairing polydimethylsiloxane–polyurethane (PDMS–PUR) and polyethylene glycol–polyurethane (PEG–PUR) Cu-catalyzed networks. *J. Mater. Chem. A*, 2, 37, 15527–15534, 2014.

32. Pu, W., Fu, D., Wang, Z., Gan, X., Lu, X., Yang, L., Xia, H., Realizing crack diagnosing and self healing by electricity with a dynamic crosslinked flexible polyurethane composite. *Adv. Sci.*, 5, 5, 1800101, 2018.
33. Tee, B.C.K., Wang, C., Allen, R., Bao, Z., An electrically and mechanically self-healing composite with pressure-and flexion-sensitive properties for electronic skin applications. *Nat. Nanotechnol.*, 7, 12, 825, 2012.
34. Yang, L., Zhang, G., Zheng, N., Zhao, Q., Xie, T., A metallosupramolecular shape memory polymer with gradient thermal plasticity. *Angew. Chem. Int. Ed.*, 56, 41, 12599–12602, 2017.
35. Blaiszik, B.J., Kramer, S.L.B., Grady, M.E., McIlroy, D.A., Moore, J.S., Sottos, N.R., White, S.R., Autonomic restoration of electrical conductivity. *Adv. Mater.*, 24, 3, 398–401, 2012.
36. He, Y., Liao, S., Jia, H., Cao, Y., Wang, Z., Wang, Y., A self healing electronic sensor based on thermal sensitive fluids. *Adv. Mater.*, 27, 31, 4622–4627, 2015.
37. Huang, Y., Zhong, M., Huang, Y., Zhu, M., Pei, Z., Wang, Z., Xue, Q., Xie, X., Zhi, C., A self-healable and highly stretchable supercapacitor based on a dual crosslinked polyelectrolyte. *Nat. Commun.*, 6, 1, 1–8, 2015.
38. Cao, J., Lu, C., Zhuang, J., Liu, M., Zhang, X., Yu, Y., Tao, Q., Multiple hydrogen bonding enables the self healing of sensors for human–machine interactions. *Angew. Chem. Int. Ed.*, 56, 30, 8795–8800, 2017.
39. Li, Y., Chen, S., Wu, M., Sun, J., Polyelectrolyte multilayers impart healability to highly electrically conductive films. *Adv. Mater.*, 24, 33, 4578–4582, 2012.
40. Wang, H., Zhu, B., Jiang, W., Yang, Y., Leow, W.R., Wang, H., Chen, X., A mechanically and electrically self healing supercapacitor. *Adv. Mater.*, 26, 22, 3638–3643, 2014.
41. Huynh, T.P. and Haick, H., Self healing, fully functional, and multiparametric flexible sensing platform. *Adv. Mater.*, 28, 1, 138–143, 2016.
42. Yang, Y. and Urban, M.W., Self repairable polyurethane networks by atmospheric carbon dioxide and water. *Angew. Chem. Int. Ed.*, 53, 45, 12142–12147, 2014.
43. Wang, S., Liu, N., Su, J., Li, L., Long, F., Zou, Z., Jiang, X., Gao, Y., Highly stretchable and self-healable supercapacitor with reduced graphene oxide-based fiber springs. *ACS Nano*, 11, 2, 2066–2074, 2017.
44. Guo, K., Zhang, D.L., Zhang, X.M., Zhang, J., Ding, L.S., Li, B.J., Zhang, S., Conductive elastomers with autonomic self healing properties. *Angew. Chem. Int. Ed.*, 54, 41, 12127–12133, 2015.
45. Cao, Y., Morrissey, T.G., Acome, E., Allec, S., II, Wong, B.M., Keplinger, C., Wang, C., A transparent, self-healing, highly stretchable ionic conductor. *Adv. Mater.*, 29, 10, 1605099, 2017.
46. Cao, J., Lu, C., Zhuang, J., Liu, M., Zhang, X., Yu, Y., Tao, Q., Multiple hydrogen bonding enables the self-healing of sensors for human–machine interactions. *Angew. Chem. Int. Ed.*, 56, 30, 8795–8800, 2017.

47. Liu, X., Su, G., Guo, Q., Lu, C., Zhou, T., Zhou, C., Zhang, X., Hierarchically structured self-healing sensors with tunable positive/negative piezoresistivity. *Adv. Funct. Mater.*, 28, 15, 1706658, 2018.
48. Liu, X., Lu, C., Wu, X., Zhang, X., Self-healing strain sensors based on nanostructured supramolecular conductive elastomers. *J. Mater. Chem. A*, 5, 20, 9824–9832, 2017.
49. Boydston, A.J., Williams, K.A., Bielawski, C.W., A modular approach to main-chainorganometallic polymers. *J. Am. Chem. Soc.*, 127, 36, 12496–12497, 2005.
50. Boydston, A.J., Rice, J.D., Sanderson, M.D., Dykhno, O.L., Bielawski, C.W., Synthesis andstudy of bidentate benzimidazolylidene—Group 10 metal complexes and related main-chainorganometallic polymers. *Organometallics*, 25, 26, 6087–6098, 2006.
51. Khramov, D.M., Boydston, A.J., Bielawski, C.W., Synthesis and study of Janus bis(carbene)s and their transition-metal complexes. *Angew. Chem. Int. Ed.*, 45, 37, 6186–6189, 2006.
52. Boydston, A.J. and Bielawski, C.W., Bis(imidazolylidene)s as modular building blocks formonomeric and macromolecular organometallic materials. *Dalton Trans.*, 34, 4073–4077, 2006.
53. Meyer, C.D., Joiner, C.S., Stoddart, J.F., Template-directed synthesis employing reversibleimine bond formation. *Chem. Soc. Rev.*, 36, 11, 1705–1723, 2007.
54. Fukuda, K., Shimoda, M., Sukegawa, M., Nobori, T., Lehn, J.M., Doubly degradable dynamers: dynamic covalent polymers based on reversible imine connections and biodegradablepolyester units. *Green Chem.*, 14, 10, 2907–2911, 2012.
55. Ro, S., Rowan, S.J., Pease, A.R., Cram, D.J., Stoddart, J.F., Dynamic hemicarcerands and hemicarceplexes. *Org. Lett.*, 2, 16, 2411–2414, 2000.
56. Bozdemir, O.A., Barin, G., Belowich, M.E., Basuray, A.N., Beuerle, F., Stoddart, J.F., Dynamiccovalent templated-synthesis of [c2] daisy chains. *Chem. Commun.*, 48, 84, 10401–10403, 2012.
57. Belowich, M.E. and Stoddart, J.F., Dynamic imine chemistry. *Chem. Soc. Rev.*, 41, 6, 2003–2024, 2012.
58. Janeliunas, D., van Rijn, P., Boekhoven, J., Minkenberg, C.B., van Esch, J.H., Eelkema, R., Aggregation-driven reversible formation of conjugated polymers in water. *Angew. Chem. Int. Ed.*, 52, 7, 1998–2001, 2013.
59. Zhao, D.H. and Moore, J.S., Folding-driven reversible polymerization of oligo (m-phenyleneethynylene) imines: solvent and starter sequence studies. *Macromolecules*, 36, 8, 2712–2720, 2003.
60. Kovaricek, P. and Lehn, J.M., Merging constitutional and motional covalent dynamics inreversible imine formation and exchange processes. *J. Am. Chem. Soc.*, 134, 22, 9446–9455, 2012.
61. Kim, D.Y., Sinha-Ray, S., Park, J.J., Lee, J.G., Cha, Y.H., Bae, S.H., Ahn, J.H., Jung, Y.C., Kim, S.M., Yarin, A.L., Yoon, S.S., Self-healing reduced graphene

oxide films by supersonic kineticspraying. *Adv. Funct. Mater.*, 24, 31, 4986–4995, 2014.
62. Um, J.E., Chung, C.H., Lee, D.C., Yoo, P.J., Kim, W.J., Restoration of the genuine electronicproperties of functionalized single-walled carbon nanotubes. *RSC Adv.*, 4, 81, 42930–42935, 2014.
63. Benight, S.J., Wang, C., Tok, J.B.H., Bao, Z.A., Stretchable and self-healing polymers anddevices for electronic skin. *Prog. Polym. Sci.*, 38, 12, 1961–1977, 2013.
64. Bubel, S., Menyo, M.S., Mates, T.E., Waite, J.H., Chabinyc, M.L., Schmitt trigger using a self-healing ionic liquid gated transistor. *Adv. Mater.*, 27, 21, 3331–3335, 2015.
65. Xu, F. and Zhu, Y., Highly conductive and stretchable silver nanowire conductors. *Adv. Mater.*, 24, 5117–5122, 2012.
66. van de Groep, J., Spinelli, P., Polman, A., Transparent conducting silver nanowire networks. *Nano Lett.*, 12, 3138–3144, 2012.
67. Gong, C., Liang, J., Hu, W., Niu, X., Ma, S., Hahn, H.T., Pei, Q., A healable, semitransparent silver nanowire-polymer composite conductor. *Adv. Mater.*, 25, 4186–4191, 2012.
68. South, A.B. and Lyon, L.A., Autonomic self-healing of hydrogel thin films. *Angew. Chem. Int. Ed.*, 49, 767–771, 2010.
69. Li, Y., Chen, S., Wu, M., Sun, J., Polyelectrolyte multilayers impart healability to highly electrically conductive films. *Adv. Mater.*, 24, 4578–4582, 2012.
70. Li, Y., Chen, S., Wu, M., Sun, J., Rapid and efficient multiple healing of flexible conductive films by near-infrared light irradiation. *ACS Appl. Mater. Interfaces*, 6, 16409–16415, 2014.
71. Tee, B.C.K., Wang, C., Allen, R., Bao, Z., An electrically and mechanically self-healing composite with pressure- and flexion-sensitive properties for electronic skin applications. *Nat. Nanotechnol.*, 7, 825–832, 2012.
72. Odom, S.A., Chayanupatkul, S., Blaiszik, B.J., Zhao, O., Jackson, A.C., Braun, P.V., Sottos, R.N., White, S.R., Moore, J.S., Self-healing: A self-healing conductive ink. *Adv. Mater.*, 24, 2509–2509, 2012.
73. Lai, G.W., Chang, S.J., Lee, J.T., Liu, H., Li, C.C., Conductive microcapsules for self-healing electric circuits. *RSC Adv.*, 5, 104145–104148, 2015.
74. Chiechi, R.C., Weiss, E.A., Dickey, M.D., Whitesides, G.M., Eutectic gallium–indium (EGaIn): A moldable liquid metal for electrical characterization of self-assembled monolayers. *Angew. Chem. Int. Ed.*, 120, 148–150, 2008.
75. Blaiszik, B.J., Kramer, S.L.B., Grady, M.E., McIlroy, D.A., Moore, J.S., Sottos, N.R., White, S.R., Autonomic restoration of electrical conductivity. *Adv. Mater.*, 24, 398–401, 2012.
76. Li, J., Shklyaev, O.E., Li, T., Liu, W., Shum, H., Rozen, I., Balazs, A.C., Wang, J., Self-propelled nanomotors autonomously seek and repair cracks. *Nano Lett.*, 15, 7077–7085, 2015.

77. Yang, W., Song, J., Wu, X., Wang, X., Liu, W., Qiu, L., Hao, W., High-efficiency self-healing conductive composites from HPAMAM and CNTs. *J. Mater. Chem. A*, 3, 12154–12158, 2015.
78. Kowalski, D., Ueda, M., Ohtsuka, T., Self-healing ability of conductive polypyrrole coating with artificial defect. *ECS Trans.*, 16, 177–181, 2009.
79. Bailey, B.M., Leterrier, Y., Garcia, S.J., van der Zwaag, S., Michaud, V., Electrically conductive self-healing polymer composite coatings. *Prog. Org. Coat.*, 85, 189–198, 2015.
80. Niratiwongkorn, T., Luckachan, G.E., Mittal, V., Self-healing protective coatings of polyvinyl butyral/polypyrrole-carbon black composite on carbon steel. *RSC Adv.*, 6, 43237–43249, 2016.
81. Paliwoda-Porebska, G., Stratmann, M., Rohwerder, M., Potje-Kamloth, K., Lu, Y., Pich, A.Z., Adler, H.J., On the development of polypyrrole coatings with self-healing properties for iron corrosion protection. *Corros. Sci.*, 47, 3216–3323, 2005.
82. Vimalanandan, A., Lv, L.-P., Tran, T.H., Landfester, K., Crespy, D., Rohwerder, M., Redox-responsive self-healing for corrosion protection. *Adv. Mater.*, 25, 6980–6984, 2013.
83. Kowalski, D., Ueda, M., Ohtsuka, T., Self-healing ion perm selective conducting polymer coating. *J. Mater. Chem.*, 20, 7630–7633, 2010.

12

Self-Healable Artificial Skin

Younus Raza Beg, Gokul Ram Nishad and Priyanka Singh*

Department of Chemistry, Govt. Digvijay PG Autonomous College, Rajnandgaon, Chhattisgarh, India

Abstract
Artificial skin is an emerging field of research. Electronic skins (e-skins) developed so far has great potential to mimic human skin. But, improving mechanical stretchability, flexibility and self healing ability are some hurdles to overcome. Improvement of self healing ability is done by introducing dynamic bonding in hybrid materials. Pressure sensor, strain sensor, motion sensor, ECG, temperature sensor and face sensor devices etc. can be prepared by these e-skins. Improving electrical conductivity of the materials may improve the sensitivity of these devices. Electrical conductivity may improve with incorporation of some electrically conducting polymer network, nanomaterials and ions. This book chapter deals with the preparation of some important hybrid materials for e-skin along with their applications.

Keywords: Self-healable, artificial skin, artificial electronic skin, pressure sensor, strain sensor, hydrogel

12.1 Introduction

Skin is the largest organ of the body. Skin not only protects the body but it also senses touch and heat. Some camouflage protects prey from predator through change of skin color. Flexibility, stretchability, and self-healing ability are some of the important characters of skin. Artificial skin is an electronic skin which may mimic skin characteristics. To mimic characteristics of skin worldwide research work is being carried on, but self healing ability, mechanical stretchability, and flexibility are some of the major issues relevant to mimicking skin as artificial skin or electronic skin. To overcome

**Corresponding author*: priyankasingh121@yahoo.com

Inamuddin, Mohd Imran Ahamed, Rajender Boddula and Tariq Altalhi (eds.) Self-Healing Smart Materials and Allied Applications, (321–344) © 2021 Scrivener Publishing LLC

these problems self-healable gels, eutectic metal alloys [1], composites [2], elastomers, polymers, nanocomposites, and supramolecules [3] are being used. Self healing ability endows new dimensions to the artificial or electronic skin. To achieve the new horizon of self healing ability reversible hydrogen bonding, dynamic chelation, Diels–Alder reaction, carboxyl-metal coordination bonds, boroxines bondings, urea bonds, reversible cross-linking, supramolecular complexation are being utilized as powerful tools. Gels are modified to improve self healing ability by hybridizing it with another hydrogels, metal ions, polymers, elastomers, cross-lnking agents and various supramolecules. Diels–Alder reaction and retro Diels–Alder reaction may be also be carried out in these hybrid gels. Self healing ability may be achieved autonomously or by inducing it in presence of moisture, heat, light or by other way. Artificial skin can be classified into two categories of electronic skin and ionic skin. Human skin senses touch and pressure in the range of 10–100 kPa for medium pressure as well as it may sense below 10 kPa of pressure for a gentle touch. Pressure and strain sensitive electronic skin can be designed generally by sandwiching dielectric layers between two layers of gel (hybrid material), whereas, strain sensor electron skin is generally designed by using very high band tape congregated between layers of gel (hybrid material). To improve conductivity of the e-skin carbon nanotube (CNT), reduced graphen oxide (rGO), conductive polymers, nanomaterials etc. can be incorporated into e-skin. Capacitance change may be recorded as the effective area and thickness of skin changes due to applied pressure or strain because current collector ions are incorporated in hydrogel to construct ionic skin.

12.2 Preparation and Properties of Artificial Skin

This section deals with the preparation and properties of various materials used to prepare artificial skin. Various hydrogels, composite materials etc. can be prepared considering dynamic interactions for self healing ability. Sohn *et al.* (2019) prepared electronic skin by combining Ga–In alloy with Ag nanowire (AgNW). Eutectic Ga–In (EGaIn) liquid metal with combination of AgNW exhibits mechanical flexibility and electrical sustainability. Hybrid conductor is prepared through drop casting of AgNW ink onto polyimide film followed by Copper electrode (Cu) attachment subsequently then printing of EGaIn onto the electrode is done. Formation of fresh surface area owing to self healing nature is achieved while bending and stretching [1]. Sim *et al.* [2] prepared self-healing composite electrode for electronic skin through liquid crystal graphene oxide (LCGO) and AgNW.

Conductivity and mechanical strength were found to be 17,800 S/cm and 4.2 MPa, respectively. Functional group interaction between LCGO and AgNW is the main reason for bonding. Solution of LCGO and AgNW exhibits high viscosity and can be printable in different shapes (Figure 12.1) [2]. Self healing is feasible due to new hydrogen bond formation and reconnection between damaged portion. Conductive composite can be used as electrode and connected to the circuit of LED (Figure 12.1b) [2].

Han *et al.* [3] coated polypyrrole (PPy) on CNC (coated cellulose nanocrystal) and CNF (cellulose nanofiber) to form self-healable network with PVA. The composite exhibits better mechanical strength and toughness than the pure PVA. Adhesive nature was also found to be 9,670 times higher with respect to its own weight. Adhesive performances were measured in air as well as water for various substances. Weak interaction and iron chelation between PVA matrix and nanocellulose filleres are responsible for flexibility, self-healing ability, and better sensitivity. CNCs and CNF were used for biotemplates for PPy as well as reinforcement for the PVA matrix. Polymerization of pyrrole was done on CNF and CNC by using oxidatant. CNF_{Fe} is prepared by treating CNF/pyrrole with $FeCl_3$ solution. The nanocomposites of CNC/CNF-PPY were dispersed in aqueous solution of PVA to form nanocomposite of PVA. Grafting of PPy on CNF and CNC was proved by FTIR studies (Figure 12.2) [3]. Major peaks related to PPy were observed for all the prepared samples. Peaks at 1,630 cm^{-1} and 1,540 cm^{-1}

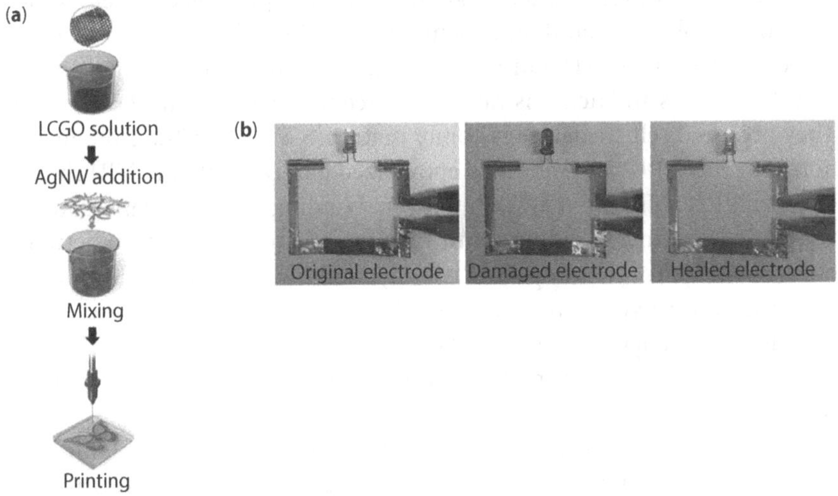

Figure 12.1 (a) Self-healing electrode made by LCGO and AgNW solution. (b) Photographs of the electrode before and after damage [Reprinted with permission from ACS publication] [2].

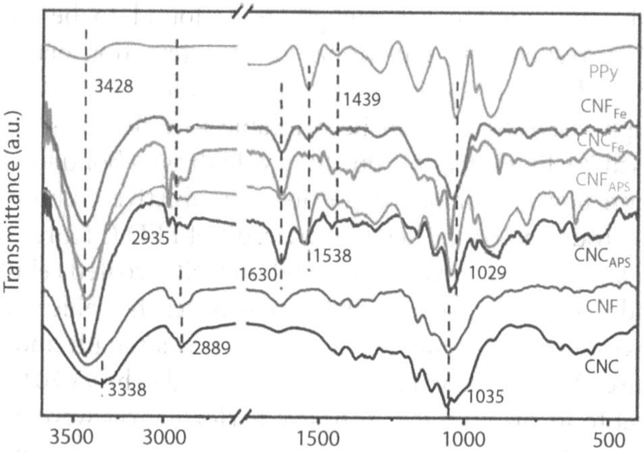

Figure 12.2 FTIR spectra of grafted CNC and CNF and various composites [Reprinted with permission from ACS publication] [3].

were due to H-bonding between pyrrole NH group and hydroxyl group of CNF/CNC. Tensile strength for CNF_{Fe}/PVA was observed to be 168 and 120.8 MPa before and after self-healing whereas pure PVA exhibited tensile strength of 32.7 MPa. They performed conductivity test and revealed that, when CNF_{Fe}/PVA film was cut, bulb extinguished immediately. After 30 min, the bulb illuminated again due to self-healing, henceforth conductivity remains unchanged after self healing [3].

Guo *et al.* [4] prepared skin like material by supramolecular elastomer composites. Skin of squid and chemopods exhibits fluorescence. Authors successfully mimicked (Figure 12.3a) [4] the skin like pattern of those squid and chemopods and demonstrated fluorescence via mechanical stretching. They prepared self-healable elastomer matrix (S-ENR) by 3-aminophenylboronic acid (APB) and modified epoxidized natural rubber (ENR) chains. The S-ENR exhibits dynamic hydrogen bonding and boroxines bonds. Dynamic boraxines bonding forms during dehydration reaction between boron and hydroxyl groups whereas dynamic hydrogen bonding forms between boron-hydroxyls and pendent secondary amines (Figure 12.3b) [4]. Fluorescent agent containing layer was wrapped between CNC/CNTs and S-ENR layers. Micro crack structure is generated on CNC@CNTs layer when strain is applied [4].

Lin *et al.* [5] deposited Tanic acid (TA) on CNC and then tanic acid coated CNCs(TA@CNC) was chelated with Ag^+ ion to form Ag/TA@CNCs. Furthermore, Ag/TA@CNCs nanocomplexes were cross-linked with polyvinyl alcohol in presence of borax to form hierarchical 3D network PB-Ag/TA@CNC nanocomposite hydrogel. TA with CNC consequently provided

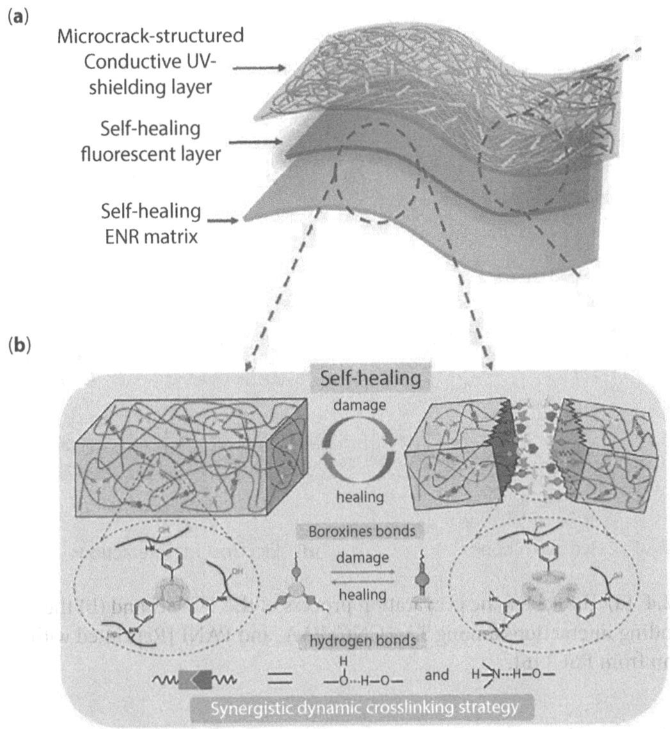

Figure 12.3 (a) Compositional schematic multifunctional material. (b) The self-healable supramolecular elastomer with synergistic dynamic crosslinking network design [Reprinted with permission from RSC] [4].

catechol groups as active sites for dynamic reversible cross-linking to facilitate self healing ability. Ag nanoparticles and catechol groups enhance conductivity and stretchability to 4.61 S m^{-1} and >4,000% respectively [5]. Superior mechanical performance (strain 4,700%) and reliable autonomous self-healing capability (HE 490%) was observed when tannic acid is coated on cellulose nanocrystals (TA@CNCs). TA@CNCs serves as highly conductive (0.13 S cm^{-1}) and highly sensitive material (GF = 11.2). TC-Gel was prepared by two step method with the help of acrylic acid (AA) monomer and N,N_0-methylenediacrylamide as a chemical cross-linking agent. Polymerization was done by UV-radiation to get T-Gel. Furthermore, oxidation of aniline is done with TA@CNC to obtain TC-Gel (Figure 12.4a) [6]. Comparative study was done for TC-gel and T-gel. TC-Gel exhibits 26 times higher conductivity (0.13 S cm^{-1}) than T-gel (0.005 S cm^{-1}). TA@ CNCs endows dual cross-linking attributed by physical as well as chemical forces. Various interactions for the material have been shown in Figure 12.4b.

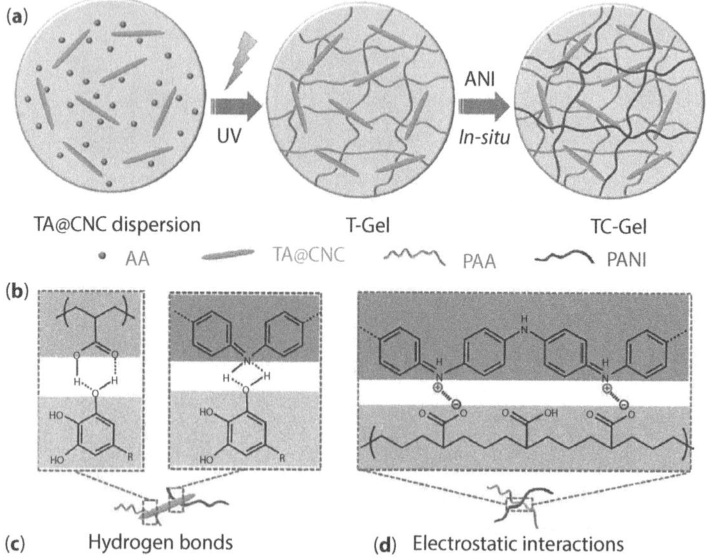

Figure 12.4 (a) The schematic preparation process of the TC-Gel and (b) the corresponding interactions among TA@CNC, PAA, and PANI [Reprinted with permission from RSC] [6].

TA@CNCs combine to PAA and PANI network with hydrogen bonding, consequently it serves as reinforcing agent. Without TA@CNCs, C-gel was prepared with aniline as the same procedure TC-gel was prepared. TC-Gel performs well as it exhibits better tensile fracture strength (0.384 MPa) and higher compression strength (3.63 MPa) than the T-Gel (Figures 12.5a, b) [6].

Park *et al.* [7] designed three dimensional artificial skins using 3D printing technique. Ring shaped and figure tip shaped devices were prepared and studied by them. They successfully demonstrated self-healing and touch sensing properties of these devices. The device may sense touching by generating 5.4 times greater electronic signal contrast. Self healing ability is attributed by secondary dynamic bonding due to ionic interaction between carboxylate anion of polyacrilic acid chain and Fe^{+3} of hydrogel (Figure 12.6a) [7]. Self healing ability was demonstrated by cut and contact method. The hydrogel healed after 30 min from cutting. After healing of the hydrogel, when strain is applied in it torn in different direction (Figure 12.6b) [7].

Kim *et al.* [8] prepared thermally healable polydimethylsiloxane (PDMS) by introducing reversible cross-linking moieties of Diels–Alder reactive species. Solvent vapor treatment for self-healing is applied. A dynamic Diels–Alder reaction may heal the substance when its retro products again

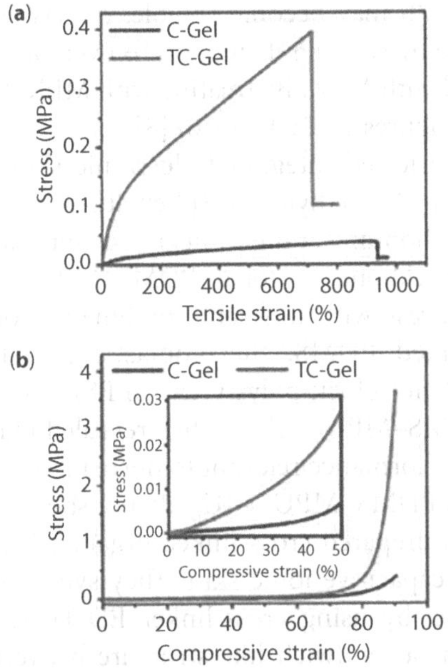

Figure 12.5 (a) Typical tensile and (b) compression stress–strain curves of TC-Gel and C-Gel [Reprinted with permission from RSC] [6].

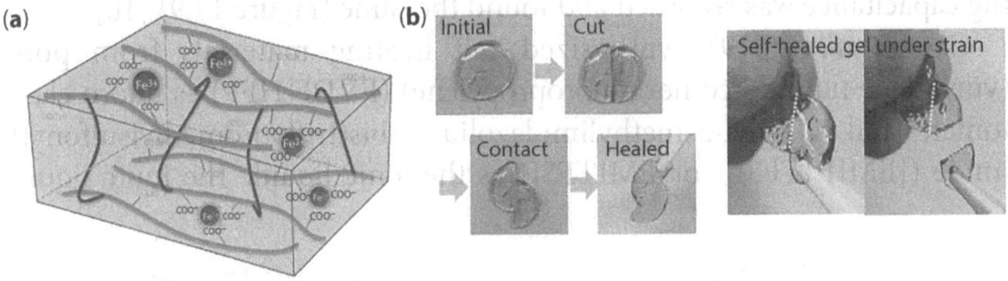

Figure 12.6 (a) Ion conductive self-healing hydrogel. (b) Cut-and-contact self-healing experiment [Reprinted with permission from ACS] [7].

take part in Diels–Alder reaction. To accomplish this follow-up reaction, methyl ethyl ketone (MEK) is used for dissolving retro-products (furan and maleimide) so that subsequent cross-linking takes place between maleimide and furan containing silocone polymer. Excellent self healability under milder condition is exhibited, as simple treatment with MEK allowed cross-linking without additional heating which may deteriorate electronic skin. 1–3 min of vapor treatment provides milder condition. They fabricated healable PDMS through face to face lamination of PDMS and AgNW on krapton. Prior to lamination PDMS was treated with MEK vapor for

10 min so that PDMS may become soluble. AgNW is easily removable from PDMS surface by rubbing it slightly. To overcome this problem, the film is again treated with MEK. By treating with MEK AgNW increasingly fused into PDMS (Figures 12.7a-1 to a-6) [8].

Son *et al.* [9] fabricated integrated electronic skin. Strong and weak dynamic unit of (4,4′-methylenebis(phenylurea) [MPU], isophorone bisurea unit (IU) bonding units integrated into a poly(dimethylsiloxane) (PDMS) backbone to form PDMS–$MPU_{0.4}$–$IU_{0.6}$. They studied self healing process with different combination of polymer matrix. Covalent cross-linked PDMS, thermoplastic polystyrene-block-poly (ethylene-ran-butylene)-block-polystyrene(SEBS) and PDMS–IU were compared with PDMS–$MPU_{0.4}$–$IU_{0.6}$. They revealed that PDMS–$MPU_{0.4}$–$IU_{0.6}$ exhibits best performance relevant to normalized resistance. Polymer matrices other than PDMS–$MPU_{0.4}$–$IU_{0.6}$ do not show self healability [9].

Zhang *et al.* [10] prepared cross-linked organogels and applied it as a dielectric layer for capacitive ionic skin. They synthesized laurylacrylate polymeric organogel by using cross-linker EDDMA. These organogels were congregated as a sandwich like structure between two conductive hydrogel layers. High sensivity of about 0.293 kPa^{-1} was measured. They demonstrated self-healing nature of this intelligent ionic skin by pressing two organogel layer (Figure 12.8) [10]. Before fracture and after self healing capacitance was recorded and found the same (Figure 12.9) [10].

Cao *et al.* (2019) synthesized self healing materials from poly (vinylidene fluoride-co hexafluoropropylene) (PVDF-HFP) as fluoro elastomer, and 1-ethyl-3-methylimidazolium bis(trifluoromethylsulfonyl) imide ($[EMI]^+[TFSI]^-$ or EMITFSI) as the ionic liquid. The ionic liquid

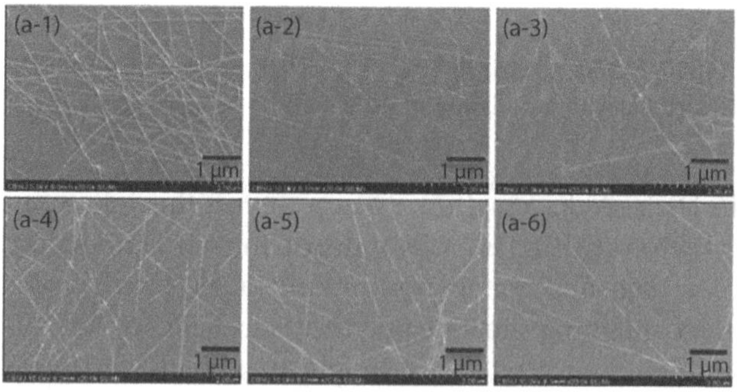

Figure 12.7 Effects of MEK vapor treatment on Ag NWs/h-PDMS. (a) SEM micrographs of the Ag NW/h-PDMS surface. Figures (a-1), (a-2), (a-3), (a-4), (a-5), and (a-6) correspond to treatment times of 0, 1, 3, 5, 10, and 20 min, respectively [Reprinted with permission from ACS] [8].

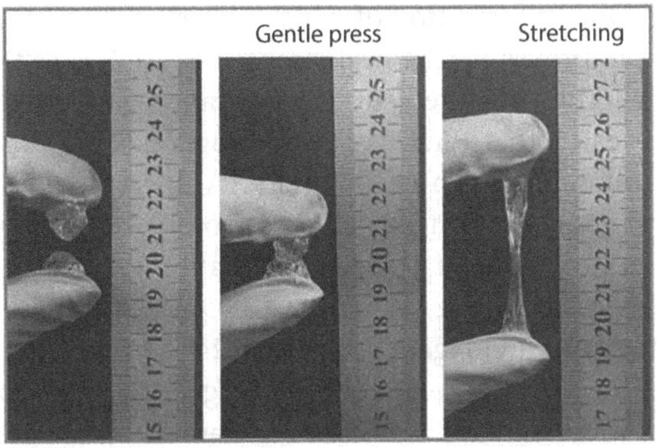

Figure 12.8 Two pieces of organogel were placed together and stretched without fracture after self-healing [Reprinted with permission from ACS] [10].

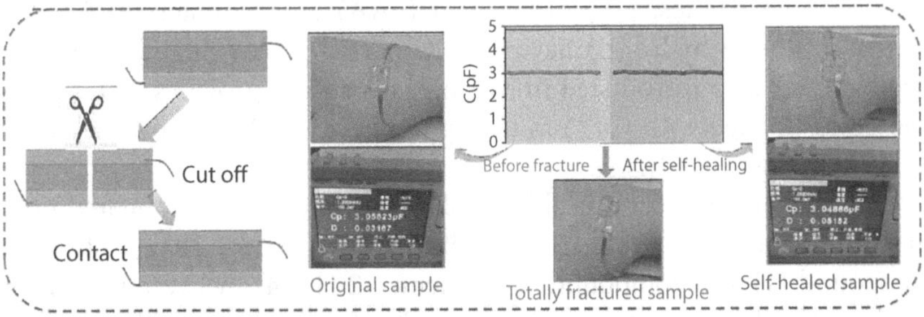

Figure 12.9 Schematic illustration of self-healing of the capacitive device and capacitance measurements before fracture/after healing [Reprinted with permission from ACS] [10].

exhibits negligible vapor pressure. They successfully demonstrated that the film is self-healable under dry and wet conditions as well as it exhibited extraordinary ionic conductivity of 10^{-3}S cm^{-1}.The film not only exhibited excellent conductivity but it also performed well under strain of 2,000%. Carbon–fluorine bond serve as a dipole so it can interact with the ionic liquid. Ion-dipole interaction between C–F bonds of elastomer and ionic liquid is very high so the water molecule cannot penetrate the ion–dipole interaction between them, this makes it different from those which interact from water. In a simple way the materials act as hydrophobic since it interacts very weakly with water molecules. Consequently this leads to self healing underwater. Conductivity of the material was found tunable. By increasing concentration of ionic liquid, conductivity of the material increased by sixfold [11].

Dai *et al.* [12] prepared hydrogel by using pectin, poly(acrylic acid) and metal ions through free radical polymerization reaction and ionic cross-linking. Due to formation of dynamic carboxyl-metal coordination bonds, pectin improves mechanical strength as well as self-healing ability of hydrogel. They analyzed surface by scanning electron microscope (SEM) and revealed that PAA-Pe/Fe^{+3} hydrogel has 3D porous network due to cross-linking between pectin and PAA. Increase of pectin content increases number of pores. At wt. percentage of 6.2% average pore size was found to be 20 μm. Content of pectin also affects the mechanical properties, which is easily visible. When pectin content increased from 1.55 to 6.2 wt%, maximum elongation up to 1,124% with respect to 200% is observed due to more complex cross-linking between pectin and PAA. Similarly with increase of Fe^{+3} content mechanical properties of hydrogel improves. Without Fe^{+3} content hydrogel exhibits fracture stress of 6.9 kPa. However, after addition of Fe^{+3} it exhibits fracture stress of 67.8 kPa [12]. Lei *et al.* [13] prepared hydrogel based on amorphous calcium carbonate (ACC) nanoparticles, polyacrylic acid (PAA) and alginate (ACC/PAA/alginate mineral hydrogel) having ACC particle cross-linked with PAA. Alginate was also introduced in hydrogel because it provides better physical cross-linkage with Ca^{2+} ion owing to chelation. Henceforth alginates enhance elasticity [13].

Jun *et al.* [14] prepared healable e-skin with the help of UV-cured polymer (urethane acrylate-based) hybridized with AgNWs. They synthesized compound (1) and compound (2) [Scheme 12.1] [14] by several steps and fabricated healable pressure sensitive capacitor (Figure 12.10) with the help of these compounds 1 and 2 and AgNWs. Dynamic urea bonds are responsible for healing ability of this e-skin material. This e-skin sensor is healable when AgNWs are heated through intense pulsed light irradiations. They demonstrated that capacitance increases when applied pressure is increased [14].

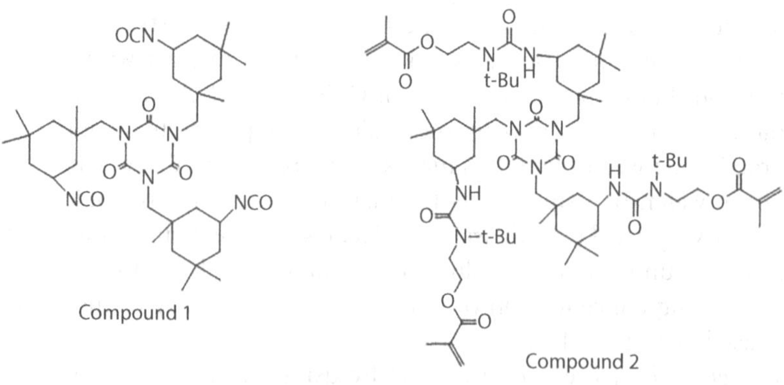

Scheme 12.1 Structure of compound 1 and compound 2 [14].

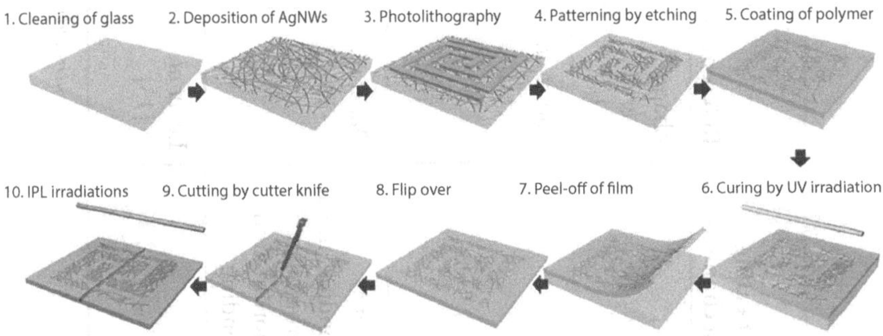

Figure 12.10 AgNW/polymer-based pressure sensor fabrication (steps 1–7) and the healing test (steps 9 and 10) [Reprinted with permission from RSC] [14].

Double network hydrogel is prepared by using polyvinyl alcohol acetoacetate (PVAA), polyacrylamide (PAM) hydrogel and Fe^{3+} ions. PAM hydrogel does not exhibit self-healing ability because it is chemically cross-linked. To improve self-healing ability, PVAA and Fe^{3+} are incorporated so that second cross-linking contributes towards self healing ability. Self healing ability is achieved due to supramolecular complexation between Fe^{3+} and PVAA, which is dynamic reversible interaction in nature. Fast chelation is possible between Fe^{3+} ion and carbonyl groups of acetoacetyl on PVAA, hence Fe^{3+}/PVAA-PAM hydrogel exhibits excellent self healing ability. Self healing is simply achieved by contacting two pieces of the gel [15]. Ternary polymer composite by PANI, PAA, and phytic acid (PA) is prepared by easy solution processable approach. Excellent self healing efficiency of 99% in 24 h as well as excellent stretchability up to 500% has been observed for the composite. PANI/PAA/PA composite exhibits conductivity of 0.12 S cm^{-1}. Aromatic backbone of PANI makes the composite stiff. To reduce the stiffness PAA is incorporated whereas PA serves as dopant to improve electrical conductivity up to 600 times [16].

Bis (3-aminopropyl)-terminated poly-(dimethylsiloxane) and 2,4-tolylene diisocyanate is reacted in presence of Al^{3+} ion to prepare self healing elastomer. Quadruple hydrogen bonding facilitates self healing ability as well as higher stretchability. Self healing ability and mechanical strength can be manipulated by stronger coordination bonds with Al and weaker quadruple hydrogen bonding. Novel elastomer displays high tensile stress (2.6 MPa), exceptional toughness (about 14.7 MJm^{-3}), high stretchability (about 1,700%), and high self-healing efficiency (90%). Smart self-healable and highly stretchable (1,250%) skin can be prepared when Au thin film is embedded in this elastomer [17]. PEDOT:PSS thin film (about 1 μm) exhibits self-healing ability in the presence of distilled water. Electrical properties were restored 100%. Electrical conductivity of PEDOT:PSS as

Table 12.1 Summary of self-healing materials for artificial skin.

SI. No.	Material	Conductivity	Stretchability	Self-healability efficiency	Pressure sensivity	Self-healing mechanism	References
1	Lauryl acrylate-based polymeric organogels	–	>500%	–	0.293 kPa^{-1}	–	Zhang *et al.* [10]
2	Fluorocarbon elastomer and a fluorine-rich ionic liquid (EMITFSI)	10^{-3} Scm^{-1}	–	–	–	–	Cao *et al.* [11]
3	PAA-Pe/Fe^{3+} hydrogel		1,214%	94%	0.65 kPa^{-1}	hydrogen bonding, metal-carboxylate coordination bonding	Dai *et al.* [12]
4	ACC/PAA/alginate mineral hydrogel	–	–	–	1 kPa		Lei *et al.* [13]

(*Continued*)

Table 12.1 Summary of self-healing materials for artificial skin. (*Continued*)

SI. No.	Material	Conductivity	Stretchability	Self-healability efficiency	Pressure sensivity	Self-healing mechanism	References
5	PB-Ag/TA@CNC nanocomposite hydrogel	4.61 Sm^{-1}	>4,000%	98.6%		dynamic borate ester bonding, hydrogen bonding	Lin *et al.* [5]
6	TC-Gel	0.13 S cm^{-1}		90%		hydrogen bonding, electrostatic interactions	Shao *et al.* [6]
7	Fe^{3+}/PVAA-PAM hydrogel	–	>700%	80%	–	PVAA-Fe^{3+} Chelation, Supramolecular complexation	Tie *et al.* [15]
8	PANI-PAA-PA composite	0.12 S cm^{-1}.	≈500%	99%	37.6 kPa^{-1}	hydrogen bonding, electrostatic interactions	Wang *et al.* [16]

(*Continued*)

Table 12.1 Summary of self-healing materials for artificial skin. (*Continued*)

SI. No.	Material	Conductivity	Stretchability	Self-healability efficiency	Pressure sensivity	Self-healing mechanism	References
9	PDMS-TDI-Al elastomer		1,700%	90%	–	quadruple hydrogen bonding, coordination bonding	Wu *et al.* [17]
10	Ag NWs-doped PAAM–PVP organogel		>22,000%			Reversible Hydrogen Bonding, chemical and physical cross-linking of PAAM–PVP double-networks	Zhang *et al.* [19]

a self-healable material is observed to be 500 S cm^{-1}, which is highest so far observed among the all self-healable materials [18]. AgNWs-doped PAAM–polyvinylpyrrolidone (PVP) double-network ethylene glycol organogel is prepared. Outstanding stretchability up to 22,000% is exhibited by the gel. Dynamic hydrogen bonding between $-NH_2$ group and carbonyl group in PVP is responsible for self healing ability. AgNWs improve tensile strength of the gel up to 50 kPa as compared to the pure organogel [19]. MXene (Ti_3C_2Tx)/polyvinyl alcohol (PVA) hydrogel electrode is used as capacitive sensor for e-skin. MXene improves electrical conductivity as well as self healability of the gel. Since MXene has various hydrophilic groups, it improves self-healing ability of the PVA-gel [20]. Bionic tactile PC/rGO/PVA hydrogel is prepared by incorporation of proanthocyanins/reduced graphene oxide (PC/rGO) composite and glycerol-plasticized PVA-borax hydrogel system. Outstanding stretchability (>5,000%), and excellent self-healing (3 s, 95.73%) ability is demonstrated by the hydrogel [21]. Summary of the properties of various materials have been discussed in Table 12.1.

12.3 Applications of Electronic Skin

This section deals with the application part of the artificial skin. Various sensing devices can be prepared by using these artificial skins. Capacitive pressure sensor was constructed by assembling dielectric layer of polyethylene between two mineral hydrogel layer. Free Ca^{2+} ions and ACC act as the ion carrier. For parallel plate configuration relationship between capacitance and deformation can be expressed as $C = \varepsilon S/4\pi kd$ (where C stands for capacitance, ε stands for dielectric constant of the dielectric layer, k stands for electrostatic constant, S stands for effective area of the conducting layer, and d stands for thickness of the dielectric layer). Capacitance changes, when pressure or strain is applied. Ionic skin exhibits high pressure sensitivity of 1 kPa. A mild touch, motion and water droplets were successfully recorded by the ionic skin. Pressure sensitivity can be expressed as the slope of graph between capacitance change versus pressure. Personal health care wearable devices for the measurement of blood pressure can be formed by using the ACC/PAA/alginate hydrogel-based ionic skin. When effective area increases, capacitance also increases. This phenomena was easily demonstrated by contacting two separate parts of hydrogel (first part was colored red by rhodamine B and second part was colored blue by methylene blue). When these two parts come into contact, effective area of hydrogel increases due to autonomously self healing process. Henceforth, capacitance also increases. The healed hydrogel is also able to monitor finger motion successfully. Finger motion was recorded when finger was bent cyclically [13].

Polyethylene terephthalate as a delectric layer was sandwiched between two conductive layers of polyvinyl alcohol/cellulose nanofibril (PVA/CNF) hydrogel to prepare pressure sensor (Figure 12.11) [22]. Conductive layers of the hydrogel were connected with electrodes. To stop water evaporation conductive hydrogel was covered with polyethylene. Similarly to prepare strain sensor VHB was used as stretchable dielectric layer and congregated between two conductive PVA/CNF hydrogel layers which was connected to metallic electrode. Dehydration of hydrogel can be stopped by covering it with VHB tapes. Sensor can stretched up to 500% (Figure 12.12d) [22]. Relative capacitance changes when strain is applied (Figures 12.12e, f) [22]. Relative capacitance change was demonstrated by applying pressure upon the sensor (Figures 12.12a, b). Water dropping on sensor was successfully monitored by the sensor (Figure 12.12c) [22]. Pressure sensor was prepared by assembling PAA-Pe/Fe^{+3} hydrogel on PDMS dielectric layer and attaching it with copper (Cu) wire of diameter 0.130 mm, followed by covering entire arrangement with another self healing hydrogel. Furthermore, authors prepared strain sensor by placing capacitance strain sensor between two hydrogel layers [12].

PDMS–$MPU_{0.4}$–$IU_{0.6}$ shows excellent stretchability of 1,600% strain and excellent fracture toughness of about 12,000 Jm^{-2} which makes it an excellent stretchable electrode. Conductive elastic network of CNT was incorporated with PDMS–$MPU_{0.4}$–$IU_{0.6}$ to improve conductivity. While embedding process resistance of CNT increases up to 10 times. To reduce resistance, gold (Au) salt is doped. Similarly, AgNWs network was embedded with PDMS–$MPU_{0.4}$–$IU_{0.6}$ to form stretchable electrode. Authors demonstrated

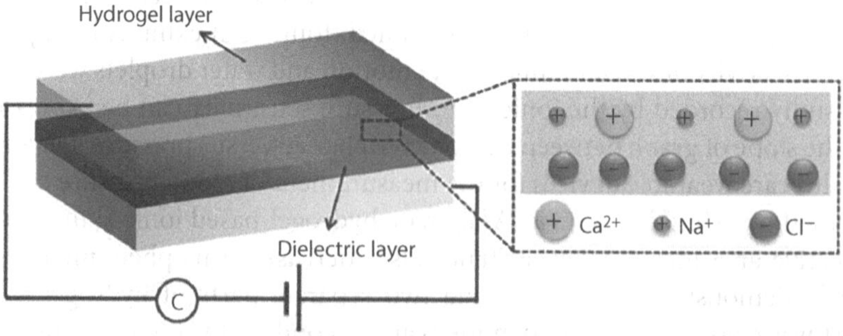

Figure 12.11 Illustration of pressure sensor made by polyvinyl alcohol/cellulose nanofibril (PVA/CNF) hydrogel [Reprinted with permission from Elsevier] [22].

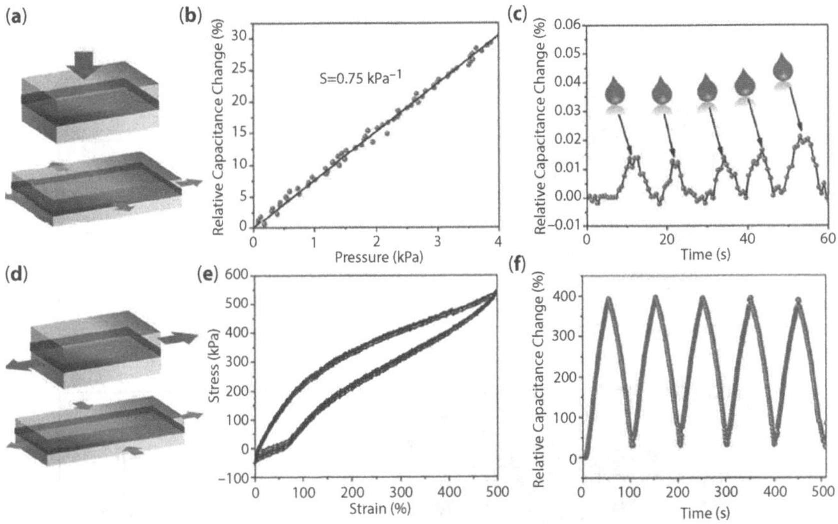

Figure 12.12 (a) Illustration of Pressure sensor (b) capacitance change of sensor upon applying pressure (c) detection of water droplets falling on sensor (d) Strain sensor (e) cyclic tensile test on sensor. (f) Relative capacitance change upon applying strain on sensor periodically [Reprinted with permission from Elsevier] [22].

that both the electrodes are self-healable autonomously and exhibit self healing. Resistance decreases due to self-healing process when two pieces of CNT and AgNWs electrode are made to come into contact. AgNW-PDMS–$MPU_{0.4}$–$IU_{0.6}$ was used in flexible circuit with inorganic LEDs. Self-healing material was employed in wearable ECG sensor on human arm. Self healing ECG sensor was recovered within a few seconds when it was damaged [9]. Self healing efficiency for MXene/PVA electrodes and capacitive strain sensitivity were found to be almost 97.4 and 97.5%, respectively [20].

AgNWs-PAAM–PVP-EG organogel can be used to detect various finger motions by adhering it to the fingers (Figures 12.13a, b, c) [19]. Directions of the wrist, different figure motions, respirations, and Parkinson's tremble can be successfully detected by the e-skin (Figure 12.14) [19].

Polyacrylic acid/reduced graphene oxide (PAA/rGO) nanocomposite as a conductive hydrogel was assembled with VHB tape as elastic substrate and sealant to prepare strain sensor. Water evaporation was minimized by putting hydrogel between VHB tapes to detect human motion. As-prepared sensor was attached with human modal [23]. Conductive functionalized single-walled carbon nanotube (FSWCNT), PVA and polydopamine were used to prepare self-healable hydrogel. The hydrogel was

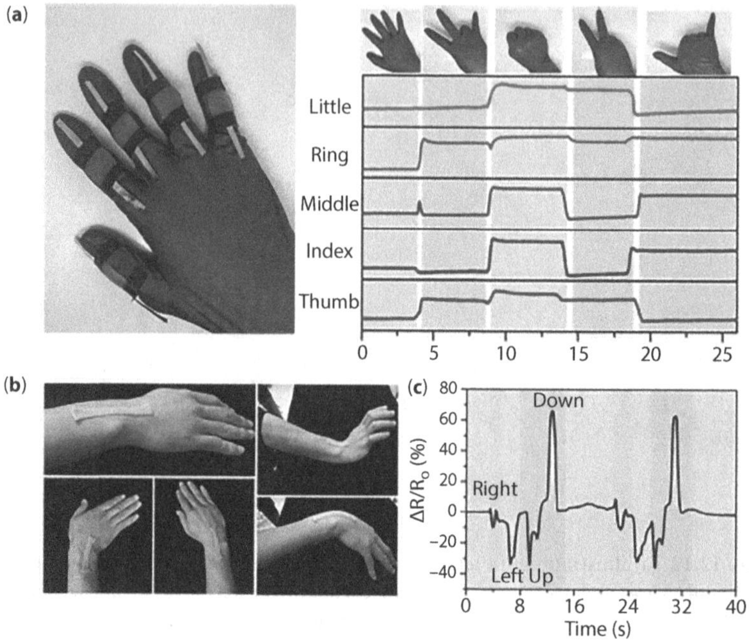

Figure 12.13 (a) The five independent sensors conformally adhered to five fingers. (b) Photographs and (c) relative resistance change of e-skin sensor tracking movement directions of the wrist. Relative resistance change of e-skin used for monitoring [Reprinted with permission from ACS] [19].

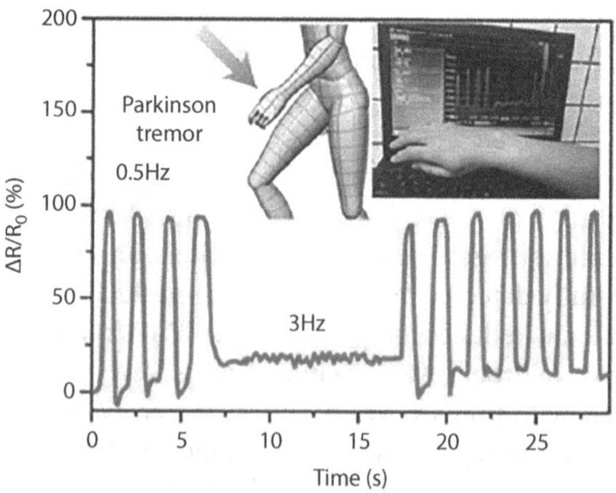

Figure 12.14 application for early diagnosis of Parkinson's disease. [Reprinted with permission from ACS] [19].

used to detect human motions. Reversible supramolecular cross-linking, π–π stacking and reversible hydrogen bonding facilitate self healing ability of this hybrid hydrogel [24].

PC/rGO nanonetwork in hydrogel acts as nerves and detects external signal owing to change in its structure and conductivity. It can detect facial expression (Figure 12.15) [21]. To be used in electrocardiograph (ECG) and electromyography (EMG) (Figure 12.16) [21], PC/rGO/PVA must have adhesive property. To improve adhesiveness metal-phenolic networks of polyphenols can be used. For this purpose PC/rGO/PVA hydrogel was treated with Aluminum(III) Chloride ($AlCl_3$) solution. PC/rGO/PVA hydrogel demonstrated excellent performance as ECG and EMG signal monitoring device [21].

Pressure sensor ionic skin was prepared with the help of PDMS (polydimethylsiloxane), sodium dodecylbenzene sulfonate (SDBS) and bimetallic ion (Fe^{3+}, Al^{3+}). The ionic skin may autonomously self heal easily within 2 s. Self healing efficiency was found to be 95%. SDBS endows

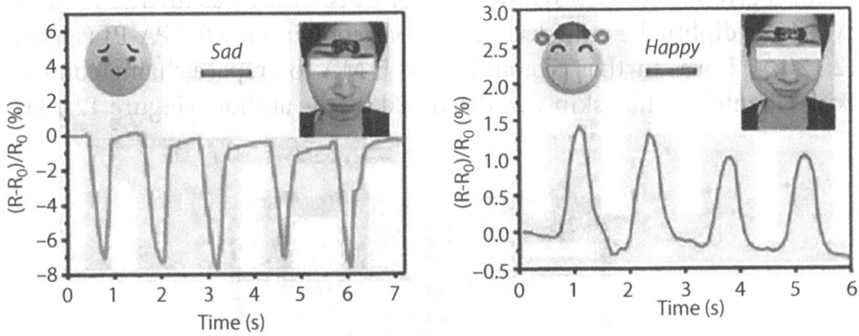

Figure 12.15 Resistance changes of the electronic skin in response to facial expressions [Reprinted with permission from Elsever] [21].

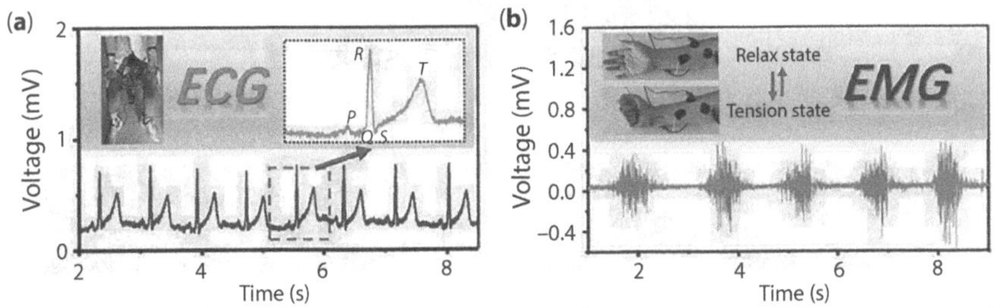

Figure 12.16 PC/rGO/PVA hydrogel electrode used to detect signals for the (a) ECG and (b) EMG [Reprinted with permission from Elsevier] [21].

porous structure to PDMS/SDBS hydrogel which further facilitates the transportation of ions within the hydrogel. Henceforth, pulse rhythm and sound vibration can be easily monitored [25].

Fluorescent agents containing layer was wrapped between CNC@CNTs layer and S-ENR layer. Micro crack structure is generated on CNC@CNTs layer when strain is applied. These microcrack structures allow UV light to pass trough it to the fluorescent layer and fluorescence is observed (Figure 12.17) [4]. When tensile strain is increased up to 50%, crack size also increases to be 50 μm hence strong fluorescence is observed. When tensile strain is relaxed the microcrack closes and consequently fluorescence stops [4].

Self-healable and stretchable electroluminescent skin was prepared by Cu-doped ZnS microparticles embedded in the PDMS–$MPU_{0.4}$–$IU_{0.6}$ polymer matrix which serves as emission layer, which is incorporated between two CNT or AgNW electrodes. CNT based Light emitting capacitor (LEC) was found stretchable up to 250%. Whereas, AgNW based LEC was found to be stretchable up to 50% without degradation in light intensity [9]. *N,N*-bis(4-nitrophenyl)benzene-1,4-diamine was treated with furfuryl glycidyl ether to synthesized DFTPN which was further reduced by 4,4-oxydiphthalic anhydride to prepare polyimide DFTPA-PI (Figure 12.12(b)). It was further crosslinked with MA to prepare chameleon like skin. Chameleon like skin was discussed by the authors (Figure 12.18a).

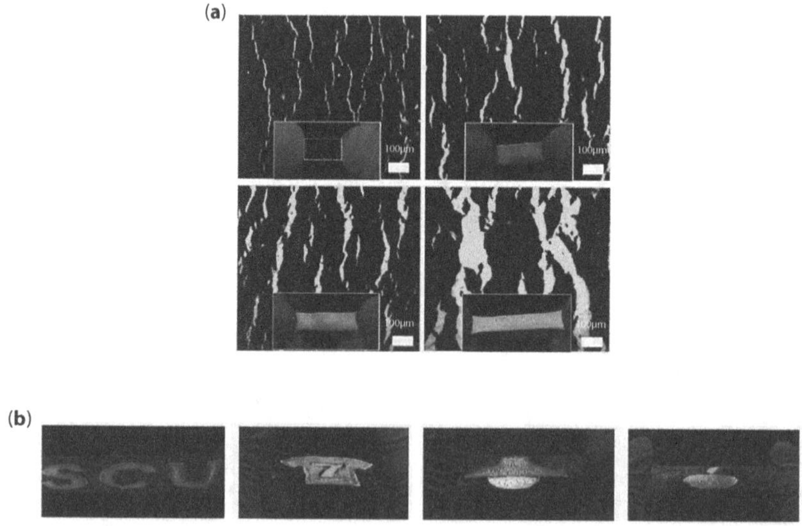

Figure 12.17 (a) LSCM images for of microcracks with a tensile strain of 0, 10, 20, 50%. luminescence images under ultraviolet light with an excitation wavelength of 365 nm. (b) The encrypted patterns [Reprinted with permission from RSC] [4].

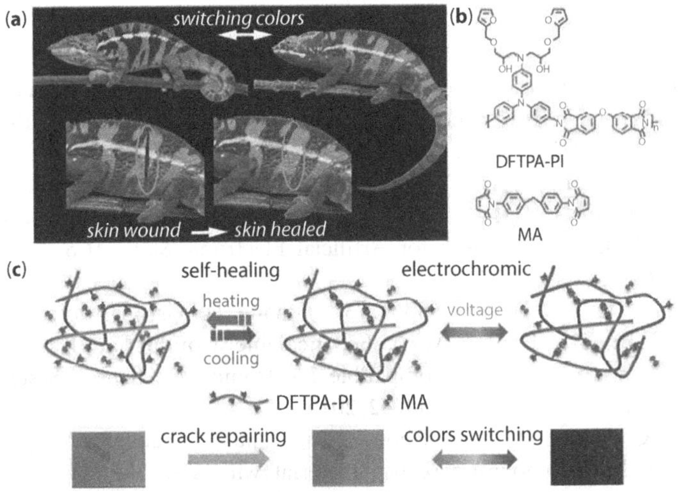

Figure 12.18 (a) The chameleon images (from Guangzhou Yingmeiqing Information Technology Co., Ltd.). (b) Structure of monomer DFTPA-PI and MA. (c) Self-healing and electrochromic function of copolymer DFTPA-PI-MA [Reprinted with permission from ACS] [26].

Color of TPA changes due to intervalence charge transfer process (Figure 12.18c). At 0.49 V oxidation leads to the formation of $DFTPA^{+}$. Further, it forms $DFTPA^{2+}$ at 1.21 V. Second step of oxidation completes at 1.5 V and color changes from yellow to olive green during this process reversibly [26].

12.4 Conclusion

Various composites, like hydrogel and organogel composites, are able to mimic human skin. Lauryl acrylate-based polymeric organogels, fluorocarbon elastomer a, PAA-Pe/Fe^{3+} hydrogel, ACC/PAA/alginate mineral hydrogel, PB Ag/TA@CNC nanocomposite hydrogel, Fe^{3+}/PVAA-PAM hydrogel, PANI-PAA-PA composite etc. have great potential for self healing ability. These materials can be used to prepare artificial skin to monitor facial expressions, body motions. These artificial skin are capable of self-healing under various conditions. They can be used to diagnosis Parkinsons disease and blood pressure, being attached to the human body. In the future with the help of these materials, minute changes on signal will be measured to diagnosis skin disease and skin cancer in early stages.

References

1. Sohn, Y. and Chu, K., Flexible hybrid conductor comprising eutectic Ga–In liquid metal and Ag nanowires for the application of electronic skin. *Mater. Lett.*, 265, 127223, 2020.
2. Sim, H.J., Kim, H., Jang, Y., Spinks, G.M., Gambhir, S., Officer, D.L., Wallace, G.G., Kim, S.J., Self-Healing Electrode with High Electrical Conductivity and Mechanical Strength for Artificial Electronic Skin. *ACS Appl. Mater. Interfaces*, 11, 46026, 2019.
3. Han, L., Cui, S., Yu, H.Y., Song, M.L., Zhang, H., Grishkewich, N., Huang, C., Kim, D., Tam, K.M.C., Self-healable Conductive Nanocellulose Nanocomposites for Biocompatible Electronic Skin Sensor System. *ACS Appl. Mater. Interfaces*, 11, 44642, 2019.
4. Guo, Q., Huang, B., Lu, C., Zhou, T., Su, G., Jia, L., Zhang, X., A cephalopod-inspired mechanoluminescence material with skin-like self-healing and sensing properties. *Mater. Horiz.*, 6, 996–1004, 2019.
5. Lin, F., Wang, Z., Shen, Y., Tang, L., Zhang, P., Wang, Y., Chen, Y., Huang, B., Lu, B., Natural skin-inspired versatile cellulose biomimetic hydrogels. *J. Mater. Chem. A*, 7, 26442–26455, 2019.
6. Shao, C., Meng, L., Cui, C., Yang, J., An integrated self-healable and robust conductive hydrogel for dynamically self-adhesive and highly conformable electronic skin. *J. Mater. Chem. C*, 7, 15208–15218, 2019.
7. Park, S., Shin, B.G., Jang, S., Chung, K., Three-dimensional Self-healable Touch Sensing Artificial Skin Device. *ACS Appl. Mater. Interfaces*, 12, 3953, 2020.
8. Kim, S.O., Han, C.J., Kim, Y., Kim, K.S., Kim, D.U., Lee, C.R., Kim, J.W., Fabrication of a bending-insensitive in-plane strain sensor from a reversible cross-linker functionalized silicone polymer. *ACS Appl. Mater. Interfaces*, 12, 6516, 2020.
9. Son, D., Kang, J., Vardoulis, O., Kim, Y., Matsuhisa, N., Oh, J.Y., To, J.W.F., Mun, J., Katsumata, T., Liu, Y., McGuire, A.F., Krason, M., Lopez, F.M., Ham, J., Kraft, U., Lee, Y., Yun, Y., Tok, J.B.H., Bao, Z., An integrated self-healable electronic skin system fabricated via dynamic reconstruction of a nanostructured conducting network. *Nat. Nanotechnol.*, 13, 1057, 2018.
10. Zhang, Z., Wang, L., Yu, H., Zhang, F., Tang, L., Feng, Y., Feng, W., Highly Transparent, Self-Healable, and Adhesive Organogels for Bio-Inspired Intelligent Ionic Skins. *ACS Appl. Mater. Interfaces*, 12, 15657, 2020.
11. Cao, Y., Tan, Y.J., Li, S., Lee, W.W., Guo, H., Cai, Y., Wang, C., Tee, B.C.K., Self-healing electronic skins for aquatic environments. *Nat. Electron.*, 2, 75, 2019.
12. Dai, S., Wang, S., Yan, H., Xu, J., Hu, H., Ding, J., Yuan, N., Stretchable and self-healable hydrogel-based capacitance pressure and strain sensor for electronic skin systems. *Mater. Res. Express*, 6, 0850b9, 2019.

13. Lei, Z., Wang, Q., Sun, S., Zhu, W., Wu, P., A Bioinspired Mineral Hydrogel as a Self-Healable, Mechanically Adaptable Ionic Skin for Highly Sensitive Pressure Sensing. *Adv. Mater.*, 29, 1700321, 2017.
14. Jun, S., Kim, S.O., Lee, H.J., Han, C.J., Lee, C.J., Yu, Y.T., Lee, C.R., Ju, B.K., Kim, Y., Kim, J.W., Transparent, pressure-sensitive, and healable e-skin from a UV-cured polymer comprising dynamic urea bonds. *J. Mater. Chem. A*, 7, 3101, 2019.
15. Tie, J., Rong, L., Liu, H., Wang, B., Mao, Z., Zhang, L., Zhong, Y., Feng, X., Sui, X., Xu, H., An autonomously healable, highly stretchable and cyclically compressible, wearable hydrogel as multimodal Sensor. *Polym. Chem.*, 11, 1327, 2020.
16. Wang, T., Zhang, Y., Liu, Q., Cheng, W., Wang, X., Pan, L., Xu, B., Xu, H., A self-healable, highly stretchable, and solution processable conductive polymer composite for ultrasensitive strain and pressure sensing. *Adv. Funct. Mater.*, 28, 1705551, 2017.
17. Wu, X., Wang, J., Huang, J., Yang, S., Robust, Stretchable, and Self-Healable Supramolecular Elastomers Synergistically Cross-Linked by Hydrogen Bonds and Coordination Bonds. *ACS Appl. Mater. Interfaces*, 11, 7, 7387, 2019.
18. Zhang, S. and Cicoira, F., Water-Enabled Healing of Conducting Polymer Films. *Adv. Mater.*, 29, 1703098, 2017.
19. Zhang, H., Niu, W., Zhang, S., Extremely Stretchable and Self-Healable Electrical Skin with Mechanical Adaptability, an Ultrawide Linear Response Range, and Excellent Temperature Tolerance. *ACS Appl. Mater. Interfaces*, 11, 24639, 2019.
20. Zhang, J., Wan, L., Gao, Y., Fang, X., Lu, T., Pan, L., Xuan, F., Highly Stretchable and Self-Healable MXene/Polyvinyl Alcohol Hydrogel Electrode for Wearable Capacitive Electronic Skin. *Adv. Electron. Mater.*, 5, 1900285, 2019.
21. Pan, X., Wang, Q., He, P., Liu, K., Ni, Y., Chen, L., Xinhua, O., Huang, L., Wang, H., Xu, S., A bionic tactile plastic hydrogel-based electronic skin constructed by a nerve-like nanonetwork combining stretchable, compliant, and self-healing properties. *Chem. Eng. J.*, 379, 122271, 2020.
22. Jing, X., Li, H., Mi, H.Y., Liu, Y.J., Feng, P.Y., Tan, Y.M., Turng, L.S., Highly transparent, stretchable, and rapid self-healing polyvinyl alcohol/cellulose nanofibril hydrogel sensors for sensitive pressure sensing and human motion detection. *Sens. Actuators B*, 295, 159, 2019.
23. Jing, X., Mi, H.Y., Peng, X.F., Turng, L.S., Biocompatible, self-healing, highly stretchable polyacrylic acid/reduced graphene oxide nanocomposite hydrogel sensors via mussel-inspired chemistry. *Carbon*, 136, 63, 2018.
24. Liao, M., Wan, P., Wen, J., Gong, M., Wu, X., Wang, Y., Shi, R., Zhang, L., Wearable, Healable, and Adhesive Epidermal Sensors Assembled from Mussel-Inspired Conductive Hybrid Hydrogel Framework. *Adv. Funct. Mater.*, 27, 1703852, 2017.

25. Wang, S., Li, Q., Feng, S., Lv, Y., Zhang, T., A water-retaining, self-healing hydrogel as ionic skin with a highly pressure sensitive properties. *J. Taiwan Inst. Chem. Eng.*, 104, 318, 2019.
26. Zheng, R., Wang, Y., Jia, C., Wan, Z., Luo, J., Malik, H.A., Weng, X., Xie, J., Deng, L., An Intelligent Biomimetic Skin of Chameleon with Excellent Performance of Self-healing and Electrochromism. *ACS Appl. Mater. Interfaces*, 10, 35533, 2018.

13

Self-Healing Smart Composites

Sithara Gopinath*, Suresh Mathew and P. Radhakrishnan Nair

School of Chemical Sciences (SCS), Advanced Molecular Materials Research Centre (AMMRC), Mahatma Gandhi Universiy, Kotayam, India

Abstract

Since the composite materials possess attractive properties, they are used in different applications such as automobiles, coating, aerospace, etc. The biologically motivated self-healing composite materials are also established to repair the cracks propagated as a result of mechanical maintenance. These materials are designed on the basis of two approaches. They are (a) discharge of smart self-healing agents towards the generated damage and (b) alterable cross linking networks. This chapter deals with a detailed study on self-healing smart composite materials and its different self-healing mechanisms.

Keywords: Self-healing, smart composites, vascular, extrinsic, capsulation, HGF, intrinsic, healing agents, cracks

13.1 Introduction

As the name indicates, self-healing materials are materials that can repair or heal the generated crack by itself. Self-healing property is similar to the biological wound healing process in living organisms (e.g. healing capacity of a wound or cut in human skin) and its mechanism is described by the working of self-healing materials which are present inside the matrix structure. These implanted self-healing materials must have the ability to sense and respond to crack generated and restore the initial condition of the system without affecting the total system performances. Studies on this self-healing mechanism in a concrete structure are already successfully

**Corresponding author*: sureshmathewmgu@gmail.com

Inamuddin, Mohd Imran Ahamed, Rajender Boddula and Tariq Altalhi (eds.) Self-Healing Smart Materials and Allied Applications, (345–360) © 2021 Scrivener Publishing LLC

reported [1–4]. The studies show that the integration of some self-healing agents inside the vessels, which are distributed all over the concrete specimen and effectively repaired the propagated cracks, without altering its properties. Analogous repairing studies are also carried out extensively by White *et al.* on a biological system [5]. Self-healing using composites materials is also a quite famous research field of interest.

Composite materials are materials that contain two or more materials with different overall properties. This composition enhances the unique properties of the system (e.g. polymer composites) [6, 7]. A smart composite material can be defined as a material with more efficient properties and smart functionalities than its primary materials. As the usage of these composites in various critical applications increases, we must have to develop a smart system with in-built capabilities to characterize and identify its chemical-degradation as well as the self-healing and mechanical damages. These smart-systems helps to process the formed cracks effectively, which mimics a biological repair method or healing method. Self-healing is generally a self-repairing process, i.e. refurbishing something impaired to its initial condition. Thus, self-healing defines a fixing of the unique properties of the system after damaging activities of the peripheral environment [8]. Autonomic-healing can be said as a synonym to describe properties of the system [9]. The main objective is to decrease the destruction or re-establish the initial properties and period of the distorted portion of the system. Thus this is a spontaneously and independently commenced reaction to cracks [9, 10]. As already discussed, these types of self-healing materials are encouraged by biological structures [11]. This idea of healing in living structures has stimulated constant efforts to imitate living systems and to integrate self-healing capacities into different polymers and its nanomaterials incorporated, i.e. polymer nanocomposites for various engineering applications [12, 13].

13.2 Self-Healing Mechanisms and its Classifications

Based on the synthesis approaches or nature of self-healing procedure, we can classify these self-healing materials under two divisions, as shown in Figure 13.1:

- Autonomic type self-repairing materials
- Non-autonomic type self-repairing materials.

The autonomic type does not require an additional or external stimulus method to repair the cracks formed or they have the ability to repair it

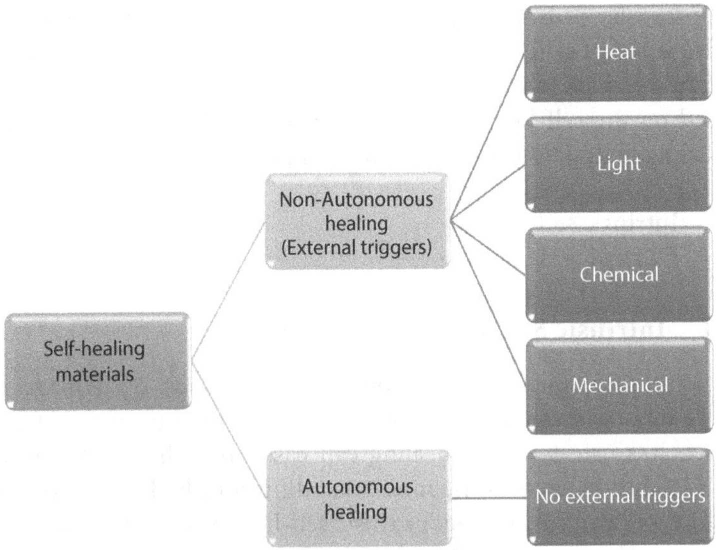

Figure 13.1 Illustration of autonomous and non-autonomous self-healing mechanisms.

by itself. Capsules or fibers which is integrated with smart healing agent can be included under this [14, 15]. The non-autonomic self-healing needs external stimulus methods to heal its damaged cracks. Additional triggering stimulus method includes light, thermal or temperature, laser beam, magnetic, chemical, etc. [11, 14].

Self-healing of polymeric material can be generally considered as a three-stage procedure. Polymeric material's self-healing process is very much similar to living material's biological self-wound healing process [16]. Let's consider an example: biological self-repair of an injured bone in a living creature. The various tissues and blood vessels in the broken bone area, jointly form a group of blood clotted cells around the injured part called hematoma, which is convenient for generating inflammatory cells. This is the first stage of self-healing procedure. This produced hematoma mass is gradually converted into fibro-cartilaginous mass, which helps to bond the occurred fracture. From this fibro-cartilaginous mass spongy type fibrous bone is formed, which is known as bone callus. These soft callus grows and is converted to hard bone callus structures and gets self-arranged as such way of the pre-injured bone matrix. This conversion or modeling step is generally known as remodeling stage [10, 17–19]. These series of stages in the fractured bone repairing process can be established in self-healing of a polymer network. In a polymeric crack self-healing procedure, first we have to stimulate the triggering action immediately as it occurred. Next, we have to transfer or transport the smart self-healing agents to the cracked area. These smart self-healing agents perform chemical repair and remodel the occurred crack in the material [19, 20].

Based on the self-healing properties, we can classify or subdivide the self-healing materials as follows, shown in Figures 13.2 and 13.3:

- Intrinsic self-healing.
- Extrinsic self-healing.

13.2.1 Intrinsic Self-Repairing Materials

Intrinsic self-healing materials accomplish their crack or damage repair through inborn reversible molecular network bonding of the structural polymer. Intrinsic type self-healing can easily be achieved by means of reversible thermal reactions, molecular diffusion, hydrogen bonding as well as by ionomeric coupling. Intrinsic self-healing or self-repairing polymers are capable of re-establishing micro-scale and macro-scale cracks via a temporary native rise in the movement of chains of polymeric materials [21, 22]. These behaviors on the polymer materials mainly depend on the specific molecular chains which enable the motion of inter-molecular

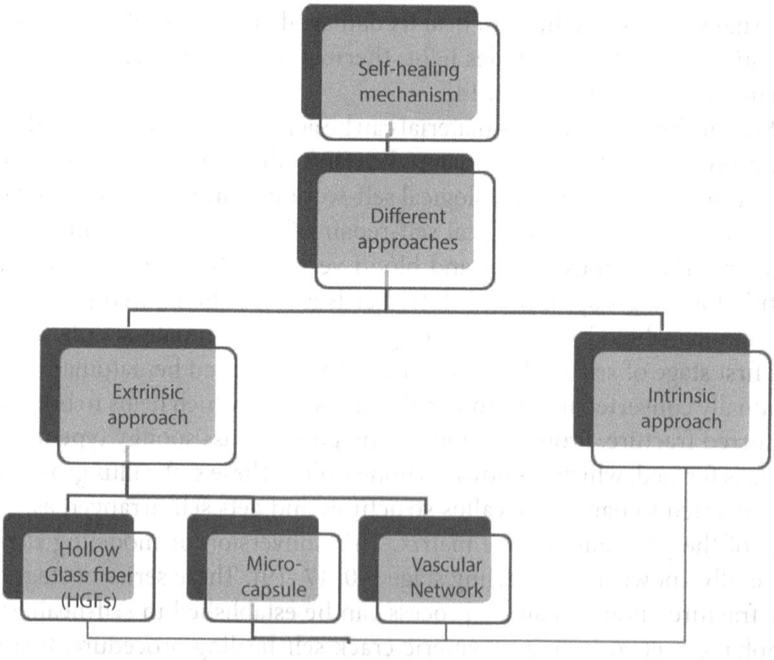

Figure 13.2 Schematic layout of self-healing mechanism.

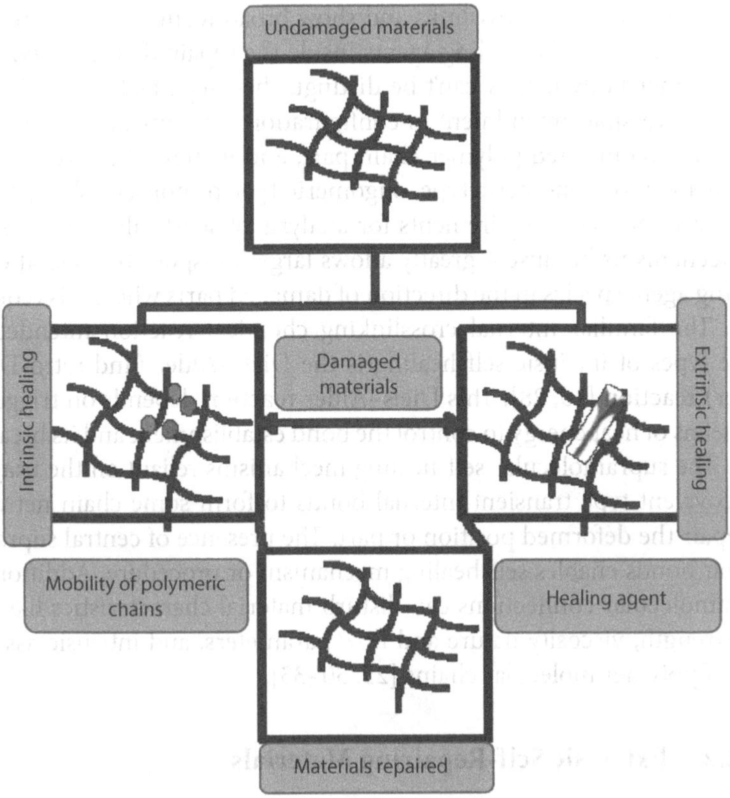

Figure 13.3 Intrinsic and extrinsic self repairing materials.

chain, with the application of some measures of energy in the form of temperature, UV, etc. results in the restoration of internal chemical and physical strength. The main advantage of this intrinsic healing is that its focuses on multiple repairments to achieve initial stage properties [22, 23]. Intrinsic healing is of two classes:

- Dynamic covalent bonds
- Supramolecular interactions.

The monomer units of the polymer networks are associated with reversible type chain network systems and the configuration can be modified either by interchange or rearrangement of chains. Based on the nature of the networks either covalent or non-covalent type, these chains may be either belong to molecular type or supramolecular type [23–25]. If they are of dynamic covalent bond type, during repairing process, they possess covalent type crosslinks and show broad formation and breaking reaction [26, 27]. The healing agents inside the repaired crack region and the polymer body matrix can't be distinguished separately [26, 27]. This type of reversible amendment in configuration consents the arrangement in the interconnected polymer chain part, and its flexibility even though they belong to monomeric type, oligomeric type or non-crosslinked type. These are the main requirements for analysis of self-healing process and its mechanisms because it greatly allows large transportation facilities for healing agent species in the direction of damaged parts where it is required [28]. The familiar internal crosslinking chemical reaction intended for these types of intrinsic self-healing is the Diels–Alder (and retro-Diels–Alder) reaction [23, 28]. This Diels–Alder reaction depends on triggering by means of heat energy to control the bond establishment and its breaking [29]. The supramolecular self-healing mechanisms reliant on the usage of noncovalent type transient internal bonds to form some chain networks to repair the deformed position or part. The presence of central supramolecular bonds enables self-healing mechanism or procedure. Additionally, supramolecular connections can disturb material characteristics like tensile strength, viscosity nature and flow parameters, and intrinsic association of polymer molecular chains [25, 30–33].

13.2.2 Extrinsic Self-Repairing Materials

Extrinsic self-healing mechanism or autonomous self-healing mechanism always uses an integrated smart liquid which is known as a smart self-healing agent to renovation the cracks. The extrinsic type self-healing system consists of three-stage or three variety of self-healing mechanisms [34, 35], shown in Figure 13.4. They are

- Capsule-type
- Vascular type
- Hollow glass fiber type.

Capsule-based self-healing materials have been developed for some of the most commonly used synthetic polymers and elastomers using a variety of capsule-based sequestration schemes. Each scheme sequesters a healing agent in a discrete capsule until damage triggers release. Capsule-based self-healing materials belong to the group of autonomic self-healing type materials. In this the smart self-healing representatives are compressed until the crack stimulates fractures and releases the healing agent. A number of methods such as interfacial method, *in situ* method and meltable

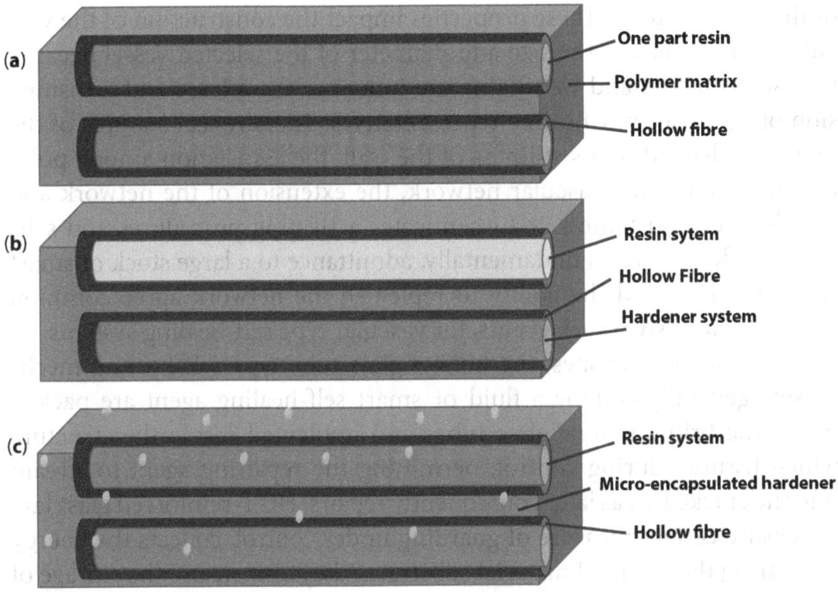

Figure 13.4 Self-healing mechanism (a) Capsule-based, (b) Vascular, (c) Hollow glass fiber.

dispersion method are commonly used [36] for the encapsulation of these responsive materials. Studies on materials integrated microcapsules have been done extensively [37]. Among that polymer structure integrated with micro-encapsulated dicyclopentadiene used along with Grubb's catalyst is quite famous for curing damage. The microcapsules are integrated with some healing agents captured inside the structural polymer matrix. When the damage or crack occurs, these microcapsules help the self-healing agent to flow into location of the propagated crack. The integrated catalyst formulates some polymerization reaction of the healing agent matrix up to about 75% crack recovery mechanism is already reported [38].

Vascular self-healing materials may quite analogous to the capsule-based method. In this, the smart self-healing agent is completely integrated with complex network of hollow canals or vessels and this interconnected network is maintained till the damages or cracks are healed or repaired. Here it is important to accomplish some better interaction among the polymer matrix substrate, the smart self-healing representative agents and the network materials, for better self-healing mechanism [39]. The vascular networks thus formed may be of one, two or three-dimensional network type. The selection of smart self-healing agents mainly depends on some characteristics such as surface wettability, viscosity and chemical reactivity of the material used. These properties impact the construction of the vascular system, mainly the size and diameter of the selected vessel because its viscous nature and wetting nature influence the release and transmission of the smart self-healing representatives. The physical features of the networks depend on its stiffness of the wall, the association among polymer matrix and the vascular network, the extension of the network and the channel distribution. It authenticates activating procedures and self-healing effectiveness. Fundamentally, admittance to a large stock of smart repairing units and the ability to replenish the network agree common repair of successive crack events, for vascular type self-healing systems.

As the name indicates, the hollow glass tube type self-healing mechanisms generally employ a fluid or smart self-healing agent are packed inside the brittle hollow glass tubes and implanted inside the structure which fissured during control, permitting the repairing agent to stream into the cracked areas later consequently repairs. HGT-reinforced glass layers ensure certain measure of guarding under control, collects the energy and caring the essential material substrate. They also ensure the storage of these self-healing smart materials. In HGF, the mechanism of self-healing depends on (a) Position and amount of the crack occurred; (b) the abilities of the representative healing agents or healing resins; (c) environmental circumstances. We can place the smart self-healing fibers in any location within the composite's lamina at a defined distance. HGT can be efficiently used as a self-healing mechanism if the correct nature of crack is made to know inside composite laminate matrix. Dry [1] initially started the research focus on hollow glass tubes based smart self-healing composite mechanism and proposed that it contains the ability to re-join the cracks which are occurred inside the lamina.

13.3 Self-Healing of Thermoplastic Materials

Thermoplastic materials are generally materials behaving as plastic nature, i.e. when heated, it becomes more soft and hardens upon cooling as depicted in Table 13.1. Crack or damage healing research studies in thermoplastic materials or polymers have been broadly began at the start of the 1980s. These thermoplastic polymers include fiber-reinforced composites, amorphous nature type, block copolymers and semi-crystalline nature type. If we connect two same types of polymer materials close together at a temperature above its transition temperature (e.g. Glass transition temperature, T_g), the phase transition occurs at its interface. This phase transition enables disappearing of the crack, by molecular diffusion process at

Table 13.1 Developments in self-healing polymer composites.

Host material	Healing material	Stimulus	Best efficiency acheived	Healing condition	Test method	Reference
Thermosetting and/or thermosetting composites	Hollow glass fiber	Mechanical	93%	24 h at ambient atm.	Flexure Strength	[43–52]
	Microencapsulation approach		80–93%	48 h at 80 °C 24 h at Ambient then 24 h at 80 °C	Fatigue resistance Fracture toughness Tensile strength	[53–54]
	Microvascular network		60–70% 7–30 cycles	12 h at ambient atm.	Fracture toughness	[55–56]
	Thermoplastic additives		30–100%	10 min at 120 °C 1–2 h at 130–160 °C	Flexure strength Tensile strength Impact strength	[57–58]
	Shape memory alloy	Electrical	77%	24 h at ambient atm.	Fracture toughness	[59]
	Carbon fiber		46%	1–20 min at 70–120 °C	Impact strength	[60]
Elastomeric	Silicone rubber	Mechanical	70–100%	48 h at ambient atm.	Tear strength	[61]
Thermoplastic	Molecular diffusion		100%	5 min at 60 °C.	Fracture toughness	[62]
	Photo-induced healing	Photo	26%	10 min at 100 °C	Flexure strength	[63]

the interface and hence enhances the mechanical strength. The research studies on crack healing in PMMA (Poly Methyl Methacrylate) by using methanol is already reported, as an example of the thermoplastics chain motion at lower temperature (from 40–60 °C) [40]. The interesting fact is that after the crack healing of PMMA the tensile strength is completely recovered to that of the initial state of the material. The self-healing using light of PMMA is also already reported [41]. Upon the exposure of light of a particular wavelength on the crack occurred area, the chain motion initiated and healing occurs. But the healing process happens on the light exposure surface, means the internal crack healing is not processed. Hence process of self-healing in these type of thermoplastic polymer materials occurs in following mechanisms [42]:

- Recombination of polymers and healing agents at chain molecular ends
- Reversible type chemical bonding based self-healing
- Active polymer method
- Nanoparticles based self-healing
- Inter-diffusion process of molecules
- Light (Photo) induced healing mechanism.

13.4 Self-Healing of Thermosetting Materials

Thermoset materials are a category of materials, which experience internal cross-linking chemical reactions with the help of thermal methods which results in the formation of a final infusible state, as depicted in Table 13.1. The self-healing thermosetting materials find many research interests nowadays and these studies resulted in real-time applications. In order to achieve more efficiency in these self-healing engineering applications the rigidity and thermal stability of the structural materials is an important factor, which is absent in most of the thermoplastics materials. The cross-linking procedure occurs inside the molecular structure of the materials during its synthesis stage determines the rigidity and thermal stability of the matrix, which means that the chain mobility is absent which is effectively discussed in thermoplastic materials. We have to adopt separate approach for the self-healing mechanism in thermosetting materials since they possess diversity in its chemical and molecular construction. The best collective methods designed for autonomic type based on thermosetting self-healing materials include integration of smart repairing self-healing agents inside a fragile container before accumulation of the containers into the polymeric substrate matrix. These fragile containers rupture upon heaping of the matrix polymer, liberating the less-viscous smart self-healing representatives to the broken locations for following curative and filling these agents in the produced cracks for repair.

The precise behavior of the self-healing methodology depends on

(i) Nature and position of the crack propagated
(ii) kind of self-healing resins used
(iii) Impact of the operative atmosphere.

13.5 Conclusions and Future Study

The research area of self-healing materials is developing day by day and it finds many applications in real engineering applications. As the interest in self-healing materials and its application increases, they find many improvements in their characteristics such as long and better lifetime, reduction in cost of maintenance and replacement and progress in product security. Mainly, polymers and metals are widely used for the production of self-healing materials. The major conclusions we reached from this study is given as follows:

- Self-healing is considered as one of the most advantageous technology for improving the lifecycle of different composite matrices.
- The structure-wise power or the structural strength of the smart composite materials can be improved by using hollow glass tubes.
- Self-healing mechanism based on capsule based system consumes revealed enhanced signs of progress in their robustness.
- For extremely smaller cracks, self-healing mechanisms such as hollow glass fiber (HGF)s and capsule types are widely used and vascular type mechanism is advantageous for larger cracks.
- In vascular type mechanism, the smart healing agents are released through more than one channel at desired area, which helps to happen healing multiple healing occurs from different directions on the crack at a time. While in HGF and capsule-based healing mechanisms there happens only single healing process per crack.

Even though, the recent research studies mainly focus on smart design, fabrications, applications in specific area etc. seldom, is this idea being exploited for industrial application and commercialization. In the future, they may find better and major applications in smart sensors and actuators inside the smart composites to recognize the occurred damages or cracks effectively and to boost the procedure proficiency.

References

1. Dry, C., Alteration of matrix permeability, pore, and crack structure by the time release of internal chemicals, in: *Proceedings: Advance in Cementitious Materials*, American Ceramic Society, Gaithersbury, Maryland, pp. 729–68, 1990.
2. Dry, C., Matrix cracking repair and filling using active and passive modes for smart timed release of chemicals from fibres into cement matrices. *Smart Mater. Struct.*, 3, 2, 118–123, 1994.
3. Dry, C. and McMillan, W., Three-part methylmethacrylate adhesive system as an internal delivery system for smart responsive concrete. *Smart Mater. Struct.*, 5, 3, 297–300, 1996.
4. Dry, C., Three design for the internal release of sealants, adhesives and waterproofing chemical into concrete to release. *Cem. Concr. Res.*, 30, 1969–1977, 2000.
5. White, S.R., Sottos, N.R., Geubelle, P.H., Moore, J.S., Kessler, M.R., Sriram, S.R., Brown, E.N., and Viswanathan, S., Autonomic Healing of Polymer Composites. *Nature*, 409, 794–797, 2001.
6. Blaiszik, B.J., Kramer, S.L.B., Olugebefola, S.C., Moore, J.S., Sottos, N.R., White, S.R., Self-healing polymers and composites. *Annu. Rev. Mater. Res.*, 40, 179–211, 2010.
7. Blaiszik, B.J., Baginska, M., White, S.R., Sottos, N.R., Autonomic recovery of fiber/matrix interfacial bond strength in a model composite. *Adv. Funct. Mater.*, 20, 3547–3554, 2010.
8. Filipovic, N., Petrovic, D., Jovanovic, A., Jovanovic, S., Balos, D., Kojic, M., DPD simulation of self-healing process of nanocoating. *J. Serbian Soc. Comput. Mech.*, 2, 42–50, 2008.
9. Ghosh, S.K., Self-Healing Materials: Fundamentals, Design Strategies, and Applications, in: *Self-Healing Materials*, pp. 1–28, Wiley-VCH Verlag GmbH & Co. KGaA, Weinheim, Germany, 2009.
10. Fischer, H., Self-repairing materials systems—A dream or a reality? *Nat. Sci.*, 2, 873–901, 2010.
11. Hager, M.D., Greil, P., Leyens, C., van der Zwaag, S., Schubert, U.S., Self-healing materials. *Adv. Mater.*, 22, 5424–5430, 2010.

12. Trask, R., Williams, H., Bond, I., Self-healing polymer composites: Mimicking nature to enhance performance. *Bioinspir. Biomim.*, 2, 1, P1, 2007.
13. Wang, Y., Pham, D.T., Ji, C., Self-healing composites: A review. *Cogent. Eng.*, 2, 1075686, 2015.
14. Puretskiy, N., *Design of Self-Repairable Superhydrophobic and Switchable Surfaces Using Colloidal Particles*, Technische Universitat Dresden, Dresden, Germany, 2013.
15. Guimard, N.K., Oehlenschlaeger, K.K., Zhou, J., Hilf, S., Schmidt, F.G., Barner-Kowollik, C., Current trends in the field of self-healing materials. *Macromol. Chem. Phys.*, 213, 131–143, 2012.
16. Wool, R.P., Self-healing materials: A review. *Soft Mater.*, 4, 400–418, 2008.
17. Doblaré, M., García, J.M., Gómez, M.J., Modelling bone tissue fracture and healing: A review. *Eng. Fract. Mech.*, 71, 1809–1840, 2004.
18. Kyllönen, L., D'Este, M., Alini, M., Eglin, D., Local drug delivery for enhancing fracture healing in osteoporotic bone. *Acta Biomater.*, 11, 412–434, 2014.
19. Sangadji, S. and Schlangen, E., Mimicking none healing process to self repair concrete structure novel approach using porous network concrete. *Proc. Eng.*, 54, 315–326, 2013.
20. Thomas, G.P., Self-Healing Materials—The Future of Repairs Azom Publisher, 2012.
21. Hager, M.D., Greil, P., Leyens, C., van der Zwaag, S., Schubert, U.S., Self-healing materials. *Adv. Mater.*, 22, 5424–5430, 2010.
22. Blaiszik, B.J., Sottos, N.R., White, S.R., Nanocapsules for self-healing materials. *Compos. Sci. Technol.*, 68, 978–986, 2008.
23. Garcia, S.J., Effect of polymer architecture on the intrinsic self-healing character of polymers. *Eur. Polym. J.*, 53, 118–125, 2014.
24. Burattini, S., Greenland, B.W., Chappell, D., Colquhoun, H.M., Hayes, W., Healable polymeric materials: A tutorial review. *Chem. Soc. Rev.*, 39, 1973–1985, 2010.
25. Lehn, J.-M., Dynamers: Dynamic molecular and supramolecular polymers. *Prog. Polym. Sci.*, 30, 814–831, 2005.
26. Chao, A., Negulescu, I., Zhang, D., Dynamic covalent polymer networks based on degenerative imine bond exchange; tuning the malleability and self-healing properties by solvent. *Macromolecules*, 49, 6277–6284, 2016.
27. Zhang, L. and Rowan, S.J., Effect of sterics and degree of cross-linking on the mechanical properties of dynamic poly(alkylurea-urethane) networks. *Macromolecules*, 50, 5051–5060, 2017.
28. Billiet, S., Hillewaere, X.K.D., Teixeira, R.F.A., Du Prez, F.E., Chemistry of crosslinking processes for self-healing polymers. *Macromol. Rapid Commun.*, 34, 290–309, 2013.

29. Deng, G., Li, F., Yu, H., Liu, F., Liu, C., Sun, W., Jiang, H., Chen, Y., Dynamic hydrogels with an environmental adaptive self-healing ability and dual responsive sol–gel transitions. *ACS Macro. Lett.*, 1, 275–279, 2012.
30. Zhang, L., Chen, L., Rowan, S.J., Trapping dynamic disulfide bond in the hare segments of thermoplastic polyurethane elastomers. *Macromol. Chem. Phys.*, 218, 1, 160030, 2017.
31. Zhang, L., Brostowitz, N.R., Cavicchi, K.A., Weiss, R.A., Perspective: Ionomer research and application. *Macromolecules*, 8, 81–99, 2014.
32. Herbst, F. and Binder, W.H., Self-Healing Polymers via Supramolecular, Hydrogen-Bonded Networks, in: *Self-Healing Polymers: From Principles to Applications*, Wiley-VCH Verlag GmbH & Co. KGaA, Weinheim, Germany, 2013.
33. Heinzamann, C., Weder, C., de Espinosa, L.M., Supramolecular polymers adhesive: Advanced materials inspired by nature. *Chem. Soc. Rev.*, 45, 342–358, 2016.
34. Grewell, D. and Benatar, A., Welding of Plastics: Fundamentals and New Developments. *Int. Polym. Process.*, 22, 43–60, 2007.
35. Caminero, M.A., Pavlopoulou, S., Lopez-Pedrosa, M., Nicolaisson, B.G., Pinna, C., Soutis, C., Analysis of adhesively bonded repairs in composites: Damage detection and prognosis. *Compos. Struct.*, 95, 500–517, 2013.
36. Blaiszik, B., Kramer, S., Olugebefola, S., Moore, J.S., Sottos, N.R., White, S.R., Self-healing polymers and composites. *Annu. Rev. Mater. Res.*, 40, 179–211, 2010.
37. Varley, R.J. and van der Zwaag, S., Towards an understanding of thermally activated self-healing of an ionomer systems during ballistic penetration. *Acta Mater.*, 56, 5737–5750, 2008.
38. White, S.R., Sottos, N.R., Geubelle, P.H., Moore, J.S., Kessler, M.R., Sriram, S.R., Brown, E.N., Viswanathan, S., Autonomic healing of polymer composites. *Nature*, 409, 794–797, 2001.
39. Chotard, T.J., Pasquiet, J., Benzeggagh, M.L., Residual performance of scarf patch-repaired pultruded shapes initially impact damaged. *Compos. Struct.*, 53, 317–331, 2001.
40. Lin, C.B., Lee, S., Liu, K.S., Methanol-induced crack healing in poly(methyl methacrylate). *Polym. Eng. Sci.*, 30, 21, 1399–1406, 1990.
41. Chung, C.-M., Roh, Y.-S., Cho, S.-Y., Kim, J.-G., Crack healing in polymeric materials via photochemical [2+2] cycloaddition. *Chem. Mater.*, 16, 21, 3982–3984, 2004.
42. Wu, D.Y., Meure, S., Solomon, D., Self-healing polymeric materials: A review of recent developments. *Prog. Polym. Sci.*, 33, 5, 479–522, 2008.
43. Pang, J.W.C. and Bond, I.P., A hollow fibre reinforced polymer composite encompassing self-healing and enhanced damage visibility. *Compos. Sci. Technol.*, 65, 11–12, 1791–1799, 2005.

44. Pang, J.W.C. and Bond, I.P., Bleeding composites—Damage detection and self-repair using a biomimetic approach. *Compos. Part A: Appl. Sci. Manuf.*, 36, 2, 183–188, 2005.
45. Dry, C., Procedures developed for self-repair of polymer matrix composite materials. *Compos. Struct.*, 35, 3, 263–269, 1996.
46. Li, V.C., Lim, Y.M., Chan, Y.W., Feasibility study of a passive smart self-healing cementitious composite. *Compos. Part B*, 29, 6, 819–827, 1998.
47. Dry, C. and McMillan, W., Three-part methylmethacrylate adhesive system as an internal delivery system for smart responsive concrete. *Smart Mater. Struct.*, 5, 3, 297–300, 1996.
48. Dry, C., Matrix cracking repair and filling using active and passive modes for smart timed release of chemicals from fibers into cement matrices. *Smart Mater. Struct.*, 3, 2, 118–123, 1994.
49. Motuku, M., Vaidya, U.K., Janowski, G.M., Parametric studies on self-repairing approaches for resin infused composites subjected to low velocity impact. *Smart Mater. Struct.*, 8, 5, 623–638, 1999.
50. Bleay, S.M., Loader, C.B., Hawyes, V.J., Humberstone, L., Curtis, P.T., A smart repair system for polymer matrix composites. *Compos. Part A*, 32, 12, 1767–1776, 2001.
51. Jang, B.Z., Chen, L.C., Hwang, L.R. *et al.*, The response of fibrous composites to impact loading. *Polym. Compos.*, 11, 144–157, 1990.
52. Trask, R.S. and Bond, I.P., Biomimetic self-healing of advanced composite structures using hollow glass fibres. *Smart Mater. Struct.*, 15, 3, 704–710, 2006.
53. Sanada, K., Yasuda, I., Shindo, Y., Transverse tensile strength of unidirectional fibre-reinforced polymers and self-healing of interfacial debonding. *Plast. Rubber Compos.*, 35, 2, 67–72, 2006.
54. Kessler, M.R. and White, S.R., Self-activated healing of delamination damage in woven composites. *Compos. Part A: Appl. Sci. Manuf.*, 32, 5, 683–699, 2001.
55. Toohey, K.S., Sottos, N.R., Lewis, J.A., Moore, J.S., White, S.R., Self-healing materials withmicrovascular networks. *Nat. Mater.*, 6, 8, 581–585, 2007.
56. Williams, H.R., Trask, R.S., Bond, I.P., Self-healing composite sandwich structures. *Smart Mater. Struct.*, 16, 4, 1198–1207, 2007.
57. Zako, M. and Takano, N., Intelligent material systems using epoxy particles to repair microcracks and delamination damage in GFRP. *J. Intell. Mater. Syst. Struct.*, 10, 10, 836–841, 1999.
58. Hayes, S.A., Zhang, W., Branthwaite, M., Jones, F.R., Self-healing of damage in fibre-reinforced polymer-matrix composites. *J. R. Soc. Interface*, 4, 13, 381–387, 2007.
59. Kirkby, E.L., Rule, J.D., Michaud, V.J., Sottos, N.R., White, S.R., Manson, J.-A.E., Embedded shape-memory alloy wires for improved performance of self-healing polymers. *Adv. Funct. Mater.*, 18, 15, 2253–2260, 2008.

60. Murphy, E.B., Bolanos, E., Schaffner-Hamann, C., Wudl, F., Nutt, S.R., Auad, M.L., Synthesis and characterization of a single-component thermally remendable polymer network: staudinger and stille revisited. *Macromolecules*, 41, 14, 5203–5209, 2008.
61. Keller, M.W., White, S.R., Sottos, N.R., A self-healing poly(dimethyl siloxane) elastomer. *Adv. Funct. Mater.*, 17, 14, 2399–2404, 2007.
62. Lin, C.B., Lee, S., Liu, K.S., Methanol-induced crack healing in poly(methyl methacrylate). *Polym. Eng. Sci.*, 30, 21, 1399–1406, 1990.
63. Chung, C.-M., Roh, Y.-S., Cho, S.-Y., Kim, J.-G., Crack healing in polymeric materials via photochemical [2+2] cycloaddition. *Chem. Mater.*, 16, 21, 3982–3984, 2004.

14

Stimuli-Responsive Self-Healable Materials

G. Jerald Maria Antony[1], S. Raja[1] and S.T. Aruna[2]*

[1]*Structural Technologies Division, CSIR-National Aerospace Laboratories, Bangalore, India*
[2]*Surface Engineering Division, CSIR-National Aerospace Laboratories, Bangalore, India*

Abstract
Self-healing (S-H) is a bio-motivated conception because nature exhibits a plethora of self-healable materials. The concept of S-H materials (SHM) emanated two decades ago and it progresses to captivate the interest of researchers. Such materials are in great demand for real engineering applications such as electronics and robotics, aerospace and sporting goods. S-H systems can be made from a variety of polymers, ceramics and metallic materials. S-H is a bottom-up technique on a microscale that demonstrates the ability to repair mechanically damaged materials and heals these cracks and also retains the requirement of structural strength. The S-H property can reduce the replacement costs, and enhance the durability and product safety and reliability. The structural polymer composites are exposed to the possibility of the formation of cracks, failure, and degradation. The S-H materials are categorized into autonomic and non-autonomic systems. As the name indicates, the former type of systems requires no stimulus for healing and the latter type of systems require a stimulus for healing. This chapter discusses about S-H materials, stimuli-responsive S-H materials with emphasis on stimuli responsive S-H polymers from very recent published literature.

***Keywords*:** Self-healing, damage, repair, stimuli, heat, light

**Corresponding author*: aruna_reddy@nal.res.in

Inamuddin, Mohd Imran Ahamed, Rajender Boddula and Tariq Altalhi (eds.) Self-Healing Smart Materials and Allied Applications, (361–378) © 2021 Scrivener Publishing LLC

14.1 Self-Healing Materials

In the past one decade, there has been a great demand for self-healing (S-H) materials as polymers and composites are being progressively used in structural applications including space, automobile, defense, and construction industries. As the name "S-H materials" indicates, these materials can repair the damage themselves and recover functionality after degradation, damage and failure [1]. The idea of S-H materials has been motivated by the restoration functionality observed in biological systems, which includes microvascular constituents as an approach to imitate the autonomic healing abilities of biological organisms, but the mechanism in S-H materials is not identical to the S-H mechanism prevailing in biological systems. These materials are an important brand of imminent smart materials that when damaged can autonomously repair or can be restored to their original form in presence of outside triggers like heat, light, pH, magnetic field, etc. [2]. Thus these materials are potential futuristic materials as they impart longer lifetimes, reduced maintenance time and cost and also increase the safety of systems. The research on S-H materials began in early 2000 and the first S-H material based on polymers was announced in 2001 by Scott R.W., and co-workers from the University of Illinois [1]. The S-H system was made of a dicyclopentadiene microcapsule incorporated in an epoxy matrix containing Grubbs' catalyst [1]. Whenever a crack was formed, the dicyclopentadiene was set free and the catalyst initiated ring-opening metathesis polymerization reaction to make a tough and extremely cross-linked network, which immediately resulted in crack closure and such healing process occurred only once. This was followed by a flurry of research activities on S-H materials as evident from the first international conference held in 2007 on S-H materials [2]. In recent years, there are serious efforts to develop materials with S-H properties with the aim of (i) facilitating remotely controlled S-H properties that completely reestablish the mechanical properties of the original material upon damage/fracture; (ii) repairing micrometer-sized cracks or cuts with negligible outside aid; and (3) to lengthen the life of materials [3].

There is a great requirement for such materials in applications such as electronics and robotics, aerospace and sporting goods, healthcare, etc. S-H materials are going to revolutionize bioinspired device technologies such as electronic (e) skin, e-nose, and e-tongue or a combination of all. The concept of S-H is very important for the following applications: (i) systems such as deep-sea wind farms and underground pipes where the refurbishment is intrinsically very expensive and the requirements are very high on reliability; (ii) tunnels and large infrastructural works wherein a greater than forty-year service has to be assured; (iii) aircraft and nuclear storage systems wherein very high reliability is required; and (iv) automobiles and optical systems where high surface quality is of utmost importance.

S-H materials encompass all classes of materials (Figure 14.1) and the most prevailing types of S-H materials are polymers or elastomers [1]. Each of these materials has its own S-H mechanisms. Basically, S-H materials can be classified into autonomous and non-autonomous materials. The differences between these S-H materials are summarized in Table 14.1. The usual features of healing agents are low monomer viscosity, low volatility, quick polymerization in ambient conditions and a lower rate of shrinkage during polymerization. Low viscosity in the monomer facilitates easy unloading of healing materials from the fractured capsules to fill scratches and be effective in wetting the developed surface. Thus it is evident that the stimuli-responsive S-H materials are basically intrinsic and mostly they

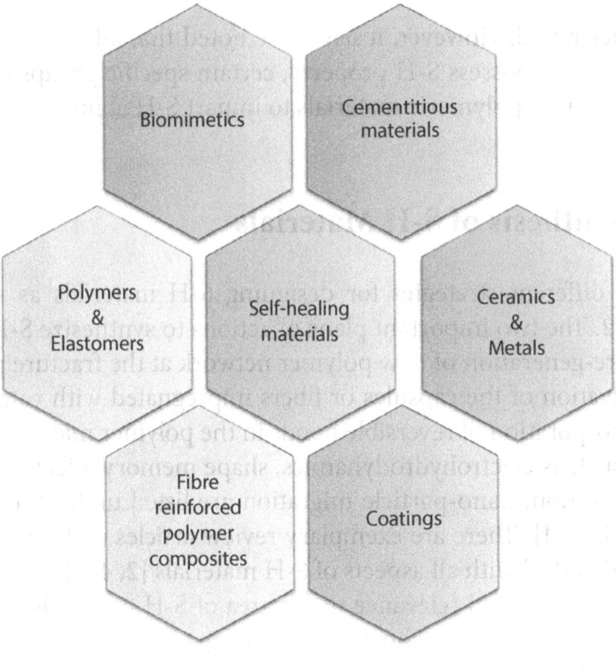

Figure 14.1 Classification of S-H materials.

Table 14.1 Differences between extrinsic and intrinsic S-H materials.

Extrinsic S-H materials (Autonomous)	Intrinsic S-H materials (Non-autonomous)
Healing agent is embedded inside the matrix	S-H is due to the incorporation of noncovalent bonds into the polymeric matrix. The covalent bonds include hydrogen bond, ionic interactions or metal–ligand interactions
Healing agent is released when damage occurs	Healing agent is not stored inside the material
Allows only a single time healing at a definitive location or a limited number of healing cycles	Theoretically heal endlessly
The healing process occurs autonomously	The healing process requires an outside trigger such as light or heat

are polymer-based. However, it has to be noted that all the conventional polymers do not possess S-H property; certain specific groups have to be incorporated into polymeric materials to impart S-H ability.

14.2 Synthesis of S-H Materials

There are different strategies for designing S-H materials as shown in Figure 14.2. The two important plans of actions to synthesize S-H materials are (i) re-generation of new polymer network at the fracture point due to the activation of the capsules or fibers impregnated with catalysts and (ii) the incorporation of reversible bonds in the polymer matrix. The other methods such as electrohydrodynamics, shape memory effect, conductivity, co-deposition, nano-particle migration are listed under the miscellaneous section [4]. There are exemplary review articles and books on S-H materials that deal with all aspects of S-H materials [2, 4, 5].

The importance and relevance of the area of S-H materials are evident from Figure 14.3 (a) which shows an increasing trend in the publication of papers over the years on S-H materials and also it is apparent that umpteen papers have been published on S-H topic in the journals related to materials science (Figure 14.3b).

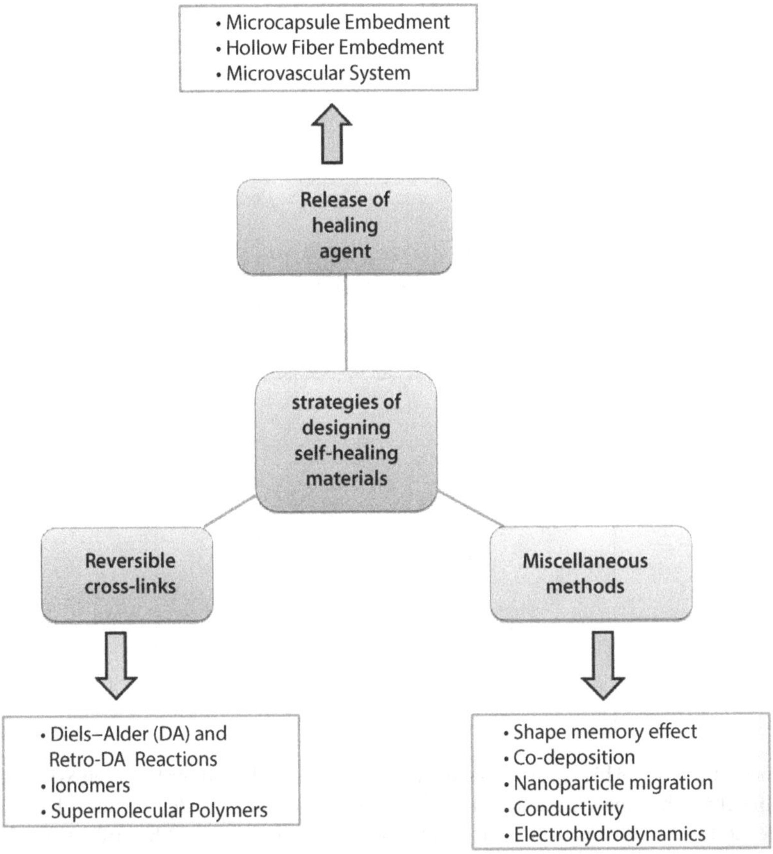

Figure 14.2 Flowchart depicting the different strategies for designing S-H materials.

14.3 Types of S-H Materials

There are four main kinds of S-H materials viz., (i) materials containing embedded healing agents, (ii) materials containing internal vascular circulation similar to blood, (iii) S-H materials based on shape memory effect, and (iv) reversible polymers. In the case of S-H materials containing embedded healing agents, the microcapsules containing an adhesive release that just fills the crack and binds the cracked portions by the polymerization and heals the damaged surface. However, the main disadvantage with this method is that due to the smaller size of the capsules they limit the amount of damage they can rectify. The other disadvantage is that it can heal only once.

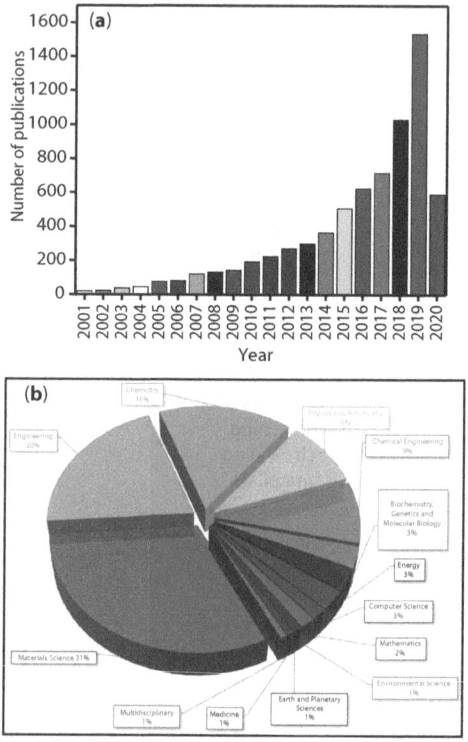

Figure 14.3 (a) The histogram showing the trend in the publications based on S-H and (b) pie-chart showing the publications related to S-H published under various subjects since 2001, based on data collected from the Scopus analysis as on 10th April 2020.

Regarding microvascular S-H materials, similar to the human body, there exists a comprehensive vascular system containing a network of tubes similar to blood vessels of different sizes that can pump healing materials to the repair spot only when they are needed. The main advantage of this process is that cracks up to ten times the size that of the microcapsules can be sealed and repaired. However, it could pose a threat if a crack is progressing faster than it is being repaired as the repair material has to travel further.

Regarding S-H materials based on shape memory effect, there is a need for delivering the right kind of stimuli (like heat) to the area where damage has occurred; the material reverts to its original form. Such systems were pioneered by Henry Sodano and are known as autonomous adaptive structures, e.g., steel rebar rods in reinforced concrete systems.

In the case of reversible polymers based S-H materials, some of the broken segments of the polymer, upon energizing with stimuli like heat or light, the broken segments join up again naturally and rebind themselves to other molecules in its vicinity and thereby excellently overturning the damage and repairing the material. Thermoplastics are an important class of polymers as they can be designed such that if they're cracked or damaged, when they are heated, the constituents of the thermoplastic will break down into their monomers and when they cool down, the original polymer reforms and switch the damage. Such polymeric material has been found to completely recover the impact of a firing bullet up to 9 mm diameter by utilizing the localized heat due to the impact.

14.4 Need for Stimuli-Responsive Shape Memory (S-RSM) Materials

S-RSM materials possess the capability to perceive and respond to outside or inner triggers like stress, temperature, pH, electric/magnetic energy, and change in size, shape, color, and properties like conductivity, permeability, etc. These materials are being used in various applications ranging from aerospace to biomedical. Among the different varieties of S-H materials, S-H polymers are in great demand as there is an ever-increasing demand for viable materials with extended lifetimes [5]. Shape memory polymers (SMPs) exhibit faster shape recovery, a high strain recovery, low density, economical, processed easily, chemically stable, good recyclability along with biocompatibility, and biodegradability. Because of their extraordinary shape varying properties, SMPs are appealing for various uses apart from S-H. There is a need for not only extremely engineered materials possessing self-repair, but also S-H grade polymers that can heal without interference under typical situation. An SMP consists of two structural elements: (a) a hard segment or the permanent net–points that generate 3-D network design that facilitates for permanent shape memory by stopping chain slippage/flow/creep on impairment and (b) a soft segment or the reversible molecular switching segments, that fix or release the interim shape when provoked by the trigger and utilizing the glass transition temperature (Tg), crystallization or melting temperature, etc. [6–11]. There are excellent review articles on stimuli-responsive polymers and hence details of stimuli-responsive materials are not presented in this chapter.

There has been a greater impetus to develop stimuli-responsive S-H materials which have been driven by the difficulties in generating mechanically potent S-H materials with high strength and stiffness at low temperatures.

14.5 Stimuli-Responsive or Nonautonomous S-H Materials

Stimuli-responsive S-H materials undergo S-H in the presence of various stimuli. For a large fraction of non-autonomous S-H materials, always an external stimulus such as heat [9], light [11], water [12–14], magnetic field/electricity [15–17] and solvent [18–20] (Figure 14.4) is required for S-H. The water and solvent may be combined as chemical stimuli. Also, some polymers can utilize multi-stimuli response for S-H. The shape memory-assisted S-H strategy was suggested by Li *et al.* [21, 22]. It was evident from the study, that for S-H of a damaged area, it suffices to activate only the damaged area without disturbing the remaining sample. From this perspective, light-induced S-H seems to be a potential choice for the stimuli-responsive S-H strategy than the bulk heating because of its on-demand treatment and remote activation. There has been an increased zeal in the development of stimuli-responsive S-H materials as evident from the increased publications on this subject (Figure 14.5).

A large number of review articles and books have been published on S-H materials [2, 3]. The review article by Amaral and Pasparakis [23] discusses the interplay of key chemical reactions that stimulate the healing mechanisms in a precise way. A few illustrative upcoming S-H elastomers, rigid polymers, hydrogels and membranes that can find real-world applications are discussed. The latest review delves on the role of shape memory effect

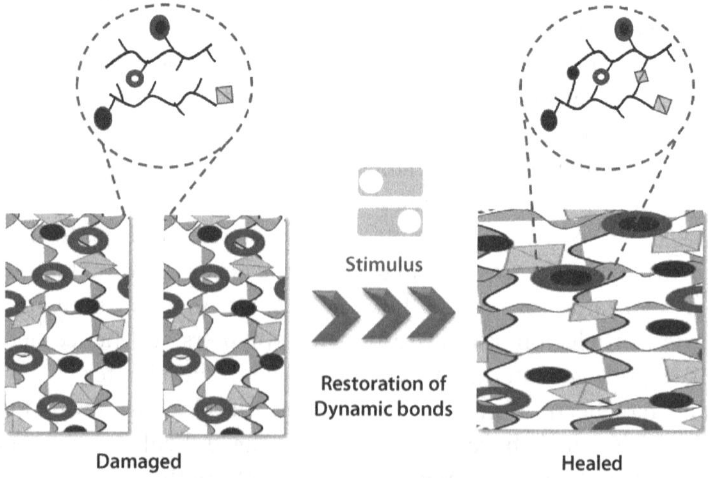

Figure 14.4 Schematic illustration of S-H in stimuli-responsive S-H polymers.

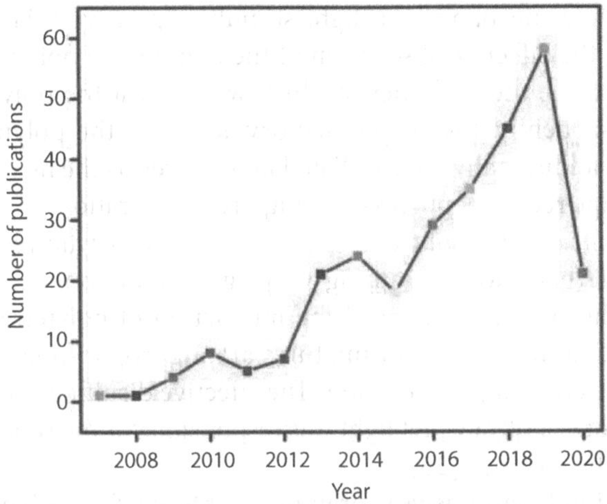

Figure 14.5 The plot depicting the trend in the publications based on stimuli-responsive S-H materials [Based on the Scopus database as on April 2020].

(SME) in S-H phenomenon [24]. Beginning from constructural needs and thermodynamics, mensurable conditions of the SME are deliberated in this review. Also, the physical factors that control the S-H process and the mechanisms of strength regain subsequent to shape recovery are explained in detail [24]. The present chapter mainly focuses on typical examples of S-H stimuli-responsive polymers from very recent published literature and it does not include gels. The systems include those which exhibit S-H in response to external trigger such as light, heat, electricity/magnetic field, etc.

14.5.1 Light Stimuli-Responsive S-H Materials

In the case of light stimuli-responsive S-H materials, the healing process is triggered through exposure to light irradiation. Light is the most widely used stimulus because it can be incident on the desired area without influencing the remaining area and it can also be incident on inaccessible damaged areas thus providing remote access to initiate S-H. Apart from this, to suit a real application, the wavelength of the light can be tuned to selectively match the specific reaction conditions. The light sources include ultra-violet (UV) lamps, near infra-red (NIR) lamps, lasers, etc. Among these sources, laser stimuli based S-H materials can find diverse uses in biomedicine, biosensors and bioelectronics, drug delivery systems, etc. The healing mechanism in light stimuli-responsive S-H materials is attributed to the effective absorption of the light by the polymer and converting it into the thermal energy which would heat the polymer film to the T_g. Consequently, the shape memory action of the polymers can be initiated and additionally the irradiated light as well as the heat resulting in the rapid shape recovery process and high recovery ratio.

A set of ultra-violet light curable polyurethane acrylic resins exhibiting good S-H behavior containing varying disulfide-bond contents was produced. The synthesis involved the interaction of polytetrahydrofuran diol and 2, 2′-dithiobisethanol mixtures among isophorone diisocyanate ensued by the end-capping reaction. The effective disulfide exchange reaction in presence of ultra violet light was responsible for the repetitive repair of cracks in the coatings [25].

A multi-stimuli-responsive supramolecular tough S-H polymer network consisting of two varieties of crosslinks: (i) chemical crosslinks made by the free-radical polymerization of crystalline poly(ethylene glycol) (PEG) and poly(ε-caprolactone) (PCL) and (ii) dynamic crosslinks made by hydrogen bonding of ureido-pyrimidinone (UPy) motifs was developed. The UPy units effectively absorbed the UV light and converted it into the thermal energy (~63 °C) culminating in the healing of the flaw with 86% toughness restoration. By forming a conductive layer onto the polymer, a prototype strain sensor was displayed [26]. It is evident that there is a need for tailoring the polymer to achieve ~100% restoration in the toughness.

Light stimuli-responsive, S-H and multi-functionalized graphene oxide (mfGO) sheathed with N-, P- and Si-moieties was synthesized through *in-situ* polymerization [27]. The cracked surface healed due to SME; explicitly, the graphitic regions of graphene could absorb energy from the VIS-NIR light and transfer the energy to the diselenide-containing polyurethane (dPTD) matrix culminating in the melting of the crystals and closing of cracked surface. This successively promoted the interfacial chain diffusion and therefore considerably reclaiming the innate properties. The studies showed that 2 wt% mfGO/dPTD matrix exhibited a synergistic combination of enhanced toughness, SME, S-H, flame retardancy, and water repellency. The development of such multifunctional stimulus-responsive composite materials in an economical way is the need of the hour and researchers have to evolve new strategies to achieve this.

14.5.2 Thermal Stimuli-Responsive S-H Materials

Thermal stimuli-responsive S-H systems are the most explored healing systems and they can refurbish the microcracks and salvage most of the initial properties of the material under heat stimuli. Invernizzi *et al.* [28] reported a unique thermal stimulus-responsive 4D-printable SMP possessing self-repairing capabilities, furnished by self-assembling UPy units resulting in supramolecular structures. This novel method employed 4D printing by utilizing a cost-effective printer. To print the methacrylated PCL (PCLDMA) + ureidopyrimidinone methacrylate (UPyMA) formulation at 45 °C, a customized hot-plate was introduced in the printer during the whole printing procedure. This enabled the best UPyMA diffusion into the PCLDMA. The 4D-objects in the form of a hockey stick exhibited good S-H behavior and phenomenal shape memory properties and its mechanical properties were akin to polycaprolactone (PCL)-based cast specimens. Along with this, the printed and refurbished specimens exhibited a healing efficiency of tensile maximum stress suitable for use as actuator devices.

A new thermal responsive, S-H and highly transparent thermoreversible cross-linked polymer coating has been obtained from furan-functionalized acrylic copolymer and an aliphatic bismaleimide and is shown in Figure 14.6 [29]. The polymer system was synthesized by the Diels–Alder (DA) reaction and a luminescent solar concentrator (LSC) with 4.9% optical efficiency was designed. The functionality and S-H behavior of DA-based polymer matrix is shown in Figure 14.6. The healing capability of this polymer was evaluated on a mechanically scratched LSC and a complete recovery of surface scratches was achieved after sufficient thermal treatment (Figure 14.6b). This work may lead to the fabrication of remarkably efficient multivalent thermoresponsive ingenious LSC systems.

14.5.3 Chemical Stimuli-Responsive S-H Materials

Under the influence of chemical stimuli such as pH, humidity, water, etc., many non-autonomous S-H materials undergo reversible bond formation. Pan *et al.* developed a superhydrophobic surface containing low surface energy fluoroalkylsilane (FAS) encapsulated microcapsule which was responsible for repairing superhydrophobicity with respect to the crushed surface microstructures and damaged surface chemistry [30]. The authors have shown the feasibility of adorning micropillars with microspheres. The surface was prepared through molding process on a pH-responsive microcapsule–precoated poly(dimethylsiloxane) (PDMS) micropillar array mold. Microcapsules decorated superhydrophobic shape memory micropillar array with a water contact angle (WCA) of 153° was attained by external heating and pressing; when the micropillar structure crumple, the crushed surface becomes hydrophobic with a WCA of 100°. When the surface is

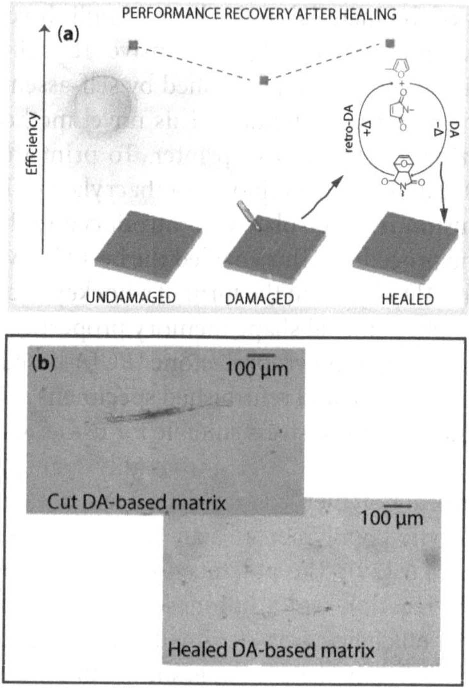

Figure 14.6 (a) Schematic representation depicting the S-H capability of DA based polymer matrix (b) Optical microscope images depicting the thermal stimuli-responsive S-H of the DA-based matrix after damage [Reprinted (adapted) with permission from Ref. [30]) Copyright (2020) American Chemical Society].

further reheated, due to the release of encapsulated FAS under pH stimuli, the surface recovers to its original state and regains superhydrophobicity (Figure 14.7).

14.5.4 Electric/Magnetic Stimuli-Responsive S-H Materials

In electric/magnetic stimuli-responsive polymers, highly dielectric receptive materials like carbon nanotubes, nanosized magnetite, etc., are incorporated in a polymer matrix which helps in inducing a shape transition by the utilization of electric/magnetic energy. These nanoparticles generate inductive heating by converting electromagnetic energy from an external high-frequency field to heat [31]. Also, some of the magnetic oxides like magnetite are approved for biomedical use due to their non-toxicity, bio-degradability (half-life of ~45 days) and the metabolism happens through the normal iron pathway [32]. In magnetic-stimuli responsive

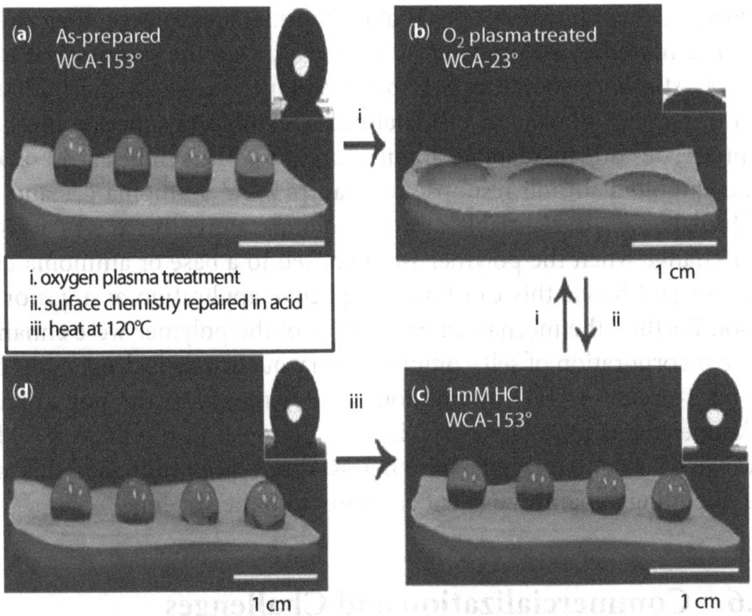

Figure 14.7 Microcapsules decorated superhydrophobic shape memory micropillar array surface with various surface modifications [Reprinted (adapted) with permission from Ref. [30] Copyright (2020) American Chemical Society].

S-H polymers, the S-H is triggered via electromagnetic induction. During the healing process, the polymer chain diffusivity increases by utilizing the heat generated during the oscillation of the nanoparticle magnetic moment known as Neel relaxation [33, 34].

Urban *et al.* developed a S-H p-methyl methacrylate/n-butyl acrylate/heptadecafluorodecyl methacrylate SMP comprising of superparamagnetic iron(III) oxide (γ-Fe_2O_3) nanoparticles employing a magnetic field [35]. When the polymeric film was physically separated, it self-healed and formed a mono block post 120 minutes and the same was repeated for several cycles. This was attributed to the fluctuation of uniformly spread 12 nanometer-sized nanoparticles at the frequency of 300 kHz, culminating in a confined heat and consequential interfacial dispersal to the impaired regions, which restored the properties [35].

14.5.5 Multi-Stimuli Responsive S-H Material

In the fields of aerospace, electronics and civil engineering, there are considerable requirements for effective polymeric materials possessing a synergistic combination of S-H ability, stimuli-responsive, high toughness and modulus, etc. Yang *et al.* have described the synthesis of novel multiple-stimuli responding S-H polyurethane via the Michael addition reaction among thiols and bismaleimides, exhibiting superior mechanical properties due to adaptive architecture [36]. The synthesized polyurethane exhibited stimuli response to heat, pH and ammonia gas and was self-healable and reclaimable at a high temperature. There was a visible color change when the polymer was exposed to a base or ammonia environment and hence this can find prospective application as a gas or pH sensor. Further, the mechanical properties of the polymer were enhanced by the incorporation of self-complementary quadruple hydrogen bonding among 2-ureido-4-[1H]-pyrimidione (UPy) motifs in the polyurethane backbone chains [36]. However, chemically crosslinked polymers seldom be repaired when damaged. Nonetheless, the authors addressed this issue by embedding dynamic covalent crosslink into the system.

14.6 Commercialization and Challenges

At present, S-H materials advancement is mostly in the initial or product stage, and hence these materials are not feasible for different uses. The scanty uses being advanced up to now persist in the automotive, aerospace, and construction sectors. Arkema (France) began the industrial production in 2009 and has developed an exhaustive range of self-repairable products tailored to several industrial applications. Autonomic Materials, Inc. (AMI, U.S.) is offering innovative S-H product technologies that extend coating lifetime and performance by up to 4 times longer than existing coatings. The microencapsulated additives developed by AMI are compatible with most solvent, water and powder-based coatings and provide protection to all kinds of substrates including metals, wood, concrete, plastic, composite, etc. minimizing upkeep and repair. The world's initial S-H clear coat by the brand name "Scratch Guard Coat" for car surfaces has been commercialized by Nissan Motor Co. Ltd. Similarly, in this category, Bayer Materials Science has also developed two-component polyurethane clear coats and this coating cure scratches in the presence of sunlight. The very recently discovered S-H rubber may be used in the toy industry. The other industrial sections where these materials use are forecasted in the aviation industry, wherein hollow fibers strengthened composites are a viable remedy to regain damaged surfaces/components. S-H polymers can also be used soon in space applications. The S-H concretes may become an everyday commodity in the construction industry. Biocompatible S-H composite may prolong the life of biomedical implants.

One of the most predominant challenges in making S-H materials is the clear paradoxical requirement of high stiffness and high molecular dynamics, which gives rise to the S-H process. Artificial self-healed materials fulfill lesser functional demands set for the application as the number of repairs is restricted and the self-healed material will not exactly possess the same properties as that in the pre-damage situation. Also, many of the raw materials used are not readily available and are not environmentally friendly. Thus there is a need to create S-H polymers/coatings based on commonly used monomers/oligomers using simple and economical fabrication methods. Research should be driven to better discern the role of shape memory in the S-H process and optimization to provide the most healing potential.

14.7 Conclusions

S-H materials are very much essential for the future and they are going to play a decisive role in all spheres of life starting from biomedical to aerospace. In particular stimuli-responsive S-H polymers would play a pivotal role in the future. Among the different stimuli-responsive S-H materials, light and heat stimuli-responsive S-H materials are advantageous and will find a plethora of applications. However, more research should be focused on developing such materials using cost-effective and commonly available S-H materials.

References

1. https://en.wikipedia.org/wiki/self-healing_material, 2020.
2. Ghosh, S.K. (Ed.), *Self-healing Materials: Fundamentals, Design Strategies, and Applications*, WILEY-VCH Verlag GmbH & Co. KGaA, Weinheim, 2009.
3. *First International Conference on Self-Healing Materials*, 12 April 2007, Delft University of Technology.
4. Hager, M.D., van der Zwaag, S., Schubert, U.S. (Eds.), *Self-healing Materials*, Springer, Cham, Switzerland, 2016.
5. Sumerlin, B.S., Next-generation self-healing materials. *Science*, 362, 150–151, 2018.
6. Rousseau, I.A. and Mather, P.T., Shape memory effect exhibited by smectic-C liquid crystalline elastomers. *J. Am. Chem. Soc.*, 125, 15300, 2003.

7. Qin, H. and Mather, P.T., Combined one-way and two-way shape memory in a glass-forming nematic network. *Macromolecules*, 42, 273, 2008.
8. Lendlein, A., Jiang, H., Jünger, O., Langer, R., Light-induced shape-memory polymers. *Nature*, 434, 879, 2005.
9. Aoki, D., Teramoto, Y., Nishio, Y., Self-healing containing cellulose acetate derivatives: preparation and characterization as a shape memory-recovery material. *Biomacromolecules*, 8, 3749, 2007.
10. Kumpfer, J.R. and Rowan, S.J., Thermo-, photo-, and chemo-responsive shape-memory properties from photo-cross-linked metallo-supramolecular polymers. *J. Am. Chem. Soc.*, 133, 12866, 2011.
11. Li, J., Viveros, J.A., Wrue, M.H., Anthamatten, M., Shape-memory effects in polymer networks containing reversibly associating side-groups. *Adv. Mater.*, 19, 2851, 2007.
12. Iacono, S.D., Amendola, E., Martone, A., Pastore, A., Filippone, G., Acierno, D., Zarrelli, M., Giordano, M., Thermally activated multiple S-H Diels–Alder epoxy system. *Polym. Eng. Sci.*, 57, 674, 2017.
13. Lee, K.M., Koerner, H., Vaia, R.A., Bunning, T.J., White, T.J., Light-activated shape memory of glassy, azobenzene liquid crystalline polymer networks. *Soft Matter*, 7, 4318, 2011.
14. Yang, B., Huang, W.M., Li, C., Lee, C.M., Li, L., On the effects of moisture in a polyurethane shape memory polymer. *Smart Mater. Struct.*, 13, 191, 2003.
15. Huang, W.M., Yang, B., An, L., Li, C., Chan, Y.S., Water-driven programmable polyurethane shape memory polymer: Demonstration and mechanism. *Appl. Phys. Lett.*, 86, 114105/1–3, 2005.
16. Qi, X., Yao, X., Deng, S., Zhou, T., Fu, Q., Water-induced shape memory effect of graphene oxide reinforced polyvinyl alcohol nanocomposites. *J. Mater. Chem. A*, 2, 2240, 2014.
17. Mohr, R., Kratz, K., Weigel, T., Lucka-Gabor, M., Moneke, M., Lendlein, A., Initiation of shape memory effect by inductive heating of magnetic nanoparticles in thermoplastic polymers. *Proc. Natl. Acad. Sci. U.S.A.*, 103, 3540, 2006.
18. Schmidt, A.M., Electromagnetic activation of shape memory polymer networks containing magnetic nanoparticles. *Macromol. Rapid Commun.*, 27, 1168, 2006.
19. Yakacki, C.M., Satarkar, N.S., Gall, K., Likos, R., Hilt, J.Z., Shape-memory polymer networks with Fe_3O_4 nanoparticles for remote activation. *J. Appl. Polym. Sci.*, 112, 3166, 2009.
20. Lv, H., Leng, J., Liu, Y., Du, S., Shape-memory polymer in response to solution. *Adv. Eng. Mater.*, 10, 592, 2008.
21. Li, G. and Nettles, D., Thermomechanical characterization of shape memory polymer based self-repairing syntactic foam. *Polymer*, 51, 755–762, 2010.
22. Li, G. and Uppu, N., Shape memory polymer based self-healing syntactic foam: 3-D confined thermomechanical characterization. *Compos. Sci. Technol.*, 70, 1419, 2010.

23. Amaral, A.J.R. and Pasparakis, G., Stimuli responsive self-healing polymers: Gels, elastomers and membranes. *Polym. Chem.*, 8, 6464, 2017.
24. Hornat, C.C. and Urban, M.W., Shape memory effects in self-healing polymers. *Prog. Polym. Sci.*, 102, 101208, 2020.
25. Zhao, D., Liu, S., Wu, Y., Guan, T., Sun, N., Ren, B., S-H UV light-curable resins containing disulfide group: Synthesis and application in UV coatings. *Prog. Org. Coat.*, 133, 289, 2019.
26. Salzano de, L.M., Marturano, V., Manganelli, M., Santillo, C., Ambrogi, V., Filippone, G., Cerruti, P., Light-responsive and self-healing behavior of azobenzene-based supramolecular hydrogels. *J. Colloid Interface Sci.*, 568, 16, 2020.
27. Du, W., Jin, Y., Lai, S., Shi, L., Shen, Y., Yang, H., Multifunctional light-responsive graphene-based polyurethane composites with shape memory, self-healing, and flame retardancy properties. *Compos. Part A: Appl. Sci. Manuf.*, 128, 105686, 2020.
28. Invernizzi, M., Turri, S., Levi, M., Suriano, M., 4D printed thermally activated S-H and shape memory polycaprolactone-based polymers. *J. Electrochem. Soc.*, 166, B3274, 2019.
29. Tatsi, E., Fortunato, G., Rigatelli, B., Lyu, G., Turri, S., Evans, R.C., Griffini, G., Thermoresponsive host polymer matrix for self-healing luminescent solar concentrators. *ACS Appl. Energy Mater.*, 3, 1152, 2020.
30. Chen, M. and Wu, L., Top of Form Bottom of Form Smart superhydrophobic surface with restorable microstructure and self-healable surface chemistry. *ACS Appl. Mater. Interfaces*, 12, 5157, 2020.
31. Schmidt, A.M., Electromagnetic activation of shape memory polymer networks containing magnetic nanoparticles. *Macromol. Rapid Commun.*, 27, 1168, 2006.
32. Weissleder, R., Stark, D.D., Engelstad, B.L., Bacon, B.R., Compton, C.C., White, D.L., Jacobs, P., Lewis, J., Superparamagnetic iron oxide: pharmacokinetics and toxicity. *Am. J. Roentgenol.*, 152, 167–173, 1989.
33. Ahmed, T.J., Stavrov, D., Bersee, H.E.N., Beukers, A., Induction welding of thermoplastic composites—An overview. *Compos. Part A*, 37, 1638, 2006.
34. Ahmed, A.S. and Ramanujan, R.V., Curie temperature controlled self-healing magnet–polymer composites. *J. Mater. Res.*, 30, 946, 2015.
35. Corten, C.C. and Urban, M.W., Repairing polymers using oscillating magnetic field. *Adv. Mater.*, 21, 5011, 2009.
36. Yang, Z., Wang, F., Zhang, C., Li, J., Zhang, R., Wu, Q., Chen, T., Sun, P., Bioinspired S-H polyurethanes with multiple stimulus responsiveness. *Polym. Chem.*, 10, 3362, 2019.

15

Mechanically-Induced Self-Healable Materials

M. Ramesh[1]*, L. Rajeshkumar[2] and R. Saravanakumar[3]

[1]*Department of Mechanical Engineering, KIT-Kalaignarkarunanidhi Institute of Technology, Coimbatore, India*
[2]*Department of Mechanical Engineering, KPR Institute of Engineering and Technology, Coimbatore, India*
[3]*Department of Mechanical Engineering, VSB Technical Campus, Coimbatore, Tamil Nadu, India*

Abstract

Design and fabrication of materials with self-healing ability has been in interest recently and this nature is induced in the materials through incitements like humidity, temperature and pH. Development of such materials has very large practical applicability. In case of extrinsic self-healable materials, the healing agent has to be encapsulated within the matrix material and has to be placed within the material in advance at the time of fabrication. Here the matrix which encloses the healing resin is not inherently healable. As the material starts to develop cracks, the material within the matrix flows into those cracks by capillary action and the healing action is done. Healing materials loaded within the host material can have healing action in two modes such as: healant loaded in microcapsules and pipelines. Mechanical stimulus for inducing the healing agent to flow within the damaged area is much easier than other incitements methods. This makes the self-healing materials to be more adaptable and better performing materials in various technical and numerous practical applications. Considering these facts, the materials that requires manual interference for inducing the self-healing nature may be discontinued for practice. This chapter deals with the various mechanical methods of inducing self-healing nature of a material.

***Keywords*:** Self-healing, mechanically induced, hydrogel, corrosion, cellulose nano crystals

**Corresponding author*: mramesh97@gmail.com

Inamuddin, Mohd Imran Ahamed, Rajender Boddula and Tariq Altalhi (eds.) Self-Healing Smart Materials and Allied Applications, (379–404) © 2021 Scrivener Publishing LLC

15.1 Introduction

Healing system with external excitation depends on vascular network or capsule as the resin holding medium [1–3]. Design of materials that actuates themselves automatically is based on the catalyst to be used and the healing agent employed in extrinsic self-healing [4–10]. Two widely used approaches for designing those extrinsic materials include micro-vascular network and micro-encapsulation. In these design methods, the process of self-healing is initiated by rupture or internal damage caused due to various reasons in vascular networks and capsule surfaces [1]. Mostly the shell of the capsules were manufactured using some compounds like polymelamine urea formaldehyde (PMUF), poly-urea formaldehyde (PUF), poly-melamine formaldehyde (PMF) and polyurethane and various polymerization techniques were followed for the fabrication of capsules [4–9, 11]. Yet some of the common problems faced by micro-encapsulation approach are the cost of catalyst which is very high and the emptying of entire encapsulated healing agent at the damage location all at once for the process of self-healing. Many researchers took the aforementioned issues and addressed them through their researches so as to astound them in near future [12–16]. This chapter deals with the mechanically induced self-healing materials that are based on the different types of materials used. Initially gel based, crystal based and composite based mechanically induced materials are discussed followed by specific healing for the corrosion based on capsule and fiber has been discussed.

15.2 Mechanically-Induced Self-Healing Based on Gel

Research on soft materials depends greatly on the gel molecules which has very less molecular weight. Because of enormous water content and wide variety of applications in bio-engineering, hydrogels are considered most important among the gels. Many researchers had studied various types of hydrogels including self-healing organogels with low molecular weight, polymer based hydrogels and nanocomposite based hydrogel self-healing materials and quoted many elements as its examples. Though these materials have wide applicability in the medical field such as in muscle thixotropy and nerve fiber regeneration, none of these soft materials with low molecular weight self-healing ability could be used as forerunner for developing them further. Yet, recent focus has been given to supramolecular soft smart-materials which can be incorporated with the ability of self-repairing. This forces the researchers to develop a low molecular weight self-healing hydrogel which could perform the above said medical functions such as thixotrophy. Any developed material, when it replicated the performance and properties of low molecular weight hydrogel (LMHG) could be potentially applied in self-repairing operations or some other challenging applications in medical field [9].

LMHGs are mostly sensitive towards external mechanical stress or strain and when such stress or strain is applied on to the material they release the binder solvent encapsulated within them and makes its own self-integrated network system to collapse irrevocably. Self-healing agents could be added within these gels which acts as solid suspensions and the elastic properties of gels were lost if the applied stress is retracted back. Yet, thixotropic gels could be an exception as they could recuperate their elastic properties when the applied stress or strain is removed from them and these gels regain their properties by disintegrating themselves into a viscous solution upon the application of mechanical stresses. Even nerve damages could be eradicated with these hydrogels if they are given with electroactive properties. Various researchers have contributed in analyzing the properties of shear-thinning, self-healing and thixotropic hydrogels. Very few of none of the researchers have still reported about the discovery of amino acid based self-healing, thixotropic or LMHGs or developing a self-healing material based on carbon nanocomposites. Only stated amino acid based hydrogel is the 1-Aminoundecanoic-acid-containing gels by few researchers [10].

Stimuli based non-covalent bonded supramolecular polymer or chemically bonded hydrogels were developed recently which could be triggered by peripheral stimulus like the presence of ions, pH or temperature. Due to the irresponsiveness to the stimuli by covalent bonded hydrogels and due to the absence of sol–gel transition characteristics, non-covalent bonded supramolecular self-healing materials were developed which has enhanced response to any stimulus and better mechanical stability [10]. Gels developed by crosslinking of reversible covalent bond with functional group of polymers and metals were in the research focus lately. Like chemical gels, these organic gels exhibit better mechanical properties due the above strengthening mechanism and also respond better to the stimuli thus enabling reversibility.

Response to the stimulus by a supramolecular polymeric gel was governed by the nature of non-covalent bond formation mechanism that takes place for the building block structuring in supramolecular polymer gel systems. If the selected non-covalent bond formation mechanisms were selected properly, then the supramolecular hydrogels which actively respond to stimuli could be manufactured in an easy way. Researchers in the field of traditional polymer science and engineering could make well use of this material and has wider scope of application in many science and engineering fields. Commonly used procedures include end-functionalized material synthesis and polymerization of functional group encompassed co-monomer with the key monomer molecule. After certain development, self-healing materials were fabricated by using micro-encapsulated polymeric resins and other reagents which induce the healing property in supramolecular gels. Process of self-healing initiates when a crack is formed in polymer composites followed by the release of monomer into the crack and then by occurrence of polymerization thus retarding the crack growth. Advanced systems like multi stimuli responsive self-healable materials, self-healing polymers whose molecules were formed by non-covalent bonds and self-assembled supramolecular gels for self-repairing are to be developed in future which could warrant for newer applications. Figure 15.1 depicts the structure of monomers, copolymers and supramolecular polymer hydrogel.

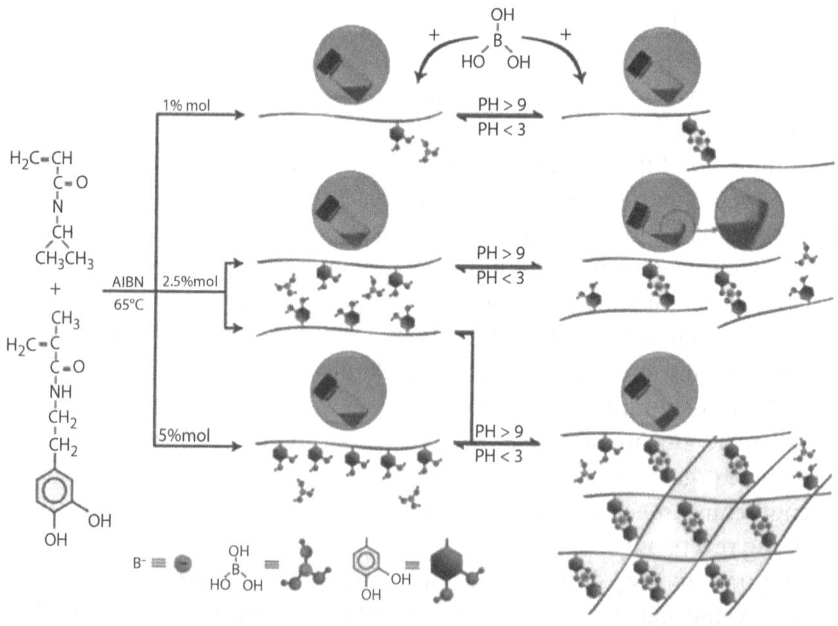

Figure 15.1 Schematic of (a) monomers, (b) copolymers, and (c) supramolecular polymers and gels [17].

Self-healing mechanism is dominant in mussel adhesive proteins and byssus. Byssal proteins interact and improve the coordination with the metals by bond formation in such a way that they develop a self-healing ability even beyond the irreparable damage occurred to them after many mechanical deformations. By utilizing this metal-protein interaction mechanism, a novel group of self-healing hydrogel polymeric material with controllable properties and molecular structure can be evolved. For making the polymeric chain of protein molecules to interact with metal ions in a better way numerous procedures were established from which precise polymer structures as per the requirement could be developed. Recent researchers tend to synthesize the mussel byssal protein self-healing mechanism artificially to open new era in self-healing researches [17, 18].

Ligands play a major role in encompassing the solvent and facilitating the gelation with the polymer. Previous researchers focused mainly in special ligands like polyethylene glycol ligands added with a metal–ligand, but nowadays small molecular ligands carry a high demand because of their large metallic density and diverse exhibition of properties. In supramolecular chemistry, adenine which is a pure nucleobase is considered to be an imperative ligand. Adenine along with its derivative compounds, when used as a ligand, may form metal-organic framework, metal ion coordination compounds, nanoparticles and organic molecules with hydrogen bonding. Currently a research outcome highlighted the adenine gelation with tricarboxylic acid derivatives where they formed p–p stacking and hydrogen bonding. Interests have grown in finding out the possibilities of accomplishing adenine derivative gelation with the help of metal coordination and bonding [18, 19]. However, gel based quadraplex compounds were formed from guanine and their derivatives by homogenizing them with monovalent metal ions while the same process has been little explored for adenine and its derivatives. Though nanoparticles were the most common products out of the above reactions, comparatively hydrogels would serve better for the development of stimulus-responsive materials. This could possibly be done by repairing of materials and guest encapsulation of molecules which was reported in some works.

Apart from using adenine nucleoside and nucleobase as ligands, developing the coordination between nucleotide molecules influences the reaction significantly. Few researchers considered this as an important area of research and used Zn^{2+} as a coordinator metal ion due to its diversity of coordination with various ligands, low toxicity, behavior as a better Lewis acid and comparatively biocompatible. Initial experiments focused on the reaction between three majorly used nucleotides such as adenosine monophosphate (AMP), guanosine monophosphate (GMP) and cytosine monophosphate (CMP) along with Zn^{2+}metal ions. These nucleotides have a very strong ability to coordinate even at neutral pH while the other nucleotide namely Thymidine monophosphate (TMP) has weaker coordination capacity. During the experimental procedure, the above said three nucleotide solutions were used to titrate Zn^{2+} and by using UV visible spectroscopy wavelength of 500 nm the titrants were captured. At the end no coloured products were observed from the titration and the magnification of light waves showed the precipitation of metal ions and light scattering through them. UV visible spectroscopy results illustrated that all the combined samples of nucleotide and Zn^{2+} intensely scattered light and the tendency increased with increase in content of Zn^{2+}. At the maximum content of Zn^{2+}, a saturated state was attained and no scattering was observed. At a content ratio of 1:1 between Zn^{2+} and AMP, a stoichiometric saturation and coordination was attained. Bond established by Zn^{2+} with AMP was much stronger when compared with other nucleotides, since the content of Zn^{2+} required to establish a strong bond with other nucleotides, CMP and GMP, was very high [19].

This ratio of combination of nucleotide with metal ion resulting in gel formation could be easily compared with other gel types (Figure 15.2). For such comparison the characteristics of freshly formed Zn^{2+}/AMP gel was carried out by varying the temperature, pH and ionic strength. Initially, Zn^{2+}/AMP gel was made with the buffer solution of

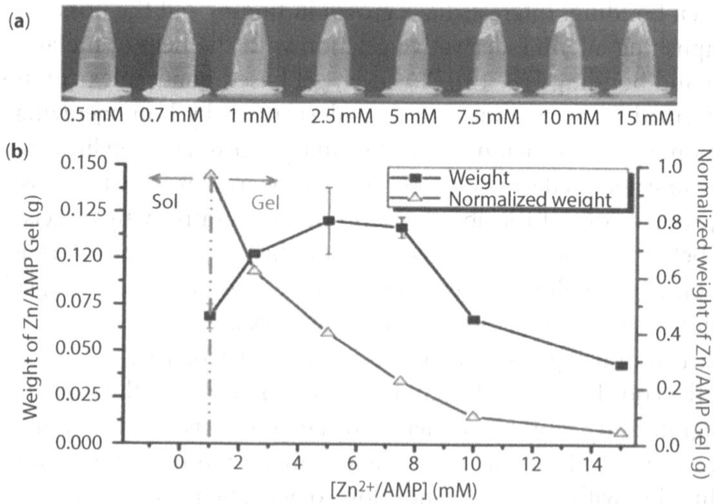

Figure 15.2 (a) Photograph of gels formed by the 1:1 Zn^{2+}:AMP mixture at different monomer concentrations, (b) The final gel weight and normalized gel weight of the Zn/AMP gels prepared with different concentrations of Zn^{2+} and AMP [20].

4-(2-hydroxyethyl)-1-piperazineethanesulfonic acid (HEPES) at a pH of 7.4 and at a concentration of 10 mM. Then the gels attained were disintegrated by mechanical vortexing after the buffer solution was suitably modified as per the condition requirements. Gel was again made to undergo centrifugal action preceded by the quantification of gel by its weight as shown in Figure 15.3(a). Due to centrifugal action, few gel particles were suspended at the bottom of the tube making, the solution visibility unclear after the disintegration of gel. Disintegrated gel rendered cloudy solution. After repeated washing of the gel, it became clear and this clarity has much influence over the ionic strength as depicted in Figure 15.3(b). It was also concluded that the gel formation was not observed below a buffer solution concentration of 1 mM [20, 21].

It was also noted that irrespective of the concentration of salt used for washing, all the gels lost about 30% of its weight when all the washes were completed. Hence, it could be stated that the stability of Zn^{2+}/AMP gel was not governed by the ionic strength. When the pH influence over the gel was tested, it could be seen that the sol-gel transition, as shown in Figure 15.3(d), occurred at a pH of 4 and 13 while the strength of the gel was maximum at neutral pH value. This reveals that the presence of adenine in

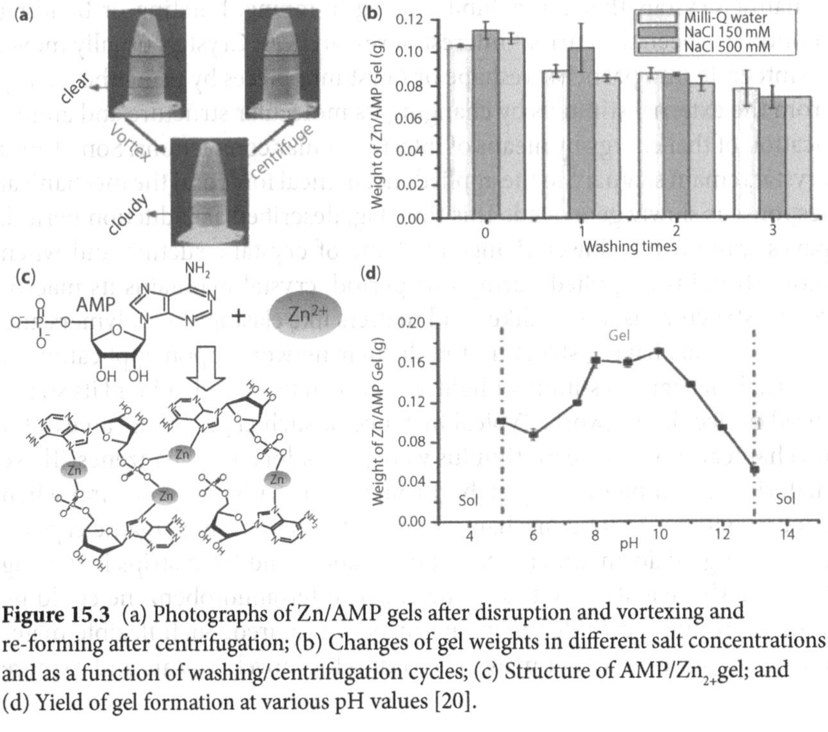

Figure 15.3 (a) Photographs of Zn/AMP gels after disruption and vortexing and re-forming after centrifugation; (b) Changes of gel weights in different salt concentrations and as a function of washing/centrifugation cycles; (c) Structure of AMP/Zn_{2+}gel; and (d) Yield of gel formation at various pH values [20].

the gel rendered this result and at a lower pH of 3.5 it can be protonated. High pH values require more amount of metal ions to precipitate the solution while it may be available for gelation with AMP. Finally the gel was observed to be very stable at a temperature of 100 °C, when an attempt has been made to analyze the effect of temperature on the gel. In terms of rheological property measurements, a frequency sweep experiment was conducted. As a result of the experiment, the loss modulus value of the gel was much lower than the storage modulus. Frequency response plots exhibited a linear range in which the gel properties were found to be elastic which is the typical frequency range (0.03–30 Hz) of supramolecular gels. Figure 15.3(c) depicts a tentative structure of the Zn^{2+}/AMP gel formed [20, 21].

15.3 Mechanically-Induced Self-Healing Based on Crystals

Emergence of efficient nano or micro-energy converters of kinetic energy in the form of mechanical movement from light or heat energy by the dynamic crystals that can exhibit curling, jumping, bursting or bending motion has been the current interest of researchers. Crystals usually move, disintegrate into particles, reshape or burst into pieces by using the energy from the external stimulus by changing its molecular structure and amplification of that energy by means of intermolecular connections. Sometimes crystal remain stagnant to the applied mechanical force and the mechanical response is slowly generated. This time lag, described as induction period, paves way for the slower change of shape of crystal structure and when more stimulus is applied during this period, crystal maintains its macroscopic structure as it is. Unlike solid matters like elastomers, polymers and liquid crystals having structured molecular network, upon application of external mechanical stimuli or light, crystal can twist, coil or bend its structured molecular network. Typical instance of such crystal that can exhibit mechanical compliance to stimulus were polyhalogenated benzenes. These materials could bend completely through 360° without deforming when applied with a confined mechanical stress. Similarly some organic crystals can undergo deformation at many points and could form strips in zig-zag fashion. The acicular crystals of the classical tetrafluorobenzene could be wound around a cylinder as many times as required. Such flexible materials may serve as model materials to develop novel crystals and explore the hidden facts of yet to discover molecular solids. They also close the gap between soft unstructured polymer materials and ductile metals or metal alloys. Stochastic event triggered fast actuation of crystals by means of photosalient and thermosalient process always renders disintegrated results while this is not the case with slow crystal deformation effects like elastic and plastic bending.

Most of the defect location cracks of the crystals actuated by thermosalient process rendered explosion or splitting up as an outcome which makes them unfit for most of the real world applications. This case has been reported with an exception in which the crystals of pyroglutamic acid could be persistently actuated for more than 15 times without undergoing much deformation or disintegration. When the elastic energy is suitably combined with external stimulus, it may provoke the reversible reaction within short period of time and hence the response time by soft material actuators could be very minimum. Such combination may also give rise to many interesting properties like superelasticity along with its existing mechanical properties. Yet slower response times do have its own demerits like less effective energy transducing ability while faster response time ends up n disintegration of elements. The balance between the time of response and the maintenance of integrity of crystal remains as the major hindrance n developing actuator materials for many practical applications. Development of dynamic molecular crystals with reversible reaction and faster actuation are in the light of research currently.

Even though the mechanical and thermal induced transitions have a common mechanism for its deliberation, they occur on different scales of time to change from one form to the other as shown in Figure 15.4. Surface incisions are the places where the transformation usually commences and particularly it happens at the (010) plane of the crystal which resembles the crystal glide plane. Front phase of the crystal vestiges as straight phase and the transition in this phase advances at a faster rate. In this phase the transition would be stopped by the front phase of another incision and here the transition of single field of crystal incision will be accomplished. These occurrences were experimentally induced in the terephthallic acid by some researchers and were confirmed by atomic force microscopy (AFM). AFM images shown in Figure 15.5 depicts the incisions on (010) plane of the crystal which are in between molecular surfaces like the shape of stairs with a dimension of 35–45 nm. The surfaces of stairs were observed to be parallel to the molecules of the crystal and were noted to be interior crystal slip planes.

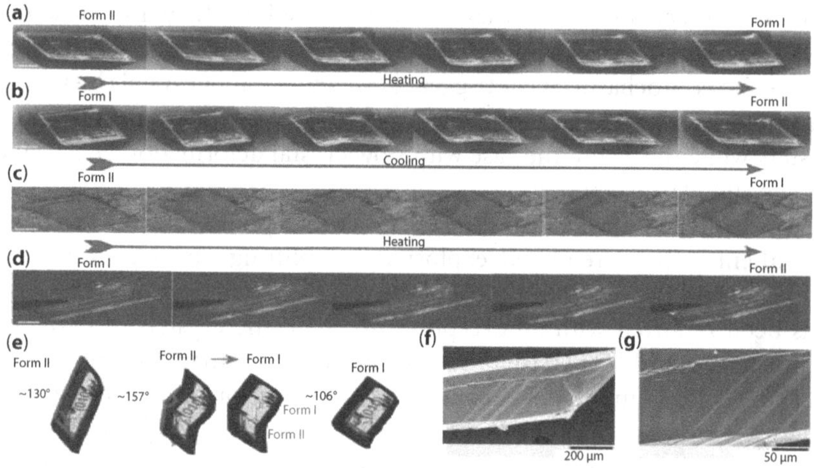

Figure 15.4 Mechanical reconfiguration of crystals during the thermally and mechanically induced phase transitions between forms I and II; (a, b); (c) Pictographs recorded using variable-temperature serial hot-stage scanning electron microscopy; (d) Snapshots of the mechanically induced phase transition recorded using optical microscope; (e) Changes in angles between the crystal faces during mechanical reconfiguration of a single crystal; (f, g) Scanning electron micrographs showing alternating layers of form II and form I [22].

Figure 15.5 AFM images of form II terephthalic acid at (010) plane; (a, b) Topography image (a) and the corresponding phase image (b); (c) Topography scan recorded on a 10 μm × 10 μm area; (d) 3D images showing terraces [22].

15.4 Mechanically-Induced Self-Healing Based on Composites

Polymer matrix composites find variety of applications in electronics, transportation, and civil construction and defense vehicles. Reliability of polymeric materials hangs highly when they undergo failure by fracture which also reduces their lifetime and sustainability. In contrast to the natural materials, these synthetic polymeric materials were not gifted with the ability of self-healing after a damage which is the most required property in the current scenario. Researches has given prime importance in developing such self-healing polymeric materials, which could repair themselves after a mechanical injury, during the past two decades.

Mechanical stimulus induced chemical interactions between molecules are in light these days due to the distinct properties resulting from those interactions. Self-healing materials developed by the materials which are sensitive to mechanical stimulus (mechanopores) by incorporating them into polymeric materials has kindled a unique reaction which could not be obtained by the materials induced by other stimuli in the field of polymer composites. Even identification of a repair and eliciting of some other captivating properties could be made possible with the aid of such reactions in polymers. Reactions that are induced mechanically in monochain of polymers, bulk materials and solutions are the research focused areas in mechano-chemistry of polymeric materials. Due to a wide range of applicability, occurrence of mechano-chemical reactions at the fiber–matrix interface of the nano composites needs better research depths which are very meager to date [19].

Few reports on the scissile of cellulose nano crystals (CNCs) being activated by mechanical stimulus yet recombinable reversibly through mechanopore with dynamic covalent bond are available. During the experiment, the characteristics of the modified mechanopore were observed by mechano-transduction over the surface of the element and that adopted nanofiller were used to manufacture a mechanopore based self-healing materials. CNCs evolution is due to many interesting properties like high-aspect-ratio, high tensile stiffness, low toxicity, abundance of cellulose in nature, low density and ease of access while these materials are being used as model materials to develop future self-healing composites. Some researchers experimented the fabrication of CNCs segregated from cotton fibers and they were functionally modified on their periphery with diarylbibenzofuranone (DABBF) for developing them as mechanopore modified self-healing material with dynamic covalent bond [20]. DABBF forms two colored radicles upon the application of mechanical force through reversible dissociation. Due to the low radicle concentration, they again recombine automatically at room temperature equilibrium conditions. As DABBF has a tendency of reversible radicle recombination property, this material is used to manufacture self-healing polymers.

Bulk display of CNCs whose periphery was attached with the covalent bonded modified mechanopore which was mechanically activated by using a DABBF-based scissile element was carried out. Recombination of DABBF radicles occurred at a very slow pace than formation in DABBF polymers and monomers in the backgrounds other than CNCs such as gels, bulk materials and solutions. These effects are normally due to the constrained movement of molecules which eventually renders mechanical stress sensitive mechanopores. Self-healing ability of the CNCs developed significantly through these mechanopore modification carried out at the surface by using DABBF which also acted a potential reinforcements to CNCs and the self-healing ability retention is more owing to reversible linkages. Self-healing composite materials with enhanced properties and the damage detection design could be more prominently done as the mechanopore modified nano CNC materials have promising applications in the field of mechano-chemistry [23].

A hypothesis states that the radicals arose due to mechanical activation of DABBF could be recombined at a faster rate by the application of a suitable solvent as it mobilizes the radicles quickly. Results of such experiments on recombination characteristics of radicles revealed that the rate of recombination was governed by the homogeneity of DABBF in CNC and not by the solubility of DABBF into the solvent used. It was also understood that DABBF possess high solubility in dimethylformamide (DMF) anddichloromethane (DCM) but poor solubility in water while CNCs exhibited better solubility in water but poor solubility in the two former solutions [22]. Experiments revealed that the DABBF radicles recombined faster in water while the recombination was very slow in DCM even DABBF has good solubility in it. From these observation it could be inferred that the radicles at CNC surfaces has no agility for recombining with other radicles within the same molecule of CNC but could easily recombine with the molecules from various CNCs. Figure 15.6 illustrates the various stages of radicle recombination through self-healing phenomenon of broken DABBF on the surface of CNCs.

Apart from mechanical stimulus, thermal stimulus combined with mechanical action could be used to stimulate the self-healing process in a polymer composite material. Few researchers carried out experiments on Diels–Alder (DA) thermally reversible reaction for developing thermally

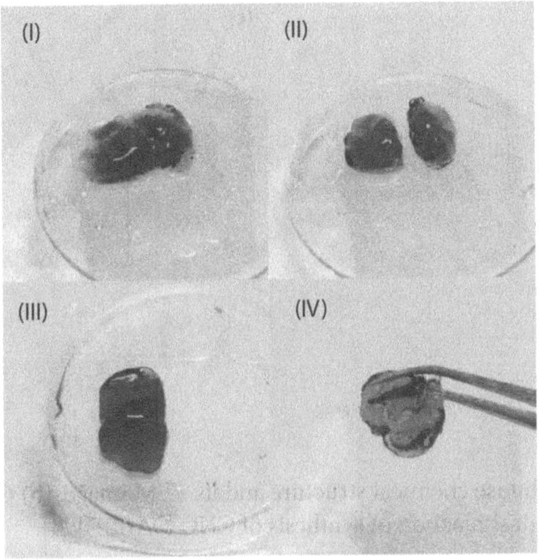

Figure 15.6 Visualization of self-healing in CNCs: (I) initial sample, (II) cut sample, (III) sample after joining the pieces together, and (IV) proof of self-healing [23].

responsive self-healing materials. DA reaction has diene and dienophile as initial elements reacting through cycloaddition. As the products of DA reaction reforms the diene and dienophile compounds at elevated temperatures due to the instability of formed products (DA Adducts) it is said to be a reversible reaction (retro DA reaction). Thermally activated reversible crosslink self-healing polymers and some other new polymer materials were manufactured through the thermal reversible phenomenon of DA reaction. Optimum process route was experimented for the production of thermally reversible gel polymers by many researchers [24].

This process can also be used for developing reversible crosslinking polymers that could be used for the removal of thermosetting polymers (Figure 15.7). Polymer encapsulants and microelectronics dielectrics could be developed by such removal materials. Thermo-responsive micelles can be manufactured from the DA adducts which were based on furan–maleimide surfactants that were cleaved thermally. Such adducts could also be used for the fabrication of nanoscale probe lithography, smart coatings, thermo-responsive optical polymers, modular polymeric color switches and materials for data storage.

Furan–maleimide compounds were fabricated by using tiny molecules of 4 furan and 3 maleimide elements that increases the mechanical properties and self-healing efficiency of polymer materials. These elements are used to form a macromolecular structure with DA adducts through the thermally reversible reactions. DA adducts disintegrate at a temperature

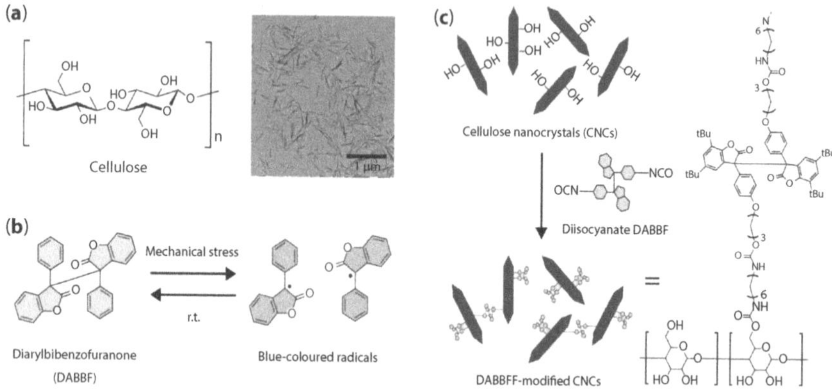

Figure 15.7 (a) Cellulose chemical structure and its TEM image, (b) Chemical structure of DABBF, (c) Chemical reaction of synthesis of CNC-DABBF [24].

of 150 °C for 15 min by following retro DA reaction thus rendering 30% disintegration. After this reaction, the disintegrated furan and maleimide elements could perform the self-healing action by mobilizing themselves and establishing a structural relinking. Hence at the above said temperature the healing efficiency may almost be 50%. If the DA adducts break into two pieces then the self-healing phenomenon rebinds them by the above process. Some authors measured the mechanical properties of such rejoined elements and found that its tensile strength value was 68 MPa which is comparable with epoxy resin whose tensile strength value falls in between 27 and 88 MPa [24]. Few other mechanical characteristics such as flexural strength and modulus, compression strength and modulus of 4F-3M structural network were also comparable with the epoxy resin. In order to increase the self-healing efficiency of such polymers to further high values, monomer substitutes with low melting point can be used and these materials possess a better agility over the cracks of crosslinked matrix materials commendably.

Graphene is an obvious choice among the materials that could be used as a filler in polymers as they have better compatibility and huge π–conjugate system. By using graphene as a filler, mechanically enhanced composite material with an enormous tensile strength could be fabricated. Graphene possess some noteworthy characteristics including thermal conductivity, super chemical stability, electrical conductivity and better IR and microwave absorbability due to which graphene responds to electrical and electro-magnetic signals better. Aforementioned characteristics could serve as a better evidence for the belief that a better and novel self-healing polymeric materials could be developed in integration with graphene and may render superior self-healing in multiple channels along with better mechanical properties. Researchers are at the nuptial stage as far as this integration is considered. Figure 15.8 shows the schematic structure of graphene based self-healing materials [25].

Some experiments were conducted for testing the mechanical properties of functional graphene (FG) dispersed in polyurethane (PU) matrix. For an optimum volume fraction of FG In PU matrix, the tensile strength increased by a considerable magnitude due to the homogenization of graphene in the PU matrix. But the self-healing materials without graphene exhibited very poor or low mechanical characteristics because of their limited option towards healing processes which in turn states the importance of incorporating FG–PU composites for self-healing applications due to its superior tensile strength [26]. Repeated healing is also a most important concept in the self-healing phenomenon where a self-healing polymer material manufactured by reversible physical or mechanical reactions or by microvascular materials embedded on the crack interfaces could induce repeated healing on the polymers. Healing repeatability is very difficult to be incorporated while manufacturing microvascular materials and a

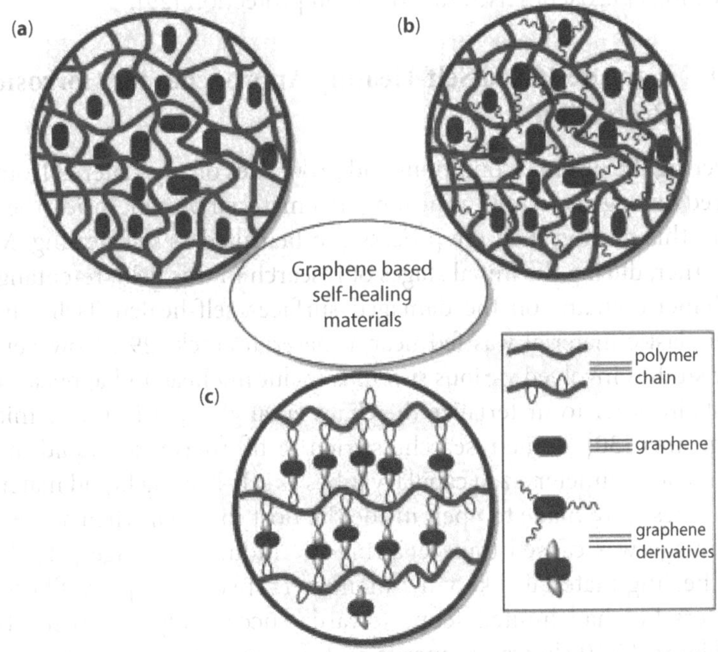

Figure 15.8 Graphene based self-healing materials [25].

simple method is yet to be developed by the researchers. Meanwhile, only limited materials could be manufactured by reversible mechanical reactions for self-healing and healing repeatability of those materials works in nuptial fractures only [27, 28].

15.5 Mechanically-Induced Self-Healing for Corrosion

Corrosion is the process of depletion of the top surface of any materials by means of internal or external forces. Since the presently used corrosion inhibition methods involve the use of costly consumables like frequent change of sacrificial metals or unavoidable use of poisonous chemicals which are costly, self-healing process have taken as a method for eradicating corrosion in recent times by researchers. When self-healing method of corrosion protection is adopted, it may eliminate the usage of toxic chemicals and other costlier consumables use. Various researches have been carried out in newer methods of extrinsic self-healing for corrosion prevention through fiber-based and capsule-based methods. Further development of methods may lead to still cheaper process and a novel one. Figure 15.9 depicts the various self-healing approaches used for corrosion protection [29].

15.5.1 Capsule-Based Self-Healing Approaches for Corrosion Protection

Recovering the original conditions and properties of any material from the incurred damage is termed as healing. If a material present inside the host induces this healing then the process can be called as self-healing. Many days earlier, during the initial stages of research in this field, re-entangling of polymeric chains on the damaged surfaces self-healed itself when a thermoplastic material was fed near a material crack [29]. However, the earlier studies involved various stimulus including heat and application of solvents in order to materialize the chain entanglement in the vicinity of damage area [30]. Some researchers tried to fill thermoset liquid monomers into a millimeter sized capillary tubes as self-healing liquid materials. These tubes were made to open inside the host material when a mechanical damage was caused thus repairing the induced damage [31]. Many liquid healing materials like vinyl monomers, two-part epoxy, fibers and polyesters but had limited scope towards successfully employing them [30]. Figure 15.10 shows the merits and demerits of various methods of self-healing in the form of a pictograph.

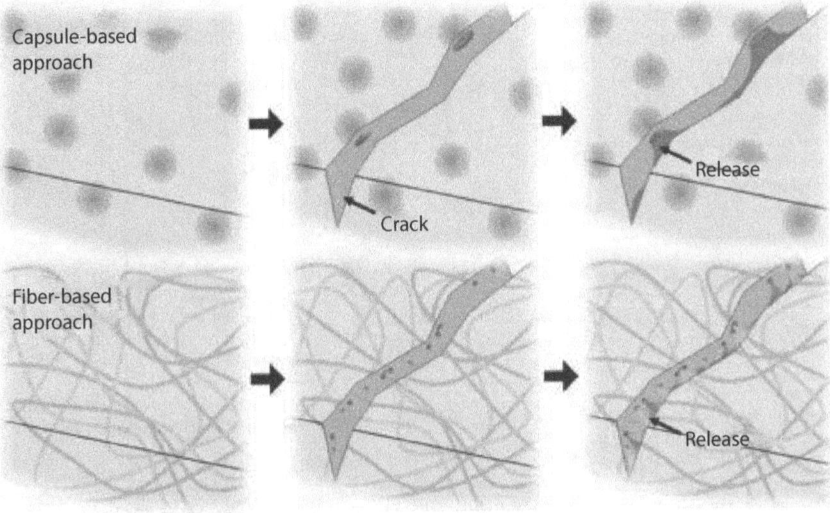

Figure 15.9 Capsule and fiber based approaches for self-healing [26].

Then the size of the capsule used to carry the resin and hardener was reduced to microscale and these hollow tubes were transported to damage locations for self-healing [32]. When the resin and hardener mix within the damage site, the repairing phenomenon happens through the thermoset network formation. If an incandescent dye was used for damage detection, then a visual image of the crack and its location could easily be obtained [33, 34]. It was later found that olefin-based monomer dicyclopentadiene (DCPD) from ring-opening metathesis polymerization (ROMP) under the catalyst action of Grubbs' first generation catalyst serves as a better chemical reaction for healing the crack damage that occurred in epoxy matrix [27, 35, 36]. Such system with Grub's catalyst encapsulated in capsules along with DCPD as a primary material are fed into epoxy based composites for a better self-healing of corrosion problems. Figure 15.11 depicts the reaction of DCPD, optical micrograph of the specimen and the mechanism of self-healing.

When damage occurs in the epoxy based composites the microcapsules respond to the mechanical stimulus by exterior wall rupturing. After rupturing, the crack on the epoxy based composite was filled by the self-healing material as it oozes out through capillary action. This leads to the ROMP reaction as the DCPD encounters the incorporated catalyst directly in the damage site. From the experiments it could be inferred that the strength of epoxy composite after healing had attained 75 to 90% of its original value

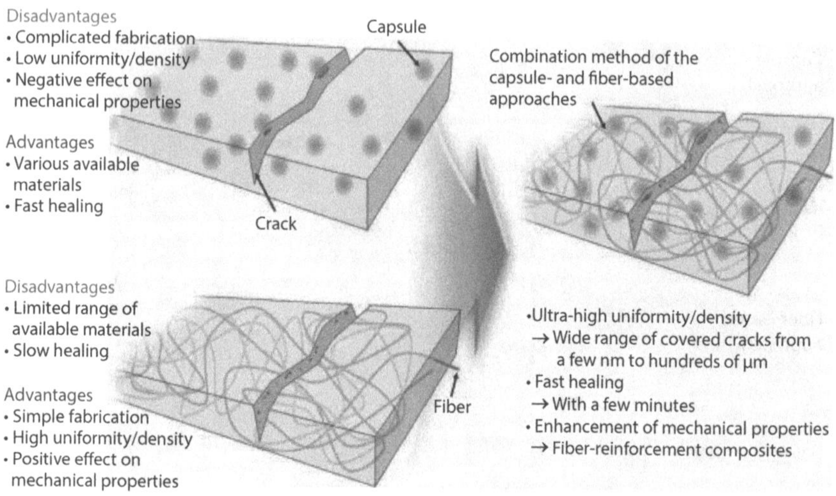

Figure 15.10 Schematic of self-healing methods for corrosion protection [26].

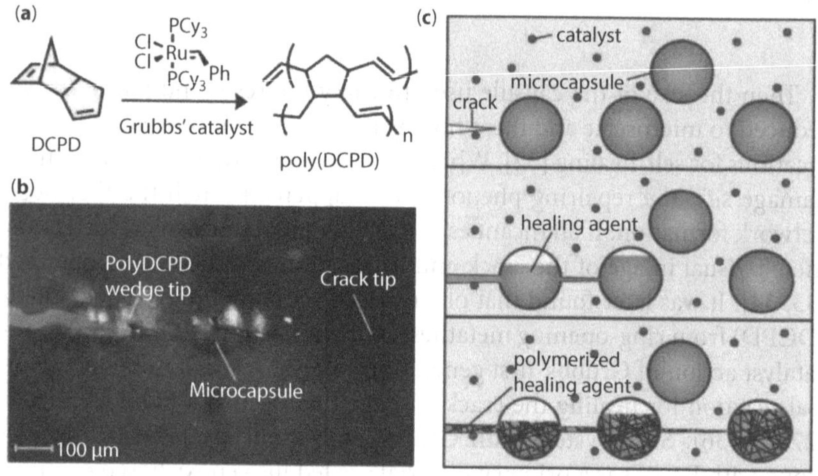

Figure 15.11 (a) Reaction system of the monomer DCPD enduring ring-opening metathesis polymerization to form poly (DCPD) [27], (b) Optical micrograph of an *in situ* fatigue specimen showing poly (DCPD) [28], (c) Mechanism of autonomic self-healing system [27, 35].

[27, 37–39]. Though the strength of the composites almost recovers within an hour, complete recovery takes upto 12 to 14 h [28, 39]. When compared with exothermic DCPD reaction at low temperatures of 78 °C which had lower time for polymerization reaction to take place, endothermic DCPD has slower polymerization kinetics [40]. Figure 15.12 depicts the schematic of hybrid silica–polyurea gel in self-healing microcapsules. Factors affecting the efficiency of healing in capsule based self-healing includes solidity of monomer without catalyst, chemical compatibility of matrix and catalyst, adhesion between matrix and a different polymer and reaction kinetics of polymerization [41].

As DCPD has better strength when compared with vinyl based monomers, it is mostly preferred as compatible monomer for all ROMP reactions. As stated above, poly DCPD formation in the damage site remains segregated with the epoxy matrix due to the lack of compatibility with the polar group of epoxy matrix. Thus to improve the intermolecular interaction of new polymer with the epoxy matrix, additional healing agents with polar groups were deliberated [41]. A catalyst is said to function appropriately when it reacts and dissolves within the crack site in a swift pace immediately when the monomer got released from the microcapsule. In recent times, at the time of microencapsulation the catalyst was separately fortified in wax microspheres which in turn enhance the new polymer adhesion with the epoxy matrix [43, 44]. Figure 15.13 illustrates the corrosion self-healing with various sized capsules containing the healing agents.

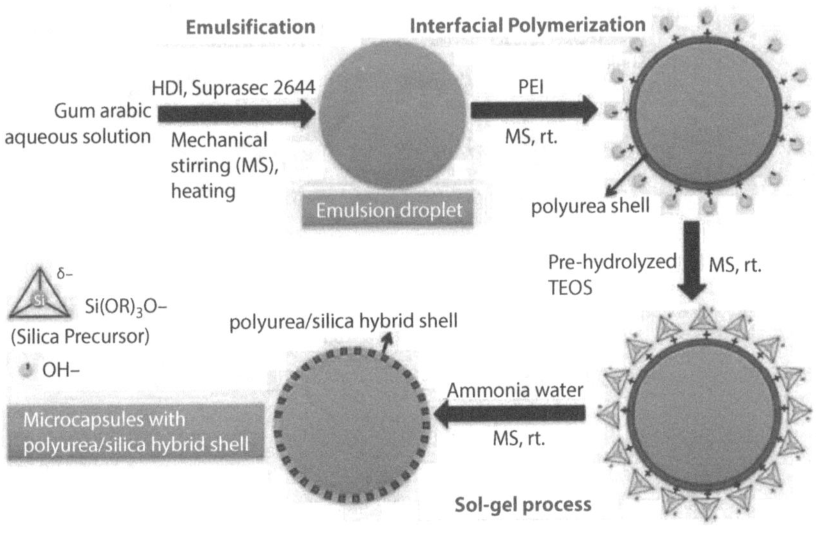

Figure 15.12 Schematic of process for fabricating polyurea/silica hybrid self-healing microcapsules [42].

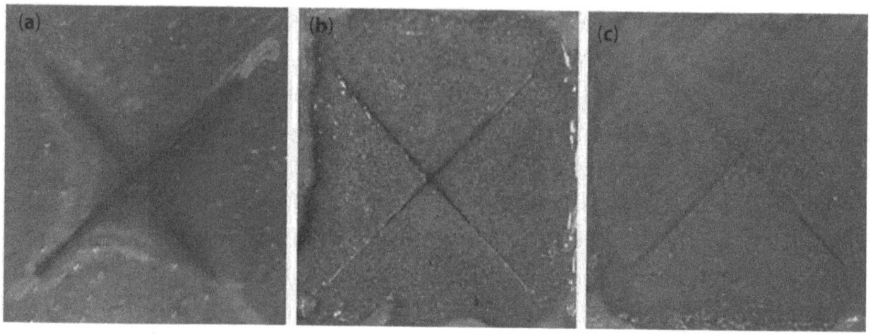

Figure 15.13 Corrosion healing of samples (a) without capsules and (b) with micrometer- and (c) nanometer-scale capsules [45].

15.5.2 Fiber-Based Self-Healing Approaches for Corrosion Protection

Few researchers employed a distinct method for processing self-healing materials inculcated within natural fiber reinforced composites. They mixed a resin monomer with capsule-less micro sized droplets of healing agent and siloxane based curing liquid mixed homogenously with the natural fibers processed by emulsion-electrospinning [45, 46]. Initially the

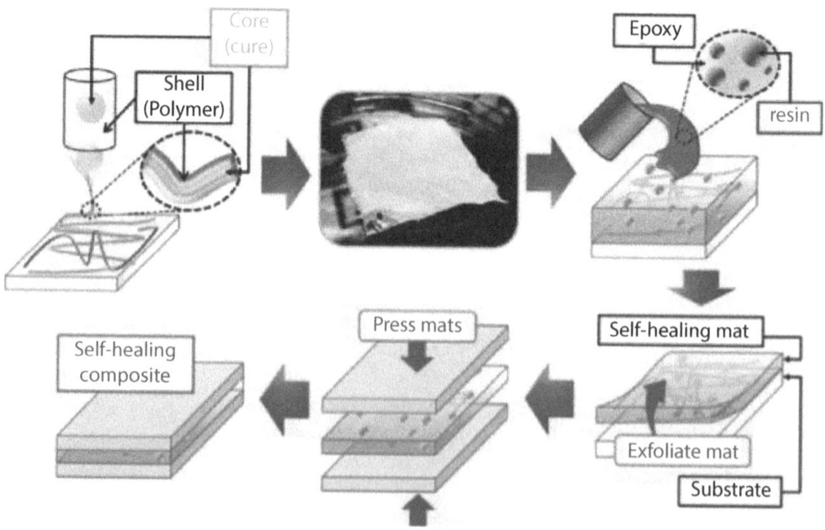

Figure 15.14 Schematic representation of fabrication of core–shell self-healing composites [46].

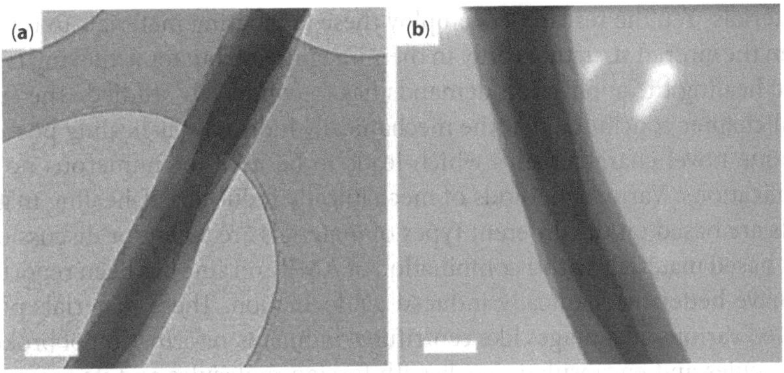

Figure 15.15 TEM images of (a) resin-infused core–shell NFs and (b) curing-liquid-infused core-shell NFs [47].

core and shell made up of curing liquid and poly-acrylonitride were mixed by electrospinning followed by the dispersion of epoxy matrix and resin upon the natural fibers that resulted in capsule-less micro-droplets formation in the epoxy matrix. When the natural fibers were electrospun for a time period of 5 min, the transparency of the natural fibers were as high as 93% which was considered to be a desirable magnitude. This formed an assorted mixture of core–shell natural fibers and micro-droplets without capsule as shown in Figure 15.14.

Figure 15.15 shows the transmission electron microscopic images of core–shell healing agents with resin infused natural fibers and curing liquid infused. Images illustrated that the siloxane resin along with the curing agent were encompassed within the core and were concealed by solidified poly-acrylo-nitride shells. Though the quantity of natural fibers used for the production of self-healable materials was very meager, their anti-corrosion properties, when employed, stood above the other materials used for the same purpose. This was validated by performing corrosion tests on the materials incorporated with such materials.

15.6 Conclusion

Interest towards any self-healing material with any stimulus induced has grown enormously in many researchers which could be seen from the number of publications made during each year. Researchers are much indulgent not only towards the modes of failure of a composite material but also towards the technique of inducing repair and choice of self-healing materials. Yet, the researchers employ these fascinating materials to repair even the aircraft structures only in their imagination but for achieving realistic healing the application demands has to completely studied. The current chapter concluded that the mechanically induced self-healing possess unique novel characteristics which leads to be used for numerous novel applications. Various methods of mechanically induced self-healing materials are based on the different types of materials are taken for discussion. Gel based materials with a combination of AMP and zinc has been reported to have better mechanically induced gel formation. These materials possessed various advantages like centrifugal inducing, reformation of broken molecules and encapsulating gels with foreign molecules and its removal. Due to these advantages gel based self-healing materials can easily be triggered by a simple centrifugal action. In crystal based self-healing materials in order to induce a thermosalient transition for ploymorphing the crystal, mechanical stimulus has to be given to make the crystal to move under mechanical motion. Polymer materials can also be induced by mechanical stimulus and the polymers' response will be through their molecular misalignments. Mechanically induced self-healing materials can also be employed in corrosion healing based on capsule and fiber which are quite interesting that it reveals complete healing is achieved in both the cases. With all the above facts, this chapter provides an inclusive consolidation of various works carried out in the field of mechanically induced self-healing materials so far and also paves way to orient the researchers' in future possible direction in this field.

References

1. White, S.R., Sottos, N.R., Geubelle, P.H., Moore, J.S., Kessler, M.R., Sriram, S.R. *et al.*, Autonomic healing of polymer composites. *Nature*, 409, 794–797, 2001.
2. Billiet, S., Hillewaere, X.K.D., Teixeira, R.F.A., Du Prez, F.E., Chemistry of crosslinking processes for self-healing polymers. *Macromol. Rapid Commun.*, 34, 290–309, 2013.
3. Olugebefola, S.C., Hamilton, A.R., Fairfield, D.J., Sottos, N.R., White, S.R., Structural reinforcementof microvascular networks using electrostatic layer-by-layer assembly with halloysite nanotubes. *Soft Matter*, 10, 544–548, 2014.
4. Brown, E.N., Kessler, M.R., Sottos, N.R., White, S.R., *In situ* poly (urea-formaldehyde) microencapsulation of dicyclopentadiene. *J. Microencapsul.*, 20, 719–730, 2003.

5. Blaiszik, B.J., Caruso, M.M., McIlroy, D.A., Moore, J.S., White, S.R., Sottos, N.R., Microcapsulesfilled with reactive solutions for self-healing materials. *Polymer*, 50, 990–997, 2009.
6. Keller, M.W., White, S.R., Sottos, N.R., A self-healing poly (dimethyl siloxane) elastomer. *Adv. Funct. Mater.*, 17, 2399–2404, 2007.
7. Yuan, Y.C., Rong, M.Z., Zhang, M.Q., Preparation and characterization of microencapsulated polythiol. *Polymer*, 49, 2531–2541, 2008.
8. Liu, X., Sheng, X., Lee, J.K., Kessler, M.R., Synthesis and Characterization of Melamine-Urea Formaldehyde Microcapsules Containing ENB-Based Self-Healing Agents. *Macromol. Mater. Eng.*, 294, 389–395, 2009.
9. Cho, S.H., Andersson, H.M., White, S.R., Sottos, N.R., Braun, P.V., Polydimethylsiloxane-based self-healing materials. *Adv. Mater.*, 18, 997–1000, 2006.
10. Therriault, D., Shepherd, R.F., White, S.R., Lewis, J.A., Fugitive inks for direct-write assembly of three-dimensional microvascular networks. *Adv. Mater.*, 17, 395–399, 2005.
11. Mauldin, T.C. and Kessler, M.R., Self-healing polymers and composites. *Int. Mater. Rev.*, 55, 317–346, 2010.
12. Skorb, E.V. and Andreeva, D.V., Layer-by-Layer approaches for formation of smart self-healing materials. *Polym. Chem.*, 4, 4834–4845, 2013.
13. Zhu, M., Rong, M.Z., Zhang, M.Q., Self-healing polymeric materials towards non-structural recovery of functional properties. *Polym. Int.*, 63, 1741–1749, 2014.
14. Fayyad, E.M., Almaadeed, M.A., Jones, A., Abdullah, A.M., Evaluation techniques for the corrosion resistance of self-healing coatings. *Int. J. Electrochem. Sci.*, 9, 4989–5011, 2014.
15. Yu, X., Chen, L., Zhang, M., Yi, T., Low-molecular-mass gels responding to ultrasound and mechanical stress: Towards self-healing materials. *Chem. Soc. Rev.*, 43, 5346–5371, 2014.
16. Wei, Z., Yang, J.H., Zhou, J., Xu, F., Zrínyi, M., Dussault, P.H. *et al.*, Self-healing gels based on constitutional dynamic chemistry and their potential applications. *Chem. Soc. Rev.*, 43, 8114–31, 2014.
17. Sigel, R.K.O. and Sigel, H., A stability concept for metal ion coordination to single-stranded nucleic acids and affinities of individual sites. *Acc. Chem. Res.*, 43, 974–984, 2010.
18. Roy, S., Baral, A., Banerjee, A., An amino-acid-based self-healing hydrogel: Modulation of the self-healing properties by incorporating carbon-based nano-materials. *Chem.—Eur. J.*, 19, 44, 14950–14957, 2013.
19. Huang, P.-J.J. and Liu, J., Rational evolution of Cd^{2+}-specific DNAzymes with phosphorothioate modified cleavage junction and Cd^{2+} sensing. *Nucleic Acids Res.*, 43, 6125–6133, 2015.
20. Liang, H., Zhang, Z., Yuan, Q., Liu, J., Self-healing metal-coordinated hydrogels using nucleotide ligands. *Chem. Commun.*, 51, 82, 15196–9, 2015.

21. Vatankhah-Varnoosfaderani, M., Hashmi, S., Nejad, A.G., Stadler, F.J., Rapid self-healing and triple stimuli responsiveness of a supramolecular polymer gel based on boron–catechol interactions in a novel water-soluble mussel-inspired copolymer. *Polym. Chem.*, 5, 2, 512–523, 2014.
22. Karothu, D.P., Weston, J., Desta, I.T., Naumov, P., Shape-memory and self-healing effects in mechanosalient molecular crystals. *J. Am. Chem. Soc.*, 138, 40, 13298–306, 2016.
23. Imato, K., Irie, A., Kosuge, T., Ohishi, T., Nishihara, M., Takahara, A., Otsuka, H., Mechanophores with a reversible radical system and freezing-induced mechanochemistry in polymer solutions and gels. *Angew. Chem. Int. Ed.*, 54, 6168–6172, 2015.
24. Capadona, J.R., Shanmuganathan, K., Trittschuh, S., Seidel, S., Rowan, S.J., Weder, C., Polymer nanocomposites with nanowhiskers isolated from microcrystalline cellulose. *Biomacromolecules*, 10, 712–716, 2009.
25. Imato, K., Natterodt, J.C., Sapkota, J., Goseki, R., Weder, C., Takahara, A., Otsuka, H., Dynamic covalent diarylbibenzofuranone-modified nanocellulose: Mechanochromic behaviour and application in self-healing polymer composites. *Polym. Chem.*, 8, 13, 2115–22, 2017.
26. An, S., Lee, M.W., Yarin, A.L., Yoon, S.S., A review on corrosion-protective extrinsic self-healing: Comparison of microcapsule-based systems and those based on core–shell vascular networks. *Chem. Eng. J.*, 344, 206–20, 2018.
27. White, S.R., Sottos, N.R., Geubelle, P.H., Moore, J.S., Kessler, M.R., Sriram, S.R., Brown, E.N., Viswanathan, S., Autonomic healing of polymer composites. *Nature*, 409, 794–797, 2001.
28. Brown, E.N., White, S.R., Sottos, N.R., Retardation and repair of fatigue cracks in a microcapsule toughened epoxy composite—Part II: *In situ* self-healing. *Compos. Sci. Technol.*, 65, 2474–2480, 2005.
29. Wool, R.P., *Adhesion and Absorption of Polymers, Part A*, vol. 12A, L.-H. Lee, (Ed.), Plenum Publishing Corp, New York, 197, 1980.
30. Wool, R.P., Self-healing materials: A review. *Soft Matter*, 4, 400–418, 2008.
31. Dry, C., Procedures developed for self-repair of polymer matrix composite materials. *Compos. Struct.*, 35, 263–269, 1996.
32. Bleay, S.M., Loader, C.B., Hawyes, V.J., Humberstone, L., Curtis, P.T., A smart repair system for polymer matrix composites. *Compos. Part A*, 32, 1767–1776, 2001.
33. Pang, J.W.C. and Bond, I.P., Bleeding composites—damage detection and self-repair using a biomimetic approach. *Compos. Part A*, 36, 183–188, 2005.
34. Pang, J.W.C. and Bond, I.P., A hollow fibre reinforced polymer composite encompassing self-healing and enhanced damage visibility. *Compos. Sci. Technol.*, 65, 1791–1799, 2005.
35. Trask, R.S. and Bond, I.P., Biomimetic self-healing of advanced composite structures using hollow glass fibres. *Smart Mater. Struct.*, 15, 704–710, 2006.

36. Trask, R.S., Williams, G.J., Bond, I.P., Bioinspired self-healing of advanced composite structures using hollow glass fibres. *J. R. Soc. Interface*, 4, 363, 2007.
37. Brown, E.N., Sottos, N.R., White, S.R., Fracture testing of a self-healing polymer composite. *Exp. Mech.*, 42, 372–379, 2002.
38. Kessler, M.R., Sottos, N.R., White, S.R., Self-healing structural composite materials. *Compos. Part A*, 34, 743–753, 2003.
39. Kessler, M.R. and White, S.R.J., Cure kinetics of the ring-opening metathesis polymerization of dicyclopentadiene. *Polym. Sci., Part A: Polym. Chem.*, 40, 2373–2383, 2002.
40. Mauldin, T.C., Rule, J.D., Sottos, N.R., White, S.R., Moore, J.S., Self-healing kinetics and the stereoisomers of dicyclopentadiene. *J. R. Soc. Interface*, 4, 389, https://doi.org/10.1098/rsif.2006.0200, 2007.
41. Wilson, G.O., Caruso, M.M., Reimer, N.T., White, S.R., Sottos, N.R., Moore, J.S., Evaluation of ruthenium catalysts for ring-opening metathesis polymerization-based self-healing applications. *Chem. Mater.*, 20, 3288–3297, 2008.
42. Wu, G., An, J., Sun, D., Tang, X., Xiang, Y., Yang, J., Robust microcapsules with polyurea/silica hybrid shell for one-part self-healing anticorrosion coatings. *J. Mater. Chem. A*, 2, 11614–11620, 2014.
43. Rule, J.D., Brown, E.N., Sottos, N.R., White, S.R., Moore, J.S., Wax-protected catalyst microspheres for efficient self-healing materials. *Adv. Mater.*, 17, 205–208, 2005.
44. Caruso, M.M., Davis, D.A., Shen, Q., Odom, S.A., Sottos, N.R., White, S.R., Moore, J.S., Mechanically-induced chemical changes in polymeric materials. *Chem. Rev.*, 109, 11, 5755–98, 2009.
45. Boura, S.H., Peikari, M., Ashrafi, A., Samadzadeh, M., Self-healing ability and adhesion strength of capsule embedded coatings—Micro and nano sized capsules containing linseed oil. *Prog. Org. Coat.*, 75, 292–300, 2012.
46. Lee, M.W., An, S., Lee, C., Liou, M., Yarin, A.L., Yoon, S.S., Hybrid self-healing matrix using core–shell nanofibers and capsule less microdroplets. *ACS Appl. Mater. Interfaces*, 6, 10461–10468, 2014.
47. An, S., Liou, M., Song, K.Y., Jo, H.S., Lee, M.W., Al-Deyab, S.S., Yarin, A.L., Yoon, S.S., Highly flexible transparent self-healing composite based on electrospun core–shell nanofibers produced by coaxial electrospinning for anti-corrosion and electrical insulation. *Nanoscale*, 7, 17778–17785, 2015.

16

Self-Healing Materials in Robotics

Sunny Kumar

Department of Petroleum Engineering, Roorkee College of Engineering, Roorkee, Uttrakhand, India

Abstract

There are many self-healing materials reported such as polymer, resins and liquid metals. These materials are very important for soft robotics. While performing their tasks, robots can damage their parts. So if they are prepared from self-healing materials, they will be utilized for longer-term (time span will be improved) for applications. Here, we have described the different self-healing materials, their synthesis methods and applications in robotics. We have described the chemistry of the self-healing materials and how the self-healing materials work. There are different applications in robotics such as actuator, skin, electronics and structure. This chapter also explains the approaches of the self-healing as capsules, vascular and intrinsic self-healing system. The potential applications with damage resilience mechanism are also explained. Most importantly, self-healing materials will save time and money in soft robotics field.

***Keywords*:** Self-healing, soft robotics, bioinspired, mechanism, applications, approaches

16.1 Introduction

For next generation soft robots, advanced new synthetic materials [1] have reduced in weight and size for next generation robotics to perform complex tasks [2]. Manufactured machineries and electronics are smaller, lighter and less stiff aid them to better fit into the environment and inter-mingle with humans [3]. Most healing materials are ceramics, metal, polymers, ceramics and their composites. Generally, materials damage occurs

Email: chemicalsunnyraj@gmail.com

Inamuddin, Mohd Imran Ahamed, Rajender Boddula and Tariq Altalhi (eds.) Self-Healing Smart Materials and Allied Applications, (405–414) © 2021 Scrivener Publishing LLC

by electrical, mechanical, thermal, ballistic and simultaneously material should have the capability to heal and regain the materials' original properties [4]. During material application, there may consequences of damage, degradation, and failure [5, 6]. In robotics application, the robots can be breaking at any instant in time. They break, like, literally all the time. The breakage can happen in both hardware and software sometime, while a person is working with a robot on an important task. Suddenly, when any period of time breakage happens, so there will be no other way to complete the task but to invest on a new machine and repair the breakage for a long time. Resolving these physical robot breaking problems can be through the usage of structural materials which can heal themselves automatically many times [7]. A thermo-reversible polymer is covalent network of Diels–Alder polymers. This polymer were commonly applied to synthesis for three applications such as soft pneumatic, self-healing actuators (artificial muscles and soft gripper [8]). During action, soft pneumatic actuators involve the leaks and perforations because of extreme pressures or wear [9, 10].

16.2 Chemistry of Self-Healing (S-H) Materials

There are two ways to define the Self-Healing mechanisms which are autonomic and nonautonomic [11] as illustrated in Figure 16.1. When there is no stimulus required for operation that is called autonomic system. This system is similar to the biological system where healing compound transports to the damage area.

When there is requirement to heal the damaged area by external energy such as heat, light mechanical or chemical that is called non-autonomic system. So, the chemistry is bonding of covalent and non-covalent in between the material during healing.

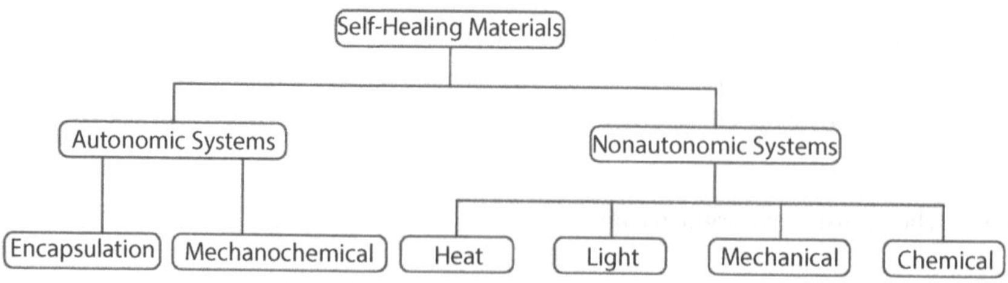

Figure 16.1 Organization of self-healing materials based on their chemistry [11].

16.3 Working of Self-Healing (S-H) Material

Previously we have defined the two types of healing mechanisms. We consider the nonautonomous system where damage is repaired by the external energy. For this case, during the handling of robot, they can be damaged. The damaged area has covered with dirt in surrounding environment which is initially cleaned and applied with the external treatment. So that damage bonding shows the high strength. In another autonomous system, the damage is recovered by the room temperature not requiring the external energy. So in this system, soft robotics damage does not require extra effort for heating. But it takes a long time to complete the bonding between the damaged part of materials. The required time may be a few days like 3 to 15 days depending upon the material and recovery percentage of the area. The size of damage or cut is also a parameter which affects the time of healing immediately or for a long time. Mostly this system is used in the soft pneumatics hands.

16.4 Application of Self-Healing Robots

Mechanical properties require to customization, so materials are mixed together and the structural part will become flexible. In the robotics field, advanced sensors require flexibility and capability to sense the damage. There has been enormously progress occurring in previous years in the fields of electronics, structure [12], actuators, and skins.

16.4.1 Self-Healing Electronics for Soft Robotics

For soft robotics, systems require large amount of electronics [13] which are stretchable and flexible [14]. Presently, there are few soft materials in electronics which are liquid metal, polyelectrolytic gel and elastomers. During the application of the soft robotics, there are issues on the damages which are tearing and puncture. So, we have required the specific type of construction for flexible circuit in the presence of loading environments. Generally, liquid metal is a relevant material during rapturing the circuit and the rejoining of the broken part. In these systems, there is no requirement for external heat or mechanical effort. This liquid metal-based material shows the effective signal transfer through the circuit during the severe rapture or damage. The current passing through the circuit does not show the losses during soft robotics operation after reconnecting the damage. Instead of the circuit area, the remaining area was insulated because of fabrication by the elastomers. There is no loss observed during the stretching of the circuit in soft robotics. The circuit line can be created in few microns for self-healing electronics. This self-healing electronics is also a potential application in the soft biological organisms.

16.4.2 Self-Healing Electrostatic Actuators

Actuators [15] are commonly used applications for soft robotics as transducers for prevention of the damage and flexibility of robotics that the soft actuator requires. Soft actuators are a combination of hydraulic system and electronics parts which provides the hydraulic and electrostatic force. This combination is called hydraulically amplified self-healing electrostatic actuators [16]. This system shows the different mode of actuation. There are so many materials available for actuating which are capable for using in multiple time self-heal during damage. The common fabrication techniques for actuators include Poly Multi-User MEMS Processes (MUMP), Silicon-On-Insulator multiuser mems process, and Metal MUMPs [17, 18]. These actuating materials are used for handling the objects and sensing muscles for soft robotics.

16.4.3 Self-Healing Skin for Robotics

For fabricating of the soft robotics, the skin is required to cover the body of robots that is generally soft and flexible as our human skin. The flexible skin provides the required motion and displacements for complete the job [19]. The robotics skin is required to perform sensing and actuation in given environments [20]. If we use the skin in different pattern than the behavior motion of the skin will be different. Using this soft robotic skin helps to perform different multifaceted tasks and motions. There are some skin materials which are self-healing properties that have advantages for damaging during performance of task. Some common skin materials are used for soft robotics such as poly(N,N-dimethylacrylamide)-poly(vinyl alcohol)/reduced graphene oxide (PDMAA-PVA/rGO) hybrid film [21] and single wall nanotube (SWNT) coated on PDMS and electrochromic pigments based film [22].

16.5 Approaches to Self-Healing

In the previous section, we have studied about the self-healing materials that can be used for different applications for soft robotics. But there are different methods for healing the damages which are capsule based, vascular and intrinsic methods. These methods are different from each other during the damage part.

When a system is damaged during an operation and it heals by the stored material in capsules, this healing phenomenon is called capsule based self-healing approach. Some common examples are given Table 16.1. In the healing mechanism, a capsule released the healing material and reaction occurs in the damage area.

Instead of previous method, when the damage part of the soft robot is healed by flowing the stored healing agent from the hollow channels or capillaries. This method is known as vascular materials-based approach. Some common examples are given in Table 16.2. After one-time damage repairing, the vasculars are empty which are further filled with healing materials.

Table 16.1 Capsule self-healing systems.

Capsule-based self-healing systems			
Matrix materials	**Healing agent**	**Catalyst/ polymerization**	**References**
Epoxy	DCPD	Grubbs' catalyst	Brown *et al.* [23], Rule *et al.* [24]
Epoxy composite	Epoxy resin	Latent CuBr2(2-MeIm)4 catalyst	Yin *et al.* [25]
PDMS	Silanol-terminated PDMS resin/PDES	DBTL	Cho *et al.* [26]
Epoxy-amine paint	Epoxy-amine adduct	Environmental stimulus	Sauvant *et al.* [27]

Table 16.2 Vascular self-healing systems.

Vascular self-healing systems			
Materials	**Vascular connectivity**	**Healing chemistry**	**References**
Epoxy/glass pipettes	1D	Cyanoacrylate	Dry *et al.* [28]
Epoxy/plain-weave S2 glass fabric/glass pipettes	1D	Two-part epoxy	Motuku *et al.* [29]
Brittle epoxy coating/ ductile epoxy substrate	3D	DCPD-Grubbs' catalyst	Toohey *et al.* [30]

Table 16.3 Intrinsic self-healing systems.

Intrinsic self-healing systems		
Materials	**Healing mechanism**	**References**
Furan-maleimide derivative (2M4F)	DA/rDA reaction	Chen *et al.* [31]
Epoxy/poly(bisphenol-A-co-epichlorohydrin) thermoplastic phase	Meltable secondary additive	Hayes *et al.* [32]
Polystyrene	Molecular diffusion	McGarel *et al.* [33]
Poly(ethylene-co-methacrylic acid) copolymer	Noncovalent bonding: ionomeric	Kalista *et al.* [34]

In another approach, sometimes a damage or tear occurs during the operation then the damage is healed by the external energy which creates the bond such as a hydrogen or ionic interaction between the damaged area. In this case reverse polymerization reaction occurs. So, this phenomenon is called intrinsic material healing approach. Some examples are presented in Table 16.3. This approach can be used multiple times in healing the damaged area.

16.6 Material Application and Damage Resilience Mechanism

Self-healing materials are utilized as actuators, structures, and electronics (sensing) and are commonly used active materials in silicon oil, liquid metal wax, elastomers and hydrogels for various applications. The properties of such material which is very important for self-healing include area strain, conductivity, compressive strength curvature change. The material systems are categorized as physically and chemically self-healing. A brief study shown in Table 16.4 presents the system, application, active materials, stimulus, recovery time and properties of these materials [35].

16.7 Conclusion

There is tremendous progress in the improvement of the self-healing material system in soft robotics that helps the mechanically vigorous structural,

Table 16.4 Literature study of damage and recovery properties of materials in soft robotics.

Damage resilience system	Application	Active material	Self-healing stimulus	Recovery time scale	Recovered material property	Reference
Physical self-healing (fluidic reflow)	Actuator	Silicon oil	Voltage reduction	0–5 s	Area strain	Hunt *et al.* [36]
	Electronics	Liquid metal	thermal	0–10 s	conductivity	Lie *et al.* [37]
	Structures	wax	thermal	10–20 min	Compressive strength	Cheng *et al.* [38]
Chemical Self-Healing (Solid bonding)	Actuator	S-H elastomer	UV	30 s	Curvature change	Wallin *et al.* [39]
	Electronics	S-H elastomer	thermal pressure	15–30 s	conductivity	Tee *et al.* [40]
	Electronics	S-H hydrogel	thermal	30–60 s	conductivity	Darabi *et al.* [41]
	Actuator	S-H elastomer	Thermal	72 h	Area strain	Li *et al.* [42]

electronics and robotic portions. The smart materials include the ability to detecting damage, and healing or correcting materials properties and geometry to avoid damage failure. The healing response can be controlled by the repair time, the repaired system properties and percentage of material heal. The repeatability (fatigue) and reversibility (recovery) are essential to be explored more in soft robotic applications. At some point, damage occurs many times during the performing the task by robotics which decrease the material properties. So the challenge is to improve the material properties of damage area such as strength and hardness of the section. The improvement for these areas can utilize stiffness and damage tolerance of various soft robotics for health care and human machine interface application. The smart self-healing material in soft robotics can reduce the response time and money.

References

1. Yang, G.-Z., Fischer, P., Nelson, B., New materials for next-generation robots. *Sci. Robot.*, 2, 1–2, 2017.
2. Laschi, C., Mazzolai, B., Cianchetti, M., Soft robotics: Technologies and systems pushing the boundaries of robot abilities. *Sci. Robot.*, 1, 1–11, 2016.
3. Bartlett, M.D., Dickey, M.D., Majidi, C., Self-healing materials for soft-matter machines and electronics. *NPG Asia Mater.*, 11, 1–4, 2019.
4. Wool, R.P., Self-healing materials: A review. *Soft Matter*, 4, 400–418, 2008.
5. Blaiszik, B.J. *et al.*, Self-Healing Polymers and Composites. *Annu. Rev. Mater. Res.*, 40, 179–211, 2010.
6. Martin, P., Wound Healing—Aiming for Perfect Skin Regeneration. *Science*, 276, 75–81, 1997.
7. Shepherd, R.F., Stokes, A.A., Nunes, R.M.D., Whitesides, G.M., Soft Machines That are Resistant to Puncture and That Self Seal. *Adv. Mater.*, 25, 6709–6713, 2013.
8. Roels, E., Terryn, S., Brancart, J., Van Assche, G., Vanderborght, B., A Multi-Material Self-Healing Soft Gripper, in: *2019 2nd IEEE International Conference on Soft Robotics (RoboSoft)*, pp. 316–321, 2019.
9. Terryn, S., Brancart, J., Lefeber, D., Assche, G.V., Vanderborght, B., Self-healing soft pneumatic robots. *Sci. Robot.*, 2, 1–12, 2017.
10. Terryn, S., Brancart, J., Lefeber, D., Van Assche, G., Vanderborght, B., A Pneumatic Artificial Muscle Manufactured Out of Self-Healing Polymers That Can Repair Macroscopic Damages. *IEEE Robot. Autom. Lett.*, 3, 16–21, 2018.
11. Williams, K.A., Dreyer, D.R., Bielawski, C.W., The Underlying Chemistry of Self-Healing Materials. *MRS Bull.*, 33, 759–765, 2008.

12. Nakashima, S. *et al.*, Resistance-based self-sensing system of active self-melting bolt towards autonomous healing structure, in: *2018 IEEE International Conference on Soft Robotics (RoboSoft)*, pp. 88–93, 2018.
13. Cao, Y. *et al.*, Self-healing electronic skins for aquatic environments. *Nat. Electron.*, 2, 75–82, 2019.
14. Markvicka, E.J., Bartlett, M.D., Huang, X., Majidi, C., An autonomously electrically self-healing liquid metal–elastomer composite for robust soft-matter robotics and electronics. *Nat. Mater.*, 17, 618–624, 2018.
15. Miriyev, A., Trujillo, C., Caires, G., Lipson, H., Rejuvenation of soft material–actuator. *MRS Commun.*, 8, 556–561, 2018.
16. Acome, E. *et al.*, Hydraulically amplified self-healing electrostatic actuators with muscle-like performance. *Science*, 359, 61–65, 2018.
17. Cauchi, M., Grech, I., Mallia, B., Mollicone, P., Sammut, N., The Effects of Structure Thickness, Air Gap Thickness and Silicon Type on the Performance of a Horizontal Electrothermal MEMS Microgripper. *Actuators*, 7, 38, 2018.
18. Iamoni, S. and Somà, A., Design of an electro-thermally actuated cell microgripper. *Microsyst. Technol.*, 20, 869–877, 2014.
19. Booth, J.W. *et al.*, OmniSkins: Robotic skins that turn inanimate objects into multifunctional robots. *Sci. Robot.*, 3, 1–9, 2018.
20. Yu, K., Xin, A., Du, H., Li, Y., Wang, Q., Additive manufacturing of self-healing elastomers. *NPG Asia Mater.*, 11, 1–11, 2019.
21. Hou, C. *et al.*, A strong and stretchable self-healing film with self-activated pressure sensitivity for potential artificial skin applications. *Sci. Rep.*, 3, 1–6, 2013.
22. Chou, H.-H. *et al.*, A chameleon-inspired stretchable electronic skin with interactive colour changing controlled by tactile sensing. *Nat. Commun.*, 6, 1–10, 2015.
23. Brown, E.N., Kessler, M.R., Sottos, N.R., White, S.R., In situ poly(urea-formaldehyde) microencapsulation of dicyclopentadiene. *J. Microencapsul.*, 20, 719–730, 2003.
24. Rule, J.D., Brown, E.N., Sottos, N.R., White, S.R., Moore, J.S., Wax-Protected Catalyst Microspheres for Efficient Self-Healing Materials. *Adv. Mater.*, 17, 205–208, 2005.
25. Yin, T., Zhou, L., Rong, M.Z., Zhang, M.Q., Self-healing woven glass fabric/epoxy composites with the healant consisting of micro-encapsulated epoxy and latent curing agent. *Smart Mater. Struct.*, 17, 015019, 2007.
26. Cho, S.H., White, S.R., Braun, P.V., Self-Healing Polymers: Self-Healing Polymer Coatings (Adv. Mater. 6/2009). *Adv. Mater.*, 21, 645–649, 2009.
27. Sauvant-Moynot, V., Gonzalez, S., Kittel, J., Self-healing coatings: An alternative route for anticorrosion protection. *Prog. Org. Coat.*, 63, 307–315, 2008.
28. Dry, C.M. and Sottos, N.R., Passive smart self-repair in polymer matrix composite materials, in: *Smart Structures and Materials 1993: Smart*

Materials, vol. 1916, pp. 438–444, International Society for Optics and Photonics, Bellingham, Washington USA, 1993.
29. Motuku, M., Vaidya, U.K., Janowski, G.M., Parametric studies on self-repairing approaches for resin infused composites subjected to low velocity impact. *Smart Mater. Struct.*, 8, 623–638, 1999.
30. Toohey, K.S., Sottos, N.R., Lewis, J.A., Moore, J.S., White, S.R., Self-healing materials with microvascular networks. *Nat. Mater.*, 6, 581–585, 2007.
31. Chen, X. *et al.*, A Thermally Re-mendable Cross-Linked Polymeric Material. *Science*, 295, 1698–1702, 2002.
32. Hayes, S.A., Zhang, W., Branthwaite, M., Jones, F.R., Self-healing of damage in fibre-reinforced polymer-matrix composites. *J. R. Soc. Interface*, 4, 381–387, 2007.
33. McGarel, O.J. and Wool, R.P., Craze growth and healing in polystyrene. *J. Polym. Sci. Part B Polym. Phys.*, 25, 2541–2560, 1987.
34. S.J.K., Jr., Ward, T.C., Oyetunji, Z., Self-Healing of Poly(Ethylene-co-Methacrylic Acid) Copolymers Following Projectile Puncture. *Mech. Adv. Mater. Struct.*, 14, 391–397, 2007.
35. Bilodeau, R.A. and Kramer, R.K., Self-Healing and Damage Resilience for Soft Robotics: A Review. *Front. Robot. AI*, 4, 1–13, 2017.
36. Hunt, S., McKay, T.G., Anderson, I.A., A self-healing dielectric elastomer actuator. *Appl. Phys. Lett.*, 104, 113701, 2014.
37. Lei, Z., Wang, Q., Sun, S., Zhu, W., Wu, P., A Bioinspired Mineral Hydrogel as a Self-Healable, Mechanically Adaptable Ionic Skin for Highly Sensitive Pressure Sensing. *Adv. Mater.*, 29, 1700321, 2017.
38. Cheng, N.G., Gopinath, A., Wang, L., Iagnemma, K., Hosoi, A.E., Thermally Tunable, Self-Healing Composites for Soft Robotic Applications. *Macromol. Mater. Eng.*, 299, 1279–1284, 2014.
39. Wallin, T.J. *et al.*, Click chemistry stereolithography for soft robots that self-heal. *J. Mater. Chem. B*, **5**, 6249–6255, 2017.
40. Tee, B.C.-K., Wang, C., Allen, R., Bao, Z., An electrically and mechanically self-healing composite with pressure- and flexion-sensitive properties for electronic skin applications. *Nat. Nanotechnol.*, 7, 825–832, 2012.
41. Darabi, M.A. *et al.*, Skin-Inspired Multifunctional Autonomic-Intrinsic Conductive Self-Healing Hydrogels with Pressure Sensitivity, Stretchability, and 3D Printability. *Adv. Mater.*, 29, 1700533, 2017.
42. Li, C.-H. *et al.*, A highly stretchable autonomous self-healing elastomer. *Nat. Chem.*, 8, 618–624, 2016.

17

Self-Healing Materials in Aerospace Applications

M. Harikrishna Kumar[1], C. Moganapriya[1], A. Mohan Kumar[1], R. Rajasekar[1] and V. K. Gobinath[2]*

[1]*Department of Mechanical Engineering, Kongu Engineering College, Erode, India*
[2]*Department of Mechatronics Engineering, Kongu Engineering College, Erode, India*

Abstract
Fiber reinforced composites (FRCs) were used in various applications due to its advantages of high strength to weight ratio. So, FRC was found to be one of the most preferred material in aerospace applications. But, one major limitation of FRC was the formation of micro-cracks when the material is subjected to impact load. It reduces the life span of the material and creates riskiness to humans when used as engineering materials. In-order to overcome the limitation of FRC, many research works have been focused on development of self-healing composites. Self-healing composite materials are seen as a potential material in engineering applications. Healing of materials has been categorized into intrinsic and extrinsic materials based on the mechanism of healing the damaged region of materials during operating conditions. Extrinsic mechanism of healing approach is preferred in aerospace applications. This chapter discusses about the different types extrinsic self-healing materials used in aerospace application.

Keywords: Composites, polymer, self-healing materials, aerospace applications

17.1 Introduction

Composite materials are prepared by using polymers, metals and ceramics as base materials. Self-healing materials belong to a type of composite materials which belong to a term called smart materials. Self-healing

**Corresponding author*: rajasekar.cr@gmail.com

Inamuddin, Mohd Imran Ahamed, Rajender Boddula and Tariq Altalhi (eds.) Self-Healing Smart Materials and Allied Applications, (415–434) © 2021 Scrivener Publishing LLC

materials can be defined as the power to heal itself by means of recover or repair damages automatically and autonomously with or without any external intervention [1]. When self-healing materials are damaged by ballistic or thermal or mechanical or any other form, self-healing materials possess the power to regain its functionality, fully or partially [2, 3]. Very few materials have the ability of self-healing itself during its functionality. When a material got damaged during its operation, self-healing property of a material increase the life time of the product and enduring strength which enhances the economic and human safety aspects [2, 4]. This characteristic of the material has to be considered as a major factor while selecting a material for an application. Because the product has been designed to perform intended application where repairing of the material is not possible. Self-healing man made materials can regain their structural integrity in case of a failure [1]. Self-healing property in materials can be incorporated. Self-healing materials was used in various applications such as automotive, aerospace and building industries.

This chapter discusses about the self-healing materials used in aerospace applications. Over the past decades, many metal parts used in aerospace are replaced by polymer matrix composite (PMC) due to its advantages in terms of characteristics such as high specific strength and stiffness with low weight ratio, good vibration damping capacity, corrosion resistance and electromagnetic transparency [5, 6]. Carbon fiber composites is also one of the alternative material used in aerospace applications instead of metal parts. Aircraft say Boeing Dreamliner, employ fiber reinforced composites which at last decreases the fuel consumption as well as environmental pollution [7]. It also extends its advantages in terms of manufacturability, flexibility in design and part count reduction. Still, one of the major limitation of PMC which was manufactured from brittle polymers is subjected to micro-cracking due to impact and cyclic (mechanical/thermal) loads [5, 6]. De-bonding of fiber-matrix was also one of the major failure modes which routinely found in the applications of composite materials [8]. Aircraft may undergo damage due to variety of impact loads such as ground service equipment, runway debris, bird strike, hail ice prior to take off and during landing [7]. Composite materials possess a major challenge in maintenance and inspection of structural parts such as wings and fins of aircrafts [8]. Composite damages occur at sub-laminate locations which is invisible to naked eyes. These damages can be detected by non-destructive techniques such as ultrasonic, infrared thermography, X-ray tomography, and computerized vibro thermography. Due to high maintenance and inspection cost significant damages have been undetected [7, 9]. Hence, self-healing technology in composite materials seems to be promising to heal or repair fracture or resistance to damage initiation in structural components of aircraft. One effective solution to solve this issue, by usage of self-healing polymeric materials which arrests the damage and micro-crack formation at an early stage there by preventing catastrophic failures [7].

17.2 Classification of Self-Healing Materials

Self-healing materials are classified according to the mechanism of self-healing property in materials, during the material failure. Self-healing materials can be classified as (i) autonomic—without human intervention triggering of self-healing property take place (ii) non-autonomic—needs human intervention for triggering of self-healing property [1, 10]. In both autonomic and non-autonomic self-healing materials, healing agents are mixed with base materials while preparing composites. When external damage occurs on composites, healing agent will become active and repair the external damages [11]. Garcia *et al.* have classified self-healing polymer materials as extrinsic (discontinuous) and intrinsic (continuous) healing mechanism. Figure 17.1 represents the classification of self-healing materials based on mechanism of healing [10, 12].

17.2.1 Intrinsic Mechanism

Intrinsic self-healing materials are a continuous process. Intrinsic self-healing materials work on the mechanism of healing due to physical interaction between molecules or chemical interactions such as reversible reactions due to heat, hydrogen bonding, ionomers, dissolving thermoplastics and

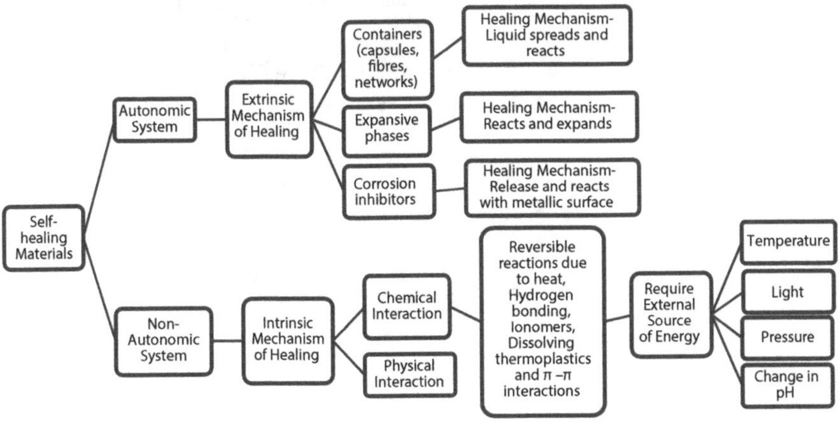

Figure 17.1 Classification based on mechanism of self-healing [10, 12].

π–π interactions. Chemical reaction between molecules in self-healing materials was activated by external source such as change in pH value, light, temperature, pressure, etc. [11, 13].

17.2.2 Extrinsic Mechanism

Extrinsic self-healing materials depend on external healing agent which is available in the form of capsules or vascular networks. Healing agents in the form of capsules or vascular networks along with catalyst is embedded in to the base polymer matrix during the manufacture of the composite material. Extrinsic self-healing mechanism can be called as autonomic system. Since, catalyst embedded in to the base matrix of the composite initiate the healing process along with the healing agent. So, internal or external damages or rupture of is healed by (i) microencapsulation and (ii) microvascular network [11, 13].

17.2.2.1 Microencapsulation

Healing agent is incorporated in the form of microcapsules along with catalyst in the base polymer matrix. Embedding of microcapsules in to the polymer matric is shown in Figure 17.2. This is called as an extrinsic approach of healing. During the formation of internal or external damage in composite, rupture of microcapsule takes place through which releases the healing agent to the damaged part. Healing agent reacts with the catalyst through which polymerization process is initiated which repairs the

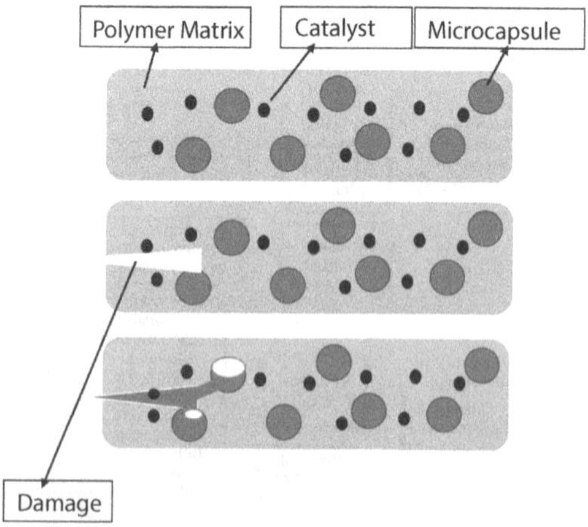

Figure 17.2 Concept of microencapsulation [12].

damaged part [6, 8]. Embedding of healing agent and catalyst into polymer matrix can be carried out by four methods, such as (i) both healing agent and catalyst can be dispersed in polymer matrix [14], (ii) both healing agent and catalyst can be embedded in different capsules [15], (iii) healing agent alone embedded in polymer matrix and polymerization should be initiated with external source of energy [16] and (iv) healing agent or catalyst can be kept in a separate phase.

17.2.2.2 Microvascular Network

Over the past decades, vascular based self-healing materials have been extensively studied by the researchers. Since, a large number of damages can be healed; therefore larger damages can be healed. In microencapsulated technique repeated healing at same region is only possible if healing liquid is available in the capsule. Embedding microvascular fibers or hollow tubes were seen as a different approach which overcomes the limitation of microencapsulated technique. Microvascular network approach is an extrinsic approach which is similar to microencapsulated technique. Microvascular network self-healing materials was like blood vessels seen in living beings. Microvascular network consist of hollow tubes or fibers or a network of capillaries which contains healing agent that may be used to heal the damaged parts shown in Figure 17.3 [3, 6]. Micro vascular network

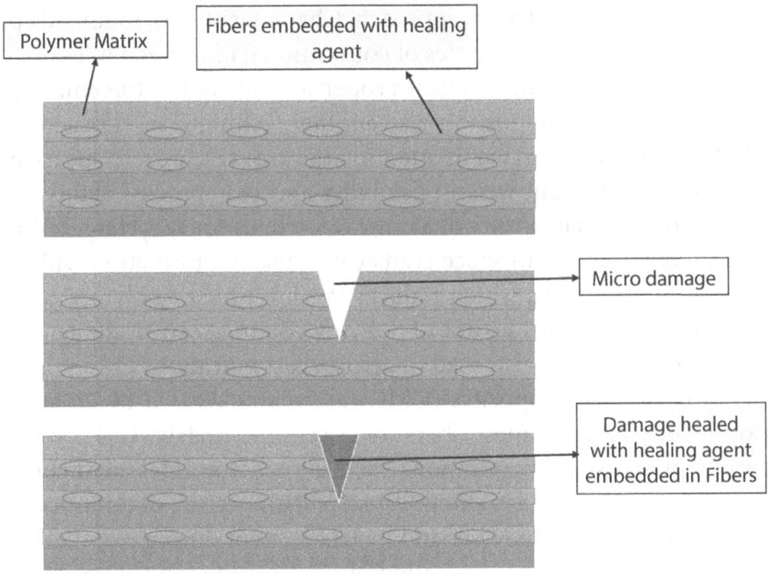

Figure 17.3 Concept of microvascular network [12].

technique was used to manufacture fiber reinforced polymer (FRP) composite materials. Since, fibers or hollow tube can be interconnected to form a network which efficiently delivers the healing agent. So, a large damaged surface area can be healed. Also, the network of fibers or hollow tubes influence on the physico-mechanical properties of composite [8].

17.3 Self-Healing Materials in Aerospace Applications

Carbon fiber reinforced polymers (CFRP) and Glass fiber reinforced polymers (GFRP) were most commonly used materials in aerospace applications. CFRP and GFRP materials were selected as a material for structural applications in aerospace applications. Since CFRP and GFRP material can provide high strength to low weight ratio which minimizes the fuel consumption in aircrafts. The sensitivity of advance composites to impact load is high which results in high maintenance and inspection cost. Number of research works has been carried out using self-healing techniques in fiber reinforced composites to repair the internal or external damages which occurred due to impact load [7, 9].

During manufacturing of CFRP and GFRP composites used in aerospace applications employ epoxy resin. Some research works has also been carried out to convert a conventional epoxy resin in to a self-healing epoxy resin which works on intrinsic approach. Characteristics of modified epoxy resin matches with the properties of conventional one which does not create impact on physico-mechanical properties during reinforcement [17]. Coating of polymer composite is also another technique used in aerospace applications and it has drawn its attention towards researchers due to many advantages such as light weight, flexibility and easy process ability of the material. Long term exposure of atomic oxygen, thermal cycling, UV radiation and low pressure in space crafts can cause the formation and propagation of micro-cracks which reduces the life time of polymer coating. In-order to enhance the life time of polymer coatings, self-healing polymer coatings technique was also used in aerospace applications [18].

Ceramic matrix composites (CMC) are manufactured by embedding carbon or SiC or aluminum fibers in to ceramic matrix. CMC are used in variety of aerospace applications such as rocket nozzles, aeronautic jet engines, heat engines and braking systems. Brittle nature of ceramic can be reduced by optimizing the fiber and matrix content during manufacturing of CMCs. One of the major limitation of CMCs is life time of the composite reduced due to oxidation. One effective way to increase the life time of

CMCs are by filling the damaged region with healing agent or by giving protective coating at the time of oxygen diffusion [19].

17.3.1 Fiber Reinforced Polymers

Williams *et al.* have investigated the significant effect on reinforcement of hollow glass fibers in to CFRP. Hollow glass fiber (HGF) which is embedded in to CFRPs will provide self-healing effect. Epoxy resin was used as base matrix to manufacture CFRPs. Hand lay-up technique has been employed to manufacture CFRP composites. Specimens were manufactured by maintaining distance between the HGF as 70 and 200 μm. Figures 17.4 and 17.5 show the arrangement of HGF and carbon fiber in base matrix. Laminates manufactured with 200 μm distance was better when compared to laminates manufactured with 70 μm distance. Since, uniform distribution of HGFs was affected in 70 μm distance. Williams *et al.* have demonstrated the concept of self-healing CFRPs with four point flexural testing and damage analysis has also been made. Laminates prepared with 70 μm distance of HGF shows lower flexural strength at initial condition when compared to the laminates with 200 μm of HGF. However, laminates prepared with 70 μm distance of HGF shows better healing properties when compared to laminates prepared with 200 μm distance [9].

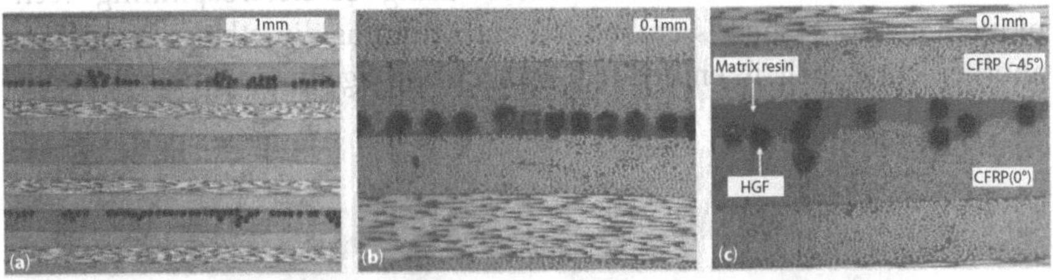

Figure 17.4 (a) HGF spaced at 70 μm, (b) Bonding of HGF with base matrix, (c) Rearrangement of HGF [9].

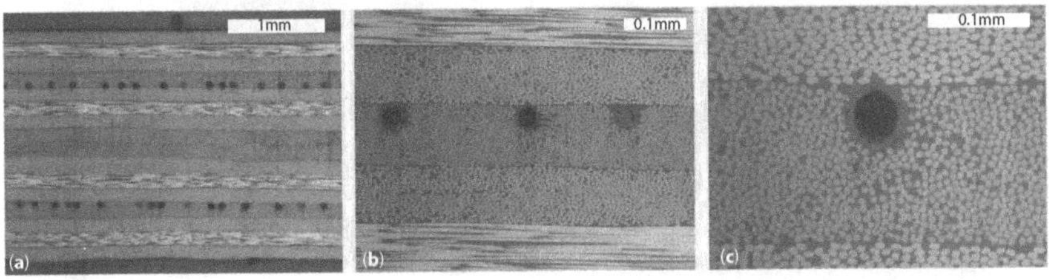

Figure 17.5 (a) HGF spaced at 200 μm, (b) uniform spacing of HGF, (c) Bonding of HGF with base matrix [9].

Another work has been reported using carbon fiber self-healing composites. 2D self-healing composites has been manufactured using metathesis polymerization technique which employs blend of two olefins (ENB/DCPD blend) as a healing agent. Metathesis healing agent is associated with degree of cross-linking which is used in aerospace applications. Since, healing of damaged structures will take place at very low temperatures at −50 °C. In this work, blending of polymers was to done in-order to enhance the degree of cross-linking pf the metathesis polymer. Self-healing efficiency of manufactured composite is analyzed using Atomic Force Microscope (AFM) and Field Emission Scanning Electron Microscope (FESEM) images of fractured surface. Fourier Transform Infrared Spectroscopy (FTIR) helps to identify the activity of catalyst that helps in healing the damaged structures. AFM images were captured on the fractured surfaces. Figure 17.6 indicates the distribution of microcapsules along the carbon fibers in the composites. With the help of FESEM delamination of composites has been studied. From Figure 17.7 it is observed that delaminated regions have been healed by polymerization which is indicated in red arrows. Author suggest that it is possible to use self-healing epoxy resin in FRP to heal the impact damage and the same can be used aerospace applications [7].

Lee *et al.* have developed a composite comprising two core-shell nanofibers which is embedded in PDMS (Polydimethylsiloxane) epoxy matrix. Core–shell fibers were manufactured using co-electrospinning technique. One core–shell is embedded with resin as dimethyl siloxane and another core–shell is embedded with curing agent as dimethyl-methyl

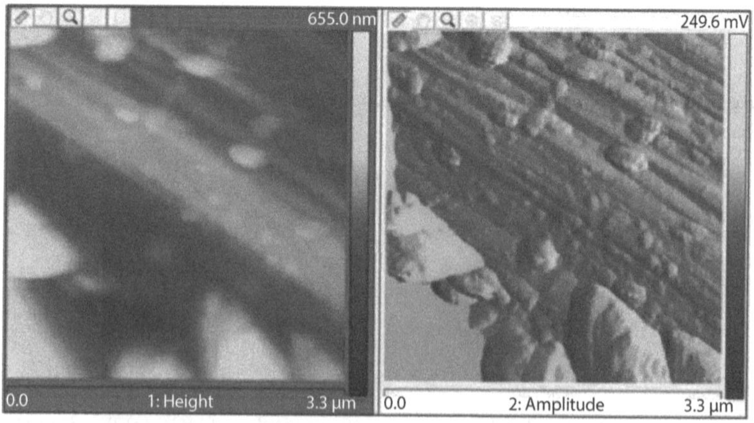

Figure 17.6 AFM images of fracture surfaces [7].

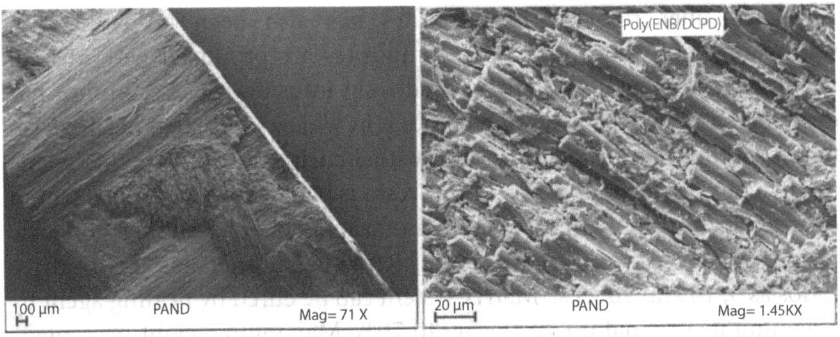

Figure 17.7 FESEM images of fracture surface of delaminated region [7].

hydrogen-siloxane. Both the core–shells are surrounded using PAN shell in separate nano-fibers (PRC) and these nano-fibers act as a healing medium during damage of the composites. Figure 17.8 represents the experimental procedure of electrospinning and composite embedded with two core nano-fibers in PDMS epoxy matrix. Author has analyzed the mechanical properties of material under the prolonged cyclic loading condition which is more commonly observed in aerospace applications. Mechanical properties of the composites are analyzed using tensile test by applying 15% strain which is irreversible plastic deformation. Beyond the 15% strain, breakage of nano-fibers occurs which helps in initiation of self-healing property. During the breakage of nano-fiber resin and curing agent will get released in the damaged region and the healing process take place. Composite prepared with PRC nano-fiber embedded in PDMS matrix shows good self-healing properties during the testing procedure [20].

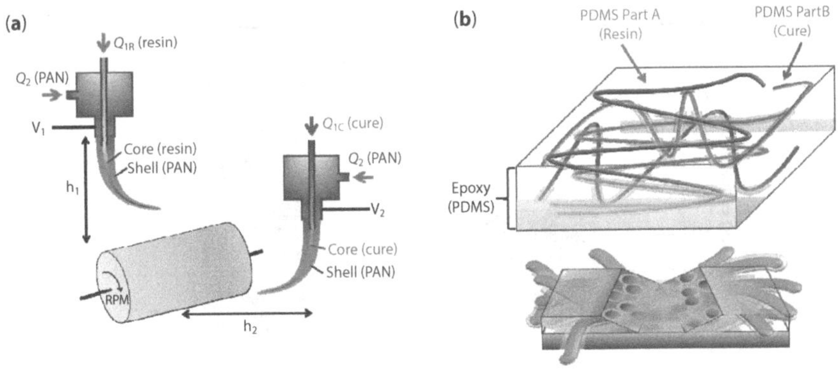

Figure 17.8 (a) Experimental procedure of electrospinning, (b) Composite embedded with two core nano-fibers in PDMS epoxy matrix [20].

E-glass fiber reinforced composites has also been prepared with self-healing property. Glass fiber composites are produced with 10% of healing agent using a vacuum assisted resin infusion process. Author has described a novel self-healing technique in which solid-state healing system for a thermosetting epoxy resin with a curing agent nadic methylene methylene anhydride (NMA). Epoxy resin is blended with curing agent in different proportions. Diglycidyl ether of bisphenol-A based thermoplastic resin was added in to the base epoxy blend as healing material with three ratios as 5, 10 and 20 wt%. Matrix system can be cured by healing agent by regaining its original properties up to 70%. Maximum weight percentage of healing agent helps in recovery of maximum weight percentage of original properties shown in Figure 17.9 [21].

Woven glass fabric embossed in epoxy matrix. Self-healing glass composites are manufactured by incorporating epoxy microcapsules and $CuBr_2$(2-MeIm)$_4$ hardener in base epoxy matrix. Healing efficiency was analyzed by conducting compression after impact (CAI) test. Healing efficiency depends on various factors such as impact energy, repair pressure, size and content of the epoxy microcapsule. Increase in impact energy decreases the healing efficiency. Healing efficiency of the composite depends on the content of healing agent in microcapsule. However, micro-cracks created in the composite during impact load has been healed effectively which prevents further damage of the material during operation. In this system, polymerization will take place at 140 °C [22].

Norris *et al.* have first manufactured FRP composite incorporating microvascular networks to furnish self-healing property in the material. Lost wax process were used for embedding microvascular networks in to FRP composites. Solder wire was used as the microvascular material to

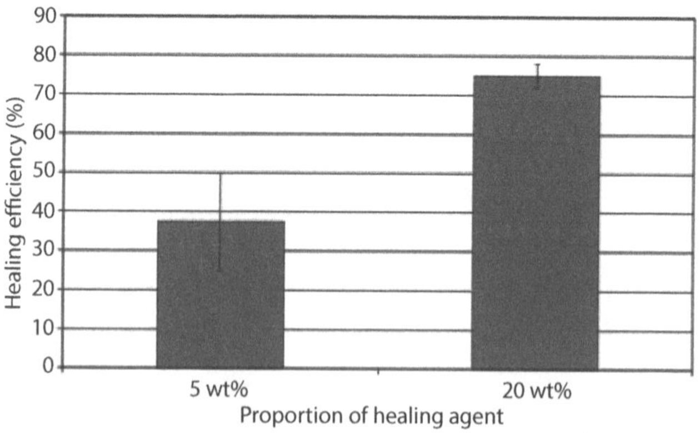

Figure 17.9 Healing efficiency with weight percentage of healing agent [21].

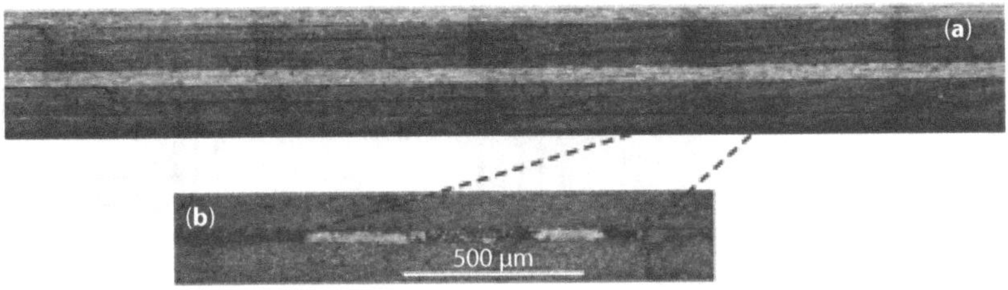

Figure 17.10 Microscopic image (a) Damaged region, (b) Self-healed region [23].

perform its operation. Initially manufactured laminates were subjected to low-velocity impact load before filling uncured epoxy resin in to vascular networks. Microvascular networks filled with healing agent provided >97% of healing efficiency after impact compression test. Author has performed various evaluation procedures such as NDR C-scan testing and microscopic testing to prove healing process happened in damaged region. Figure 17.10 shows the microscopic images of partially infused self-healed region. The process of healing has also been compared self-healing functions of bone in human beings. This process of self-healing should be a highly suitable desirable characteristics for engineering materials [23].

17.3.2 Modified Epoxy

Researchers have also found an alternative approach for deriving self-healing materials is that by modifying the properties of base matrix (epoxy) to self-healing nature. Coope *et al.* conducted a study by altering conventional epoxy amine system to self-healing epoxy using an external source of energy as heat for polymerization process. Epoxy amine was combined with furfuryl and maleimide functional groups in a two-step process and synthesized to form epoxy which contains Diels–Alder (DA) furan and maleimide moieties. Epoxy is modified with DA since it provides thermo-reversible reaction. Figure 17.11 represents the schematic view of DA self-healing process. Thermo-reversible reaction was required in epoxy to balance the properties of thermoset and thermoplastic materials. Since, polymerization could take place using external source of energy this process becomes intrinsic self-healing approach. This approach is also preferred load bearing aerospace structures where healing agent will be placed in damages regions such as skin-stringer run outs, ply drops and around drilled holes. Bismaleimide content is varied in furan functionalized epoxy resins and the performance has been evaluated by the author. 20 pph of bismaleimide content in furan functionalized epoxy resins provided most

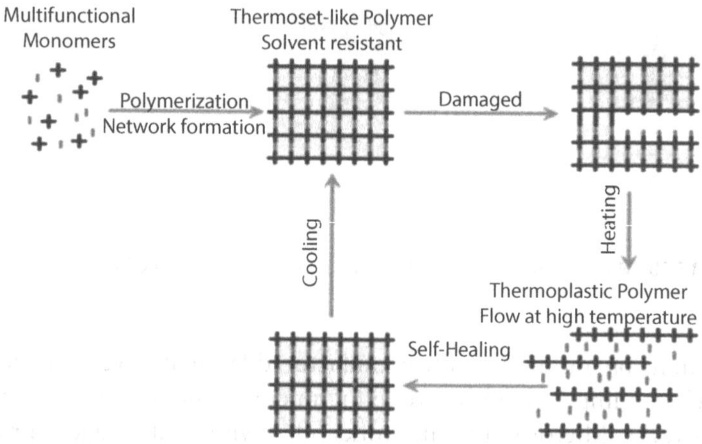

Figure 17.11 Schematic view of DA self-healing process [17].

prominent healing efficiency when heated is applied at 150 °C for 5 min. Modified epoxy resin shows better healing properties in FRP composites also [17].

Coope *et al.* have also derived a system of material in self-healing concept by embedding in FRP composite. Diglycidyl ether bisphenol A (DGEBA) epoxy monomer was encapsulated using poly (urea-formaldehyde) (pUF) shell. DGEBA epoxy was taken as healing agent and scandium (III) triflate was taken as catalyst for curing of healing agent. During initiation and propagation of crack microcapsules get ruptured, healing agent reacts with locally dispersed catalyst which promotes the polymerization healing process. Analysis were carried out using tapered double cantilever beam (TDCB) and the consequences of microcapsule content and healing temperature and time were considered as a factor for deriving healing efficiency. Author was able to observe that 80% of recovery of properties is achieved through healing system. Healing performances of the catalyst (scandium(III) triflate) is comparable with Grubbs catalyst which is more commonly used [24]. Kamphaus *et al.* derived a cost-effective self-healing materials using tungsten (VI) chloride (WCl6) as a catalyst for ring-opening metathesis polymerization (ROMP) of exo-dicyclopentadiene (exo-DCPD) healing applications. Environmental stability of WCl_6 using three different delivery methods. WCl_6 were prepared in different forms as WCl_6 (Aldrich) form, recrystallized form and in wax-protected form to use in self-healing epoxy. WCl_6 used as Aldrich form and recrystallized form resulted in 100% recovery in damaged regions [25]. Zhang *et al.* used two healing agent carriers in the form of microcapsules and hollow glass bubbles (HGB). Microcapsules contains epoxy solutions (EP capsules) and HGB is filled with amine solutions (AM-HGBs). Samples are prepared using those two healing agents in self-healing epoxy system. About 80 wt% of epoxy is filled in EP capsules and around 33 wt% of amine solution is filled in AM-HGBs through *in situ* polymerization and vacuum assisted method. EP capsules and AM-HGBs were incorporated in to epoxy matrix as shown in Figure 17.12. AM-HGBs showed high performance at ambient temperatures but performance has reduced at 80 °C. Weight and mass ratio of the two healing agents in damaged region were compared considering the factors such as size, concentration and weight percentage of healing agent in carriers. Considering a simplified cubic array model, it is derived that longest diffusion of healing agent is inversely proportional to the cubic root of the size, concentration and weight percentage of healing agent in carriers [26]. Guo *et al.* have manufactured a UV-triggered self-healing material through incorporation of Sio_2 in the form microcapsules. A single Sio_2 microcapsule comprises of SiO_2 precursor, epoxy resin and cationic photo-initiator manufactured through interfacial/*in-situ* polymerization technique shown in Figure 17.13. Photo-initiators can get deactivated due to the presence of oxygen in space environment. But entrapment of oxygen in to microcapsule take place at elevated temperatures or strong solvent will lead to degradation of polymer shell ultimately lead to spreading of healing species in base matrix. Incorporated microcapsules show solvent-resistance and thermal stability. In addition to this 89% of the curing efficiency of epoxy resin is achieved during UV exposure. This material has a valid potential application in aerospace industries. Polymer-based coatings are well suitable in aerospace applications due to accurate stoichiometric ratio [18].

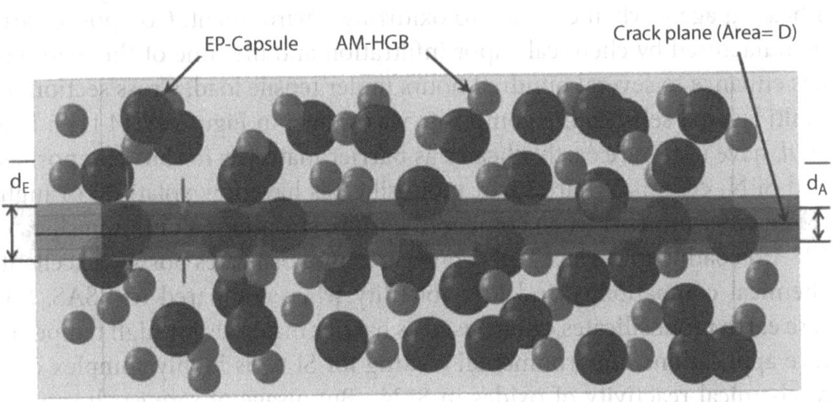

Figure 17.12 Randomly distributed EP Capsules and AM-HGB in epoxy matrix [26].

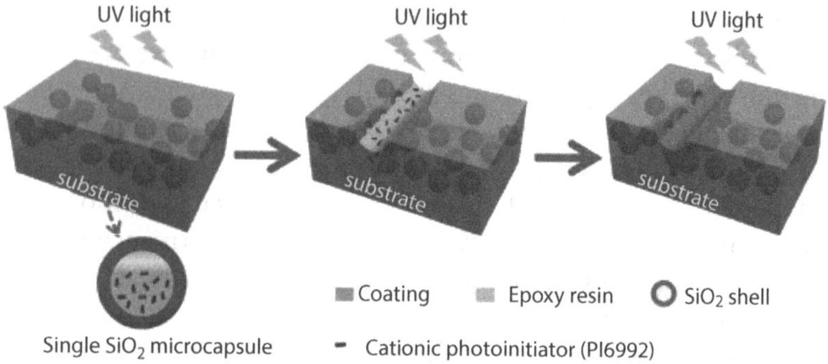

Figure 17.13 Incorporation of UV-triggered self-healing of SiO_2 microcapsules embedded into epoxy resin coating [18].

17.3.3 Ceramic Matrix Composites

Naslain *et al.* have introduced boron-bearing refectory material in ceramic matrix composites in-order to improve the oxidation resistance of the composites at high temperature applications. Aerospace application is one of the predominant application where oxidation resistance is considered as one of the major property to improve the life span of the material. In this work, instead of using pure pyrocarbon as interphase material, B-doped pyrocarbon and hexogonal-Boron Nitride (hex-BN) was used to enhance the oxidation resistance. B-doped pyrocarbon and hex-BN was used as homogeneous or multilayered fiber coatings to arrest the cracks formed under thermos-mechanical load which increases toughness of the material. A self-healing composite has been manufactured by combining B-doped pyrocarbon and Si-bearing compound which forms B_2O_3 act as a healing agent when exposed to oxidizing environment. Composites are manufactured by chemical vapor infiltration and life time of the material has enhance to several hundred hours under tensile load. Cross section of multi layered self-healing composite was shown in Figure 17.14 [19]. Lee *et al.* have used rare earth silicates as barrier materials for SiC composites and Si_3N_4 ceramics. Rare earth monosilicates have low volatility in high steam environments when compared barium strontium aluminum silicate (BSAS) coating materials. Since, rare earth monosilicates possess excellent chemical compatibility and low volatility when compared to BSAS. So, rare earth monosilicates can be used as barrier materials for high temperature applications. Environmental coating for Si_3N_4 is highly complex due to chemical reactivity of oxides in Si_3N_4. But, usage of rare earth monosilicates as a top most coating in composites possess excellent superior

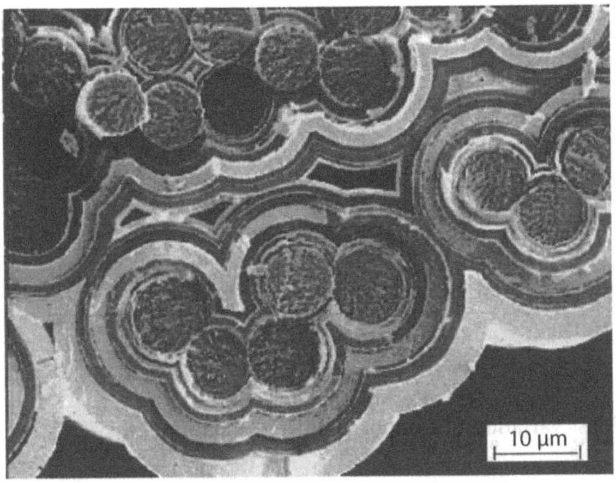

Figure 17.14 Cross section of multi layered self-healing composite.

temperature capability and extended life span when compared to BSAS coating of composite. So, rare earth silicates can reduce oxidation of the ceramic composite which extends the life time of the material [27].

Ceramic matrix composites (CMC) are most preferable material used in advanced aeronautic engines. In recent days CMC are used as an alternative material for nickel-based alloys. CMC are manufactured by reinforcing SiC fibers in to SiC matrix. SiC matrix consist of SiC(f)/Si–B–C fibers and commercial name of the material used is Cerasep A40C. Multi scale approach was used to understand the behavior of Cerasep A40C reinforced composite under dry air at 550 °C. First oxidation rate of fiber and boron carbide was studied and in the secondary phase, degradation rate of whole composite were studied. Matrix crack are sealed by the oxide which is formed by boron carbide. Mass variation of the composite during oxidation under dry air at 550 °C is shown in Figure 17.15. During the self-healing process, B_2O_3 preferentially concentrates on the fibers and tends to flow along the multilayered seal-coat [28].

Quemard *et al.* have also investigated about non-oxide ceramic matrix composite incorporating the CERASEP A410 fibers in SiC matrix. This work is almost similar to the type of work carried out by Nualas *et al.* In this case durability and self-healing efficiency of the composite are examined under 600 °C for the duration of 600 h under high pressure steam environments. Degradation of composite was determined from measuring residual strength, strain to failure, weight changes and oxides formed. Weight changes of the composite under 600 °C for 600 h were shown in Figure 17.16. Incorporation of small amount of SiC fiber in boron oxide in

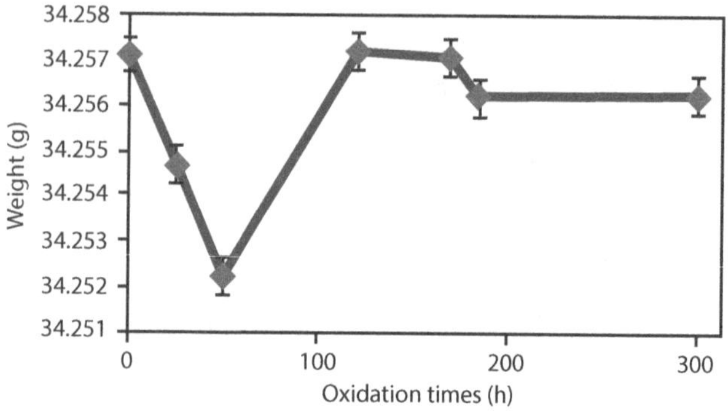

Figure 17.15 Mass variation of composite during oxidation [28].

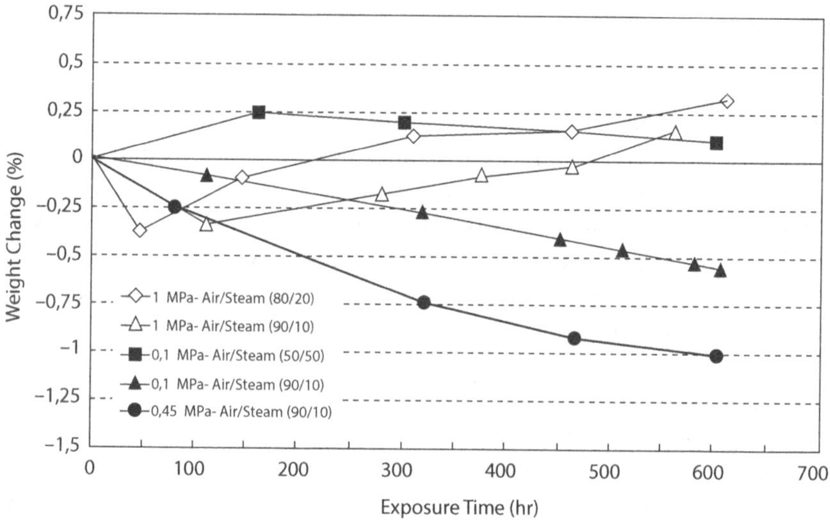

Figure 17.16 Weight variation of composite [29].

high pressure environment shows better material characteristics at 600°C [29].

Carbon fiber reinforced in SiC (C/SiC) matrix was manufactured. C/SiC composites were manufactured with Si–B–C coatings under different combinations. Manufactured C/SiC composites were oxidized for 50 h and their weight loss, residual strength and self-healing efficiency were analyzed. C/SiC composite with Si–B–C coatings has shown excellent oxidation resistance. C/SiC composite with Si-B-C coating on external surface shows superior mechanical properties below 1,000 °C. C/SiC composite with Si–B–C coatings on intermediate surface shows better oxidation

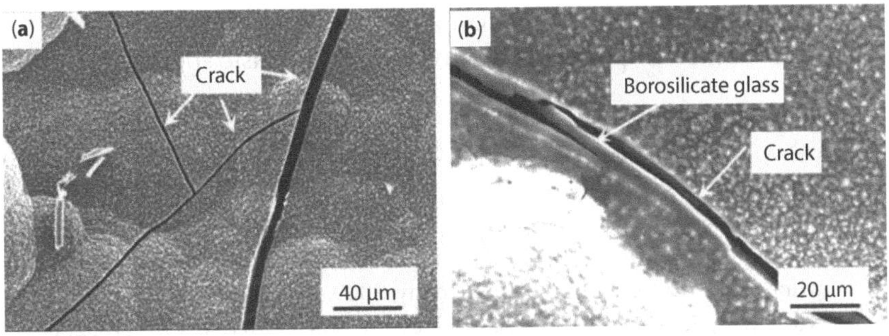

Figure 17.17 SEM images (a) surface morphology of specimen oxidized at 700 °C, (b) surface morphology of specimen oxidized at 1,300 °C [30].

properties above 1,000 °C. Oxidation of Si–B–C forms boron silicate glass fills the cracks of the composite which provides excellent oxidation resistance. Formation of boron silicate glass is observed through SEM images shown in Figure 17.17. As a result, Si–B–C coating on C/SiC composite acts as a good self-healer by increasing the oxidation resistance in the temperature ranging from 700 to 1,300 °C [30].

17.4 Conclusion

This chapter extensively discuses about the various approaches of developing self-healing material and different types of materials used in aerospace structural and high temperature applications.

- Intrinsic mechanism of self-healing approach requires external source of energy such as light, temperature, pressure, etc., for initiation of healing mechanism in the external and internal damaged region.
- Extrinsic self-healing approach gets initiated when healing agent (resin) and catalyst (curing agent) reacts with each other. Resin and curing agent comes in to contact with each during voids or micro-cracks formed due to impact and cyclic load.
- Extrinsic mechanism of healing was seen as a preferred approach in the materials used in aerospace applications. Since, micro-cracks or laminar damages are healed autonomically by polymerization of resin and curing agent.
- Application of self-healing mechanism in various materials such as glass fiber reinforced polymer composites, carbon

fiber reinforced polymer composites, carbon fiber reinforced ceramic composite are discussed with extrinsic approaches.

References

1. Ghosh, S.K., *Self-healing materials: fundamentals, design strategies, and applications*, Wiley Online Library, Germany, 2009.
2. Wool, R.P., Self-healing materials: A review. *Soft Matter*, 4, 400–418, 2008.
3. Bekas, D., Tsirka, K., Baltzis, D., Paipetis, A., Self-healing materials: A review of advances in materials, evaluation, characterization and monitoring techniques. *Compos. Part B: Eng.*, 87, 92–119, 2016.
4. Wool, R.P., *Polymer interfaces: Structure and strength*, Hanser, New York, 1995.
5. Nairn, J.A., Matrix microcracking in composites. *Polym. Matrix Compos.*, 2, 403–432, 2000.
6. Kessler, M., Self-healing: A new paradigm in materials design. *Proc. Inst. Mech. Eng. Part G: J. Aerosp. Eng.*, 221, 479–495, 2007.
7. Raimondo, M., De Nicola, F., Volponi, R., Binder, W., Michael, P., Russo, S., Guadagno, L., Self-repairing CFRPs targeted towards structural aerospace applications. *Int. J. Struct. Integrity*, 7, 5, 656–670, 2016.
8. Das, R., Melchior, C., Karumbaiah, K., Self-healing composites for aerospace applications, in: *Advanced composite materials for aerospace engineering*, pp. 333–364, Elsevier, England, 2016.
9. Williams, G., Trask, R., Bond, I., A self-healing carbon fibre reinforced polymer for aerospace applications. *Compos. Part A: Appl. Sci. Manuf.*, 38, 1525–1532, 2007.
10. Williams, K.A., Dreyer, D.R., Bielawski, C.W., The underlying chemistry of self-healing materials. *MRS Bull.*, 33, 759–765, 2008.
11. Fainleib, A. and Purikova, O., Self-healing polymers: Approaches of healing and their application. *Полімерний журнал*, 41, 1, 4–18, 2019.
12. Garcia, S. and Fischer, H., Self-healing polymer systems: properties, synthesis and applications, in: *Smart Polymers and their Applications*, pp. 271–298, Elsevier, England, 2014.
13. Thakur, V.K. and Kessler, M.R., Self-healing polymer nanocomposite materials: A review. *Polymer*, 69, 369–383, 2015.
14. White, S., Sottos, N., Geubelle, P., Moore, J., Kessler, M., Sriram, S., Brown, E., Viswanathan, S., Autonomic healing of polymer composites. *Nature*, 409, 794, 2001. 415, 817–817, 2002.
15. Keller, M.W., Encapsulation-based self-healing polymers and composites, in: *Healable Polymer Systems*, pp. 16–61, The Royal Society of Chemistry, Cambridge, UK, 2013.

16. Yin, T., Rong, M.Z., Zhang, M.Q., Yang, G.C., Self-healing epoxy composites—Preparation and effect of the healant consisting of microencapsulated epoxy and latent curing agent. *Compos. Sci. Technol.*, 67, 201–212, 2007.
17. Coope, T., Turkenburg, D., Fischer, H., Luterbacher, R., Van Bracht, H., Bond, I., Novel Diels–Alder based self-healing epoxies for aerospace composites. *Smart Mater. Struct.*, 25, 084010, 2016.
18. Guo, W., Jia, Y., Tian, K., Xu, Z., Jiao, J., Li, R., Wu, Y., Cao, L., Wang, H., UV-triggered self-healing of a single robust SiO2 microcapsule based on cationic polymerization for potential application in aerospace coatings. *ACS Appl. Mater. Interfaces*, 8, 21046–21054, 2016.
19. Naslain, R., Guette, A., Rebillat, F., Pailler, R., Langlais, F., Bourrat, X., Boron-bearing species in ceramic matrix composites for long-term aerospace applications. *J. Solid State Chem.*, 177, 449–456, 2004.
20. Lee, M.W., An, S., Jo, H.S., Yoon, S.S., Yarin, A.L., Self-healing nanofiber-reinforced polymer composites. 1. Tensile testing and recovery of mechanical properties. *ACS Appl. Mater. Interfaces*, 7, 19546–19554, 2015.
21. Hayes, S., Jones, F., Marshiya, K., Zhang, W., A self-healing thermosetting composite material. *Compos. Part A: Appl. Sci. Manuf.*, 38, 1116–1120, 2007.
22. Yin, T., Rong, M.Z., Wu, J., Chen, H., Zhang, M.Q., Healing of impact damage in woven glass fabric reinforced epoxy composites. *Compos. Part A: Appl. Sci. Manuf.*, 39, 1479–1487, 2008.
23. Norris, C.J., Meadway, G.J., O'Sullivan, M.J., Bond, I.P., Trask, R.S., Self-healing fibre reinforced composites via a bioinspired vasculature. *Adv. Funct. Mater.*, 21, 3624–3633, 2011.
24. Coope, T.S., Mayer, U.F., Wass, D.F., Trask, R.S., Bond, I.P., Self-healing of an epoxy resin using scandium (III) triflate as a catalytic curing agent. *Adv. Funct. Mater.*, 21, 4624–4631, 2011.
25. Kamphaus, J.M., Rule, J.D., Moore, J.S., Sottos, N.R., White, S.R., A new self-healing epoxy with tungsten (VI) chloride catalyst. *J. R. Soc. Interface*, 5, 95–103, 2008.
26. Zhang, H. and Yang, J., Development of self-healing polymers via amine–epoxy chemistry: I. Properties of healing agent carriers and the modelling of a two-part self-healing system. *Smart Mater. Struct.*, 23, 065003, 2014.
27. Lee, K.N., Fox, D.S., Bansal, N.P., Rare earth silicate environmental barrier coatings for SiC/SiC composites and Si3N4 ceramics. *J. Eur. Ceram. Soc.*, 25, 1705–1715, 2005.
28. Nualas, F. and Rebillat, F., A Multi-Scale Approach of Degradation Mechanisms inside a SiC (f)/Si–B–C (m) Based Self-Healing Matrix Composite in a Dry Oxidizing Environment. *Oxid. Met.*, 80, 279–287, 2013.

29. Quemard, L., Rebillat, F., Guette, A., Tawil, H., Louchet-Pouillerie, C., Self-healing mechanisms of a SiC fiber reinforced multi-layered ceramic matrix composite in high pressure steam environments. *J. Eur. Ceram. Soc.*, 27, 2085–2094, 2007.
30. Zuo, X., Zhang, L., Liu, Y., Cheng, L., Xia, Y., Oxidation behaviour of two-dimensional C/SiC modified with self-healing Si–B–C coating in static air. *Corros. Sci.*, 65, 87–93, 2012.

18

Bio-Inspired Self-Healable Materials

Archita Sharma and Shailendra Kumar Arya*

University Institute of Engineering and Technology (UIET), Panjab University (PU), Chandigarh, India

Abstract

Nature renders tremendous broad-scoped elements with diverse functions as bio-inspiration for the researchers dealing with such materials. With an immense power of biological systems to perceive, respond, maintain, flourish and heal, majorly the focus is on bio-inspired self-healable materials. With the advancements in material chemistry, microscale and nanoscale fabrication approaches, it is now possible to fabricate bio-inspired systems with a potential to mimic numerous aforementioned functions. Such bio-inspired systems have given birth to self-healing materials in which it is autonomous and site-specific healing of damage. This autonomous healing is precisely a dynamic attribute of material that resembles a specific natural material of living systems. Every year much work has been done to establish varied self-healable bio-inspired systems along with their production at large scale keeping in mind the ratio of property to cost. This chapter focuses significantly on self-healing materials, properties, and principle behind the self-healing process. This discussion briefly gives an insight of self-healing polymers and their types. Furthermore, it covers case studies inspired by nature along with few examples to shed more light on the self-healable materials. The research on biomimetic materials is flourishing rapidly at large platform and thus a promising field of futuristic research front.

Keywords: Autonomous repair, bio-inspired, biomaterials, biomimetics, polymers, self-healing, supramolecular self-assembly, tissue engineering

**Corresponding author*: skarya_kr@yahoo.co.in

Inamuddin, Mohd Imran Ahamed, Rajender Boddula and Tariq Altalhi (eds.) Self-Healing Smart Materials and Allied Applications, (435–474) © 2021 Scrivener Publishing LLC

18.1 Introduction

Plants and animals majorly consist of biological materials with the potential to regulate the function of cells, render eyes to capture, coordinate light and allow plants to adapt a light and allow animals to show movement. Such potential functions have encouraged humankind to fabricate materials, devices, systems, etc. with a notion to streamline our daily activities. Such bio-based natural materials have a few number of elements that can be exploited to harmonize numerous polymers and minerals. Whereas on the other hand, history has exploited a variety of elements to fabricate a particular system or structure and this allows mankind to invent such materials that have not been used by nature. With the advent of the industrial revolution, technical materials have been fabricated but such materials need extremely high temperatures that are not accessed by biological organisms. There is an intricate and hierarchical structuring of the natural materials which leads to the growth of such materials according to the genetic material instead of fabrication under the desired design (Figure 18.1) [1–3].

Due to natural selection, living organisms have enhanced functional systems with evolution. One of the most remarkable characteristics is the potential to self heal the damaged tissues which can be imposed on applications like cuts, bruises, external damages, etc. This remarkable self-healing potential obliges them to withstand ruptured or completely lost body parts [4, 5]. Self-healing is a big issue for artificial materials as there is an absence of self-healing characteristics to repair inevitable degradation and unanticipated damage during exploitation. To sort this out, fabrication of self-healing materials is grabbing a lot of attention nowadays for practical applications by mimicking the function and behavior of living organisms with a notion to recover the structure and characteristics in a spontaneous manner [6–8]. The basic purpose of self-healing materials is to impede the damage caused by external or internal physical and chemical stimuli. Like living organisms, such self-healable materials have the capacity to stretch the shelf life of material remarkably by averting possible failures due to microcracks accumulated in the materials [9]. Till now, in the majority the research regarding self-healing materials are basically focused on restoring mechanical damage to bulk materials along with improvements in shelf life through numerous methods, for example, encapsulation, reversible chemistry, micro-vascular networks, phase separation of nanoparticles, hollow fibers, monomer phase separation, etc. [4, 5, 10].

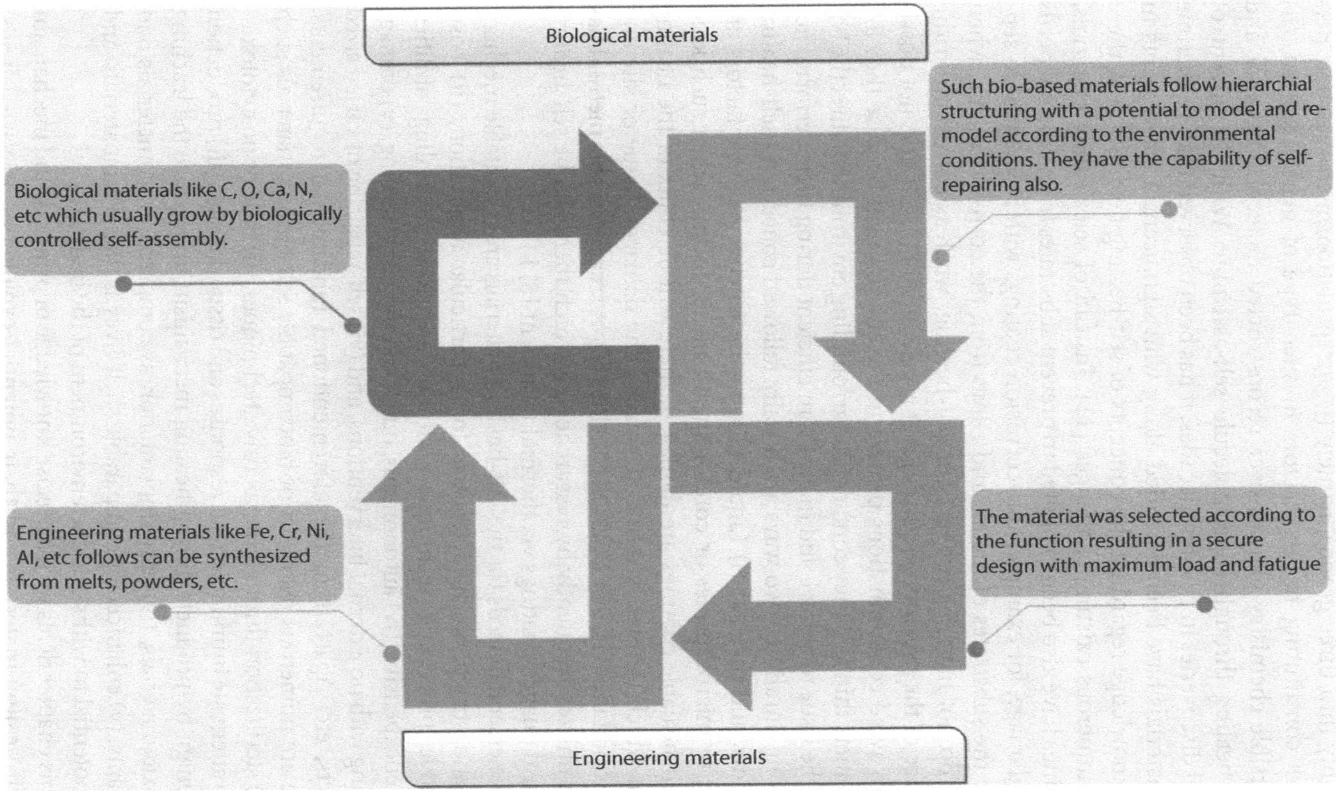

Figure 18.1 Fabrication of biological and engineering materials through different strategies to accomplish desired functionality [3].

To synthesize self-healing materials the approaches are categorized roughly into three groups, namely (i) encapsulation/micro-vascular networks constituting healing factors in such type of materials itself, (ii) reversible chemistry planted over reconstructed molecular network and (iii) healing through supermolecule self-assembly. With the advent of novel discoveries in past few decades, it has been observed that the lifetime of materials have been extended along with significant improvements in the knowledge regarding the concepts of self-healing field. But still, there are numerous existing challenges [11]. Instead of self-healing materials, coatings have also been exploited to prevent substrates from external perils and damage, for example, anti-corrosion coatings with a potential to separate the substrates of metals and alloys from the corrosive media, when they do not have the property of self-healing, whereas corrosion on the surface of the substrate requires great acceleration [12, 13]. It is not possible to avoid contaminations present in the environment because they are extremely thin and they contact the surrounding environment directly. For this the issue to resolve and make them efficient during their performance for long durations, two ways are usually followed concerning self-healing coatings and they are (i) Liquid filling—the reactive healable factors are filled in microcapsules or core or sheath fibers. In this, when such kind of self-healable coatings are damaged, there will be efflux of the healing agent which ultimately results in healing the damaged tissue or site and (ii) exploitation of reversible Diels–Alder (D–A) reaction. This method has been utilized frequently by researchers in which there will be self-healing of the damaged coatings via thermal treatment [5, 14–16].

The essential needs for the self-healing materials are (i) suitable mechanical and structural characteristics for a particular application, (ii) cost-effective processing techniques without any loss of healing abilities, (iii) enexpensive, (iv) autonomous, quick and repetitive healing, (v) damage sensing with no external interventions, and (vi) self-repairing of varied sized defects, etc. Apart from the aforementioned fundamental requirements, there are numerous restraints of biocompatible self-healing materials such as (i) sterilization through inexpensive techniques, (ii) little or no cytotoxicity to attenuate immunological responses and tissue damage, (iii) production of benign by-products during healing mechanism to decrease the leaching of toxic materials, (iv) easy implantation as compared to biomaterials and implants, (v) enhanced or comparable shelf-lives as that of biomaterials, and (vi) potential to withstand the environment of the body [17].

This chapter deals with a precise knowledge of principal and mechanism of self-healing materials and their potential exploitation in varied fields. There will be a section in this chapter discussing the new and novel class of self-healing polymers and their types. At the end of the chapter, there will be numerous case studies regarding self-healable materials following few examples to shed more light on bio-inspired self-healable materials.

18.1.1 Self-Healable Materials and Coatings

Within nature, the process of self-healing occurs one at the level of single molecules like deoxyribose nucleic acid (DNA) repair and another at the macroscopic level such as fusion of broken bones, closure, and healing of blood vessel injuries, etc. Such kind of mechanisms is well-known and apparent like the healing of an injured finger. Nevertheless, engineered or artificial materials lack the ability of self-healing. They are fabricated with a notion to prevent damage instead of the management of damage [18, 19].

Self-healable materials are categorized into two major classes (i) external stimulus-responsive self-healable materials. This class has the ability to reinstate their indigenous physical and mechanical characteristics with respect to extrinsic stimulus and (ii) automatic self-healable materials. This category of self-healing materials repairs the damage or fracture in the absence of external interference. Concerning the need for external stimulus, the self-healing materials are divided into three methods: (a) capsulate and microvascular, (2) reversible covalent bond and (3) supramolecular self assembly [15].

18.1.1.1 The Process of Self-Healing Through the Exploitation of Micro-Capsule and Micro-Vascular Method

In this method, the healing agent/factor, the catalyst and initiator are packed individually in the microcapsules inside the materials (Figure 18.2) [20, 21]. Due to the fracture in the materials, there will be prompt mixing of the healing agents, initiators, and catalysts via capillary action at the damaged site. After that, there will be an initiation of the healing process and result in the formation of novel polymeric materials leading to regeneration. This micro-capsule self-healing method is suitable particularly for the cross-linked materials [22]. A team of researchers investigated this method of self-healing and they exploited (a) the reaction called Ring Opening Metathesis Polymerization (ROMP) with a notion to produce novel materials immediately, (b) the monomer namely dicyclopentadiene and (iii) the Grubbs' catalyst which was impregnated separately into the micro-capsules within the matrix of epoxy resin. The process of healing initiated during the rupturing micro-capsules due to

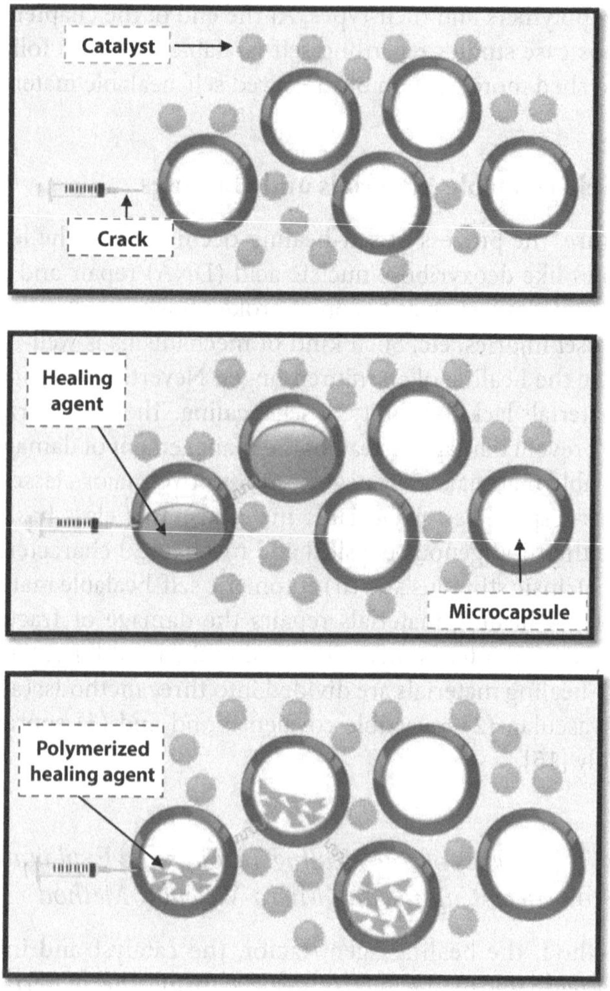

Figure 18.2 A micro-capsule self-healing process [15].

crack. Since there will be a quick release of the monomer dicyclopentadiene and the catalyst through capillary action, there will be proper mixing at the damaged site to start the ROMP reaction. After the reaction is completed, there will be a production of new cross-linked within that very damaged site [15].

Apart from efficient repairing of the damage, there are certain demerits of this approach such as the process of healing at a particular damaged are occurs only once with a healing efficiency of approximately 67%, the cost of catalyst is expensive, the stability of the catalyst inside the matrix is poor, etc. [19, 23]. To date, the research is focused significantly on how to overcome such demerits of this approach. Considering the healing efficiency, majority of the work has been done by the researchers and led to the conclusion that with the quick dissolution of the catalyst there will be an increase in the healing efficiency by many folds [24, 25]. Other than the rapid dissolution of catalysts, a distinct approach to boost the healing efficiency is by the addition of Shape Memory Alloy wires into the composites. Considering this approach, the Shape Memory Alloy wires are activated which makes the crack smaller and thus better self-healing process and thus enhanced healing efficiency [26, 27].

To surmount the issues of cost of the catalyst and its stability in the matrix, there has been bulk exploitation of catalyst based on tin and healing agents based on siloxane. The advantage is that they can tolerate extreme harsh conditions and have better stability in a humid environment because of the hydrophobic property of the siloxane-based healing agents [28]. Additionally, the tin-based catalyst has the potential to cope up with the high temperatures (more than 100 °C), but the healing efficiencies are quite poor. Concerning the cost associated with the catalysts, tungsten (VI) catalysts have been used for the ROMP reaction as tungsten is comparatively cheap and eco-friendly [29]. Instead of major advantages of tungsten, the healing efficiency is poor with limited area of applications [30].

Another issue of self-repairing of fractures at the damaged site only once can be over-powered by simulating the network of capillaries in the skin dermal layer. The researched have fabricated a 3-D microvascular network consisting of coatings of the epoxy resin which constitutes catalysts. These networks can be re-filled with healing agents for its repetitive utilization in repairing the damage (Figure 18.2) [30–32]. This epoxy resin coating has the potential to serve numerous healing agents individually at the site of injury. During the damage, the healing agents went into the cracks and contact the catalyst which is ingrained in the epoxy resin coating. Before the onset of next healing cycle, the healing agents are then re-filled and thus repetitive recovery of this particular self-healing system with the healing efficiency of approximately 30–70%. Another approach is the exploitation of hollow borosilicate glass fibers with curing resin filled in a polymer matrix with a thought to render the ability of self-repairing to the materials [30], for instance, a team ingrained the fibers in the polymeric matrix and load the mixture of repair agent/factor and ultra-violet fluorescent dye by vacuum-assisted infiltration method [32]. With the application of stress that breaks the matrix, the hollow fibers get fractured and discharge the two epoxy components. Furthermore, there will be successive mixing of resin and hardener which renders strength to repair the damage. The purpose of ultra-violet (UV) fluorescent dye in this approach is to provide better visualization of the self-healing process when irradiated by ultra-violet (UV) light [15].

18.1.1.2 Self-Healing Process Through Reversible Covalent Bond Formation

Materials that exploit reversible breaking and reformation of the network of polymers for restoring macroscopic fractures or to re-attach damaged or broken fragments have grabbed a lot of attention recently. The materials are devised in such a way so as to retain the shape along with the notion to boost the mechanical strength of the materials through the incorporation of the reversible chemical bonds [16]. The quintessential approaches in the self-healing system are by the exploitation of the thermoreversible Diels–Alder cycloaddition reaction. In this very approach, diene, and dienophile are considered as the prototype which react and form [4 + 2] cycloaddition [33–35]. Diels–Alder (D–A) cycloaddition reaction is a conducive way to form C–C bonds through an easy reaction called click reaction with no particular conditions. The merits of this type of reaction are that (a) there are negligible side reactions, (b) thermo-reversible reaction as there is no requirement of delivering additional chemical like an expensive catalyst, (c) easy fabrication of linear and dendritic polymeric materials, (d) high mechanical strength, (e) a better stability of such materials, and (f) potential to repair the same fractured part many times, etc. [36–38].

A group of researchers designed a self-healing system that exploits a translucent polymeric material cross-linked through the reactions of dienophile (maleimide) and diene (furan) units [34, 39]. In this particular study, they exploited the merits of thermoreversible features of the Diels–Alder cycloaddition reaction for reheating the sample and allow the re-formation of new covalent bonds to understand the materials remuneration. After that numerous approaches have been proposed [40–42] but issue like restricted thermal reversibility and expensive fabrication protocols impede their practical applications as self-healable materials. For instance, the furfuryl adduct in the reverse Diels–Alder cycloaddition reaction (D–A reaction) is thermally not stable, particularly when the temperature is 150 °C or more. Another researcher make changes in the system by cross-linking polystyrene with end groups constituting maleimide with di-2-furfuryl adipate [41, 43]. The team of researchers observed that the reverse Diels–Alder cycloaddition reaction progressed at 130°C in the absence of side reactions by altering the diene to pendant 2-furfuryloxy moieties with an MDPBM cross-linking agent/factor, and utilizing 2-methyl furan as a snare to pick up the released MDPBM [44]. Considering the aforementioned studies, a group of researchers has exploited Diels-Alder cycloaddition reaction in bis-maleimides and polyketones to accomplish the intent of reusing thermoset polymers [40]. Such reports have suggested that the self-healing systems showed 100% healing efficacy even after repetitive series. Another chemical D–A reaction in which dicyclopentadiene, a monomer is used which behaves as a diene and dieneophile both in the Diels–Alder cycloaddition reaction have been majorly investigated in the recent past [45–47]. The advantage of such type of reaction is of the monomer which has enhanced thermal stability as compared to furan-maleimide [40].

Additionally, apart from the self-healing system based on D–A reaction, there are numerous types of self-healing systems like hydrogen bonding, dynamic covalent transitions, ionomers, etc. [48–50]. Usually, the covalent bonds have relatively high energy and therefore there is a requirement of energy in the form of heat and light in bulk amounts to break and reform covalent bonds. On the other hand, non-covalent bonds like ionic bonds, hydrophobic bonds, metal-coordination bonds, hydrogen bonds, etc make use of molecules and ions to form an assemblage that have low kinetic stability. A group devised a system based on cross-linking via hydrogen bonds.

In any case, if the polymer is damaged, it is possible to repair the damage automatically by joining fractured ends concomitantly for some time at optimal temperature. Such type of approach is particularly replicable and the restored materials have preserved mechanical properties after numerous repetitions [50–52]. But the limitation of this approach is that it is very much important to join the fractured ends as quickly swiftly to gather minimal atonement as the hydrogen bonding react with the neighboring units only in their approach. Currently, the approaches to conjugate the shape memory materials (SMM) with self-healing approach have led to enhancing the self-healing process. With this, there is an elimination of the requirement of an extrinsic force to seal the crack at the time of the restoration process. Furthermore, the thermoplastic ionic polymers–ionomers have been utilized at a broad scale in the self-healing domain. Such kind of ionomers constitutes hydrocarbon chains with a carboxylic acid group which is either partially or completely neutralized with metal or quaternary, ammonium ions. The positive aspect of this approach is that it a facile and cost-effective fabrication process [53–55].

18.1.1.3 Self-Healable Systems on the Basis of Supramolecular Self-Assembly

In the majority, this particular kind of self-healing approach build on the utilization of non-covalent interactions, hydrogen bonds, and transient bonds to form structures with the potential to repair the damaged area [56–58]. In general, linear supramolecular polymers can to heal their own through thermal regimen. Mostly, the damaged supramolecular materials can be healed thermally in the case when two broken/damaged parts are joined altogether due to the thermal diffusion of polymeric chains across the interface (Figure 18.3). The heat source is crucial enough as the process of healing will proceed when the polymer is placed above its glass transition temperature (Tg) more than the repetition time (Tr). Concerning

Figure 18.3 The process of self-healing of supramolecular material: (A) when the supramolecular material is incise into two sections the fragile supramolecular bonds (above) or clusters (down) are damaged or broken and (B) the bonding motifs of supramolecular materials are present in unconnected form over the fractured surface for a particular duration, but have the potential to rearrange again; (C) again after contact the supramolecular bonds or clusters are formed again, thus healing of the damaged crack [57].

the process of self-healing a crack, macromolecules will disperse over the interface between two parts of the broken polymer when given the heat. After that, they permeate and finally re-entangle with neighboring chains, thus intertwine the two portions of the thermoplastic together. Such self-healing materials have the potential to attain the original strength again but its jade resistance of such materials has been decreased beyond its maximum value [59, 60].

Concerning such self-healable materials based on supramolecular assembly, numerous researchers engaged bulk materials with various non-covalent bonds to stick altogether a network of oligomeric or polymeric chains. The non-covalent interactions have a distinctive function in the structure of the self-healable materials [61, 62]. Majority of the investigation have been done on the self-healing systems based on the non-covalent interactions assembled on the π–π stacking interactions. From the reports, the supramolecular polymeric systems of pyrenyl terminated polyamides are embedded into the chain folds of polyimides. The polyimide has the capability of chain-folding around the π-electron rich in pyrenyl chain ends to form a complementary, π–π stacked, non-covalent polymer group. Such type of interaction resulted in the formation of the film. At the same time, the damaged film is formed again when all the broken pieces are pressed altogether at a heating temperature of 80 °C. This results in the complete and rapid recovery of modulus strength. The sectional separation of the complementary π–π stacking interactions, results in the start of the reparation process. These dissociated components flow and fill the fractured area with elevated temperature. When the temperature were decreased, the non-covalent π–π stacking interactions was again well-rooted in a progressive manner with a notion to reinstate the physical characteristics of the material [63, 64].

18.1.2 Mechanism of Self-Healing Materials

There are numerous methods and concepts for different classes of self-healing materials (Table 18.1) that describe their potential applicability (Figure 18.4) in different fields of research. The reason behind them is the special molecular architecture and respective range of temperatures at which they are exploited [19].

During the exploitation of materials, all types of natural and synthetic materials stockpile near the damage. If this particular damage zone extends the critical value, then there will be a complete component decline and hence, loss of the function of the device whereas self-healing materials are

Table 18.1 Different potential mechanisms for different self healing materials (Symbols represents the expected success level; ⊕ - potential, ◆ - unlikely, ✳ , positive [19].

Mechanism	Encapsulation	Channel transport	Expanding phases	Temperature increase	Phase separation	Reversible cross links	Biological processes	Electrochemical processes
Polymer & Composites	positive	positive	positive	positive	positive	positive	potential	unlikely
Ceramics & Concrete	positive	positive	positive	positive	potential	unlikely	positive	positive
Metals	potential	potential	positive	positive	potential	unlikely	unlikely	positive

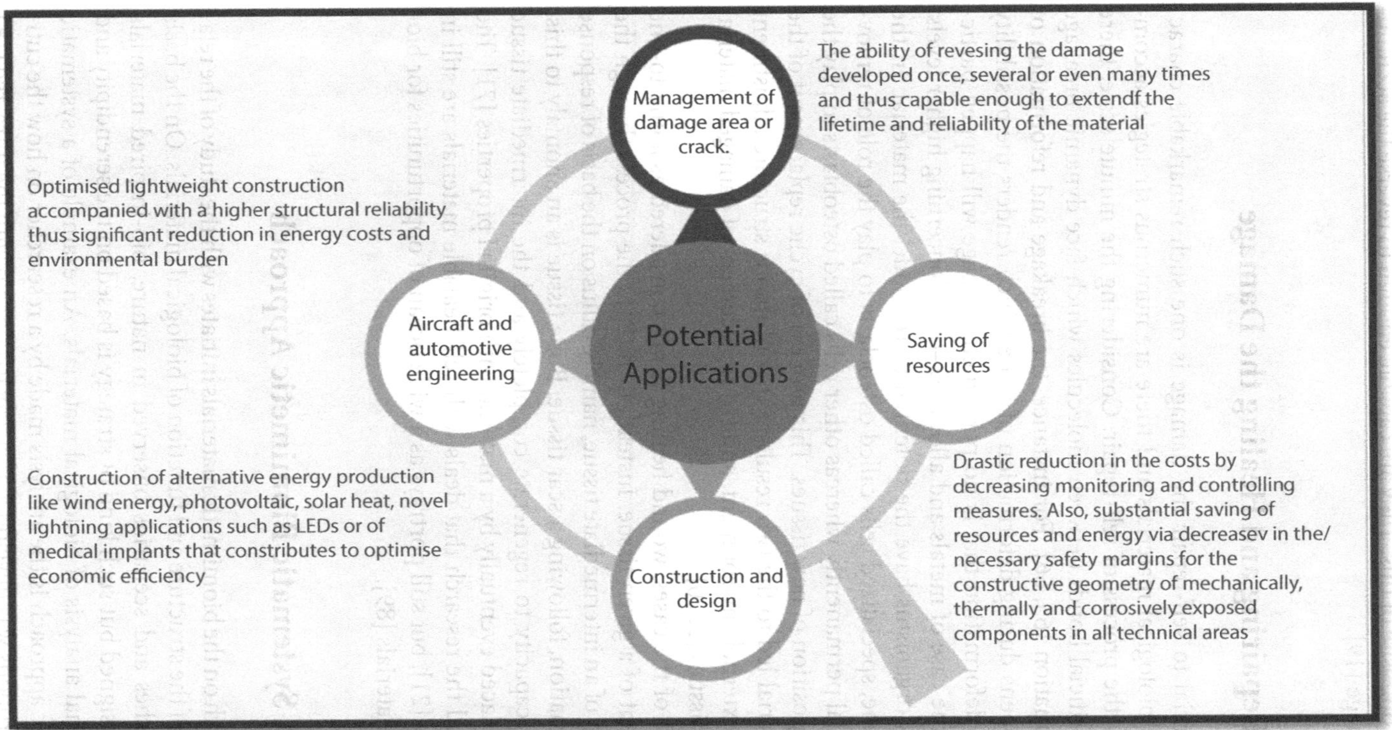

Figure 18.4 Possible application of self-healing materials in different fields of research [19].

complete opposite of this. They are capable enough to reverse the accumulated damage [19].

18.2 Repairing and Healing the Damage

The potential to self-repair the damage is one such remarkable characteristic of biological materials and there are numerous strategies accompanied by the process of self-repair. Considering the minute scale, there occur sacrificial bonds between molecules which face dynamic breakage and reformation [65, 66]. For instance, the breakage and reformation of bond happens during deformation of bone which renders the possibility of plastic deformation that is no permanent damage will happen particularly in the case of metals and alloys [67–69]. Concerning higher levels, numerous organisms have the potential to re-model the material. In the case of bone, specialized cells called osteoclasts to play the role of removing material permanently whereas other cells called osteoblasts to play the role of deposition of new tissues. This process of cyclic replacement of the bone material led to the two results: (a) continuous structure adjustment to varied external environment, and (b) replacement of damaged material with new tissues [2, 70].

In most of the cases, wound healing is not considered as a one-to-one replacement of a given tissue. Instead, it initiates the process through the generation of an intermediate tissue, namely callus on the basis of response to inflammation, following a scar tissue. Bone tissue is an anomaly to this. It has the capacity to regenerate completely and the intermediate tissue will be replaced eventually by a material with original properties [71]. The science and the research that deals with self-healable materials are still in its infancy [21], but still portrays as a wide ocean of opportunities for biomimetic materials [89].

18.3 A Systematic Biomimetic Approach

The research on the biomimetic materials initiates with the study of the relationships of the structure and function of biological materials. On the basis of approaches and scenarios observed in nature, bio-inspired materials have be designed but such kind of strategy is based on the serendipity and on the actual analysis of biological materials. An example of a systematic biomimetic approach is the analysis made by a researcher on how the cuticle of arthropods was developed in a way to cope up with the infra-red (IR) and ultra-violet (UV) irradiation along with the demands for sensory transmission, movement, etc. This analysis helped to understand the similarity between the designs of the cuticle with well-known technology is only approximately 20% [72].

Another systematic biomimetic approach is to make an inventory of biomimetic solutions. Once they are exposed while analyzing biological materials, they can be converted into large databases where it is possible to retrieve through engineers with a notion to look forward to technical solutions. Earlier, such kind of databases have been designed for selecting materials when dealing with technical developments and nowadays such type of databases have been applied for both materials and processes [73, 74]. Initially, numerous efforts have been made to develop a system in which all well-known biomimetic solutions can be placed altogether categorized on the basis of function [75, 76]. These types of tools are very much profitable for developing bio-inspired materials and processes. At last, the manufacturing verification of biological processes may result in an iterative process between biology and engineering which provides and way to gain an understanding from engineering to biology. This significant unexplored way renders the possibility to contribute towards biological science by numerous engineers [77, 78].

18.4 Self-Healable Materials: Case Studies

This section deals with the overview of numerous case studies that elaborates the mechanism of self-repair in various biological archetypes to get a better insight of such mechanisms at play [79].

18.4.1 Regrowth of Limbs

Regeneration is defined as the restoration of obsolete body parts and the repair of their structure and functions, both. The process of regeneration of the whole body from small parts takes place in both simple organisms like planarians and hydra-simple freshwater polyps and more complex organisms like urodele amphibians such as the salamander, newt, frogs like *Xenopus* spp., zebrafish, etc. The knowledge related to the lack of regeneration capacity in humans as compared to some animals is still infancy whereas in general there was a loss of regenerative capacities during evolution [80].

Here in this case study the researchers mainly focused on the re-growth of limbs in salamanders which is a well-documented and extraordinary example of regeneration [81]. Apart from the regeneration of limbs and tails, salamanders have the capacity to restore both upper and lower jaws, specific sections of the eye, the intestine, minute parts of the heart, etc. [82]. Of all above, the remarkable feature is the complete re-growth of a limb (proximo-distal axis: between shoulder and wrist) which includes complex tissue patterning, full vascularization, innervations, etc. The process of regeneration of limb is categorized into three phases [83]; (a) in the initial sealing phase, the surface of the wound is wrapped with a specialized epithelium named as wound epithelium, within a few hours after removing it surgically. With the passage of time, this wound epithelium is converted into a specialized secretory epithelium which is considered as a source of signals which is significant enough through the regeneration process, (b) this wound epithelium helps in de-differentiation through the induction of the cells in the underlying tissues which means returning to less specialized characteristics of cells following the proliferation of cells which results in the formation of blastema (which is an amalgamation of cells with a strength to generate particular organs and body parts) at the end of the amputated limb [81], (c) in the final phase, which takes place after approximately two weeks, the blastema cells get re-differentiated into an exact replica of the limb which has been removed surgically. At this very particular phase, the blastema acts similar to the newly formed embryonic limb bud in a way the routes of these two mechanisms that is restoration and development of limb are thought to assemble [83].

The process of distal transformation explains that limb elements that are more distal to its positional memory have the potential to regenerate and thus when blastema of the upper arm is transplanted onto a lower arm stump, there is a reformation of all the distal structure through the blastema from the position where the regenerated limb consists of sections of a second lower arm [84]. With the surgical removal of the forelimb of salamander via hand and later sutured to the side of the body which results in lucrative healing. Furthermore, another incision was made via the upper limb. Since both stumps had positional memory of an upper limb, there was the regeneration of two complete limbs [85].

After a clean investigation between the mechanism of restoration and embryonic formation of limbs from limb buds, yet, the query raises on the similarity between the elemental mechanisms of both the process. Contrary to the development of the embryo, numerous tissues on an amputated limb of an adult salamander is an epitome of a huge complex initial condition for the process of regeneration [81]. Another difference is the size of the plane of limb amputation which is about ten times larger than a limb bud due to diffusion of certain signaling molecules [82]. Another difference is the role of innervations in the process of regeneration. In certain conditions, the surgical interruption with a notion to divert the nerve endings to a lateral wound on the limb results in the development of a complete patterned limb from this lateral wound in developed animals [81, 83].

18.4.2 The Mechanism of Bone Healing

Bone with complex hierarchical structure show self-healing strength at numerous levels varying from the restoration of macroscopic fractures to form reversible bonds again at nanoscale [79]. First and foremost, the mechanism of healing of bone is a regenerative process as a whole and results in the complete restoration of the pre-fractured bone with no formation of scar tissue. Considering the micron scale, there will be no damage to the bone material via bone re-modeling mechanism whereas at the molecular level, due to the constant exchange of non-covalent molecular bonds, there will be an observation of plastic deformation of bone. When bones are loaded beyond their extreme strength, it results in the fracture of bone as well as damage to the nearby soft tissues. A particular condition which follows mechanical fixation of the fractured region, there will be a movement of bone fragments against each other and thus results in the process named as secondary bone healing through the generation of extra tissue as callus. This mechanism of secondary bone healing is further sub-divided into three overlying phases [86, 87]; (i) in the initial phase, there is a formation of a blood clot and further, the site with dead materials has an inflammatory response in the form of immune cells that clean up the site. At this stage, there will be an infiltration of hematoma through microscopic blood vessels and fibroblasts and is converted into granulation tissue, (ii) during the second stage, the mesenchymal progenitor cells move into the callus and further gets differentiated and form fibrous tissues, cartilage, fibro-cartilage, bone, etc. [88], (iii) during the final phase of bone re-modeling, there will be a removal of bone material which is present in excess and thus results in the formation of bone with virtual indistinguishable features when considering the structure and function from the pre-fractured initial arrangement of bone [79]. Blood clotting and scar formation both have the potential to freeze and repair the damage but do not have the potential to completely reinstate the tissue to its original non-damaged stage. There are various self-healable agents for the mechanism of self-healing bone cement (Figure 18.5) [17].

Concerning the molecular level, repair processes majorly devotes the plastic nature of bone and high toughness. Additionally, bone exploits the

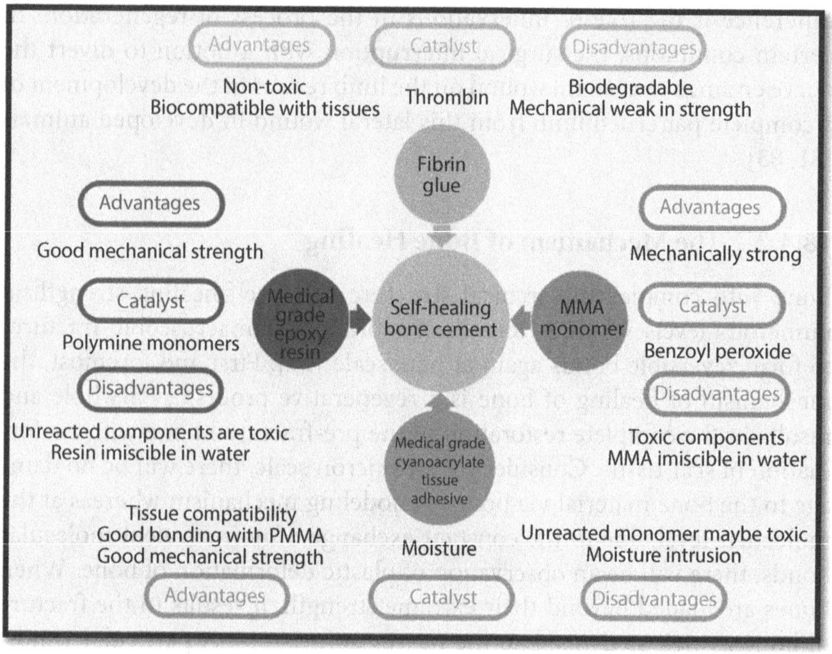

Figure 18.5 Self-healable systems for self-healing bone cement [17].

deformation process through the gliding of structural elements opposing each other with a notion to enhance the toughness of the material [89].

18.4.3 Cutaneous Wound Healing

In the majority of the human population, the wound healing of the skin such as cuts, scrapes, etc. is one of the most familiar biological mechanisms of restoration. In case of very minute cut or scrape, the cascade of wound healing of the skin is extremely complex [90, 91]. There are various prospects of cutaneous wound healing which are germane with a notion of designing self-healable materials, particularly the potential of rapid transportation of fundamental healable agents through an intricate network of the vasculature to the damaged area with a purpose to seal the wound quickly. Similar to the mechanism of healing of bone and cutaneous wound in the case of mammals is categorized into 3 discrete and overlying classes [90, 92]; (i) inflammation, (ii) formation of new tissues, and (iii) re-modeling. The very initial step is the process of restoration to tape out the wound through interim shelter to resolute the fluids and blood loss. This is achieved through the clot formation which consists of platelet aggregates entrenched in a loose fibrin matrix. This clot will serve as a scaffold for the infiltration of cells and acts as a stockpile for the signaling molecules [79].

There is a recruitment of various inflammatory leucocytes at the site of the injury to avert the onset of infection through the removal of dead tissue and bacteria. With the initiation of the second stage, the epithelium starts behaving as a barrier again and thus, keratinocytes will migrate from the edge of the wound to the injured dermis which further separates the viable tissues from the desiccated eschar, which consists of necrotic tissue along with dried blood. In the case of healing a bone, there will be an initiation of production of new blood vessels termed as angiogenesis which provides blood supply at the site of injury and thus converts the fibrin matrix into granulation tissue, which constitutes a community of thick cell in a network of collagen, fibronectin, and hyaluronic acid (HA). Considering such stage, a few of the fibroblasts gets differentiated into myofibroblasts, which are capable enough to compact the nearby area of tissue through their contractile properties and thus helps in contracting the wound. At the final stage of re-modeling, in the majority, the cells get activated when the injured site goes through the controlled cell death termed as apoptosis or abandon the wound. With time (approximately a year), there will an initiation of re-modeling of the residual acellular collagen matrix, which forms a scar with type I collagen having a tensile strength of approximately 70% of its first value as compared to the uninjured skin [90, 92]. After all the possible research, the major clinical issue is the prevention and treatment of excessive scarring of the skin especially due to burns [93].

18.5 Applications of Bio-Inspired Self-Healable Materials—Examples

18.5.1 Bio-Inspired Ionic Skin for Pressure Sensing

Human skin is soft, potent and has the ability of self-healing. Such features provoked researchers to perform numerous investigations on artificial skin-like materials such as electronic skin, soft wearable sensors, etc. [94, 95]. Such kind of devices can convert external stimulus like pressure [96], strain [97, 98], temperature [99, 100], humidity [101], twist deformation [102], etc. into detectable electronic signals such as current [102], resistance [102, 103], capacitance [96–98], etc. Due to this, there are numerous exploitations in the domain of artificial intelligence, soft robotics, personal healthcare, etc. [96, 104, 105]. Apart from the zeal of research in this domain, there are subtle obstacles while initiating the research of imitation of human skin due to the novel association of mechanical and sensory features [106].

With a notion to completely mimic the sensory features and feel of skin, it is substantial enough to design materials with features such as low elastic modulus, better stretchability, etc. [101]. Presently, there are three important approaches for designing stretchable intelligent skin: (a) buckling flexible electronic devices [107, 108], (b) imitating discontinuous stiff components, [109] and (c) designing intrinsic stretchable materials [110, 111]. Nonetheless, concerning biomimetic and medical applications, it is not sufficient to have materials with features of flexibility and stretchability. The important aspect of artificial skin apart from flexibility and stretchability is mechanical compliance and durability for practical applications like plastic deformation with similar curved and dynamic surfaces along with autonomous self-healing property to sense wound with quickly self-repair [111–113]. In the recent past, a group of researchers designed the concept of ionic skin in which there occurs ionic transduction via hydrogels or termed as ionic gels. This concept broadens the area of designing artificial intelligent skins with novel merits of hydrogel-based sensors like biocompatibility, high stretchability, etc [97] whereas exemplary ionic skins which are chemically crosslinked polyacrylamide (PAAM) hydrogels [97, 114, 115] or ionic gels [116] have inadequate self-healing capacity and less intimate interfacial contact. On the contrary, ionic skins made up of physically cross-linked supramolecular hydrogels have better stretchability, plastic deformation abilities, autonomous self-healing capacity and thus develops the next-gen soft artificial intelligent skins [106].

In this research work, researchers developed a bio-inspired mineral hydrogel composed of small amorphous calcium carbonate nanoparticles (ACC) cross-linked physically with polyacrylamide acid (PAA) and alginate chains to manufacture a unique mechanically acclimatized ionic skin sensor. Because of novel viscoelastic features, this sensor is docile and show self-healing potential along with sensing subtle pressure alterations like gentle finger touch, human movements, small water droplets. When stretched, the hydrogel has elasticity within a small range of strain and exhibits plastic deformation above 30% strain whereas, with large strain, the plastic deformation is because of the splitting of the physical crosslinks present within the hydrogel. To verify the behavior of hydrogel in comparison with the curved and dynamic surfaces, the film of hydrogel was surfaced over a prosthetic finger. The hydrogel film dynamically acclimatizes itself and accustoms the movement of the finger. The team sandwiched hydrogel between two acrylic elastomer tapes, namely Very High Bond (VHB) 4905, 3M, in close contact and then stretched. It has been observed that

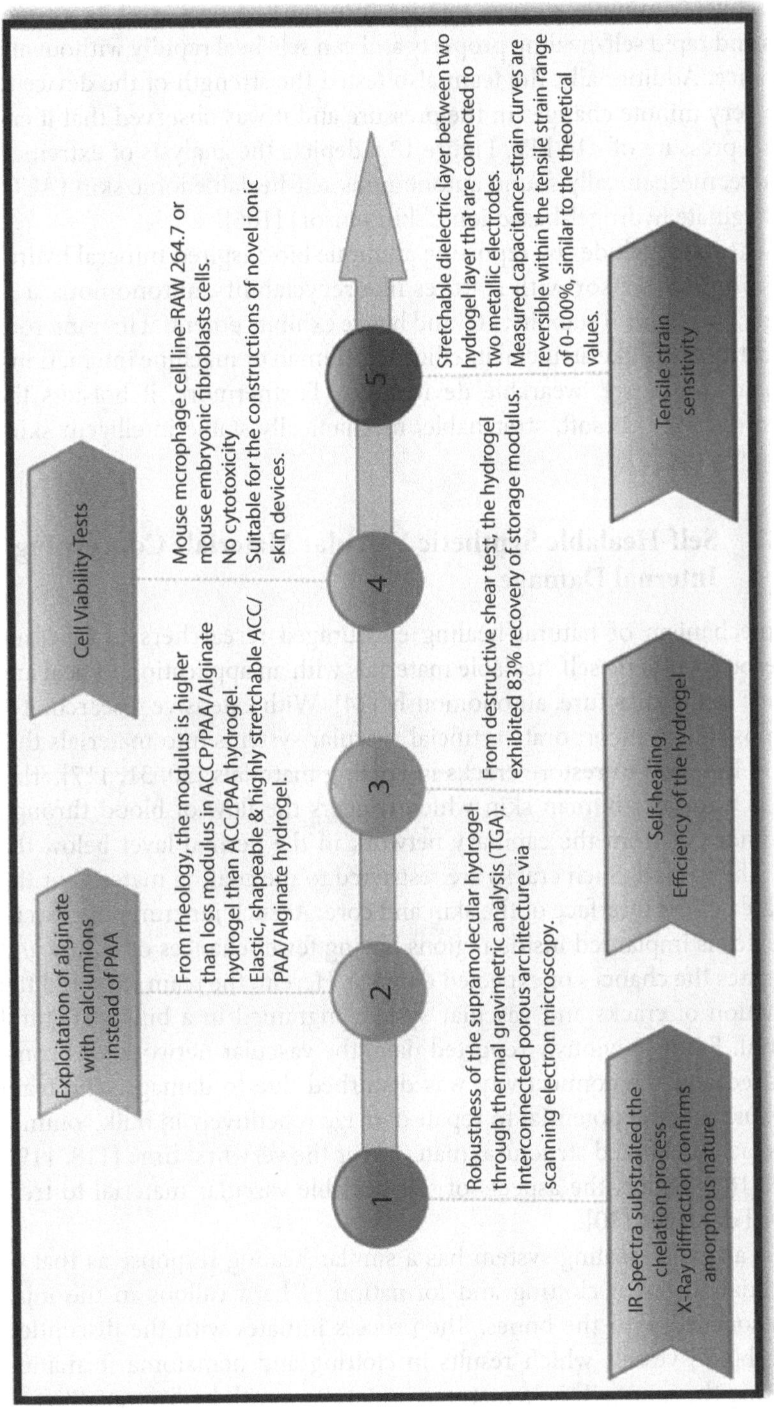

Figure 18.6 Analysis of extremely perceptive, mechanically docile, and autonomous self-healable ionic skin [106].

within twenty minutes at room temperature, the hydrogel exhibits autonomous and rapid self-healing property and can self-heal rapidly without any assistance. Additionally, the team also tested the strength of the device to sense very minute changes in the pressure and it was observed that it can sense a pressure of <10 kPa. Figure 18.6 depicts the analysis of extremely sensitive, mechanically sound autonomous self-healable ionic skin (ACC/PAA/alginate hydrogel-based ionic skin sensor) [106].

The team concludes by reporting a unique bio-inspired mineral hydrogel as an ionic sensor with features like recyclability, autonomous self-healing, mechanical stability, etc. and hence exhibits potential in numerous applicative areas like artificial intelligence, human or machine interactions, personal healthcare, wearable devices, etc. Furthermore, it bolsters the design of next-gen soft, stretchable, mechanically stable intelligent skin-like apparatus [106].

18.5.2 Self-Healable Synthetic Vascular Materials Concerning Internal Damage

The mechanism of natural healing encouraged researchers to fabricate numerous synthetic self-healable materials with an application to heal any damage to the structure, autonomously [14]. With extensive research, it is now possible to incorporate artificial vascular systems into materials that require less load to restore cracks in coating materials [30, 31, 117]. This approach mimics human skin which triggers the flow of blood through the minor cut from the capillary network in the dermal layer below the site of the wound. Such cracks are restricted to the coating material or the near area of the interface of the skin and core. At such juncture, the vascular device is implanted inside regions having fewer chances of failure and eliminates the chances of expected damage. Herein, the team explored the interaction of cracks and vascular system ingrained in a bulk structural material. From previously reported data, the vascular network was compromised and the connectivity was disturbed due to damage. The team also illustrates the potential to repair damage repetitively in bulk volumes within a vascularized structural material for the very first time [118, 119]. Figure 18.7 depicts the aspects of self-healable vascular material to treat internal damage [120].

This artificial healing system has a similar healing response as that of the initial stages of clotting and formation of hard callous in the injuries associated with the bones. The process initiates with the disruption of the blood vessels which results in clotting and hematoma formation enclosing the injury. This hematoma mineralizes with hydroxyapatite at a

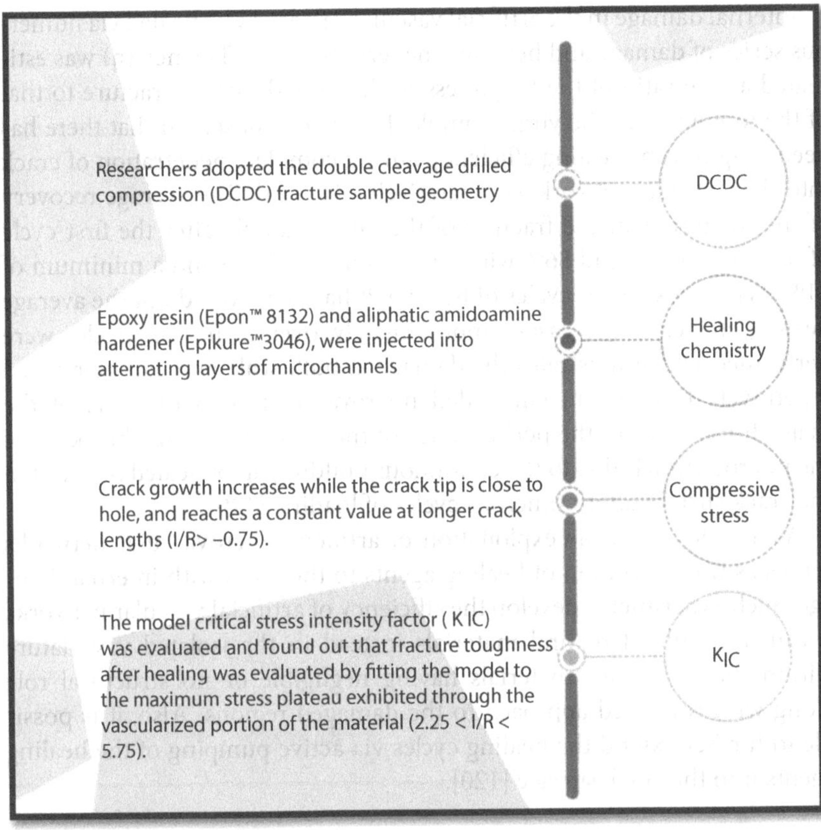

Figure 18.7 Aspects of self-healable vascular material for internal damage [120].

gradual pace and forms a hard callus of woven bone [121]. In the course of time, the lamellar bone replaces woven bone and accommodates the service loads in accordance with Wolff's law [89, 122]. Furthermore, vascular canals present in the cortical bone escalate the toughness of the bulk material through deflection of crack propagation and enhancing the bridging of crack and microcracking [123]. This healing process has two components (i) epoxy resin (Epon™ 8132) and (ii) aliphatic amidoamine hardener (Epikure™3046), there both were infused into the layers of microchannels [31]. The components exhibit less with a stoichiometric ratio of 2.2:1, which is in comparison with the microchannels via sample thickness (3:2). Since every layer of microchannels are separate, there is no chance of contact of both of the healing components as far as fracture happens and intersect both resin- and hardener-filled microchannel [120].

Internal damage in the artificial vascular material was healed via numerous series of damage and healing. The term healing efficiency (η) was estimated as the ratio of the toughness of the healed sample fracture to that of the toughness of the virgin sample. It has been observed that there has been a significant healing efficiency during complete penetration of crack into the vascular network, comparatively [31, 124]. The average recovery of the toughness of the fracture of the virgin sample after the first cycle of healing was around 86% with a maximum of 120% and a minimum of 61%. With subsequent cycles of healing, it has been found out the average healing efficiency decreases. Approximately, over 13 healing cycles were performed which illustrates the damage at bulk scale and can be revived repetitively through an embedded microvascular network. One of the major hindrances in the performance of the network was the blockage of the microchannels due to the continuous building up of healed material in the crack surface after numerous cycles of loading [120].

As a conclusion, the exploitation of artificial microvascular networks serves as transportation of healing agents to the areas with internal damage. Such experiments develop the efficiency of artificial vascular networks within the bulk structural materials instead in the isolated vasculature within the secondary materials having negligible or no structural role along with restricted approach to the damaged regions. Also, it is possible to further extend the healing cycles via active pumping of the healing agents into the crack surface [120].

18.5.3 Biobased Self-Healable Color Hydrogel

Due to interaction of light with photonic nanostructures and periodic intrinsic nanostructures result in structural colors that have grabbed a lot of attention of the researchers because of their remarkable features such as their prevalence in nature, shelf-life, biological properties, etc. [125, 126]. A lot of efforts have been made on unraveling the bio-inspired structural color materials. There are numerous applicative areas of structural color hydrogels (one of the most significant examples to be studied concerning structural color materials) like in switches [127, 128], optical devices [129, 130], sensing materials [131, 132], wearable electronics [133], etc. [22, 134]. Nonetheless, during their research from applications point of view, it is unavoidable to manage the degradation and aggregation of damaged materials with a notion to accomplish next-gen. Thus, it is important to synthesize bio-inspired structural color materials with better survival rates, which is a need of an hour. Simultaneously, to boost the survival rates and shelf-life of an organism, the function of self-healing like spontaneous healing of the injury, restoration of the function of a particular organ or material, etc. [135, 136]. Keeping such remarkable features in mind, researchers did a lot of investigation and fabricated self-healable hydrogel materials and fused them with the homogeneous hydrogels for numerous significant applications [137–139]. In spite of this, it is possible to carry forward the process of fusion of self-healable hydrogels in the adjacent nanostructures but that will result in the demolition of the periodic photonic scaffolds. Thus, a self-healable hydrogel with substantial structural color is still not documented. Also, the construction of the same is still a big challenge [140].

Herein this research work, the team illustrates the construction of nanocomposite which constitutes a self-healable structural color (Figure 18.8). This particular nanostructure consists of (a) inverse opal scaffold of methacrylate gelatin (GelMA) hydrogel, and (ii) glutaraldehyde filler which is cross-linked with BSA hydrogel through enzyme additives of glucose oxidase (GOX) and catalase (CAT). This nanocomposite ensures the stability of both inverse opal structure and its associated structural colors. On the other hand, the protein hydrogel filler renders material with seal-healable property by attaching glutaraldehyde with the lysine residues of BSA and the GOX and CAT additives through reversible covalent interaction. Since both the methacrylate gelatin (GelMA) and protein hydrogels are obtained from naturally occurring organisms, thus the nanocomposite imparts high biocompatibility and plasticity. It has been illustrated that it is possible to fabricate a chain of novel structural color materials with one-dimensional linear microfiber, two-dimensional pattern, and three-dimensional photonic path structures via assembling and healing the components of nanocomposite structural color hydrogels thus rendering fit for numerous applications like counterfeit management, integrated optics, biomedical engineering, etc. [140].

The inverse opal scaffolds of methacrylate gelatin (GelMA) hydrogel were synthesized through the replication of templates made up of silica colloidal crystals fabricated through the process of self-assembly of silica nanoparticles in silica capillaries or over the glass slides. These crystals are present in a closed packed manner and ultimately results in the formation of an ordered structure when dehydrated. This ordered packing renders the property of interconnected nanopores all through the template which allows the infiltration of methacrylate gelatin (GelMA) pre-gel suspension solution. After the penetration of the pre-gel suspension solution into the nanopores and complete filling of all the voids present in the template through capillary action, the solution was polymerized and later converted into hydrogel under the influence of ultra-violet (UV) light. Eventually, the

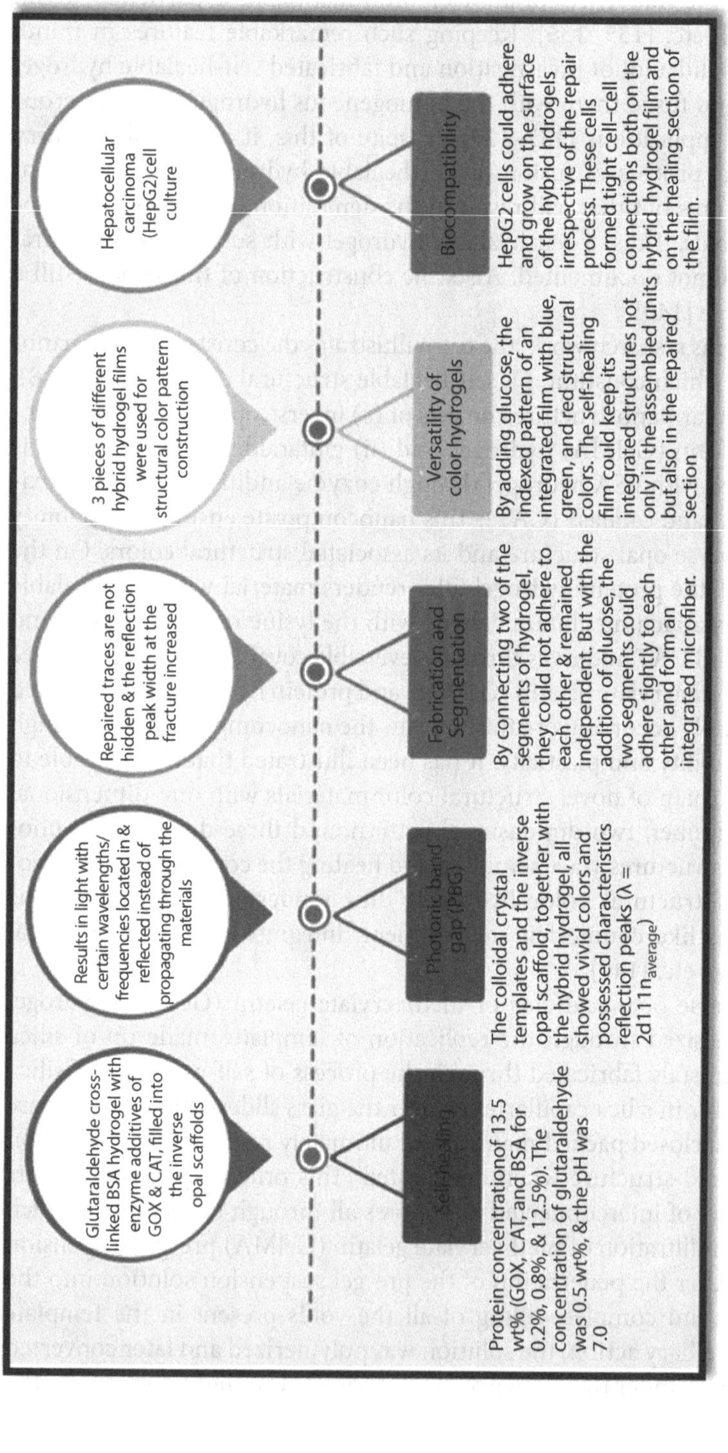

Figure 18.8 An outline of bio-based self-healable structural color hydrogels [140].

inverse opal scaffolds were procured when silica nanoparticles were etched and named as inverse opal methacrylate gelatin (GelMA) hydrogel scaffold. Such type of scaffold outlines magnificent structural colors which are considered as a significant feature of such kind of materials. The enzyme additive glucose oxidase (GOX) facilitates glucose to get oxidized into gluconolactone following hydrolysis to gluconic acid for decreasing the pH value of the protein hydrogel. The by-product of the glucose oxidation, hydrogen peroxide (H_2O_2) gets decomposed into water (H_2O) and oxygen (O_2) through the action of enzyme catalase to avoid the oxidation of the imine bonds along with providing support to the cyclic reaction of glucose oxidation. By regulating pH, imine bonds render the ability to self-heal the protein hydrogel [140].

From this illustrative research work, it is possible to even heal the joints of microfibrous segments with varied structural colors by treating with glucose. With varied structural colors, the nearby segments can still adhere together firmly and led to the formation of an integrated microfiber with enhanced elasticity when compared to the actual microfibers. This suggests that enzyme assisted hybrid structural color hydrogel was appropriate for varied types of inverse opal scaffolds with varied nanometric aperture sizes. Hence, a novel strategy of self-assembly have been devised to fabricate a multifarious structural color hydrogel [140].

To illustrate the concept of two-dimensional patterning and three-dimensional structural color, the methacrylate gelatin, and proteins hybrid yellow structural color hydrogel film was incised into pieces of varied sizes following stacking from large to the small size and can touch each other due to their porous property. Hence, this renders an integrated three-dimensional pyramid structure through the strategy of self-healing assembly. Such kind of characteristic properties of the self-healable structural color hydrogels portrays the skilled nature of hydrogels in various applicative areas [140].

18.5.4 Bio-Inspired Support for Repairing Damaged Articular Cartilage

The damaged cartilage and associated dysfunction of clinical joints alter numerous individuals. Since it is well-known that damaged articular cartilage has restricted potential to self-repair, face intense wear, age-related degeneration due to tissue vascularity, the less mitotic activity of the nearby cells, chondrocytes [141]. With the existence of numerous surgical approaches to restore cartilage defects and function of joints [142], still the restored tissue led to the generation of mechanically inferior fibrocartilage which has limited clinical efficacy. Apart from surgical procedures, additional surgery helps in regaining complete function and thus advanced partial or total knee replacement. Hence, it is a need of an hour to discover novel regenerative medicines to enhance the process of restoration to allow ample tissue regeneration and there is an overwhelming need for novel restorative medicine approaches to build up the repair mechanisms to aid sufficient tissue regeneration and durability. In recent days, scaffolds based on polymeric materials have been exploited to enhance the process of tissue regeneration through the recapitulation of numerous physical cues of nearby tissues [143]. Concerning this, various methods have been exploited to produce fibrous scaffolds such as self-assembly, phase separation, electrospinning, etc. [144].

In this research work, the team utilized a low-density, a three-dimensional nanofibrous network which constitutes poly(vinyl alcohol)-methacrylate (PVA-MA) and a composite of poly(vinyl alcohol)-methacrylate (PVA-MA)/chondroitin sulfate (CS)-chondroitin sulfate-methacrylate (CS-MA) which provides a bed for the infiltration of cells during cell seeding, designated as poly(vinyl alcohol) (PVA) fibers and chondroitin sulfate (CS) fibers, respectively (Figure 18.9). Evaluation of the nanofibers was executed to acknowledge morphology, diameter, and the content of chondroitin sulfate (CS). *In vitro* investigation was done to analyze the chondrogenesis of bone marrow-derived stem cells (bMSCs). In addition to this, they make use of rat model with an osteochondral defect to scrutinize the ramification of acellular fiber scaffolds to repair articular tissues under the influence of the physiological environment. The nanofibrous scaffolds can provide base and augment the *in-vitro* chondrogenesis along with the complete support to initial cell infiltration and tissue restoration in an *in-vivo* osteochondral defect model. Through the incorporation of the bio-signal of the chondroitin sulfate (CS) into the extracellular matrix of cartilage that gives direction for the specificity of tissues, *in-vitro* and *in-vivo* [145].

From the scanning electron microscope (SEM) imaging, the fibers of poly vinyl alcohol (PVA) showed the diameter of 500 ± 100 nm whereas the chondroitin sulfate-methacrylate (CS-MA) fibers showed a diameter of 410 ± 60 nm. To get more knowledge of the impact of the physical structure of the nanofibrous network and incorporated bio-polymer during the formation of tissue, the proliferation of the cell was monitored on the 21st and 42nd days. It has been found out that there has been an increase in the content of DNA by 2.6 ± 0.7 times and 2.03 ± 0.9 times in the chondroitin sulfate (CS) fibers and polyvinyl alcohol (PVA) fibers, respectively on day 42 as compared to 21st day. Also, at 21st day, the cells

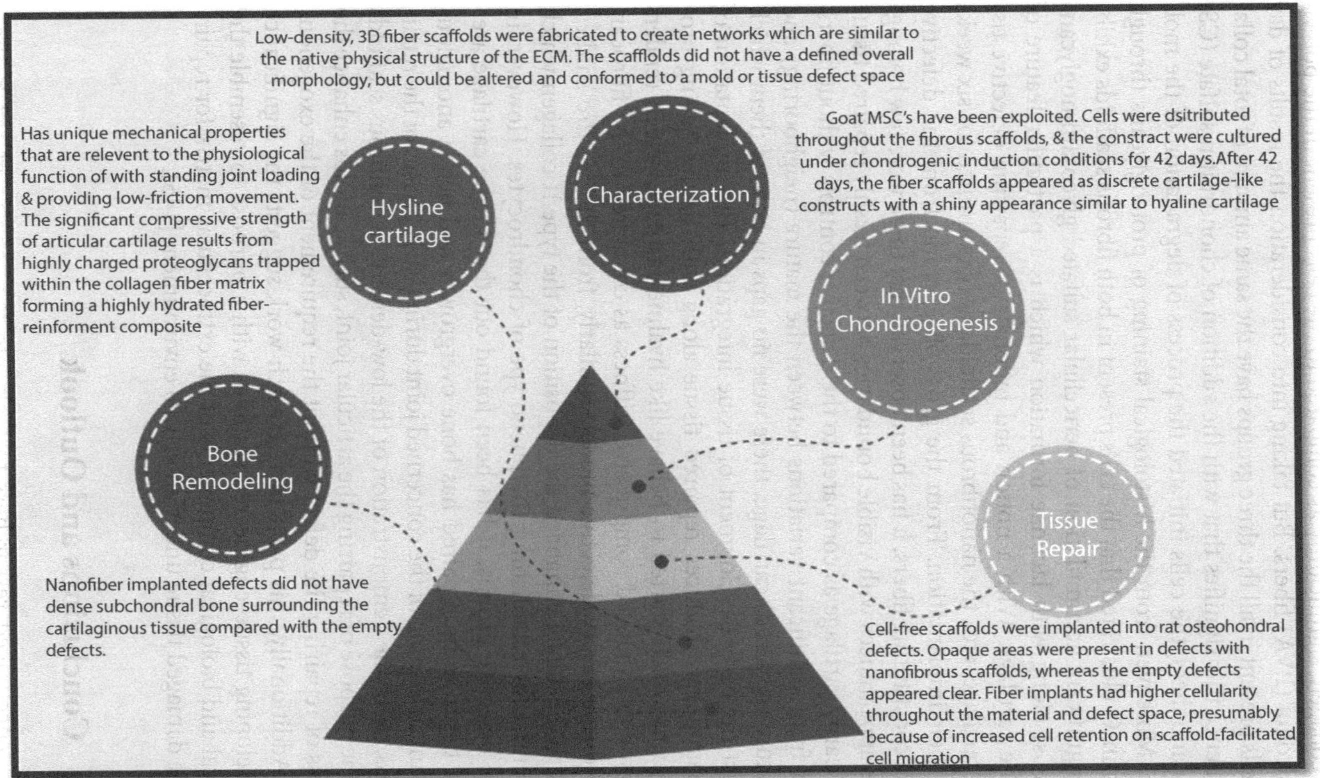

Figure 18.9 An outline of support of nanofibers for repairing articular cartilage [145].

present in the chondroitin sulfate (CS fibers) results in the production of total collagen, significantly as compared to the cells present in the polyvinyl alcohol (PVA) fibers. But taking into consideration the results of day 42, cells present in all the three groups have the same amount of total collagen, hence this signifies that with the addition of chondroitin sulfate (CS) into nanofibers, the cells initiated the process of degradation of the molecule. Moreover, from the histological staining of proteoglycans through safranin-O illustrated that the cells present in both fibrous scaffolds exhibited significant accumulation of pericellular sulfated glycosaminoglycans (sGAGs) along with lacunae formation which is a particular feature of hyaline cartilage. With a notion and rationale of repairing defective tissues, cartilage, etc., the nanofibrous scaffolds were examined for six weeks after their implantation. From the gross observation from the defective model constituting fibers, it has been observed that there were few homogenous appearances with visible boundaries between freshly repaired tissue and nearby cartilage as compared to the defective controls without fibers. Apart from significant variations between the mature (neighboring) and immature (repaired) cartilage, there were no gaps in both of them which suggest the early development of tissue integration. The implantation of the nanofibers produces repaired tissue along with tissues having proteoglycans which started to behave like hyaline cartilage, which renders tissues with enhanced compressive stiffness as compared to the fibrocartilage of the empty defective model. Similarly, from the *in-vitro* studies, chondroitin sulfate enhanced the deposition of the type II collagen which is essential for establishing the phenotype of chondrocytes. However, in all the defective groups, it has been found out that in the cartilage area of the tissue to be repaired has bone overgrowth and hence anomalous mechanical loading of the concerned joint during early phases of the repair process. With the demonstration of the low-density, nanofibrous scaffolds with the purpose of repairing the articular joint, such scaffolds enhance the process of repairing the defect without the requirement of the exogenous cells. Additionally, this particular research work suggests the significance of developing tissue-engineered scaffolds with a purpose to resemble the physical and biological components of the extracellular matrix for repairing the damaged tissue under an *in-vivo* environment [145].

18.6 Conclusions and Outlook

Nature encouraged researchers to look into the concept of self-healing and self-healable materials which has grabbed attention at a large scale in the last few years and advance the level of research. The principle of self-healable materials has extended the global scientific world and has attracted possible attention from chemistry towards engineering. The significant switch from two-dimensional to three-dimensional materials have started to revolutionize the ways to create bio-inspired micro-environment with a purpose to evaluate the response of the cell towards material cues, engineering tissues, *in-vitro* disease modeling, precision medicines, etc. for the treatment of patients. The obsolete field of biomimetic materials research has been enhanced in a dynamic way due to certain reasons like better interaction between biological materials and materials sciences. The concept of bio-inspired materials required detailed information regarding the relationship between the structure and functions of the biological material. The word self-repair is itself contradictory which means that self suggests that the process will undergo autonomously and repair suggests the exploitation of tools. This misnomer exploits the potential of biological role models for devising innovative biomimetic self-repairing materials and allows such materials to respond against alterations in the environment and the requirements of humans. In the majority of the self-repairing systems in numerous industries, it is not possible to mimic the potential of nature.

References

1. Jeronimidis, G. and Atkins, A.G., Mechanics of Biological Materials and Structures: Nature's Lessons for the Engineer. *Proc. Inst. Mech. Eng. Part C J. Mech. Eng. Sci.*, 209, 4, 221–235, 1995.
2. Currey, J.D., Materials science. Hierarchies in biomineral structures. *Science*, 309, 5732, 253–254, 2005.
3. Fratzl, P., Biomimetic materials research: What can we really learn from nature's structural materials? *J. R. Soc. Interface*, 4, 637–642, 2007.
4. Yang, Y. and Urban, M.W., Self-healing polymeric materials. *Chem. Soc. Rev.*, 42, 17, 7446–7467, 2013.
5. Syrett, J.A., Becer, C.R., Haddleton, D.M., Self-healing and self-mendable polymers. *Polym. Chem.*, 1, 7, 978–987, 2010.
6. Wool, R.P., Self-healing materials: A review. *Soft Matter*, 4, 3, 400–418, 2008.
7. Gupta, S., Zhang, Q., Emrick, T., Balazs, A., Russell, T., Entropy-driven segregation of nanoparticles to cracks in multilayered composite polymer structures. *Nat. Mater.*, 5, 3, 229–233, 2006.
8. Yang, J., Keller, M.W., Moore, J.S., White, S.R., Sottos, N.R., Microencapsulation of isocyanates for self-healing polymers. *Macromolecules*, 41, 24, 9650–9655, 2008.

9. Burattini, S., Greenland, B.W., Chappell, D., Colquhoun, H.M., Hayes, W., Healable polymeric materials: A tutorial review. *Chem. Soc. Rev.*, 39, 6, 1973–1985, 2010.
10. Cannon, S.L., McKenna, G.B., Statton, W.O., Hard-Elastic fibers. (A review of a novel state for crystalline polymers). *J. Polym. Sci. Macromol. Rev.*, 11, 1, 209–275, 1976.
11. Guimard, N.K., Oehlenschlaeger, K.K., Zhou, J., Hilf, S., Schmidt, F.G., Barner-Kowollik, C., Current trends in the field of self-healing materials. *Macromol. Chem. Phys.*, 213, 2, 131–143, 2012.
12. Cho, S.H., White, S.R., Braun, P.V., Self-healing polymers: Self-healing polymer coatings (Adv. Mater. 6/2009). *Adv. Mater.*, 21, 6, 2009.
13. Park, J.H. and Braun, P.V., Coaxial electrospinning of self-healing coatings. *Adv. Mater.*, 22, 4, 496–499, 2010.
14. Blaiszik, B., Kramer, S., Olugebefola, S., Moore, J., Sottos, N., Self-healing polymers and composites. *Annu. Rev. Mater. Res.*, 40, 5, 179–211, 2010.
15. Yin, X., Liu, Z., Wang, D., Pei, X., Yu, B., Zhou, F., Bioinspired self-healing organic materials: Chemical mechanisms and fabrications. *J. Bionic Eng.*, 12, 1, 1–16, 2015.
16. Liu, Y.L. and Chuo, T.W., Self-healing polymers based on thermally reversible Diels–Alder chemistry. *Polym. Chem.*, 4, 7, 2194–2205, 2013.
17. Brochu, A.B.W., Craig, S.L., Reichert, W.M., Self-healing biomaterials. *J. Biomed. Mater. Res. A*, 96A, 2, 492–506, 2011.
18. van der Zwaag, S., *Self Healing Materials—An Alternative Approach to 20 Centuries Materials Science*, Springer, Dordrecht, The Netherlands, 100, 2007.
19. Hager, M.D., Greil, P., Leyens, C., van der Zwaag, S., Schubert, U.S., Self-healing materials. *Adv. Mater.*, 22, 47, 5424–5430, 2010.
20. Kessler, M.R., Self-healing: A new paradigm in materials design. *Proc. Inst. Mech. Eng. Part G J. Aerosp. Eng.*, 221, 4, 479–495, 2007.
21. White, S.R., Caruso, M.M., Moore, J.S., Autonomic healing of polymers. *MRS Bull.*, 33, 8, 766–769, 2008.
22. Sottos, N. *et al.*, Autonomic healing of polymer composites. *Nature*, 409, 6822, 794–797, 2001.
23. Amamoto, Y., Kamada, J., Otsuka, H., Takahara., A., Matyjaszewski., K., Repeatable photoinduced self-healing of covalently cross-linked polymers through reshuffling of trithiocarbonate units. *Angew. Chem. Int. Ed. Engl.*, 50, 7, 1660–1663, 2011.
24. Wilson, G.O., Caruso, M.M., Reimer, N.T., White, S.R., Sottos, N.R., Moore, J.S., Evaluation of ruthenium catalysts for ring-opening metathesis polymerization-based self-healing applications. *Chem. Mater.*, 20, 10, 3288–3297, 2008.
25. Cho, S.H., White, S.R., Braun, P.V., Room-temperature polydimethylsiloxane-based self-healing polymers. *Chem. Mater.*, 24, 21, 4209–4214, 2012.
26. Neuser, S. and Michaud, V., Improving solvent-based self-healing materials through shape memory alloys. *Polymer (Guildf).*, 53, 2, 370–378, 2012.

27. Xu, W. and Li., W., Constitutive modeling of shape memory polymer based self-healing syntactic foam. *Int. J. Solids Struct.*, 47, 9, 1306–1316, 2010.
28. Cho, S.H., Andersson, H.M., White, S.R., Sottos, N.R., Braun, P.V., Polydimethylsiloxane-based self-healing materials. *Adv. Mater.*, 18, 8, 997–1000, 2006.
29. Kamphaus, J.M., Rule, J.D., Moore, J.S., Sottos, N.R., White, S.R., A new self-healing epoxy with tungsten (VI) chloride catalyst. *J. R. Soc. Interface*, 5, 18, 95–103, 2008.
30. Toohey, K.S., Sottos, N.R., Lewis, J.A., Moore, J.S., White, S.R., Self-healing materials with microvascular networks. *Nat. Mater.*, 6, 8, 581–585, 2007.
31. Toohey, K., Hansen, C., Lewis, J., White, S., Sottos, N., Delivery of two-part self-healing chemistry via microvascular networks. *Adv. Funct. Mater.*, 19, 9, 1399–1405, 2009.
32. Pang, J.W.C. and Bond, I.P., A hollow fibre reinforced polymer composite encompassing self-healing and enhanced damage visibility. *Compos. Sci. Technol.*, 65, 11–12, 2005.
33. Sanyal, A., Diels–Alder cycloaddition-cycloreversion: A powerful combo in materials design. *Macromol. Chem. Phys.*, 211, 13, 1417–1425, 2010.
34. Chen, X. *et al.*, A thermally re-mendable cross-linked polymeric material. *Science*, 295, 5560, 1698–1702, 2002.
35. Sauk, J., Byun, J., Kim, H., Composite nafion/polyphenylene oxide (ppo) membranes with phosphomolybdic acid (PMA) for direct methanol fuel cells. *J. Power Sources*, 143, 1–2, 136–141, 2005.
36. Stille, J.K. and Noren, G.K., Catenation and kinetics of the diels-alder step-growth reaction in the synthesis of phenylated polyphenylenes. *Macromolecules*, 5, 1, 49–55, 1972.
37. Stille, J.K., Rakutis, R.O., Mukamal, H., Harris, F.W., Diels–Alder polymerizations. iv. polymers containing short phenylene blocks connected by alkylene units. *Macromolecules*, 1, 5, 431–436, 1968.
38. Pratama, P.A., Sharifi, M., Peterson, A.M., Palmese, G.R., Room temperature self-healing thermoset based on the Diels–Alder reaction. *ACS Appl. Mater. Interfaces*, 5, 23, 12425–12431, 2013.
39. Goodall, G.W. and Hayes, W., Advances in cycloaddition polymerizations. *Chem. Soc. Rev.*, 35, 3, 280–312, 2006.
40. Zhang, Y., Broekhuis, A.A., Picchioni, F., Thermally self-healing polymeric materials: The next step to recycling thermoset polymers? *Macromolecules*, 42, 6, 1906–1912, 2009.
41. Canary, S.A. and Stevens, M.P., Thermally reversible crosslinking of polystyrene via the furan–maleimide Diels–Alder reaction. *J. Polym. Sci. Part A Polym. Chem.*, 30, 8, 1755–1760, 1992.
42. Liu, Y.L. and Chen, Y., Thermally reversible cross-linked polyamides with high toughness and self-repairing ability from maleimide- and furan-functionalized aromatic polyamides. *Macromol. Chem. Phys.*, 208, 2, 224–232, 2007.

43. Laita, H., Boufi, S., Gandini, A., The application of the Diels–Alder reaction to polymers bearing furan moieties. 1. Reactions with maleimides. *Eur. Polym. J.*, 33, 8, 1203–1211, 1997.
44. Goussé, C., Gandini, A., Hodge, P., Application of the Diels–Alder reaction to polymers bearing furan moieties. 2. Diels–Alder and Retro-Diels–Alder reactions involving furan rings in some styrene copolymers. *Macromolecules*, 31, 2, 314–321, 1998.
45. Grigoras, M. and Colotin, G., Copolymerization of a bisanthracene compound with bismaleimides by Diels–Alder cycloaddition. *Polym. Int.*, 50, 12, 1375–1378, 2001.
46. Syrett, J.A., Mantovani, G., Barton, W.R.S., Price, D., Haddleton, D.M., Self-healing polymers prepared via living radical polymerisation. *Polym. Chem.*, 1, 1, 102–106, 2010.
47. Stille, J.K. and Plummer, L., Polymerization by the Diels–Alder Reaction1,2. *J. Org. Chem.*, 26, 10, 4026–4029, 1961.
48. Chen, Y., Kushner, A.M., Williams, G.A., Guan, G., Multiphase design of autonomic self-healing thermoplastic elastomers. *Nat. Chem.*, 4, 6, 467–472, 2012.
49. Nakahata, M., Takashima, Y., Yamaguchi, H., Harada, A., Redox-responsive self-healing materials formed from host-guest polymers. *Nat. Commun.*, 2, 1, 511, Oct. 2011.
50. Burnworth, M. *et al.*, Optically healable supramolecular polymers. *Nature*, 472, 7343, 334–337, 2011.
51. Montarnal, D., Tournilhac, F., Hidalgo, M., Couturier, J.L., Leibler, L., Versatile one-pot synthesis of supramolecular plastics and self-healing rubbers. *J. Am. Chem. Soc.*, 131, 23, 7966–7967, 2009.
52. Montarnal, D., Cordier, P., Soulié-Ziakovic, C., Tournilhac, F., Leibler, L., Synthesis of self-healing supramolecular rubbers from fatty acid derivatives, diethylene triamine, and urea. *J. Polym. Sci. Part A Polym. Chem.*, 46, 24, 7925–7936, 2008.
53. Moore, R.B., Ionomers: Synthesis, Structure, Properties and Applications Edited by M.R. Tant (Eastman Chemical Company), K.A. Mauritz (University of Southern Mississippi), and G.L. Wilkes (Virginia Polytechnic Institute & State University). Chapman & Hall: New Yo. *J. Am. Chem. Soc.*, 120, 15, 3822, 1998.
54. Schué, F., Introduction to ionomers, A. Eisenberg and Joon-Seop Kim Wiley Interscience, New York, USA, p 327, 1998, *Polym. Int.*, 2, 235, 2000.
55. Varley, R.J. and van der Zwaag, S., Towards an understanding of thermally activated self-healing of an ionomer system during ballistic penetration. *Acta Mater.*, 56, 19, 5737–5750, 2008.
56. Stadler, R. and Freitas, L., Thermoplastic elastomers by hydrogen bonding 1. Rheological properties of modified polybutadiene. *Colloid Polym. Sci.*, 264, 9, 773–778, 1986.

57. Herbst, F., Dohler, D., Michael, P., Binder, W.H., Self-healing polymers via supramolecular forces. *Macromol. Rapid Commun.*, 34, 3, 203–220, 2013.
58. De Lucca Freitas, L.L. and Stadler, R., Thermoplastic elastomers by hydrogen bonding. 3. Interrelations between molecular parameters and rheological properties. *Macromolecules*, 20, 10, 2478–2485, 1987.
59. Boiko, Y., Guerin, G., Marikhin, V., Prud'homme, R., Healing of interfaces of amorphous and semi-crystalline poly(ethylene terephthalate) in the vicinity of the glass transition temperature. *Polymer (Guildf)*, 42, 21, 8695–8702, 2001.
60. Kim, Y.H. and Wool, R.P., A theory of healing at a polymer-polymer interface. *Macromolecules*, 16, 7, 1115–1120, 1983.
61. Hoeben, F.J.M., Jonkheijm, P., Meijer, E.W., Schenning, A.P.H.J., About supramolecular assemblies of π-conjugated systems. *Chem. Rev.*, 104, 4, 1491–1546, 2005.
62. Pérez-García, L. and Amabilino, D.B., Spontaneous resolution, whence and whither: From enantiomorphic solids to chiral liquid crystals, monolayers and macro- and supra-molecular polymers and assemblies. *Chem. Soc. Rev.*, 36, 6, 941–967, 2007.
63. Swiegers, G.F. and Malefetse, T.J., New self-assembled structural motifs in coordination chemistry. *Chem. Rev.*, 100, 9, 3483–3538, 2000.
64. Burattini, S. *et al.*, A self-repairing, supramolecular polymer system: Healability as a consequence of donor–acceptor pi–pi stacking interactions. *Chem. Commun. (Camb).*, 28, 44, 6717–6719, 2009.
65. Fantner, G.E. *et al.*, Sacrificial bonds and hidden length: unraveling molecular mesostructures in tough materials. *Biophys. J.*, 90, 4, 1411–1418, 2006.
66. Thompson, J.B., Kindt, J.H., Drake, B., Hansma, H.G., Morse, D.E., Hansma, P.K., Bone indentation recovery time correlates with bond reforming time. *Nature*, 414, 6865, 773–776, 2001.
67. Fantner, G.E. *et al.*, Sacrificial bonds and hidden length dissipate energy as mineralized fibrils separate during bone fracture. *Nat. Mater.*, 4, 8, 612–616, 2005.
68. Gupta, H., Fratzl, P., Kerschnitzki, M., Benecke, G., Wagermaier, W., Kirchner, H., Evidence for an elementary process in bone plasticity with an activation enthalpy of 1 eV. *J. R. Soc. Interface*, 4, 13, 277–282, 2007.
69. Gupta, H.S., Seto, J., Wagermaier, W., Zaslansky, P., Boesecke, P., Fratzl, P., Cooperative deformation of mineral and collagen in bone at the nanoscale. *Proc. Natl. Acad. Sci. U. S. A.*, 103, 47, 17741–17746, 2006.
70. Fratzl, P., Gupta, H.S., Paschalis, E.P., Roschger, P., Structure and mechanical quality of the collagen–mineral nano-composite in bone. *J. Mater. Chem.*, 14, 2115–2123, 2004.
71. *Skeletal function and form: Mechanobiology of skeletal development, aging, and regeneration*, Carter, Dennis R. and Beaupré, Gary S., Cambridge University Press: Cambridge, 2001. pp. xii–318, *Int. J. Osteoarchaeol.*, 19, 3, 175–176, Scheuer 2003.

72. Vincent, J.F.V., Deconstructing the design of a biological material. *J. Theor. Biol.*, 236, 1, 73–78, 2005.
73. Ashby, M.F., *Materials selection in mechanical design*, 3rd edn., Butterworth-Heinemann, Oxford, UK, 660, 2003.
74. Ashby, M.F., Bréchet, Y., Cebon, D., Selection strategies for materials and processes. *Adv. Eng. Mater.*, 4, 6, 327–334, 2002.
75. Vincent, J.F.V. and Mann, D.L., Systematic technology transfer from biology to engineering. *Philos. Trans. A. Math. Phys. Eng. Sci.*, 360, 1791, 159–173, 2002.
76. Vincent, J.F.V., Bogatyreva, O.A., Bogatyrev, N.R., Bowyer, A., Pahl, A.K., Biomimetics: Its practice and theory. *J. R. Soc. Interface*, 3, 9, 471–482, 2006.
77. Csete, M.E. and Doyle, J.C., Reverse engineering of biological complexity. *Science*, 295, 5560, 1664–1669, 2002.
78. Vincent, J., Biomimetic modelling. *Philos. Trans. R. Soc. Lond. B. Biol. Sci.*, 358, 1437, 1597–1603, 2003.
79. Harrington, M.J., Speck, O., Speck, T., Wagner, S., Weinkamer, R., *Biological archetypes for self-healing materials BT-Self-healing materials*, M.D. Hager, S. van der Zwaag, U.S. Schubert, (Eds.), pp. 307–344, Springer International Publishing, Cham, 273, 2016.
80. Alvarado, A.S., Regeneration in the metazoans: Why does it happen? *BioEssays*, 22, 6, 578–590, 2000.
81. Nacu, E. and Tanaka, E.M., Limb regeneration: A new development? *Annu. Rev. Cell Dev. Biol.*, 27, 409–440, 2011.
82. Brockes, J.P. and Kumar, A., Appendage regeneration in adult vertebrates and implications for regenerative medicine. *Science*, 310, 5756, 1919–1923, 2005.
83. Bryant, S.V., Endo, T., Gardiner, D.M., Vertebrate limb regeneration and the origin of limb stem cells. *Int. J. Dev. Biol.*, 46, 7, 887–896, 2002.
84. Butler, E.G., Regeneration of the urodele forelimb after reversal of its proximo-distal axis. *J. Morphol.*, 96, 2, 265–281, 1955.
85. French, V., Bryant, P.J., Bryant, S.V., Pattern regulation in epimorphic fields. *Science*, 123, 4257, 969–981, 1976.
86. Betts, D.C. and Muller, R., Mechanical regulation of bone regeneration: Theories, models, and experiments. *Front. Endocrinol. (Lausanne)*, 11, 5, 211, 2014.
87. Sfeir, C., Ho, L., Doll, B., Azari, K., Hollinger, J., Fracture repair, in: *Bone Regeneration and Repair: Biology and Clinical Applications*, pp. 21–44, 88, 3, 2007.
88. Vetter, A., Witt, F., Sander, O., Duda, G.N., Weinkamer, R., The spatio-temporal arrangement of different tissues during bone healing as a result of simple mechanobiological rules. *Biomech. Model. Mechanobiol.*, 11, 1, 147–160, 2012.
89. Fratzl, P. and Weinkamer, R., Nature's hierarchical materials. *Prog. Mater. Sci.*, 52, 8, 1263–1334, 2007.
90. Singer, A.J. and Clark, R.A., Cutaneous wound healing. *N. Engl. J. Med.*, 341, 10, 738–746, 1999.

91. Martin, P., Wound healing—Aiming for perfect skin regeneration. *Science*, 276, 5309, 75–81, 1997.
92. Gurtner, G.C., Werner, S., Barrandon, Y., Longaker, M.T., Wound repair and regeneration. *Nature*, 453, 7193, 314–321, 2008.
93. Aarabi, S., Longaker, M.T., Gurtner, G.C., Hypertrophic scar formation following burns and trauma: New approaches to treatment. *PLoS Med.*, 4, 9, 234, 2007.
94. Chortos, A. and Bao, Z., Skin-inspired electronic devices. *Mater. Today*, 17, 7, 321–331, 2014.
95. Wang, X., Dong, L., Zhang, H., Yu, R., Pan, C., Wang, Z.L., Recent progress in electronic skin. *Adv. Sci. (Weinheim, Baden-Wurttemberg, Ger.*, 2, 10, 1500169, 2015.
96. Mannsfeld, S.C.B. *et al.*, Highly sensitive flexible pressure sensors with microstructured rubber dielectric layers. *Nat. Mater.*, 9, 10, 859–864, 2010.
97. Sun, J.Y., Keplinger, C., Whitesides, G.M., Suo, Z., Ionic skin. *Adv. Mater.*, 26, 45, 7608–7614, 2014.
98. Cohen, D.J., Mitra, D., Peterson, K., Maharbiz, M.M., A highly elastic, capacitive strain gauge based on percolating nanotube networks. *Nano Lett.*, 12, 4, 1821–1825, 2012.
99. Webb, R.C. *et al.*, Ultrathin conformal devices for precise and continuous thermal characterization of human skin. *Nat. Mater.*, 12, 10, 938–944, 2013.
100. Yang, H. *et al.*, Soft thermal sensor with mechanical adaptability. *Adv. Mater.*, 28, 41, 9175–9181, 2016.
101. Chortos, A., Liu, J., Bao, Z., Pursuing prosthetic electronic skin. *Nat. Mater.*, 15, 9, 937–950, 2016.
102. Lai, Y.C. *et al.*, Electric eel-skin-inspired mechanically durable and super-stretchable nanogenerator for deformable power source and fully autonomous conformable electronic-skin applications. *Adv. Mater.*, 28, 45, 10024–10032, 2016.
103. Jeon, J., Lee, H.B.R., Bao, Z., Flexible wireless temperature sensors based on Ni microparticle-filled binary polymer composites. *Adv. Mater.*, 25, 6, 850–855, 2013.
104. Bauer, S., Bauer-Gogonea, S., Graz, I., Kaltenbrunner, M., Keplinger, C., Schwodiauer, R., 25th anniversary article: A soft future: from robots and sensor skin to energy harvesters. *Adv. Mater.*, 26, 1, 149–161, 2014.
105. Trung, T.Q. and Lee, N.E., Flexible and stretchable physical sensor integrated platforms for wearable human-activity monitoringand personal healthcare. *Adv. Mater.*, 28, 22, 4338–4372, 2016.
106. Lei, Z., Wang, Q., Sun, S., Zhu, W., Wu, P., A bioinspired mineral hydrogel as a self-healable, mechanically adaptable ionic skin for highly sensitive pressure sensing. *Adv. Mater.*, 29, 22, 1700321, 2017.
107. Kim, D.H. *et al.*, Stretchable and foldable silicon integrated circuits. *Science*, 320, 5875, 507–511, 2008.

108. Rogers, J., Someya, T., Huang, Y., Materials and mechanics for stretchable electronics. *Science*, 327, 5973, 1603–1607, 2010.
109. Kim, D.H. *et al.*, Materials and noncoplanar mesh designs for integrated circuits with linear elastic responses to extreme mechanical deformations. *Proc. Natl. Acad. Sci. U. S. A.*, 105, 48, 18675–18680, 2008.
110. Oh, J.Y. *et al.*, Intrinsically stretchable and healable semiconducting polymer for organic transistors. *Nature*, 539, 7629, 411–415, 2016.
111. Benight, S., Wang, C., Tok, J., Bao, Z., Stretchable and self-healing polymers and devices for electronic skin. *Prog. Polym. Sci.*, 38, 12, 1961–1977, 2013.
112. Bandodkar, A.J., Jeerapan, I., Wang, J., Wearable chemical sensors: Present challenges and future prospects. *ACS Sens.*, 1, 5, 464–482, 2016.
113. Someya, T., Bao, Z., Malliaras, G.G., The rise of plastic bioelectronics. *Nature*, 540, 7633, 379–385, 2016.
114. Yang, C.H., Chen, B., Zhou, J., Chen, Y.M., Suo, Z., Electroluminescence of giant stretchability. *Adv. Mater.*, 22, 4480–4484, 2016.
115. Larson, C. *et al.*, Highly stretchable electroluminescent skin for optical signaling and tactile sensing. *Science (80-.)*, 351, 6277, 1071–1074, 2016.
116. Nie, B., Li, R., Cao, J., Brandt, J.D., Pan, P., Flexible transparent iontronic film for interfacial capacitive pressure sensing. *Adv. Mater.*, 27, 39, 6055–6062, 2015.
117. Williams, H.R., Trask, R.S., Bond, I.P., Self-healing composite sandwich structures. *Smart Mater. Struct.*, 16, 4, 1198–1207, 2007.
118. Huang, C.Y., Trask, R.S., Bond, I.P., Characterization and analysis of carbon fibre-reinforced polymer composite laminates with embedded circular vasculature. *J. R. Soc. Interface*, 7, 49, 1229–1241, 2010.
119. Williams, H., Trask, R., Weaver, P., Bond, I., Minimum mass vascular networks in multifunctional materials. *J. R. Soc. Interface*, 5, 18, 55–65, 2008.
120. Hamilton, A.R., Sottos, N.R., White, S.R., Self-healing of internal damage in synthetic vascular materials. *Adv. Mater.*, 22, 45, 5159–5163, 2010.
121. Carano, R.A.D. and Filvaroff, E.H., Angiogenesis and bone repair. *Drug Discov. Today*, 8, 21, 980–989, 2003.
122. Chamay, A. and Tschantz, P., Mechanical influences in bone remodeling. Experimental research on Wolff's law. *J. Biomech.*, 5, 2, 173–180, 1972.
123. Koester, K.J., Ager, J.W., Ritchie, R.O., The true toughness of human cortical bone measured with realistically short cracks. *Nat. Mater.*, 8, 7, 672–677, 2008.
124. Hansen, C.J., Wu, W., Toohey, K.S., Sottos, N.R., White, S.R., Lewis, J.A., Self-healing materials with interpenetrating microvascular networks. *Adv. Mater.*, 21, 41, 4143–4147, 2009.
125. Ge, J. and Yin, Y., Responsive photonic crystals. *Angew. Chem. Int. Ed. Engl.*, 50, 7, 1492–1522, 2011.
126. Qin, M. *et al.*, A rainbow structural-color chip for multisaccharide recognition. *Angew. Chem. Int. Ed.*, 128, 24, 6911–6914, 2016.
127. Hou, J., Zhang, H., Yang, Q., Li, M., Jiang, L., Song, Y., Hydrophilic–hydrophobic patterned molecularly imprinted photonic crystal sensors for

high-sensitive colorimetric detection of tetracycline. *Small*, 23, 2738–2742, 11, 2015.
128. Gu, H. *et al.*, Tailoring colloidal photonic crystals with wide viewing angles. *Small*, 9, 13, 2266–2271, 2013.
129. Hou, J., Zhang, H., Yang, Q., Li, M., Song, Y., Jiang, L., Bio-inspired photonic-crystal microchip for fluorescent ultratrace detection. *Angew. Chem. Int. Ed. Engl.*, 53, 23, 5791–5795, 2014.
130. Zhang, P., Zhang, F., Zhao, C., Wang, S., Liu, M., Jiang, L., Superspreading on immersed gel surfaces for the confined synthesis of thin polymer films. *Angew. Chem. Int. Ed. Engl.*, 55, 11, 3615–3619, 2016.
131. Liu, C. *et al.*, Tunable structural color surfaces with visually self-reporting wettability. *Adv. Funct. Mater.*, 26, 43, 7937–7942, 2016.
132. Kim, S.H., Shim, J.W., Yang, S.M., Microfluidic multicolor encoding of microspheres with nanoscopic surface complexity for multiplex immunoassays. *Angew. Chem. Int. Ed. Engl.*, 50, 5, 1171–1174, 2011.
133. Su, B., Tian, Y., Jiang, L., Bioinspired interfaces with superwettability: From materials to chemistry. *J. Am. Chem. Soc.*, 138, 6, 1727–1748, 2016.
134. Wang, M., He, L., Xu, W., Wang, X., Yin, Y., Magnetic assembly and field-tuning of ellipsoidal-nanoparticle-based colloidal photonic crystals. *Angew. Chem. Int. Ed. Engl.*, 54, 24, 7077–7081, 2015.
135. Griffin, D.R., Weaver, W.M., Scumpia, P.O., Carlo, D., Segura, T., Accelerated wound healing by injectable microporous gel scaffolds assembled from annealed building blocks. *Nat. Mater.*, 14, 7, 737–744, 2015.
136. Li, C.H. *et al.*, A highly stretchable autonomous self-healing elastomer. *Nat. Chem.*, 6, 8, 618–624, 2016.
137. Dong, R., Zhao, X., Guo, B., Ma, P.X., Self-healing conductive injectable hydrogels with antibacterial activity as cell delivery carrier for cardiac cell therapy. *ACS Appl. Mater. Interfaces*, 8, 27, 17138–17150, 2016.
138. Zeng, Q., Desai, M.S., Jin, H.E., Lee, J.H., Chang, J., Lee, S.W., Self-healing elastin–bioglass hydrogels. *Biomacromolecules*, 17, 8, 2619–2625, 2016.
139. Shang, L. *et al.*, Bioinspired multifunctional spindle-knotted microfibers from microfluidics. *Small*, Superwettability, 4, 13, 1600286, 2017.
140. Fu, F. *et al.*, Bio-inspired self-healing structural color hydrogel. *Proc. Natl. Acad. Sci. U. S. A.*, 114, 23, 5900–5905, 2017.
141. Mow, V.C., Ratcliffe, A., Poole, A.R., Cartilage and diarthrodial joints as paradigms for hierarchical materials and structures. *Biomaterials*, 13, 2, 67–97, 1992.
142. Kalson, N.S., Gikas, P.D., Briggs, T.W.R., Current strategies for knee cartilage repair. *Int. J. Clin. Pract.*, 64, 10, 1444–1452, 2010.
143. Deans, T.L. and Elisseeff, J.H., Stem cells in musculoskeletal engineered tissue. *Curr. Opin. Biotechnol.*, 20, 5, 537–544, 2009.
144. Vasita, V. and Katti, D., Nanofiber and their application in tissue engineering. *Int. J. Nanomedicine*, 1, 1, 15, 2006.

145. Coburn, J., Gibson, M., Monagle, S., Patterson, Z., Elisseeff, J., Bioinspired nanofibers support chondrogenesis for articular cartilage repair. *Proc. Natl. Acad. Sci. U. S. A.*, 109, 10012–10017, 2012.

19

Self-Healable Batteries

Seyyed Mojtaba Mousavi[1], Maryam Zarei[2], Seyyed Alireza Hashemi[3], Wei-Hung Chiang[1], Chin Wei Lai[4]* and Sonia Bahrani[2]

[1]*Department of Chemical Engineering, National Taiwan University of Science and Technology, Taipei, Taiwan*
[2]*Pharmaceutical Sciences Research Center, Shiraz University of Medical Sciences, Shiraz, Iran*
[3]*Department of Mechanical Engineering, Center for Nanofibers and Nanotechnology, National University of Singapore, Singapore*
[4]*Nanotechnology & Catalysis Research Centre, University of Malaya, Kuala Lumpur, Malaysia*

Abstract

In recent years the energy demands and population increase have become the major concerns of today's life. Due to destructive effects of climate change caused by using fossil fuels to answer these demands and costs of connecting remote areas to the utility grid, replacement of old conventional power supply methods with a progressive environment-friendly system is highly beneficial and nevertheless inevitable. Over time and through repeated use of electronic devices, their structure may be damaged which leads to failing their performance for a while or permanently. Self-healing materials are introduced to surmount the aforementioned drawback. This kind of material has gained substantial considerations due to its interesting characterizations. Self-healing materials are promising candidates for the fabrication of electronic devices, energy storage systems, conductive adhesives, and even sensors. The present chapter will follow several topics, first, the detailed research progress of self-healing materials is reviewed, and second, the preparation mechanisms and properties of the self-healing process are summarized. After that, self-healable electrodes and electrolytes are portrayed. Finally, to boost the improvement self-healing materials, the major opportunities and future challenges are also explained.

***Keywords*:** Self-healing, batteries, supercapacitors, conductive devices

**Corresponding author*: cwlai@um.edu.my

Inamuddin, Mohd Imran Ahamed, Rajender Boddula and Tariq Altalhi (eds.) Self-Healing Smart Materials and Allied Applications, (475–494) © 2021 Scrivener Publishing LLC

19.1 Introduction

From the environmental point of view, increasing world population and lack of renewable energy sources are main challenges of today's life. Therefore, the highlighted role of lithium-ion batteries in energy storage systems, electrical vehicles, and diverse electronic devices cannot be ignored due to its long cycle life and high energy density. These kinds of batteries possess excellent properties and more acceptable applications in comparison with previous traditional ones [1, 2]. Since batteries play a vital role in scientific experiments and industry, many efforts have been carried out to develop biocompatibility [3], anti-fogging property [4], capacity, flexibility [5] and, superomniphobic property of batteries [6]. It is necessary to mention that during long time usage, batteries may break, be damaged or be affected by external forces which can jeopardize their performance. Thence, self-healing is introduced as a potential solution for the aforementioned problems which can lead to enhancing the lifespan of batteries. The smart materials which are able to restore all or part of their structure after exposure to harsh conditions or damaging are called self-healing materials [7, 8]. High temperatures and external mechanical forces cannot jeopardize the performance of self-healing materials. The self-healing materials not only enhance the reliability of the obtained product however they are able to prolong the lifespan. Previous literature has proven that self-healing materials are valuable substances that can diminish the waste of resources. Also, several mechanisms of self-healing like intrinsic and extrinsic self-healing were reported by researchers [9–11]. As a point, it can be mentioned that dynamic reversible bonds and microcapsules are applied in intrinsic and extrinsic self-healing methods respectively. The healing intrinsic method can be applied many times while former methods were not like this [12–14]. The researchers have claimed that the advantages of self-healing materials have gradually promoted to conductive properties [15]. In today's life, the role of self-healing materials is an essential issue in terms of many applications [16, 17] (Figure 19.1). The materials with conductive properties and self-healing features have been broadly utilized in different applications like soft robotics [18], medical devices [19] and electronics [28]. So far many applicable methods are proposed by researchers to extend conductive properties to self-healable substances [20, 21]. A group of researchers has explained the improvement of self-healing materials with conductive properties [22]. Based on their report, these conductive materials can be categorized into two classes. The presence of repairing factor can make a difference between the two aforementioned classes [22].

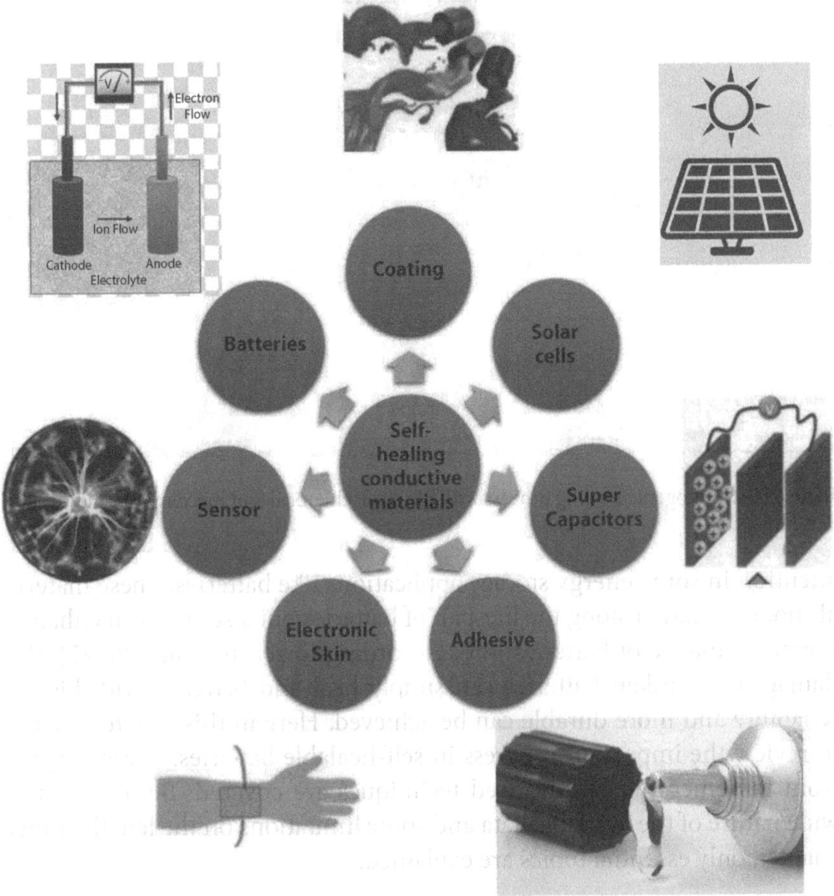

Figure 19.1 The applications of self-healing conductive materials in different areas including energy storage devices, sensor, electronic skin, adhesive, batteries and coating.

Zhang *et al.* have reviewed published papers from 2015 to 2017 to discuss the properties of self-healing conductive adhesives and coatings [23].

Another field of research in the last years is autogenic healing materials which have gained tremendous attention due to their interesting properties. As proven before, the autogenic healing mechanism can possess different causes like physical, mechanical and chemical. Figure 19.2 demonstrates the reactions of this process. The autogenic healing process can be occurred by enlargement of hydrate matrix [24], rehydration of unhydrated particles, production of calcium carbonate or degradation of particles in water. The self-healable materials have received remarkable

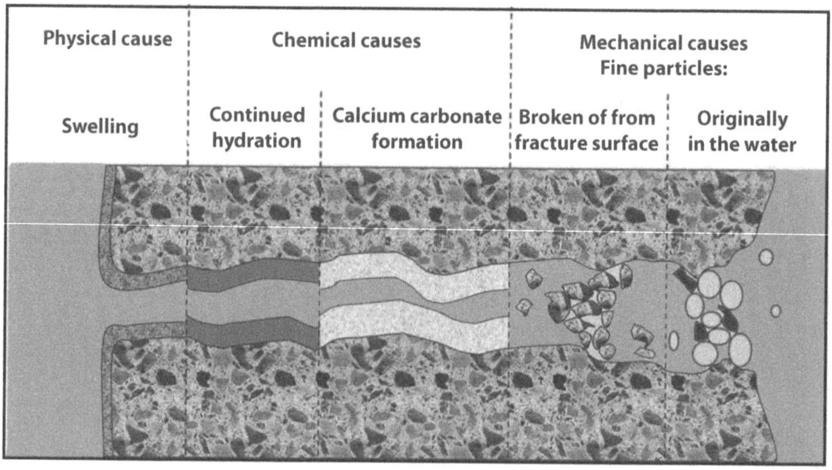

Figure 19.2 Autogenic healing process with physical, chemical and mechanical causes [28].

attention in some energy storage applications like batteries. These materials not only can prolong the lifespan of batteries but also they can enhance the performance of batteries [25]. According to self-healing [26, 27], the damaged or broken batteries can simply heal and batteries with higher reliability and more durable can be achieved. Here in this chapter, we try to review the important progress in self-healable batteries. Several topics from the emerging to advanced techniques are covered. Because of the wide nature of the available data and some limitations on the length of this chapter, only essential topics are explained.

19.2 Development of Self-Healing Materials

As we know in every biological mechanism, bones, skin or other organs of organisms are able to heal themselves after sensing damage. The same situation occurs for plants or animals and when they injure, they can repair themselves. Since the components will possess some disordered during their life cycle [29], many different composite materials are proposed to overcome damaging. From the economical point of view, the diagnosis of damage in the early stage and self-healing can reduce costs and also prevent failure. Many researchers tried to improve the self-healing based materials for conversion devices and energy storage systems. Metals with low melting point and certain polymers are proposed as two main categories of self-healing materials. To date, different self-healing mechanisms and phenomena have been revealed. White and his coworkers have developed the microencapsulation method which is one of the most known self-healing methods and it is mostly according to polymerization with the open-chain process [30]. Figure 19.3 represents the autonomic healing mechanism that by propagating the crack through the microcapsules, the healing monomer is polymerized and dispersed in the matrix. They also reported that more than 70% recovery in toughness can be achieved by the aforementioned concept. It is also claimed that this method is very applicable in both glasses and ceramics. Zhang *et al.* in their interesting work proposed that the self-healing mechanism possesses several steps like surface modification, vicinity, soaking, propagation, and randomization [31].

The researchers tried to divide self-healing materials according to chemical bonds hence single and multiple bond categories were proposed. Pu and his coworkers [32] have proposed the novel conductive composites which were self-healing and flexible. Carbon nanotubes (CNTs) and Diels–Alder bonds (PUDA) were two major parts of their proposed composite. Since CNTs could provide adequate photothermal and even photoelectrical effects, the healing rate of the composite was high (Figures 19.4A and B). The self-healing process of PUDA/CNT composite under certain conditions is illustrated in Figure 19.4(C). They finally claimed that these novel composites can be promising candidates in electronic sensors with high self-healable activity. Another group of researchers has formed polyurethane frameworks which possess self-healing ability [33]. They synthesized these networks using the one-pot method under ordinary temperature.

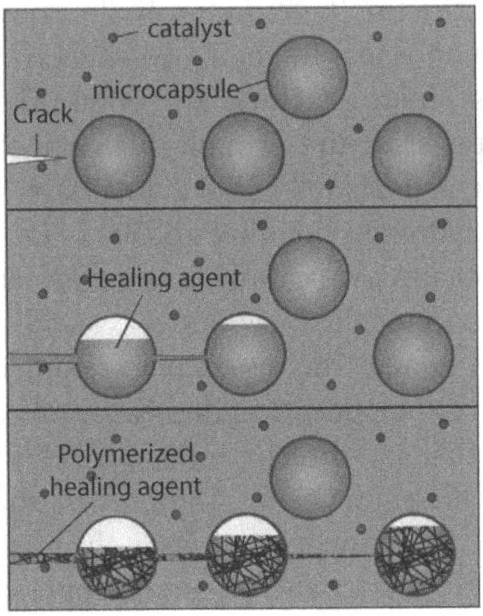

Figure 19.3 A schematic representation of autonomic healing process [30].

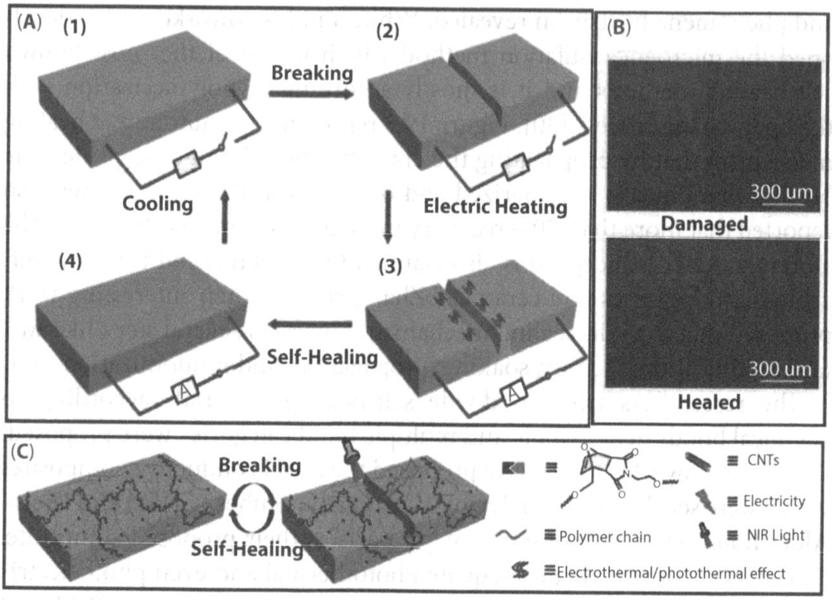

Figure 19.4 (A) Schematic of autonomic healing process of PUDA/CNTs composite as a new electronic device, (B) Microscopic images of damaged and healed samples with 1 wt% CNTs content, (C) Self-healing process of PUDA/CNTs composite under NIR light [32].

It is reported that the combination of Diels–Alder reaction and rapid shape memory effect can lead to the healing process at 50 °C. Schäfer *et al.* have offered a novel mechanism based on Diels–Alder chemistry of poly (butyl methacrylates) and structurally varied polysiloxanes for the progress of self-healing nano-composites. In their investigation silica nanoparticles were applied as cross-linking factors. The obtained results have verified that the polymer type, structure of molecules and the polymeric chains can affect the healing abilities [34]. Kang *et al.* have offered a self-healing new network using hydrogen bonding and ionic crosslinking [35]. They introduced chitosan chains as a scaffold for this formation. It is proven that the healing performance of this new network can be affected by coordination and hydrogen bonds. The researchers also demonstrated that, the prepared hydrogels were applicable in both wearable devices and artificial skin. The key factor that can directly influence healing efficiency is time. Indeed, to enhance efficiency, the healing process should occur immediately after the damage. In an interesting work, the researchers have synthesized the mixture of clay, water and organic compounds like polyethylene glycol as a self-healing hydrogel [36]. Hydrophilic effects were considered as one of the most factors of damaged blocks in the self-healing process.

Based on their results, the self-healing process could not occur if the damaged blocks were left alone for more than 60 s. Because the healed surface possesses hydrophilic effect, therefore the damaged blocks will lose their hydrophilicity which is mainly owing to the reconstruction process [36]. According to previous literature, the importance of self-healing in biological systems, electrodes, electrolytes, and other energy storage systems like capacitors and batteries must not be underestimated. Based on our best of knowledge, the commercialization and experiment on self-healable materials are anticipated to enhance quickly.

19.3 Self-Healing Batteries

It is reported that Gilbert N. Lewis was the first person who proposed lithium metal batteries and explains their properties. After him, many groups of researchers have tried to investigate and expand the properties of batteries. However, during the experiments, they realized that some obstacles like poor cycle performances and sudden breaks could affect the lifespans of batteries. Another issue after the damage was that the leakage of electrolytes which most of the electrolytes were toxic [37, 38] and could cause environmental disorders. Thence, they were looking for an alternative to overcome the aforementioned problems. Recently, self-healing materials and mechanisms have gained considerable attention caused by their interesting properties in different industries. Self-healable batteries are introduced as potential candidates in energy storage systems. Macroscopic and microscopic self-healable batteries are two major categories of self-healable batteries. If an external force causes cut or break in batteries, macroscopic self-healing occurs. Also, if during the charge and discharge processes of batteries, some micro-fractures caused by electrolytes or electrodes are formed then this kind of healing is called the micro-healing process. It is necessary to mention that to enhance cycling stability and also increase the lifespan of batteries, microscopic-healing can be an effective choice.

19.3.1 Self-Healable Electrodes

Since metallic alloy [39, 40], silicon (and its derivatives) and oxides [41, 42] can provide high energy density and they are able to deliver high theoretical specific capacities for batteries, they are widely applied as new electrodes. Short cycling life and low capacity of electrodes can be major issues in ion extract and insert processes which can limit their applications. Therefore the researchers have tried to surmount these problems by introducing new self-healable electrodes. Wang *et al.* have demonstrated by using the polymer coating (with hydrogen bonds) on inexpensive silicon microparticles, they can act as an effective anode which enhances the cycle life of batteries [43]. A schematic representation of the traditional and proposed electrode is demonstrated in Figure 19.5(a). The polymer binds the inner materials to provide the electric contact, in the traditional anodes.

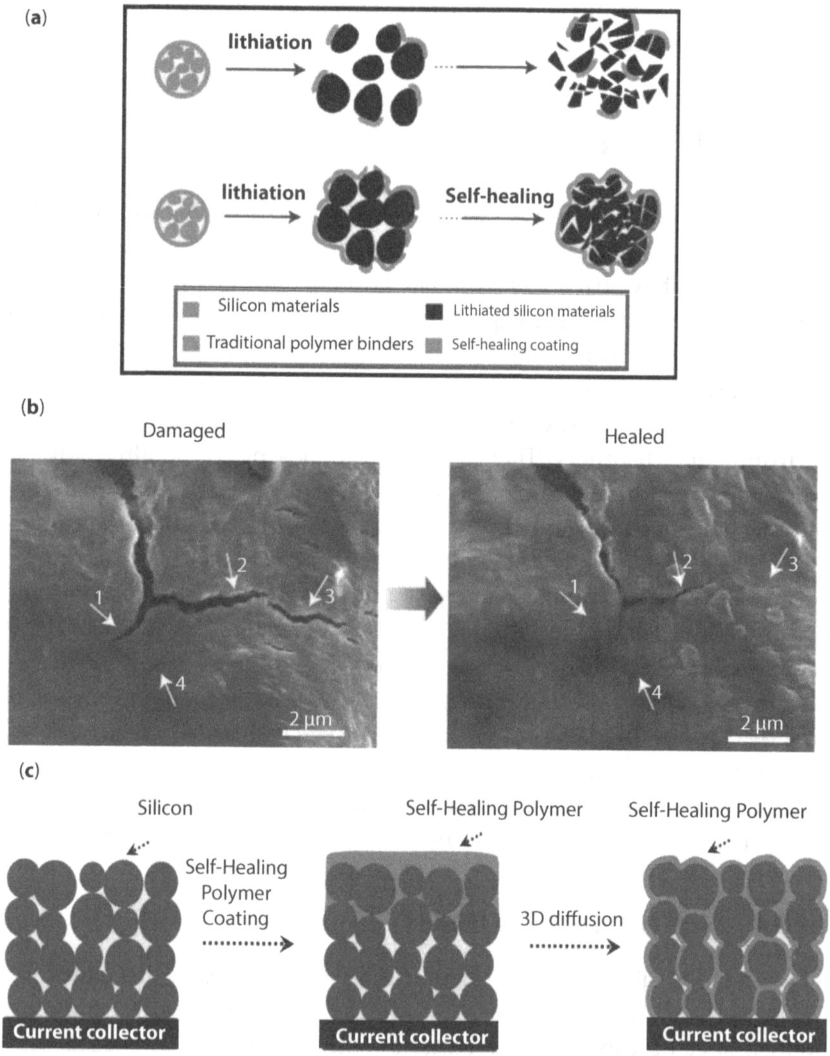

Figure 19.5 (a) Traditional and self-healing polymer electrodes in silicon based batteries [43], (b) SEM images of damaged and healed silicon based batteries [43], (c) the schematic design of spatial propagation of self-healing polymer in silicon layer [45].

While the self-healing polymers not only can bind active materials, even they are stretchable which repairs the cracks on the anode and enhance mechanical properties. The SEM images of aforementioned self-healable polymer verified that after only 5 h, the damaged parts were healed and the cracks disappeared (Figure 19.5(b)) [43]. Similarly, Chen *et al.* in their interesting experiment have designed 3-dimentioanl spatial propagation of self-healing polymers on the traditional silicon electrodes. As represented in Figure 19.5(c), the silicon particles poured on substrates of Cu. Then the framework dried under period of time and the designed self-healable polymers coated on silicon electrodes. It is proven that this novel approach can control the size of silicon particles [44] which leads to high cycling stability and acceptable mechanical properties [45].

The self-healing process is introduced as a valuable process that received remarkable consideration [46–48], hence investigating and characterizing the self-healing concept are the goal of many researchers [49, 50]. For this purpose, a group of researchers focused on the creation and growth of fractures in silicon particles under both lithiation and delithiation processes. They reported the final morphology of silicon particles during the healing process. After characterizing all the tested regions and silicon crystal, the claimed that the combination of silicon and aluminum could be the promising candidate for electrodes [51]. Hu and his group have prepared and proposed a new hydrogel with the 3D conductive framework that can be applied as self-healing anode in batteries. The *in-situ* method was utilized for polymerization of this self-healing hydrogel on the surface of silicon particles. They claimed that higher electric contacts under 3D framework can be achieved while a huge volume of damages of silicon particles restrain by using the aforementioned hydrogel [52]. Other examples and investigations of utilized materials in self-healable batteries are summarized in Table 19.1. By using the reviewed data of the table, the readers intuitively realize that the role of polymers [53], hydrogels and metallic materials are more highlighted than non-metallic [54] ones in self-healable batteries.

19.3.2 Self-Healable Electrolytes

Using self-healable electrolytes can be an impressive approach to enhancing the electrochemical properties of batteries. As we know, poor coulombic performance and low stability of cyclic are two major concerns of batteries, hence the researchers tried to surmount the aforementioned issues. Xu and his coworkers introduced a new self-healable electrolyte that could overcome the inefficiency of Li–S batteries [55]. The role of

Table 19.1 Different types of self-healable materials in batteries.

Proposed material	Type of damage	Cycle number	Healing strategy	Ref
Polysulfide	Break	50	External agent	[55]
Polymer	collapse	100	Using Hydrogen bonds	[56]
Polyvinyl alcohol	metamorphose	5	Dynamic borate ester bonding	[57]
Gallium (liquid)	Fracture	–	Conversion	[58]
Supramolecular polymer	Break	170	Using Hydrogen bonds	[59]
Lithium (liquid)	cut	1,000	Metallic dissolving	[60]
Polymer	deformation	90	Using hydrogen bonds	[43]
SnP_3/C anode	metamorphose	150	Reversible conversion	[61]
PANa cross-linked by Fe^{3+}	Break	–	Using cross-linkers	[62]
Polymerized β cyclodextrin	metamorphose	250	Using hydrogen bonds	[63]
PVDF	Pulverization	150	Metallic interactions	[45]
Ga_2O_3 nanoparticles	deformation	250	Carbon shell protection	[64]
Polyethylene glycol	Fracture	150	Using hydrogen bonds	[65]

(Continued)

Table 19.1 Different types of self-healable materials in batteries. (*Continued*)

Proposed material	**Type of damage**	**Cycle number**	**Healing strategy**	**Ref**
Polymer (liquid)	deformation	100	Chemical interaction	[66]
Fe2O3/rGO	Pulverization	300	Specific coating	[67]
Un-crosslinked ionomer	Stretchable	150	Using hydrogen bonds	[68]
Cationic additive	Dendrite	650	Electrostatic Shield	[69]
Native silanol types	Fracture	500	Ionic interactions	[70]
Network polyimines	Crack	200	Malleability, reversibility	[71]
PDMS	metamorphose	1,000	Repair cracks	[72]

self-healing materials in the solid-state generation of batteries was highlighted by a group of researchers. They prepared a solid electrolyte which possesses polymer matrix with high self-healing activity. In their strategy, polyimide can act as an effective filler for the base matrix to boost the mechanical performances of soli batteries [71]. As a note, it can be stated that, by producing self-healable electrolytes, most of the fractures and cracks will be automatically healed and therefore the performance and lifespan of batteries will be promoted. Based on previous literature, some covalent bonds like disulfide [73–75], hydrogen and as acylhydrazone bonds [76–78] can the base of self-healing mechanisms in many polymeric scaffolds. Tian *et al.* [79] in their interesting experiment have verified that poly(4-vinylpyridine)(propyl-trimethylammonium) (PVT) is a promising electrode to reduce the instability of batteries. The self-healing mechanism of Tian's experiment is illustrated in Figure 19.6(A). After the separation of polymer, the fundamental interactions between the cation (N+) and anion (TSFI−) discontinues. In the next step, the electrostatic force can repair and heal the polymer. It is reported that the above electrolyte represents efficient flexibility and it possesses acceptable ionic conductivity [79].

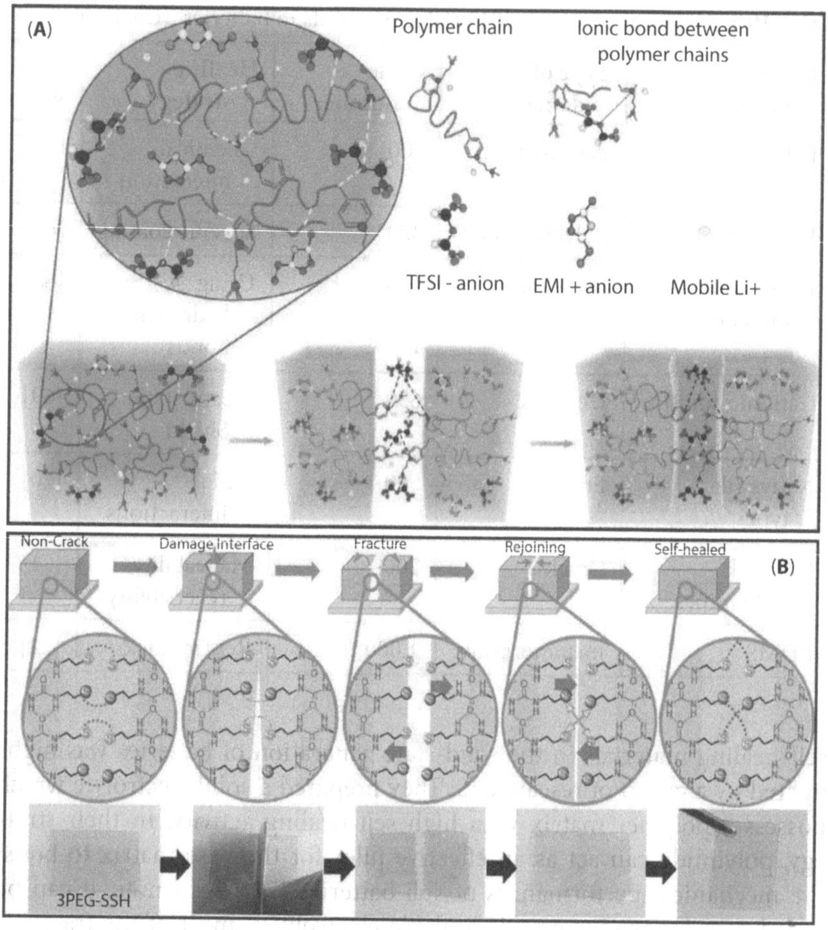

Figure 19.6 Self-healing process of (A) PVT-40%EMITFSI electrolytes [79], (B) Solid-state electrolyte [93].

Guo *et al.* in their valuable experiment, have applied the combination of ionic liquids and lithium salt hydrogen-bonded PIL-UPy supramolecular under polymer networks to fabricate highly conductive, nonflammable and a self-healable electrolyte which can broadly be utilized in the new generation of batteries. The ionic conductivity of these electrodes was 0.001 S/cm while the previous electrodes didn't have such acceptable conductivities. Also, the Coulombic efficiency after four months was about 99.7% which demonstrated that these novel electrodes were able to repair damages under 55 °C [80]. Similarly, other researchers have tried to investigate the performance of self-healing electrodes through quadruple hydrogen bonding under several temperatures [81–83]. For example, a new ion gel was fabricated through hydrogen bonds incorporated with a copolymer by Tamate *et al.* [84]. This ion gel exhibits extraordinary self-healable performance without applying any external force during 180 min. Xia and his coworkers have prepared new electrodes that possess fatty acid in their structures. It is reported that these electrodes not only promote battery life however they are efficient self-healable materials that can control the current density during charge and discharge process [85]. It is reported that the biological self-healing process can pave the route for researchers [86–89]. Hence, polysulfide with extrinsic healing properties was reported by Peng *et al.* [90] to stabilize the particles of sulfur. They reported that this self-healable material enhances the performance of batteries and can reduce their instabilities during the cyclic process. A group of researchers has claimed that without using any catalysts, the self-healing elastomer by disulfide metathesis can be achieved. This proposed elastomer will be used in different applications like electrolytes [91]. Kim and his group have fabricated a new polymer (polyurethane) through disulfide metathesis which possesses a satisfactory mechanical performance even at room temperature. This self-healable material is an efficient electrolyte with high stability and doesn't need any stimuli for its function [92]. Very recently a new self-healable electrolyte was reported by Jo *et al.* In their experiment, disulfide metathesis played a vital role as a scaffold. This proposed electrolyte has excellent mechanical and thermal properties and they can satisfy a wide range of demand in terms of electrochemical devices. Figure 19.6(B) illustrated the self-healing mechanism of solid polymer electrolyte [93]. It is necessary to mention that in all reported cases, after repeatedly usage the batteries were able to repair their structure and enhance thermal and mechanical performances.

19.4 Conclusions

So far significant efforts have been carried out to prepare and pave the route for the renovation of energy storage devices. As we know many difficulties are associated with these devices which can directly affect their efficiency and reduce their lifespan. Thence, self-healing materials are introduced to surmount the aforementioned issues. Indeed by using self-healing materials that act as electrodes and electrolytes, cut, folding fracture and deformation are not irreparable and the structure can be repaired quickly. Here, we have collected the recent strategies and preparation mechanisms of self-healing materials. In terms of self-healing batteries, we have introduced both macroscopic and microscopic self-healing materials. Also, detailed elaboration on the fabrication of self-healable electrolytes and electrodes are provided. Finally, it can be stated that self-healable materials are valuable kinds of materials that can diminish environmental contaminations and store energy in different ways hence these materials will play a fundamental role in the next generation of electronic devices.

References

1. Wang, Q. *et al.*, Progress of enhancing the safety of lithium ion battery from the electrolyte aspect. *Nano Energy*, 55, 93–114, 2019.
2. Wang, Q. *et al.*, A review of lithium ion battery failure mechanisms and fire prevention strategies. *Prog. Energy Combust. Sci.*, 73, 95–131, 2019.
3. Zhu, H. *et al.*, Wood-derived materials for green electronics, biological devices, and energy applications. *Chem. Rev.*, 116, 16, 9305–9374, 2016.
4. Li, H. *et al.*, Flexible sodium-ion based energy storage devices: Recent progress and challenges. *Energy Storage Mater.*, 26, 83–104, 2020.
5. Yao, L. *et al.*, Long-Lived Multilayer Coatings for Smart Windows: Integration of Energy-Saving, Antifogging, and Self-Healing Functions. *ACS Appl. Energy Mater.*, 2, 10, 7467–7473, 2019.
6. Schlaich, C., Wei, Q., Haag, R., Mussel-inspired Polyglycerol coatings with controlled wettability: From Superhydrophilic to Superhydrophobic surface coatings. *Langmuir*, 33, 38, 9508–9520, 2017.
7. Basu, R., Nanolayer Surface Phase Change in Self-Healing Materials. In *Nanocomposites VI: Nanoscience and Nanotechnology in Advanced Composites*, pp. 19–34, Springer, Cham, 2019.
8. Song, M.-M. *et al.*, Functional materials with self-healing properties: A review. *Soft Matter*, pp. 19–34, Springer, Cham, 2019.
9. Wool, R.P., Self-healing materials: A review. *Soft Matter*, 4, 3, 400–418, 2008.
10. Wojtecki, R.J., Meador, M.A., Rowan, S.J., Using the dynamic bond to access macroscopically responsive structurally dynamic polymers. *Nat. Mater.*, 10, 1, 14–27, 2011.
11. Bekas, D. *et al.*, Self-healing materials: A review of advances in materials, evaluation, characterization and monitoring techniques. *Compos. Part B: Eng.*, 87, 92–119, 2016.
12. Zechel, S. *et al.*, Intrinsic self-healing polymers with a high E-modulus based on dynamic reversible urea bonds. *NPG Asia Mater.*, 9, 8, e420–e420, 2017.
13. Fan, L.F. *et al.*, Repeated intrinsic self-healing of wider cracks in polymer via dynamic reversible covalent bonding molecularly combined with a two-way shape memory effect. *ACS Appl. Mater. Interfaces*, 10, 44, 38538–38546, 2018.

14. Hu, Z. *et al.*, Multistimuli-responsive intrinsic self-healing epoxy resin constructed by host–guest interactions. *Macromolecules*, 51, 14, 5294–5303, 2018.
15. Zhong, N. and Post, W., Self-repair of structural and functional composites with intrinsically self-healing polymer matrices: A review. *Compos. Part A: Appl. Sci. Manuf.*, 69, 226–239, 2015.
16. Mousavi, S.M., Hashemi, S.A., Zarei, M., Amani, A.M., Babapoor, A., Nanosensors for Chemical and Biological and Medical Applications. *Med. Chem.* (Los Angeles), 8, 205–217, 2018.
17. Mousavi, S., Zarei, M., Hashemi, S., Polydopamine for Biomedical Application and Drug Delivery System. *Med. Chem. (Los Angeles)*, 8, 218–229, 2018.
18. Stokes, A.A. *et al.*, A hybrid combining hard and soft robots. *Soft Rob.*, 1, 1, 70–74, 2014.
19. Zhang, Y. *et al.*, A magnetic self-healing hydrogel. *Chem. Commun.*, 48, 74, 9305–9307, 2012.
20. Zhu, M., Rong, M.Z., Zhang, M.Q., Self-healing polymeric materials towards non-structural recovery of functional properties. *Polym. Int.*, 63, 10, 1741–1749, 2014.
21. Kang, J. *et al.*, Tough and water-insensitive self-healing elastomer for robust electronic skin. *Adv. Mater.*, 30, 13, 1706846, 2018.
22. Kun, G. *et al.*, Research progress and applications of self-healing conductive materials. *Prog. Chem.*, 27, 6, 633–640, 2015.
23. Zhang, Q. *et al.*, Review of recent achievements in self-healing conductive materials and their applications. *J. Mater. Sci.*, 53, 1, 27–46, 2018.
24. Mousavi, S.M., Hashemi, S.A., Amani, A.M., *et al.*, Modification of Polypropylene-Starch Blend by Eggshell Nano-Particle, EVA and Maleic Anhydrideto Improve Biodegradability and Thermal Properties. *Int. J. Chem. Sci.*, 15, 4, 225, 2017
25. Cheng, X. *et al.*, Gel polymer electrolytes for electrochemical energy storage. *Adv. Energy Mater.*, 8, 7, 1702184, 2018.
26. Guo, K. *et al.*, The Deeply Understanding of the Self-Healing Mechanism for Self-Healing Behavior of Supramolecular Materials Based on Cyclodextrin–Guest Interactions. *Macromol. Chem. Phys.*, 218, 10, 1600593, 2017.
27. Nie, F.-M. *et al.*, Molecular-Level Tuning toward Aggregation Dynamics of Self-Healing Materials. *Macromolecules*, 52, 14, 5289–5297, 2019.
28. De Rooij, M, Van Tittelboom, K, De Belie, N, Schlangen, E, editors. *Self-healing phenomena in cement-Based materials: State-of-the-art report of RILEM technical committee 221-SHC: Self-Healing phenomena in cement-based materials*. Springer Science & Business Media, 2013.
29. Mousavi, S., Esmaeili, H., Arjmand, O., Karimi, S., Hashemi, S., Biodegradation study of nanocomposites of phenol novolac epoxy/unsaturated polyester resin/egg shell nanoparticles using natural polymers. *J. Mater.*, 2015, 1–6, 2015.

30. White, S.R. *et al.*, Autonomic healing of polymer composites. *Nature*, 409, 6822, 794–797, 2001.
31. Zhang, M.Q. and Rong, M.Z., Theoretical consideration and modeling of self-healing polymers. *J. Polym. Sci. Part B: Polym. Phys.*, 50, 4, 229–241, 2012.
32. Pu, W. *et al.*, Realizing Crack Diagnosing and Self-Healing by Electricity with a Dynamic Crosslinked Flexible Polyurethane Composite. *Adv. Sci.*, 5, 5, 1800101, 2018.
33. Rivero, G. *et al.*, One-pot thermo-remendable shape memory polyurethanes. *Macromolecules*, 47, 6, 2010–2018, 2014.
34. Schäfer, S. and Kickelbick, G., Self-healing polymer nanocomposites based on Diels–Alder-reactions with silica nanoparticles: The role of the polymer matrix. *Polymer*, 69, 357–368, 2015.
35. Kang, M. *et al.*, Template method for dual network self-healing hydrogel with conductive property. *Mater. Des.*, 148, 96–103, 2018.
36. Wang, Q. *et al.*, High-water-content mouldable hydrogels by mixing clay and a dendritic molecular binder. *Nature*, 463, 7279, 339–343, 2010.
37. Nam, Y.J. *et al.*, Bendable and thin sulfide solid electrolyte film: A new electrolyte opportunity for free-standing and stackable high-energy all-solid-state lithium-ion batteries. *Nano Lett.*, 15, 5, 3317–3323, 2015.
38. Wang, Y. *et al.*, Design principles for solid-state lithium superionic conductors. *Nat. Mater.*, 14, 10, 1026–1031, 2015.
39. Li, W., Sun, X., Yu, Y., Si-, Ge-, Sn-based anode materials for lithium-ion batteries: From structure design to electrochemical performance. *Small Methods*, 1, 3, 1600037, 2017.
40. Tan, H. *et al.*, Peering into Alloy Anodes for Sodium-Ion Batteries: Current Trends, Challenges, and Opportunities. *Adv. Funct. Mater.*, 29, 14, 1808745, 2019.
41. Liu, Z. *et al.*, Silicon oxides: A promising family of anode materials for lithium-ion batteries. *Chem. Soc. Rev.*, 48, 1, 285–309, 2019.
42. Wang, L. *et al.*, Synthesis, characterizations, and utilization of oxygen-deficient metal oxides for lithium/sodium-ion batteries and supercapacitors. *Coord. Chem. Rev.*, 397, 138–167, 2019.
43. Wang, C. *et al.*, Self-healing chemistry enables the stable operation of silicon microparticle anodes for high-energy lithium-ion batteries. *Nat. Chem.*, 5, 12, 1042, 2013.
44. Mousavi, S.M. *et al.*, Octadecyl amine functionalized Graphene oxide towards hydrophobic chemical resistant epoxy nanocomposites. *ChemistrySelect*, 3, 25, 7200–7207, 2018.
45. Chen, Z. *et al.*, High-areal-capacity silicon electrodes with low-cost silicon particles based on spatial control of self-healing binder. *Adv. Energy Mater.*, 5, 8, 1401826, 2015.
46. Huang, S. *et al.*, A self-healing integrated all-in-one zinc-ion battery. *Angew. Chem.*, 131, 13, 4357–4361, 2019.

47. Mi, H. *et al.*, Self-healing silicon-sodium alginate-polyaniline composites originated from the enhancement hydrogen bonding for lithium-ion battery: A combined simulation and experiment study. *J. Power Sources*, 412, 749–758, 2019.
48. Wang, D. *et al.*, A Superior δ-MnO2 Cathode and a Self-Healing Zn-δ-MnO2 Battery. *ACS Nano*, 13, 9, 10643–10652, 2019.
49. Mousavi, S. *et al.*, Modification of the epoxy resin mechanical and thermal properties with silicon acrylate and montmorillonite nanoparticles. *Polym. Renewable Resour.*, 7, 3, 101–113, 2016.
50. Mousavi, S.M. and Goudarzian, N., Hashemi, S., Unsaturated polyester resins modified with cresol novolac epoxy and silica nanoparticles: Processing and mechanical properties. *Int. J. Chem. Pet. Sci.*, 5, 1, 13–26, 2016.
51. Bhattacharya, S. and Alpas, A.T., Self-healing of cracks formed in Silicon-Aluminum anodes electrochemically cycled at high lithiation rates. *J. Power Sources*, 328, 300–310, 2016.
52. Hu, S. *et al.*, A conductive self-healing hydrogel binder for high-performance silicon anodes in lithium-ion batteries. *J. Power Sources*, 449, 227472, 2020.
53. Mousavi, S. *et al.*, Improved morphology and properties of nanocomposites, linear low density polyethylene, ethylene-co-vinyl acetate and nano clay particles by electron beam. *Polym. Renewable Resour.*, 7, 4, 135–153, 2016.
54. Mousavi, S.M. *et al.*, A conceptual review of rhodanine: Current applications of antiviral drugs, anticancer and antimicrobial activities. *Artif. Cells Nanomed. Biotechnol.*, 47, 1, 1132–1148, 2019.
55. Xu, R. *et al.*, Role of Polysulfides in Self-Healing Lithium–Sulfur Batteries. *Adv. Energy Mater.*, 3, 7, 833–838, 2013.
56. Sun, Y. *et al.*, A stretchable graphitic carbon/Si anode enabled by conformal coating of a self-healing elastic polymer. *Adv. Mater.*, 28, 12, 2455–2461, 2016.
57. Hao, Z. *et al.*, An Omni-healable and Tailorable Aqueous Lithium-Ion Battery. *ChemElectroChem*, 5, 4, 637–642, 2018.
58. Deshpande, R.D. *et al.*, Liquid metal alloys as self-healing negative electrodes for lithium ion batteries. *J. Electrochem. Soc.*, 158, 8, A845–A849, 2011.
59. Lopez, J. *et al.*, The effects of cross-linking in a supramolecular binder on cycle life in silicon microparticle anodes. *ACS Appl. Mater. Interfaces*, 8, 3, 2318–2324, 2016.
60. Ning, X. *et al.*, Self-healing Li–Bi liquid metal battery for grid-scale energy storage. *J. Power Sources*, 275, 370–376, 2015.
61. Fan, X. *et al.*, Superior Stable Self-Healing SnP3 Anode for Sodium-Ion Batteries. *Adv. Energy Mater.*, 5, 18, 1500174, 2015.
62. Huang, Y. *et al.*, A self-healable and highly stretchable supercapacitor based on a dual crosslinked polyelectrolyte. *Nat. Commun.*, 6, 1, 1–8, 2015.
63. Jeong, Y.K. *et al.*, Hyperbranched β-cyclodextrin polymer as an effective multidimensional binder for silicon anodes in lithium rechargeable batteries. *Nano Lett.*, 14, 2, 864–870, 2014.

64. Tang, X. *et al.*, High-performance Ga2O3 anode for lithium-ion batteries. *ACS Appl. Mater. Interfaces*, 10, 6, 5519–5526, 2018.
65. Munaoka, T. *et al.*, Ionically Conductive Self-Healing Binder for Low Cost Si Microparticles Anodes in Li-Ion Batteries. *Adv. Energy Mater.*, 8, 14, 1703138, 2018.
66. Ma, X. *et al.*, Three-dimensional hierarchical walnut kernel shape conducting polymer as water soluble binder for lithium-ion battery. *Electrochim. Acta*, 269, 571–579, 2018.
67. Lee, J.-G. *et al.*, Stable high-capacity lithium ion battery anodes produced by supersonic spray deposition of hematite nanoparticles and self-healing reduced graphene oxide. *Electrochim. Acta*, 228, 604–610, 2017.
68. Huang, S. *et al.*, Low addition amount of self-healing ionomer binder for Si/graphite electrodes with enhanced cycling. *New J. Chem.*, 42, 9, 6742–6749, 2018.
69. Ding, F. *et al.*, Dendrite-free lithium deposition via self-healing electrostatic shield mechanism. *J. Am. Chem. Soc.*, 135, 11, 4450–4456, 2013.
70. Kwon, T.w. *et al.*, Systematic Molecular-Level Design of Binders Incorporating Meldrum's Acid for Silicon Anodes in Lithium Rechargeable Batteries. *Adv. Mater.*, 26, 47, 7979–7985, 2014.
71. Whiteley, J.M. *et al.*, Ultra-thin solid-state Li-ion electrolyte membrane facilitated by a self-healing polymer matrix. *Adv. Mater.*, 27, 43, 6922–6927, 2015.
72. Weng, W. *et al.*, A Gum-Like Lithium-Ion Battery Based on a Novel Arched Structure. *Adv. Mater.*, 27, 8, 1363–1369, 2015.
73. Xu, Y. and Chen, D., A novel self-healing polyurethane based on disulfide bonds. *Macromol. Chem. Phys.*, 217, 10, 1191–1196, 2016.
74. Wu, X. *et al.*, Heat-triggered poly(siloxane–urethane)s based on disulfide bonds for self-healing application. *J. Appl. Polym. Sci.*, 135, 31, 46532, 2018.
75. Zhang, L. *et al.*, Self-Healing Polycaprolactone Networks through Thermo-Induced Reversible Disulfide Bond Formation. *Macromol. Rapid Commun.*, 39, 20, 1800121, 2018.
76. Tuncaboylu, D.C. *et al.*, Autonomic self-healing in covalently crosslinked hydrogels containing hydrophobic domains. *Polymer*, 54, 23, 6381–6388, 2013.
77. Zhang, D.-D. *et al.*, A self-healing PDMS elastomer based on acylhydrazone groups and the role of hydrogen bonds. *Polymer*, 120, 189–196, 2017.
78. An, Q. *et al.*, Recycling and self-healing of dynamic covalent polymer networks with a precisely tuneable crosslinking degree. *Polym. Chem.*, 10, 6, 672–678, 2019.
79. Tian, X. *et al.*, Self-healing and high stretchable polymer electrolytes based on ionic bonds with high conductivity for lithium batteries. *J. Power Sources*, 450, 227629, 2020.
80. Guo, P. *et al.*, Healable, highly conductive, flexible, and nonflammable supramolecular ionogel electrolytes for lithium-ion batteries. *ACS Appl. Mater. Interfaces*, 11, 21, 19413–19420, 2019.

81. Lai, Y. *et al.*, Colorless, Transparent, Robust, and Fast Scratch-Self-Healing Elastomers via a Phase-Locked Dynamic Bonds Design. *Adv. Mater.*, 30, 38, 1802556, 2018.
82. Zhou, B. *et al.*, A flexible, self-healing and highly stretchable polymer electrolyte via quadruple hydrogen bonding for lithium-ion batteries. *J. Mater. Chem. A*, 6, 25, 11725–11733, 2018.
83. Zhou, B. *et al.*, Self-healing composite polymer electrolyte formed via supramolecular networks for high-performance lithium-ion batteries. *J. Mater. Chem. A*, 7, 17, 10354–10362, 2019.
84. Tamate, R. *et al.*, Self-Healing Micellar Ion Gels Based on Multiple Hydrogen Bonding. *Adv. Mater.*, 30, 36, 1802792, 2018.
85. Xia, S. *et al.*, High-Rate and Large-Capacity Lithium Metal Anode Enabled by Volume Conformal and Self-Healable Composite Polymer Electrolyte. *Adv. Sci.*, 6, 9, 1802353, 2019.
86. Brown, E.N., Sottos, N.R., White, S.R., Fracture testing of a self-healing polymer composite. *Exp. Mech.*, 42, 4, 372–379, 2002.
87. George, S., Evans, D., Davidson, L., A biologically inspired programming model for self-healing systems, in: *Proceedings of the First Workshop on Self-Healing Systems*, 2002.
88. Hansen, C.J. *et al.*, Self-healing materials with interpenetrating microvascular networks. *Adv. Mater.*, 21, 41, 4143–4147, 2009.
89. Stuckrath, C. *et al.*, Quantification of chemical and biological calcium carbonate precipitation: Performance of self-healing in reinforced mortar containing chemical admixtures. *Cem. Concr. Compos.*, 50, 10–15, 2014.
90. Peng, H.-J. *et al.*, Healing high-loading sulfur electrodes with unprecedented long cycling life: Spatial heterogeneity control. *J. Am. Chem. Soc.*, 139, 25, 8458–8466, 2017.
91. Krishnakumar, B. *et al.*, Catalyst free self-healable vitrimer/graphene oxide nanocomposites. *Compos. Part B: Eng.*, 184, 107647, 2020.
92. Kim, S.M. *et al.*, Superior toughness and fast self-healing at room temperature engineered by transparent elastomers. *Adv. Mater.*, 30, 1, 1705145, 2018.
93. Jo, Y.H., Li, S., Zuo, C., Zhang, Y., Gan, H., Li, S., Yu, L., He, D., Xie, X., Xue, Z., Self-healing solid polymer electrolyte facilitated by a dynamic cross-linked polymer matrix for lithium-ion batteries. *Macromolecules*, 53, 3, 1024–1032, 2020.

20

Self-Healing in Bleeding Composites

Lutfur Rahman, Ata Ullah, Muhammad Bilal Yazdani, Muhammad Irfan, Waheed S. Khan and Asma Rehman*

National Institute for Biotechnology and Genetic Engineering, Faisalabad, Pakistan

Abstract

Engineered composite materials have been used for multifarious applications. However, significant problems that material scientists are facing include composites damage and repairing at remote locations. To overcome such problems, self-healing composite materials have been designed using cost effective methods. Such materials possess self-repairing capabilities to recover damages/cracks, through the action of healing agents, present intrinsically. For this, researchers are trying various approaches but another issue with such materials is crakes visualization. To resolve both the problems (i.e. self-repairing and crakes visualization), bioinspired and biomimetic strategies, evolved from biological system are the most appropriate approaches. Living creatures have specific mechanism for wound healing and follow steps such as inflammatory response, cell proliferation and matrix remodeling. Similar mechanism is found in engineered self-healing composite materials i.e. triggering, transport of healing agent to damage sites and finally repairing. Self-healing bleeding composite materials have been developed by using biomimetic approach. These bleeding composite materials have dual function: have UV-light fluorescent dye, which is used for tracking any damage and secondly it also releases the healing agent to perform repairing function. Current chapter emphasize self-healing in bleeding composite materials, their synthesis approaches and properties, mechanism of healing, advantages, and their limitations.

Keywords: Self-healing, biomimetic, healing agent, microencapsulation, bleeding composite, mechanical properties

**Corresponding author*: asmanano@gmail.com

Inamuddin, Mohd Imran Ahamed, Rajender Boddula and Tariq Altalhi (eds.) Self-Healing Smart Materials and Allied Applications, (495–510) © 2021 Scrivener Publishing LLC

20.1 Introduction

Naturally occurring materials used for various applications (clothing, building, etc.) may have some limitations. Thus, material scientists have been searching to synthesize state-of-the-art materials with durable properties or to modify the available materials according to their desired applications. The main purpose of fabricating novel materials is to improve their life-span and other properties such as mechanical strength, stiffness, density, sustainability as well as lowered their processing and manufacturing cost. Moreover, researchers also working to incorporate properties like self-healing, damage recognition, regeneration, adaptability according to the environmental changes, in a smart reacting manner with self-defensive mechanism [1].

In the traditional system of materials repairing, welding "hot plate" was the basic method to restore the fractured surfaces. But this system has some limitations; as the actual mechanical strength may lost, as a result of which the materials strength decreases and lead to failure. Another method is to inject resin to the site of damage but most of the time the cracks may not be easily accessible which increase the chances of failure of materials [2]. Other approaches used are the application of patch alone or in combination with resin injection. Such approaches are not ideal to restore and repair damaged materials, as these are temporary solutions and each strategy need proper monitoring and external intervention. These limitation increases material's cost by maintenances and extensive care [3].

Keeping in view material's significant features and intelligent efficient self-healing mechanistic approach, composites materials have been developed and used in different fields of research. Such self-healing composite materials have the capability of autonomic recovery and adaptability, with respect to the environmental changes, in a dynamic manner. These unique qualities can differentiate them from the traditionally and naturally occurring materials. Scientist expect that self-healing property of the fabricated materials (composites) may improve their reliability, safety and can decrease their cost of synthesis. Thus, this area of material development gaining significant attention from the last few decades [1]. Composites materials are basically consisting of two or more components in which one serve as matrix phase while the other may be present in the form of particle or fiber. In most cases naturally occurring or synthetic fibers can be used to fabricate composite materials. Fabrication of composite materials, in the field of materials sciences was a valuable development which substantially have applications in a variety of other fields such as mechanical, automobile, construction, aerospace, biomedical and marine [4].

The existing composites materials have been categories in three major groups; (1) Capsule-based, (2) vascular based and (3) intrinsically self-healing materials. The classification is based on their self-healing and biomimetic repairing mechanisms as well as on the healing agent associated with these materials [5]. The basic mechanism of capsule-based materials is the oozing of liquid from a small capsule like structure at the site of crakes present in the vicinity of the materials. It is because, when there occurs damage to the materials in the form of crakes, these cause capsule's rupturing as a result of which liquid releases and fill the gaps. In case of vascular based system, the capsules are replaced by vascular structures arranged in the form of a tunnels network with special liquids flow continuously. Similarly, when there occur crakes, they result the breakage of channels and the excretion of intrinsic fluid to fill the surrounding gap. The repairing material or fluid present inside the capsules/vascular network healing system is generally known as healing agent. Damage recovery processes and materials mechanical properties maintenance mostly depends upon healing agent's behavior and nature of healing mechanism. On the other hand, materials with intrinsic self-healing capability have no special designing but possesses integral reversibility properties, based on physical or chemical bonding [6]. Fundamentally, the healing mechanism of intrinsic self-healing composite is also different from that of capsule and vascular based composite materials. Some of the reported studies showed that such composite materials may perform their self-healing by the swelling of polymers [7], solidification and melting of thermoplastic materials or by increasing viscosities of pH-sensitive micro-gels [8].

Bleeding composite system is a novel system based on biomimetic approach. This has been developed for composite materials to perform self-healing functions. Distinct features regarding self-healing materials in bleeding composites are visualization of the cracked or damaged and restoration of mechanical strength of the damaged area. Such materials can enhance visualization of damaged site by the bleeding action of a special medium/liquid such as crack penetrating or fluorescent dyes, at the site of damage. Such composite materials also perform the action of self-healing by releasing the healing agent present in their core structure. Beside these, the fabricated materials lead to restore the mechanical strength of the bleeding composite at the point damage. To study the physical and chemical properties of self-healing composite materials, various tests have been reported. These tests can be conducted after the fabrication of materials as well as when damage occurs and healed. Pang and Bond [9], fabricated bleeding composite evaluated its strength restoration by conducting a four-point blend testing. He further reported that by self-healing effect, a significant fraction of strength i.e. approximately 97% was restored [9].

20.2 Intrinsic and Extrinsic Self-Healing Materials and Their Repairing Approaches

Based on repairing mechanism of healing agents and indigenous repairing properties of materials, self-healing materials are classified as intrinsic or extrinsic. The main difference between intrinsic and extrinsic self-healing materials is: in extrinsic healing materials, healing agents are used to facilitate structure repairing while intrinsic self-repairing materials have its own healing ability. Based on materials definition Figure 20.1 shows the proposed repairing mechanism for both intrinsic and extrinsic self-healing materials [5].

In case of intrinsic self-healing materials, indigenous polymeric structure when damaged so their occur moments in their constituents' chains. Such mobility fevers healing processing but mostly depends upon energy like static load, temperature, UV light irradiation, as well as on strategy of repairing i.e. physical or chemical [10, 11]. Based on interactions during

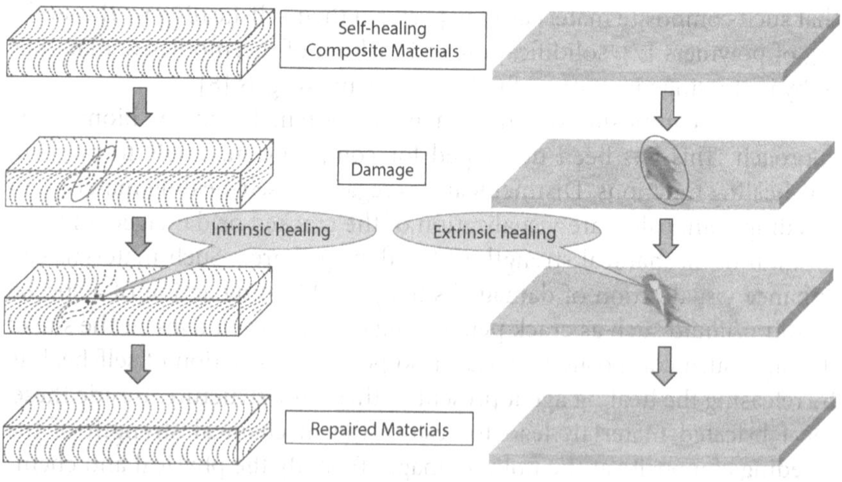

Figure 20.1 Schematic illustration showing repair mechanism in intrinsic and extrinsic self-healing composite materials.

the healing process, intrinsic materials have been classified into two major groups; dynamic covalent bonds and supra molecular interactions [12]. In dynamic covalent interactions, their occurs reshuffling and rearrangement among monomers which in turn based on their initial states. The monomeric structures have covalent cross linking, as a result of which they experienced significant bond breaking and bond forming phenomenon during healing process [13, 14]. Further self-healing in intrinsic composite materials also depends upon their monomeric nature such as the material is of which constituents (monomers, oligomers or non-crosslinked). These are the basic requirements which should be understand for composite materials the repairing process because they play very important role in the intrinsic transportation to the damaged area. Diels–Alder (and retro-Diels–Alder) reaction is the most recognized chemical reaction related to self-healing materials [12, 15].

The extrinsic self-healing composite materials are those which are basically engineered and provided with healing agents. The integrated healing agents are involved in the damage repair. Various strategies and structures are available such as capsule-based, vascular network based self-healing materials [5]. These systems are mostly derived from biological self-healing phenomenon and are known as bioinspired or biomimetic approaches.

20.3 Strategies of Self-Healing in Engineered Materials

20.3.1 Materials With Bioinspired Self-Healing Mechanism

It is understood that biological systems can overcome specific damages occurs to the organisms (plants, animals), such as wound and fractures. Living organisms possesses a vital and systematic self-healing mechanism to recover fractures and heal wounds. On the other hand, there occur damage to the materials, used in various industrial applications and it is important to recover them. Inspired from such models, materials scientists are working from the last few decades, to incorporate such self-healing capabilities to the engineered materials [1]. Under the bioinspired self-healing approach, three areas are of great interest. The very first is the recovery of failed surfaces by fusing them at the point of damage. Some materials such as polymers possess cross linking and they get damaged due to load, so the recovery of damage can be achieved by the application of heat [16]. Other materials may contain another entity in the solid form embedded in the matrix. By heating such, solid unit migrates to the area of damage and healed material's damaged portion. Although, such system works on self-healing mechanism, the basic disadvantages are: sensing of damage, no proper feedback response and heating requirements. These limitations make them imperfect for real applications [4].

Another approach is the incorporation of nanoparticles inside the multilayered polymeric structures, working on the principles of blood clotting phenomenon. Lee *et al.* [17] reported nanoparticles embedded polymeric film and studied their self-healing capabilities using integrated computer simulations. It was further described that the assembly of nanoparticles occurs at the point of nano-crake and healing occurs at the point of damage [17]. It was further reported that the repaired composite materials achieved 75 to 100% mechanical strength. Another study was also reported using fluorescent nanoparticles, but no mechanical property restoration was achieved [18]. Such type of composite materials can be used for optical, display technology and biomedical applications but due to some limitations this approach is not applicable for composite structures.

The third most interested area on materials' self-healing and repairing mechanism is "bleeding", derived from biological phenomenon, include microcapsules (Figure 20.2) and hollow fibers [16, 19].

In microencapsulation approach, the matrix is homogenously distributed with microcapsules filled with healing agent. When there occur damages/crakes, as a result of which microcapsules get damaged which in turn facilitate the release of healing agent to the surrounding. The released entity reacts with the active agent (catalyst) present in its vicinity and lead to the initiation of polymerization thus repair the damage portion as shown in Figure 20.3(A) [20–22]. The significance of microencapsulation self-healing mechanism is that the matrix can be distributed with a plenty of microcapsule. But the limitations related to such model are; microcapsule need to be fractured and the released healing materials must interact the surrounding agents for the initiation of polymerization [22]. Some additional problems may occur in the fibers reinforced polymeric composites. In such structures, microcapsules disrupt fiber morphology. In this regards excellent distribution of microcapsules and catalysts is required for proper repairing. It has been reported that, healing efficiency for composite materials depends upon the healing agent present in microcapsules [21].

Hollow fibers embedded self-healing system in engineered composite material is just like arterial network found in animals' blood circulatory system. Different arrangements and scale fibers have been investigated in a variety of engineered materials for self-healing process. Examples of such materials are bulk concrete, bulk polymer structure, polymeric composite

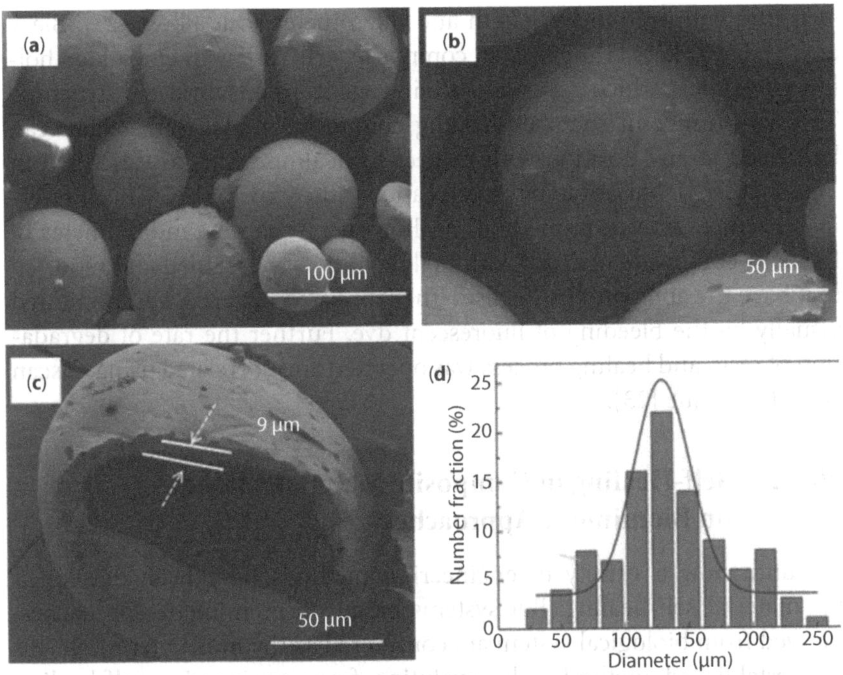

Figure 20.2 Scanning Electron Micrograph and size distribution of microcapsules. Sections labeled (a-d) representing microcapsules magnification at 100× (a) and 200× (b), cracked microcapsule (c) and size distribution. Source Li *et al.*, 2017 [19]. https://doi.org/10.1016/j.colsurfa.2017.01.046. Under License No 4785740948140.

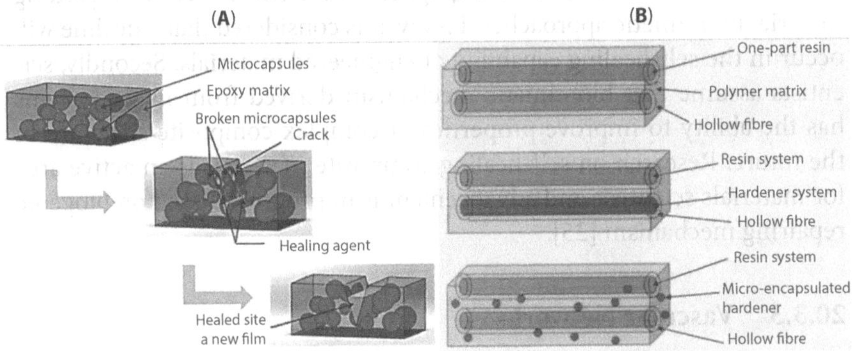

Figure 20.3 Shows schematics (A) self-healing mechanism of microcapsules filled with healing agent. Source Li *et al.*, 2017 [17]. https://doi.org/10.1016/j.colsurfa.2017.01.046. Under License No 4785740948140. (B) hollow fibers embedded self-healing composite. Source Bleay *et al.*, 2001 [16]. https://doi.org/10.1016/S1359-835X(01)00020-3. Under License No 4785740194796.

structure with millimeter length and micrometer scale fibers [23]. Bleay and his coworkers [16] prepared composite structure with embedded hollow glass fibers which act as self-healing agent in the fabricated structure. The significance of such self-healing composite model is that fiber can be placed at any position with respect to healing resins as described in Figure 20.3(B). Variation in fibers location mostly correlates with composite structure, activation method, healing agent quantity and diameter of fibers. A composite material with hollow fiber was designed and reported that have the ability to for "bruise". The fabricated material was monitored visually by the bleeding of fluorescent dye. Further the rate of degradation of resin and healing process was operated and correlated using C-scan NDE technique [23].

20.3.2 Self-Healing in Composite Materials Based on Biomimetic Approaches

Up until now a variety of engineering methods have been developed for material self-healing. But systems based on biomimetic approaches, derived from biological system are considered as the most interesting and well-established methods. The evolution from engineering self-healing strategies to biomimetic mechanism is a big challenge for the future. This biomechanistic based approach for composite materials is an emerging rea of research [24]. As discussed earlier, to date most of the research reported on self-healing engineered composite materials, is based on bioinspired mechanism not on biomimetic. Due to advancements in the field of materials sciences, the concept is changing. There are two directions regarding material biomimetic approaches. Firstly, it is considered that a decline will occur in the self-healing capability of engineered materials. Secondly, scientists assume that biomimetic mechanism derived from living systems has the ability to improve properties of complex composite structure in the future. Research on self-healing composite materials is an active area for materials scientists and advancement is in progress, based on biogenic repairing mechanism [25].

20.3.3 Vascular Networks

Living organisms possess a well-developed and multifunctional vascular system involved in various biological activities like transport of nutrients, proteins, metabolites as well as in the gracious exchange and hemostasis. Such system of tiny vessels can facilitate healing of damage parts by providing cells and other necessary constituents by the action of bleeding processes. As materials need proper monitoring which can be achieved by continuous supply of healing agents to the site of damage. In this regard the concept of vascular network system has been developed in self-healing composite materials that can delivered healing agent to the point of damage from a reservoir. Further it is also significant to provide system with frequent supply of healing agent, having the capacity to restore structure as well as physio-chemical properties of the engineered material. True vascular network system was not established initially but Toohey *et al.* [26], fabricated a polymer coating material with interconnected microvascular system for the flow of healing agent. Similarly, composite material of polymethacrylimide and glass fibers, based on simple vascular network system was fabricated and it was further reported that the system has smart delivery system for healing agent [25].

20.4 Healing Agents, Comparison With Biological Phenomenon and Bleeding Mechanism in Self-Healing Composite Materials

Blood is the key component and valuable fluid of circulatory system, involved in a variety physiological processing like transport, defense, wound healing. One of the phenomena is blood clotting which occurs at the time of fracture or injury [27]. The blood clotting mechanism was first proposed by Macfarlane, Davie and Ratnoff. Different clotting factors are involved in mammalian blood clotting. Usually these clotting factors are present an inactive form and later on activated by series of enzymes [28]. After activation, these clotting factors carryout out reaction in the form of cascade until the fibrin produced to form web like mesh in wound site to block bleeding thus lead to wound healing. Fibrin are important proteins that produce fibers in damage sites with large elasticity and extensibility. Interestingly, the blood clotting system is highly regulated. It gives quick response to injury or damage with minimal adverse effect. In healthy tissue, the endothelial cells initiate a cascade reaction which breakdown the fibrin to continues the blood circulation. Nowadays the research focusses on the development of the artificial fibers or composite materials that can mimic the natural process of blood clotting/bleeding to heal any crack or damage [29].

Materials scientists have developed biomimetic hollow fibers which can mimic the mammalian self-healing mechanism by secreting of a liquid known as healing agent. Figure 20.4 demonstrates a comparative

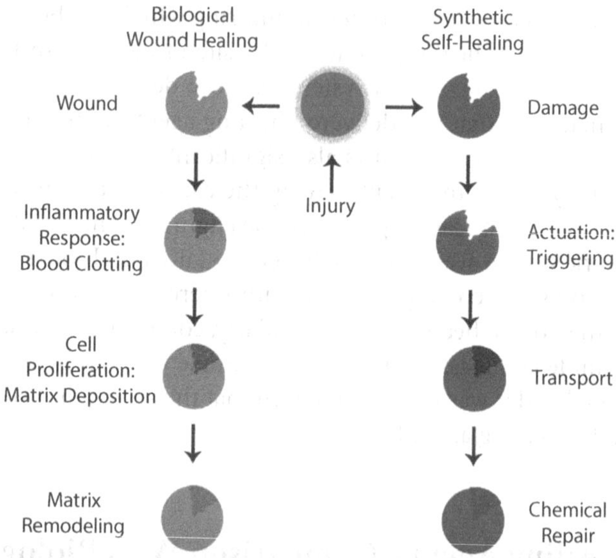

Figure 20.4 Comparation between biological and synthetic self-healing processes. Source: Garg *et al.*, 2017 [30]. https://doi.org/10.3390/biomimetics2020007. Licensed Under CC BY 4.0.

self-healing processing in living system as well repairing mechanism in chemogenic self-healing materials. Steps involved in the biological healing process after wound are inflammatory response (blood clotting), cell proliferation (matrix deposition) and matrix remodeling. On the other hand, when materials get damaged, the steps followed for repairing are triggering just like inflammatory response, transport of healing material and at last the repairing of damaged portion [30]. Although bleeding takes place when injury occurs, a significant difference in both healing mechanisms is that living system shows rapid response of hemostasis to arrest bleeding. Also, the healing process in biological systems are lengthier as compared to self-healing composite materials. Moreover, during the healing progression it is expected that the injured part will have no load lifting capability. In case of composite material's biomimetic self-healing process, a rapid and quick response is much needed to prevent crake propagation and repair damage.

Hemostasis is not a simple system, but it has been reported that it involves various biochemical reactions coupled with specific enzymes and blood cells. The main reason to mimic biological system healing phenomenon in engineered material is its degree of complexity. Keeping in view the biomimetic approach, a model was design for engineered composite materials, based on reaction with the involvement of water. Process reactions/steps are initiation, inhibition of autocatalysis and at last the precipitation reaction [31].

Materials scientists have been working for a few decades, to engineer materials with self-healing capability when they get damaged. Such system needs to use proper healing agents for example resin, to duplicate blood clotting mechanism of biological system. The best model for such engineered structures is a two-part system that is either resin with harder part or resin with catalytic portion. The working principles of both bleeding approaches are different from each other. Resin-hardener approach healing mechanism mostly depends upon proper and successful mixing of both the constituents at the point of interaction i.e. transfer of healing agent (liquid resin) to the site of damage in the composite materials. On the other hand, resin-catalyst approach significantly depends upon the reaction taking place between the healing agent and catalyst at the site of crake or damage. Both the healing agent containing approaches need initiators in the form of physical or chemical nature for example UV light irradiation, temperature and catalyst. The key point of hemostasis process, from which the scientist inspired, is that the reaction initiation in biological system occurs without inhibitor. In this regard, it is considered that the resin self-healing system might work through this mechanism [24].

Damage to the materials is a natural phenomenon and their visualization is very much important. For visualization of crakes, generally UV-light fluorescent dyes are used. Visualization through such agents performs two functions, first it helps the manufacturer to identify the damage point and secondly it monitors and insure the bleeding capability of healing agent. Fractured self-healing composite material and the bleeding action of UV-light fluorescent dye along with healing agent [9].

Based on biomimetic self-healing approach, Pang and Bond [9] developed a novel bleeding composite material. It was further reported that the system can perform dual function; visualization of damage area by the bleeding action of specific media such as fluorescent dye, and restoration through self-healing mechanism by the healing agent, stored in the composite material. The healing agent oozed out from the hollow fibers, penetrates in the area of damage and at last acting to heal the damage area and to restore mechanical properties of the material. The bleeding composite material fabricated was of borosilicate glass tubes, with external diameter of 60 μm by using bespoke fiber making facility. The final hollow fibers produced were of 1 μm which were then filled with healing agent following film infusion process. Different specimens were prepared based on repairing agent along with fluorescent dye and different treatments. Further various properties of specimens were studied. To check materials mechanical property, four-point bend flexural testing was performed using Instron 1341 servohydraulic machine. The result showed that undamaged samples possess 14% higher flexural strength as compared to damaged one. The four-point bend flexural testing results were further analyzed using Weibull analysis model to validate the mechanical strength of materials after healing processing. The result obtained through the stated model showed an increase in composite with self-healing agent as compared other samples. Moreover, all sample sets with healing agent and without healing agent were subjected to damage, by the process of indentation using the same machine and pattern of indentation behavior was obtained. No significant response in term of load-displacement was observe for both the samples.

20.4.1 Compartmentalization, Recovery After Yield and Reinforce Repair

Compartmentalization is a defensive system present in most biological systems such as animals and plants. When deep cuts occur in mammalian skin, blood clotting process starts leading finally to wound healing. Similarly, when damage occurs to plants tissues, they also a follow specific pattern by producing resistive internal boundaries at the site of damage. Compartmentalization is a significant mechanism found in biological system because with such strategy living organisms protect themselves from pathogens. There are some environmental factors such as moisture, which can damage the composite structures and decrease their strength significantly. Engineers consider mechanism followed by plants during injuries, as a form of damage tolerant model. This model may work significantly in composite materials having sandwich structure such as those use in marine operations to provide safety as well as for inspection and maintenance.

Various engineering approaches are used to design composite materials with special features. Standard fabrication approach is for the synthesis of composite structure to design fibers having sufficient strength that can response to damage. In this regard better strategy is needed to fabricate hybrid materials (i.e. composites), that mimic and adopt the mechanism (recovery) present in biological systems. Such system may be used to produce materials with elastic–plastic behavior (i.e. yielding). Although such approaches will work properly if the fabricated fibers and matrix have the capability to healed and regain their mechanical properties after. Following the stated approach and incorporation of such qualities, it could lead to a significant ameliorating advantage in the composite materials.

Till now, repairing in fibers reinforced composite structures mainly focused on the polymer matrices, because damage primarily occurs in the matrix. Though, most of the strength provided within the composite structures are due to the reinforcing phase. This strategy of reinforcement can provide significant benefits in other failure self-healing models. Inspired from biological system, bone fracture repairing in humans involved various cellular, biochemicals, hormonal, biomechanical and pathological mechanisms. Moreover, bone healing is a lengthy process of continuous deposition, remodeling and resorption. Although it is a complex phenomenon but the process can be accelerated by the application of externals forces or agents [32]. Keeping in view the biomimetic reinforced mechanism, this strategy is applicable synthetic reinforced fibrous composite materials. Further it has been suggested that additional fibers through *in-situ* growth can lead to enhance loading capacity, strength and support to the other fibers. The main issue with this model is time, as it need a long period of time for growth.

20.5 Advantages and Disadvantages of Self-Repairing Bleeding Composite Materials

Every engineered composite material has some advantages and disadvantages, in term of damage visualization and self-healing approach. Moreover, materials may possess specialty in their intrinsic, extrinsic, physical and chemical properties. Self-healing bleeding composite materials also have some advantages and limitation which are as follows:

- Most of bleeding composite materials possess a huge quantity of healing agent, for damage repair
- For such materials variety of healing agent and activation method can be used
- Using specific fluorescent dyes, visualization of crakes and damage is detectable
- Most of the hollow fibers present in such materials can be mixed easily with other i.e. conventional reinforced fibers.

The following are some disadvantages related to bleeding composite materials:

- For the release of self-healing agent, it is important that the fiber should be broken

- With the use of high viscosity resin there may occur difficulty in its infiltration, thus it must be preferring to use low viscus resin to facilitate infiltration
- The use of hollow fibers in some materials such as carbon fiber-reinforced materials may lead to mismatching due to coefficient of thermal expansion (CTE)
- Such composite materials require multistep for fabrication, as described previously [9].

20.6 Conclusion

In the recent decades, significant attention has been given to composite materials engineering through various approaches. However, use is restricted due to damage initiation and propagation, in different areas of engineering. Although, with the development of composite materials and incorporation of biomimetic approaches such as damage visualization and self-healing mechanism, the demand is increasing. Self-repairing, bleeding composite structures have been fabricated which have the capacity to visualize the area of damage by releasing UV-light fluorescent dyes at the site of damage. Additionally, such composite materials also excrete self-healing agent to repair the cracks. Some limitations have been reported but they are controllable. For example, difficulty in healing agent's infiltration is manageable by providing little bit diluted samples. The self-healing approach of bleeding composite is very much interesting and applicable, yet little literature has been reported. Therefore, extensive multidisciplinary research is required, to fabricate various kind of bleeding composites and to explore their potential uses in various fields of engineering.

References

1. Wang, Y., Pham, D.T., Ji, C., Self-healing composites: A review. *Cogent. Eng.*, 2, 1, 1075686, 2015.
2. Marioli-Riga, Z., Xenos, D., Vrettos, C., A standard analysis methodology for the stress analysis of repaired aircraft structures with the method of composite patch repair. *Appl. Compos. Mater.*, 11, 4, 191–203, 2004.
3. Lam, C., Cheng, J., Yam, C., Finite element study of cracked steel circular tube repaired by FRP patching. *Procedia Eng.*, 14, 1106–1113, 2011.
4. Kanu, N.J. *et al.*, Self-healing composites: A state-of-the-art review. *Compos. Part A Appl. Sci. Manuf.*, 121, 474–486, 2019.

5. Blaiszik, B.J. *et al.*, Self-healing polymers and composites. *Annu. Rev. Mater. Res.*, 40, 179–211, 2010.
6. Chen, X. *et al.*, New thermally remendable highly cross-linked polymeric materials. *Macromolecules*, 36, 6, 1802–1807, 2003.
7. Habault, D., Zhang, H., Zhao, Y., Light-triggered self-healing and shape-memory polymers. *Chem. Soc. Rev.*, 42, 17, 7244–7256, 2013.
8. Selver, E. *et al.*, Healing potential of hybrid materials for structural composites. *Compos. Struct.*, 122, 57–66, 2015.
9. Pang, J. and Bond, I., 'Bleeding composites'—Damage detection and self-repair using a biomimetic approach. *Compos. Part A Appl. Sci. Manuf.*, 36, 2, 183–188, 2005.
10. Hager, M.D. *et al.*, Self-healing materials. *Adv. Mater.*, 22, 47, 5424–5430, 2010.
11. Blaiszik, B., Sottos, N.R., White, S.R., Nanocapsules for self-healing materials. *Compos. Sci. Technol.*, 68, 3–4, 978–986, 2008.
12. Garcia, S.J., Effect of polymer architecture on the intrinsic self-healing character of polymers. *Eur. Polym.*, 53, 118–125, 2014.
13. Lehn, J.-M., Dynamers: Dynamic molecular and supramolecular polymers. *Prog. Polym. Sci.*, 30, 8–9, 814–831, 2005.
14. Zhang, L. and Rowan, S.J., Effect of Sterics and Degree of Cross-Linking on the Mechanical Properties of Dynamic Poly (alkylurea–urethane) Networks. *Macromolecules*, 50, 13, 5051–5060, 2017.
15. Billiet, S. *et al.*, Chemistry of crosslinking processes for self-healing polymers. *Macromol. Rapid Commun.*, 34, 4, 290–309, 2013.
16. Bleay, S. *et al.*, A smart repair system for polymer matrix composites. *Compos. Part A Appl. Sci. Manuf.*, 32, 12, 1767–1776, 2001.
17. Lee, J.Y., Buxton, G.A., Balazs, A.C., Using nanoparticles to create self-healing composites. *J. Chem. Phys.*, 121, 11, 5531–5540, 2004.
18. Gupta, S. *et al.*, Entropy-driven segregation of nanoparticles to cracks in multilayered composite polymer structures. *Nat. Mater.*, **5**, 3, 229–233, 2006.
19. Li, H. *et al.*, Preparation and application of polysulfone microcapsules containing tung oil in self-healing and self-lubricating epoxy coating. *Colloids Surf. A Physicochem. Eng. Asp.*, 518, 181–187, 2017.
20. White, S.R. *et al.*, Autonomic healing of polymer composites. *Nature*, 409, 6822, 794–797, 2001.
21. Kessler, M., Sottos, N.R., White, S.R., Self-healing structural composite materials. *Compos. Part A Appl. Sci. Manuf.*, 34, 8, 743–753, 2003.
22. Kessler, M. and White, S.R., Self-activated healing of delamination damage in woven composites. *Compos. Part A Appl. Sci. Manuf.*, 32, 5, 683–699, 2001.
23. Pang, J.W. and Bond, I.P., A hollow fibre reinforced polymer composite encompassing self-healing and enhanced damage visibility. *Compos. Sci. Technol.*, 65, 11–12, 1791–1799, 2005.
24. Trask, R., Williams, H., Bond, I., Self-healing polymer composites: Mimicking nature to enhance performance. *Bioinspir. Biomim.*, 2, 1, 1–12, 2007.

25. Williams, H. *et al.*, Biomimetic reliability strategies for self-healing vascular networks in engineering materials. *J. R. Soc. Interface*, 5, 24, 735–747, 2008.
26. Toohey, K., White, S., Sottos, N., Microvascular networks for self-healing polymer coatings, in: *Fifteenth USNC/TAM*, 25th–31st June, 2006.
27. Grazul-Bilska, A.T. *et al.*, Wound healing: The role of growth factors. *Drug Today*, 39, 10, 787–800, 2003.
28. Macfarlane, R., An enzyme cascade in the blood clotting mechanism and its function as a biochemical amplifier. *Nature*, 202, 4931, 498–499, 1964.
29. Liu, W. *et al.*, Fibrin fibers have extraordinary extensibility and elasticity. *Science*, 313, 5787, 634–634, 2006.
30. Garg, P. *et al.*, Influence of nanotechnology and the role of nanostructures in biomimetic studies and their potential applications. *Biomimetics*, 2, 2, 1–25, 2017.
31. Runyon, M.K., Johnson-Kerner, B.L., Ismagilov, R.F., Minimal functional model of hemostasis in a biomimetic microfluidic system. *Angew.*, 43, 12, 1531–1536, 2004.
32. Gittens, S.A. and Uludag, H., Growth factor delivery for bone tissue engineering. *J. Drug Target.*, 9, 6, 407–429, 2001.

21

Self-Healing Polymers

Muhammad Akram[1], Charles Oluwaseun Adetunji[2*], Mohd Imran Ahamed[3], Adrish Sohail[1], Iram Ghaffar[1], Olugbenga Samuel Michael[4], Hina Anwar[1], Musa Abidemi Muhibi[6], Juliana Bunmi Adetunji[5], Umme Laila[1] and Mathew Olaniyan[6]

[1]*Department of Eastern Medicine, Government College University Faisalabad, Faisalabad, Pakistan*

[2]*Applied Microbiology, Biotechnology and Nanotechnology Laboratory, Department of Microbiology, Edo University Iyamho, Auchi, Nigeria*

[3]*Department of Chemistry, Aligarh Muslim University, Aligarh, India*

[4]*Cardiometabolic Research Unit, Department of Physiology, College of Health Sciences, Bowen University, Iwo, Nigeria*

[5]*Nutritional and Toxicological Research Laboratory, Department of Biochemistry Sciences, Osun State University, Osogbo, Nigeria*

[6]*Department of Medical Laboratory Science, Edo University Iyamho, Auchi, Nigeria*

Abstract

There is a very high demand and search for unique material that possesses self-healing attributes. Several materials have been utilized as synthetic self-healing materials majorly from thermoplastic, elastic, shape memory polymers, polymer composites, nanocomposites and coatings. However, many of these materials still need drastic improvement in order to enhance their self-healing attributes such as strength, resistance to corrosion and conductivity. This will go a long way in improving and increasing the life span of manufactured products with self-healing characteristics. Therefore, this chapter intends to provide a general overview on numerous materials with self-healing attributes. More emphasis was also made on autonomic and non-autonomic types of self-healing materials. The mechanisms of action of these self-healing materials were also highlighted.

Keywords: Self-healing, autonomic, non-autonomic, thermoplastic, elastic, shape memory polymers, polymer composites, nanocomposites

**Corresponding author*: charliguitar@yahoo.com; adetunjicharles@gmail.com

Inamuddin, Mohd Imran Ahamed, Rajender Boddula and Tariq Altalhi (eds.) Self-Healing Smart Materials and Allied Applications, (511–530) © 2021 Scrivener Publishing LLC

21.1 Introduction

Numerous biological systems have recognized to possess numerous capabilities to play a crucial role in the process of healing wounds [1, 2]. Therefore, there is a need to search and discover numerous and novel materials that possess self-healing attributes [3–5]. Artificially synthesized self-healing materials are normally synthesized using the concept of autonomous healing that occurs in body of living organisms [6]. It has been realized that the process of self-healing ability which occurs in living organisms is vital for their survival and increased age [7–9]. Since 2001, there were about 6,000 publications on the web available that have performed several research work on self-healing ability of numerous materials [10]. These synthetic self-healing materials are chemically synthesized with materials that include thermoplastic, elastic, shape memory polymers, polymer composites, nanocomposites and coatings [11–13]. These synthetic self-healing materials possess several unique attributes such as strength, resistance to corrosion and conductivity as well as possess capability to increase the life span of manufactured products [14, 15]. There were hundreds of articles published in past years studying about polymers, chemistry and self-healing materials [5, 16]. Moreover, there is a new trend of using nanomaterials in various applications as self-healing materials. This could be linked to their unique properties [17–19]. There are two types of self-healing materials according to their nature of actions. One is autonomic and the other one is non-autonomic. In autonomic type, the healing materials are enclosed, and could automatically be release by detecting any damage and did repair while in non-autonomic type external applications are required for self-healing materials to work [20]. Specific examples of these materials in autonomic type (Figure 21.1) and non-autonomic type (Figure 21.2) are organized according to their characteristics [21, 22].

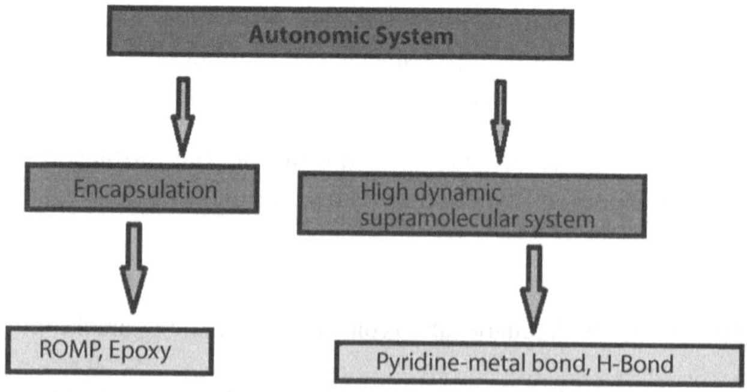

Figure 21.1 Stages involved in autonomic system during the process of self-healing.

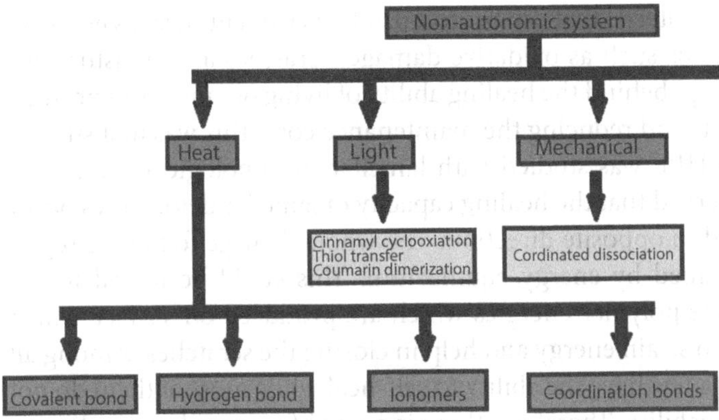

Figure 21.2 Stages involved in a non-autonomic system during the process of self-healing.

Effective chemical reactions called "click chemistry" in combination with state-of-art engineering techniques are utilized in the preparation of these efficient materials. The self-healing attribute of these materials could be linked nanoreservoir, microvascular system and microcapsule based systems [23–28]. Moreover, among all the synthetic healing materials the material containing nano polymers are considered best because of their properties like strength, conductivity, stability, lightweight and high efficiency. Also, the combination of the concept of self-healing along with nano-polymer composites gave new ideas on their safety and they could enhance the durability of these healing materials [21]. Moreover, it has been discovered that some self-healing materials like polymers have amazing healing ability but it can repair damage only a single time then after depletion of healing material at specific point the process of healing stopped. Therefore, there might be a need to produce novel self-healing polymers that could be re-mendable in nature which could be reutilized in many cycles [29]. Therefore, this chapter entails a general overview on numerous materials with self-healing attributes. More emphasis was also made on autonomic and non-autonomic types of self-healing materials. The mechanisms of action of these self-healing materials were also highlighted.

21.2 General Overview on Self-Healing Materials

It has been observed that the environment that surrounds us have various impacts on products manufactured by man. Therefore, there is a need is to make a natural material that can protect products from numerous environmental conditions such as oxidative damage, scratches and moisture etc. The main concept behind the healing ability of living organisms in protecting the products and reducing the maintenance cost. This was first studied in 1970 when HPP was studied with lamellar morphological characters, and it was observed that the healing capacity of lamellar deformities occur when stretched in opposite direction. The whole damage following repair could be explained by energy equilibrium. This could be linked to the stretching of the polymer energies which are produced on surface which are converted to strain energy and help in closing the scratches. Among all polymers only a few have the ability to self-heal while most of them do not possess this capability. Therefore, there is a need for more human discoveries and production of novel self-healing materials. Usually the methods used by humans include resins, welding and apply patches but these are not strong methods because the damage site remains weak even after repair. Also, in some cases it is impossible to reach the damaged site therefore, it might be necessary to make a material that will not only repair damage but that possess that capability to stop further damage while it will also maintains its physiochemical properties. There are two types of an ideal self-healing material. Either the healing occurs internally by repairing the bond or by externally applying any force to release the healing material that has been added previously. There are some types of healing materials that are already available in the market that melt at TG and fill the crack. The main purpose of further studies was to make a healing material that could be initiated in response to stimulus like damage [30]. These materials that cause healing in response to stimulus are called autonomic materials while there is also the other type could induce healing in case of external stimulus like chemical, thermal, optical, etc. The first designed healing material was fibers reinforced system. During this process the fibers were filled with self-healing material. Any damage or reinforcement could result the release of material from fibers and could result into healing by filling cracks [31, 32]. Thereafter, numerous numbers of researches made by the chemist to design further self-healing materials were observed with improved properties. They were succeeded in designing autonomic and non-autonomic healing materials. The autonomous healing materials were considered more useful because there is no need of human to provide external stimulus for the release of healing material however in some non-autonomous materials the natural external trigger could play a role like optical stimulus for car coating could be played by sunlight [33]. Moreover, it has been observed that when designing a self-healing model it is also necessary to observe and evaluate all of its properties including efficacy, ability to reuse at same point, physico-chemical properties and ability to maintain the properties of damaged site. That's why a variety of materials were taken under consideration [34, 35].

21.3 Design of Self-Healing

It is difficult to design a self-healing material for rigid matrices like crystalline materials, amorphous polymers and high crossed linked matrices. In University of Illinois, an architect, did experiments on designing a self-healing material using fibers for concretes and polymers [4, 31]. She made a self-healing system by using fibers filled with self-healing material. However, it was only released and triggered through stimulus like damage and heals the cracks. The evaluation showed some properties of materials that it could gradually leads to degradation while the fibers become filled with healing material which could be releases when triggered by stimulus when the fibers became coated [36]. Several experiments were done using fibers. In one experiment two types of fibers were used, one was metal fibers and others were fibers with healing material. The filled fibers had two type of materials in them, epoxy and diamine. In that experiment, the metal fibers were removed without disturbing the fluid filled fibers. It resulted into stress over materials filled fibers and the fluids were released. That reduced the stress and interacts with damaged area for healing. This design was considered easy and useful because it only required checking the design and numbers of fibers, coating material and filling materials. The type of healing materials depends on the type of choice. Atypical examples are vinyl monomers which can be used as healing agent of micro cracks which form free radicals. Those free radicals are used in providing strength to damaged area.

21.3.1 Modes of Action of Self-Healing

The analysis of healing system by solvents, fluids and catalyst has been observed that they could fit perfectly on all systems as described by Dry and Sottos [37] and Ref. [10] as well as the following model as from the following scientists [12, 26, 38], nanoscale models [39], chemical models [40], Nano segregation [24] and ballistic self-healing system [9]. Depending upon cracking and its self-healing level and molecular process, the mechanism of healing could be divided into following stages.

21.3.2 Rearrangement of Surface Dynamics

When a surface is cracked and damage is produced then the surface become rough and its characteristics are completely changed, the healing time depends upon temperature, pressure, healing materials, external stimulus and other environmental conditions. Like in case of polymer damage, the healing occurs by fibers rearrangement [41]. In case of any damage on surface, the materials were designed to move towards surface with less tension. These were usually low weight molecules, so, these can easily move towards surface. While in case of Nano fractures Nano particles move to heal [24]. This surface arrangement occurs in presence of solvents. If there is chance of any chemical reaction like oxidation, then whole process became complicated. Moreover, these solvent reactions can cause further damage by causing inflammation of micro voids.

21.3.3 Bringing the Surfaces Together

In lab experiments, this stage considered very easy because all conditions and external environment is under control, we can apply the pressure required to bring surfaces close. However, if surfaces were not pooled together then there will be no healing. It usually occurs if there is any debris causing hindrance in surface approach. Surface approach can occur either in presence of healing material or by applying pressure. However, different material surfaces have different time limit of surface approach. In case of healing material involvement, it is necessary for healing material to bind to both surfaces.

21.3.4 Wetness

After the step of surface approach, there is release of self-healing materials between two surfaces for healing process [42]. According to different models of self-healing, the wetting ability of different materials is also different. Some fluids have better speed than others. Sometimes, the chemical reactions like oxidations prevent the wetting of surfaces.

21.3.5 Diffusion

Diffusion means molecular diffusion which occurred at cracked area. It is the most valuable stage and determines the level of strength and reversal of characteristics of that damaged area. This step became easy because the area is wet, and movement of molecular chains is easy. The chain interfuses to repair area. The wetting is helpful in decreasing the friction between matrix so that chains can easily diffuses through matrix. According to [43] theory, the chains that are present at interface diffuse more thoroughly and closely, while as we move away from interface the chains of molecules are not completely diffused. In case of low weight molecules, we do not discuss about spreading of molecular chains [44] while in case of heavy weight molecules, the spread and non-connected chains are also considered important [45].

21.4 Application of Self-Healing Materials

Self-healing materials are now at preliminary level that's why these are not used in all industries, but their successful results may end at its usage in all industries. At present time, these are used in automobile, aerospace and building industries. Nissan Motors was the first company to use healing materials in car coat to protect it from scratches. They explained that the coat on car prevents scratches as a result of car wash or rough driving. The coat contains resins that prevent the deep scratches and the complete healing occurs within 1 to 7 days depending upon sunlight and temperature. Desmodur and desmophen were used in preparation of that car coat. There are two types of polymers: thermoplastic and thermosetting. In case of thermoplastic polymers, when the force is applied to create scratch then a groove is formed, and the energy used is lost that's why it didn't restore to normal. But in case of thermosetting polymer when force is applied, the groove is formed, energy stored in surrounding area. When force is removed, the energy will then be released, and polymer restore to normal shape. However, if the scratch is deep resulting into cracks then restoration may be partial (Figure 21.3).

Fibers reinforced healing system is also used in aircrafts. Now a days, industries are trying to make use of healing materials in their products like self-healing concrete, metal products and in protection of steel from corrosive damage. These materials may have a promising future if used in increasing the life of artificial bones, teeth and in other medical needs [46].

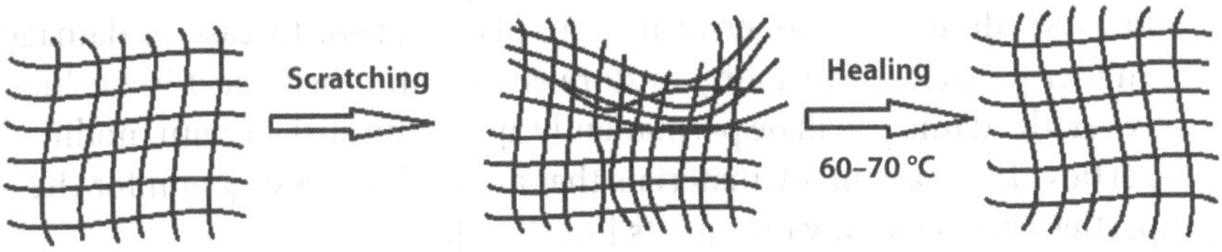

Figure 21.3 Process involved in the self-healing.

21.4.1 Properties of Self-Healing

An ideal self-healing material must have the following properties [11].

- An autonomous healing ability
- It must be re-mendable (used in multiple cycles)
- Can heal even large damaged areas
- Have fewer expenses
- Must be powerful and stiff
- Protect from deep damages
- Have ability to restore physiochemical properties of damaged area
- Must be affordable.

21.4.2 Advancement in Self-Healing

Chen and coworkers developed a healing system in which the polymer was used with ability to be use at multiple times [47]. That polymer was consisting of furan and maleimide. According to that system when damage occurs, the area was heated up to 120 °C. Such a high heat breaks down all linkages and then those linkages were formed again restoring the shape and properties of damaged area. By this process polymers can be healed multiple times.

Zhu and Wool utilized nanomers in healing system. Nanomers were made by combining the AOME with nanoclay [48]. The properties of this composition depend upon nanoclay. It was proposed that, when pressure was applied, these nanoclayparticles have ability to store that mechanical energy, later releasing that energy and restoring the properties leading to heal the damage [48]. In future there is possibility of development of self-healing materials made up of bio molecules. For example, amino acids when used in large amount have ability to resist the level of damage by resisting the stress. There is another theory about the use of healing materials that may have ability to unfold by absorbing energy produced by stress and then again fold to restore the damaged area [49–51]. Some other studies include use of rubber in healing plastic glass. In case of damage the rubber particles have ability to store energy and then use it in healing. However, the damage will be permanently appear on surface with modified properties. There are many materials that are under investigation for their properties and needs advancements [9, 52, 53].

21.4.3 Classification of Self-Healing

In self-healing process it has been discovered that different application are applicable, but before the detail explanation of types first discus about the releasing agent. During the process of healing some active agents like catalyst, dyes, monomers, hollow-fibers & micro-capsules embedded into, system which called polymeric system. Whenever the problem of cracks occurs than firstly reputing process start in these reservoirs, the releasing agent than poured into, cracks through a force of capillaries from which they solidify. It is an auto-nomic procedure not required any externally & manually intervention [46].

21.4.4 Healing Mechanism Types of Healing

Crack filling healing	this type include healing agent in hollow fiber, healing agent which is phase separated, micro-vascular network & micro-encapsulated healing agent.
Reformation of bond	ultraviolet initiated, thermally reversible & supramolecular Assembly and association of metal ligand.
Diffusion	visco-elastic healing, thermo-plastic dangling diffusion of Chain.
Strengthening property	mechano-phores.

21.4.4.1 Crack Filling Healing Process

In this mechanism of healing process mostly those polymers & composites are developed which shows or exhibit the autonomic mechanism for that purpose. Herein is the name of those releasing agent which helpful in the process of healing and these substances are solvent solution, polymers of thermo-plastic, liquid type monomers, crystal in the form of liquid & inorganic or organic type. There are many advantages of this process, by this process initially removal of cracks or closing of cracks done which is difficult later for stiff-structural composites & polymers. From this crack healing process, it is easily cure to crack & no need of external intravenous because in that process whole releasing agent is supply for the cure of these cracks, which present in the polymers [54].

In crack healing process further some steps which are involved in this procedure and the one is micro-encapsulation in that process small sized micro particle like liquid in the form of droplet & gases or solid all these en-closed into the inert-shell, & the main action of this procedure, gave protection from the environment & isolate it [55, 56]. Micro-capsule is the product which also called micro-encapsulation, composition of this capsule into shell & core two parts of these capsule, there may be irregularity in their structure e.g. in their shape either these particles possess spherical or irregular shape & size is also variable, variation of their size from Nano-micro. For self-healing purpose micro-capsule which contain catalyst & healing agent like composite are used for this purpose. From the research it was concluded that affective effect of self-healing process achieved by using those agents which present in the Matrix of polyester. The useful effect of this process is that discharge of healing agent is fast & due to that reason recovery of cracks quickly occurs. In many polymer systems the used of this microcapsule might be occurs [57]. In case of hollow embedment of fiber, some disadvantage is present in self, healing process and one is the complete utilization of the entire healing agent in case of completion of healing process. Multiple process of healing occurs only when amount of healing agent is more & this could be achieved by using micro-capsule bases self-healing which is hollow embedment fiber. Specific examples of self-healing materials and their influencing factors are shown in Table 21.1.

Table 21.1 Specific examples of self-healing materials and their influencing factors.

Parameters	**Influencing factors**
Healing	Healing process mostly faster & multiple
Catalyst	They dissolve into mono-mers & with matrix polymers no accumulation.
Polymerization	Stress relaxation, should be fast
Monomers	For polymerization enough time available when their volatility is decrease. Low viscous mono-mer for flow on the crack though the action of capillary
Microcapsule	Shelf life of capsule longer, for healing process the agent must be inert into the polymers.

21.4.4.2 Diffusion

The procedure of self-healing depends on the molecular diffusion of the species which are mobile to form an adhesive physical & chemical linkage. In this crack healing process, the role of diffusion in that is movement of mobile-species from the damage or broken area to another. Then the area which is defective might be filled with the healing agent. In case of thermo-plastic polymer, firstly with thermoset make a homogenous blend, these blends used for curing purpose before & after. In epoxy thermoset resin it is dissolved, reduction occurs in the de-lamination area. After that note the toughness of fracture, & result which obtain from that diffusion of thermo-plastic polymer occurs which filling the affected area. Another method which is used in diffusion for healing purpose is dangling chain diffusion healing process. After diffusion, by using chemical & physical interaction these dangling chain process heal the polymers.

21.4.4.3 Bond Reformation

Bond reformation shows a significant potential for healing process is shown through by using reversible polymerization of polymers. The other name of these polymers are also dynamers & mendomers and they have some specific bonds that show reversible reaction due to some external stimulus like light, heat, basic & acidic condition. Mendomer has more potential to heal itself for healing process. In bond reformation where many polymers present external heat is the main stimulus for the healing process. When used ultra violet light as an external stimulus than it effective for repairing of polymers in the bond formation. Virgin property strengthens this technique is new in field of self, healing procedure.

21.4.4.4 Application

In industry the product to be commercialized mostly depends on different mile-stones like generation of ideas (e.g. preliminary level) in laboratory implementation (e.g. at the level of product) scaling up line pilot (e.g. at the level of process) & industrial application (e.g. at level of marketing). Now days the application of self, healing material either in the level of product or it might be preliminary so, due to that reason these types of material used in different form of application. Much new advancement is made on that type of material & in future the used of these materials is more. There few applications are applicable in industries buildings, aero-space & auto-motive. According to the record it was find that first self, healing coat discovered for car & name of that car is coast guard scratch [46]. From this they not only remove the scratches of car but it also effective for 3 years, so by using this it protects the inner surface of the car. Another e.g. is urethane coats; this coat is effective and prevents the influence of light and also protects from scratching.

21.5 Specific Examples of Self-Healing Polymer

21.5.1 Intrinsic Self-Healing

The cleft healing process is initiated by specific stimulus (mostly heating) and it depends upon the functioning of polymers and polymeric matrices, known as intrinsic self-healing polymers. For the time being, autonomic healing is not possible without extraneous interference in these materials.

21.5.2 Extrinsic Self-Healing

The healing factor is discharged into the cleft planes through capillaries to initiate the healing process. So in case of extrinsic self-healing, matrix resin does not have ability to heal itself. The healing agent should be capsulized and incorporated into the materials before.

21.5.3 One Capsule System

In 2001, the development of epoxy-based material that possessed the ability of autonomic repairmen after harm, even in the absence of external interference was reported by White and co-workers [10]. In one capsule system, within an epoxy matrix autonomic healing is carried out by embedding microencapsulated healing factor (a monomer) and a chemical activator (initiator or catalyst). Microcapsules are ruptured at the site of cleft, healing agent is released into the fissure by capillary process, when this factor comes in contact with catalyst, polymerization is started and thus structural continuity is restored.

21.5.4 Self-Healing Based on Ring Opening Metathesis Polymerization

First generation Grubb's catalyst is used to initiate the ring opening metathesis polymerization (ROMP). Due to stability of monomer, it can be kept in an inactive form in a microcapsule prior to reacting with the catalyst; due to low viscosity of DCPD, it can easily move to the damaged site, minimum shrinkage usually takes place by the transition of monomer to polymer, this results into a highly cross-linked polymer "polydicyclopentadiene" (PolyDCPD) and it has prominent mechanical features. The catalyst, bis(tricyclohexylphosphine) benzylideneruthenium (IV)dichloride (first-generation Grubbs' catalyst) shows high metathesis action with DCPD monomer and possess satisfactory chemical constancy [58–60]. The ability of a catalyst to endure the treating circumstances in sample manufacturing process is assured by the chemical stability of the catalyst, and it should remain active to start ROMP throughout the healing process.

21.5.5 Solvent-Induced Self-Healing

In solvent-induced self-healing system, instead of molding, the microcapsules which consist of solvent are incorporated into the epoxy material. The rupture in epoxy material damages the microcapsules which results in the discharge of solvent into the cleft plane. During healing process, the unreacted epoxy resin as well as unreacted amines and amine end-groups are elute out by the solvent to react further. Formerly reference tests were conducted to manifest the relationship between solvent polarity and its healing ability, in these tests epoxy TDCB samples were damaged and repaired by manual injections of solvent [61]. Although the nature of this relationship was ill-defined, but five highly efficient solvents were dimethyl sulfoxide (DMSO), dimethylacetamide (DMA), nitrobenzene, dimethylformamide (DMF) and *N*-methyl pyrrolidone (NMP). Hexanes, water, cyclohexane and formamide, have shown no healing ability with both extremes of polarity and are not able to elevate recovery of mechanical properties. The five efficient solvents mentioned above, do not require the standard mechanism of urea-formaldehyde encapsulation due to their polarity. To encapsulate these solvents, various attempts were made at reverse phase encapsulation procedures, eventually (the next highly efficient solvent) was capsulized by using urea-formaldehyde mechanism. The healing power of encapsulated chlorobenzene containing samples was 82% but the healing efficiencies of nonpolar solvents, including hexanes and xylenes were 0 and 38% respectively [62].

21.5.6 Dual-Capsule Systems

Dual-capsule systems include hydroxyl-terminated PDMS condensation and hydrosilylation.

21.5.6.1 Polydimethylsiloxane Condensation

The chemically stable chemistries were realized when a substantial step of the development of PolyDiMethylSiloxane based self-healing chemistries was taken. The basis of categorization of this chemistry for self-healing functions were tin-catalyzed polycondensation of phase-separated droplets containing hydroxyl end-functionalized poly(dimethylsiloxane) (HOPDMS) a poly(diethoxysiloxane) (PDES) [63]. The catalyst, di-n-butyltindilaurate (DBTL) was incorporated in an epoxy vinyl ester matrix and it was capsulized in polyurethane microcapsules. Similar to the former systems, any crack in the matrix damaged the microcapsule that contained the catalyst, due to this fracture the constituents were discharged into cleft plane reacting with the phase-separated HOPDMS and PDES. This PDES assisted as the cross-linking factor in this poly condensation reaction. This system has more advantages than ROMP-based systems such as higher thermal stability, stability in humid and wet environment, wide availability of its components and much more economical.

21.5.6.2 Platinum-Catalyzed Hydrosilylation

The platinum-catalyzed hydrosilylation of vinyl-terminated PDMS with a methylhydrosiloxane copolymer is another PDMS chemistry used for the development of self-healing polymers. The mechanism of interaction of a crack with microcapsules in a linear elastic matrix was numerically defined by White and co-workers by using Eshelby–Mura equivalent inclusion model [10]. It was discovered by their analysis that the crack was more according to the microcapsule.

21.5.6.3 Adaptive Resistant Effect

The immune effect remains essential in vertebrates [64]. Basically leukocytes, lymphocytes and B-cells, T-cell leukocytes are the main factors that remain provided an exclusive method disinfects destructive microbes [2]. These cells can be present in a short-term response as memory cells it can be regulate through the accessory for the distinctive resistant system [65]. T-cells are innate immune system are present in the thymus and can be used as a cell-mediated. And other way B-cells initiate in bone and making antibodies, are effectively neutralizing the receptor in harmful microbes and block their active region [66]. Vaccinations are used for illness make the immunological recollection they are vital for these personalized replies. Further, dispute has remained complete to contain the natural cells killer during glowing [67]. The healing process of injury occurs in two ways: autonomically and non-autonomically, and this process activates through polymers of self-healing. Different healing mechanism are present like vascular system of healing, system of healing which is based on capsule & intrinsic system of healing. Different studies & report have been done on these mechanisms like corrosion test done, impact response & fatigue. From this study researcher focused on the intrinsic & extrinsic processes of self-healing & also their advantages, efficiency & their application in different fields like auto-mobile, coating, aero-space, energy and electronics & many more [68].

21.6 Conclusion and Recommendations

This chapter has provided comprehensive details on numerous materials with self-healing attributes. More emphasis was also made on autonomic and non-autonomic types of self-healing materials. The mechanisms of action of these self-healing materials such as rearrangement of surface dynamics: bringing the surfaces together, wetness, and diffusion were also highlighted. More information on design of self-healing, application of self-healing materials, advancement in self-healing, classification of self-healing as well as information on the various types of healing involved crack filling healing, reformation of bond, diffusion, strengthening property. There is a need for inter-multidisciplinary collaboration between material scientists, chemists, and biologist to come together and discover novel material with self-healing properties that possess very unique properties with high durability.

References

1. Holliman, J.H., Principles of Wound Healing and Tissue Repair. In: *Pathology*. Oklahoma Notes, Springer, New York, NY, 1995. https://doi.org/10.1007/978-1-4612-0813-6_5
2. Cremaldi, J.C. and Bhushan, B., Bioinspired self-healing materials: Lessons from nature. *Beilstein J. Nanotechnol.*, 9, 1, 907–935, 2018.
3. Gould, P., Smart, clean surfaces. *Mater. Today*, 6, 11, 44–48, 2003.
4. Sottos, N., White, S., Bond, I., Introduction: self-healing polymers and composites. *J. R. Soc. Interface*, 4, 13, 347–348, 2007.
5. Trask, R., Williams, H., Bond, I., Self-healing polymer composites: Mimicking nature to enhance performance. *Bioinspiration Biomimetics*, 2, 1, 1, 2007.

6. Pandey, J.K. and Takagi, H., Self-healing potential of green nanocomposites from crystalline cellulose. *Int. J. Mod. Phys. B*, 25, 31, 4216–4219, 2011.
7. Yuan, Y., Yin, T., Rong, M., Zhang, M., Self healing in polymers and polymer composites. Concepts, realization and outlook: A review. *Express Polym. Lett.*, 2, 4, 238–250, 2008.
8. Chen, Y., Kushner, A.M., Williams, G.A., Guan, Z., Multiphase design of autonomic self-healing thermoplastic elastomers. *Nat. Chem.*, 4, 6, 467, 2012.
9. Kessler, M. and Thakur, V.K., Self-healing polymer nanocomposite materials: A review. *Polymer*, 69, 369–383, 2015.
10. White, S.R., Sottos, N.R., Geubelle, P.H., Moore, J.S., Kessler, M.R., Sriram, S.R., Brown, E.N., and Viswanatham, S., Autonomic healing of polymer composites. *Nature*, 409, 6822, 794, 2001.
11. van der Zwaag, S., An Introduction to Material Design Principles: Damage Prevention versus Damage Management. In: *Self Healing Materials. Springer Series in Materials Science*, van der Zwaag S. (eds.), vol 100, Springer, Dordrecht, 2007. https://doi.org/10.1007/978-1-4020-6250-6_1
12. Bergman, S.D. and Wudl, F., Mendable polymers. *J. Mater. Chem.*, 18, 1, 41–62, 2008.
13. Michal, B.T., Jaye, C.A., Spencer, E.J., Rowan, S.J., Inherently photohealable and thermal shape-memory polydisulfide networks. *ACS Macro Lett.*, 2, 8, 694–699, 2013.
14. Zheludkevich, M., Self-Healing Anticorrosion Coatings. In: *Self-Healing Materials: Fundamentals, Design Strategies, and Applications*, Ghosh S.K. (ed), pp. 101–139, Wiley VCH Verlag GmbH & Co. KGaA, Weinheim, 2009.
15. Billiet, S., Hillewaere, X.K., Teixeira, R.F., Du Prez, F.E., Chemistry of crosslinking processes for self-healing polymers. *Macromol. Rapid Commun.*, 34, 4, 290–309, 2013.
16. Wu, X.F. and Yarin, A.L., Recent progress in interfacial toughening and damage self-healing of polymer composites based on electrospun and solution-blown nanofibers: An overview. *J. Appl. Polym. Sci.*, 130, 4, 2225–2237, 2013.
17. Barbero, E.J., Ford, K.J., Mayugo, J.A., Modeling Self-healing of Fiber-reinforced Polymer–matrix Composites with Distributed Damage, in: *Self-healing Materials*, 2009.
18. Frei, R., McWilliam, R., Derrick, B., Purvis, A., Tiwari, A., Serugendo, G.D.M., Self-healing and self-repairing technologies. *Int. J. Adv. Manuf. Technol.*, 69, 5–8, 1033–1061, 2013.
19. Kötteritzsch, J., Schubert, U.S., Hager, M.D., Triggered and self-healing systems using nanostructured materials. *Nanotechnol. Rev.*, 2, 6, 699–723, 2013.
20. Williams, K.A., Dreyer, D.R., Bielawski, C.W., The underlying chemistry of self-healing materials. *MRS Bull.*, 33, 8, 759–65, 2008.
21. Benight, S.J., Wang, C., Tok, J.B., Bao, Z., Stretchable and self-healing polymers and devices for electronic skin. *Prog. Polym. Sci.*, 38, 12, 1961–1977, 2013.

22. Herbst, F., Döhler, D., Michael, P., Binder, W.H., Self-healing polymers via supramolecular forces. *Macromol. Rapid Commun.*, 34, 3, 203–220, 2013.
23. Zako, M. and Takano, N., Intelligent material systems using epoxy particles to repair microcracks and delamination damage in GFRP. *J. Intell. Mater. Syst. Struct.*, 10, 836–41, 1999.
24. Lee, J.Y., Buxton, G.A., Balazs, A.C., Using nanoparticles to create self-healing composites. *J. Chem. Phys.*, 121, 11, 5531–5540, 2004.
25. Gutpa, S., Zhang, Q., Emrick, T., Balazs, A.C., Russell, T.P., Entropy-driven segregation of nanoparticles to cracks in multilayered composite polymer structures. *Nat. Mater.*, 5, 229–33, 2006.
26. White, S.R., Sottos, N.R., Geubelle, P.H., Moore, J.S., Kessler, M.R., Sriram, S.R., Brown, E.N., Viswanathan, S. Autonomic healing of polymer composites. *Nature*, 409, 6822, 794–797, 2001.
27. Kessler, M.R. and White, S.R., Self-activated healing of delamination damage in woven composites. *Compos. A*, 32, 683–99, 2001.
28. Kessler, M.R., White., S.R., Sottos, N.R., Self-healing of composites using embedded microcapsules: repair of delamination damage in woven composites. *10th European Conf. on Composite Materials (ECCM-10)*, Bruges, 3–7 June.
29. Toohey, K.S., Sottos, N.R., Lewis, J.A., Moore, J.S., White, S.R., Self-healing materials with microvascular networks. *Nat. Mater.*, 6, 8, 581, 2007.
30. Otsuka, H., Nagano, S., Kobashi, Y., Maeda, T., Takahara, A., A dynamic covalent polymer driven by disulfide metathesis under photoirradiation. *Chem. Commun.*, 46, 7, 1150–1152, 2010.
31. Dry, C.M., Smart building materials which prevent damage or repair themselves. *MRS Online Proceedings Library Archive*, p. 276, 1992.
32. Burattini, S., Greenland, B.W., Chappell, D., Colquhoun, H.M., Hayes, W., Healable polymeric materials: A tutorial review. *Chem. Soc. Rev.*, 39, 6, 1973–1985, 2010.
33. Holten-Andersen, N., Harrington, M.J., Birkedal, H., Lee, B.P., Messersmith, P.B., Lee, K.Y.C., Waite, J.H., pH-induced metal-ligand cross-links inspired by mussel yield self-healing polymer networks with near-covalent elastic moduli. *Proc. Natl. Acad. Sci.*, 108, 7, 2651–2655, 2011.
34. Burattini, S., Colquhoun, H.M., Fox, J.D., Friedmann, D., Greenland, B.W., Harris, P.J., Hayes, W., Mackay, M.E., Rowan, S.J., A self-repairing, supramolecular polymer system: Healability as a consequence of donor–acceptor π–π stacking interactions. *Chem. Commun.* (Cambridge, England), 44, 6717–6719, 2009.
35. Burnworth, M., Tang, L., Kumpfer, J.R., Duncan, A.J., Beyer, F.L., Fiore, G.L., Rowan, S.J., Weder, C., Optically healable supramolecular polymers. *Nature*, 472, 7343, 334, 2011.
36. Allen, K.W., *Polymer interfaces: structure and strength*, Richard P. Wool. Carl Hanser Verlag, Munich, Vienna, New York, *Polym. Inter.*, 38, 3, 305–306, 1995.

37. Dry, C.M. and Sottos, N.R., Passive smart self-repair in polymer matrix composite materials. *Smart Structures and Materials 1993: Smart Materials*, International Society for Optics and Photonics, 1993.
38. Brown, E.N., White, S.R., Sottos, N.R., Fatigue crack propagation in microcapsule-toughened epoxy. *J. Mater. Sci.*, 41, 19, 6266–6273, 2006.
39. Wool, R.P., Material response and reversible cracks in viscoelastic polymers. *Polym. Eng. Sci.*, 18, 14, 1057–1061, 1978.
40. Raghavan, J. and Wool, R., Interfaces in repair, recycling, joining and manufacturing of polymers and polymer composites. *J. Appl. Polym. Sci.*, 71, 5, 775–785, 1999.
41. Wool, R. and O'Connor, K., Time dependence of crack healing. *J. Polym. Sci.: Polym. Lett. Ed.*, 20, 1, 7–16, 1982.
42. Brochard, F., Spreading of liquid drops on thin cylinders: The "manchon/droplet" transition. *J. Chem. Phys.*, 84, 8, 4664–4672, 1986.
43. Sapoval, B., Rosso, M., Gouyet, J.-F., The fractal nature of a diffusion front and the relation to percolation. *J. Phys. Lett.*, 46, 4, 149–156, 1985.
44. Wool, R.P. and Long, J.M., Fractal structure of polymer interfaces. *Macromolecules*, 26, 19, 5227–5239, 1993.
45. Wool, R., The Importance of Interfaces in Thermoplastic Matrix Composites and their Tailoring During Manufacture. *Proceedings of International Symposium on Materials Science, Interface Design of Polymer Matrix Composites—Mechanics, Chemistry, Modeling and Manufacturing*, Risoe, Denmark, 007.
46. Ghosh, S.K., *Self-Healing Materials: Fundamentals, Design Strategies, and Applications*, pp. 307, Wiley-VCH Verlag GmbH & Co. KGaA, Weinheim, 2009.
47. Chen, X., Dam, M.A., Ono, K., Mal, A.K., Shen, H., Nutt, S.R., Sheran, K., Wudl, F., A thermally re-mendable cross-linked polymeric material. *Science*, 295, 5560, 1698–702, 2002.
48. Zhu, L. and Wool, R.P., Nanoclay reinforced bio-based elastomers: Synthesis and characterization. *Polymer*, 47, 24, 8106–8115, 2006.
49. Pochan, D. and Schneider, J., Responsive materials via designed peptides, in: *Biopolymers*, John Wiley & Sons Inc, 111 River St, Hoboken, NJ 07030 USA, 2003.
50. Patwardhan, S.V., Maheshwari, R., Mukherjee, N., Kiick, K.L., Clarson, S.J., Conformation and assembly of polypeptide scaffolds in templating the synthesis of silica: An example of a polylysine macromolecular "switch". *Biomacromolecules*, 7, 2, 491–497, 2006.
51. Pogula, S.D., Patwardhan, S.V., Perry, C.C., Gillespie, J.W., Yarlagadda, S., Kiick, K.L., Continuous silica coatings on glass fibers via bioinspired approaches. *Langmuir*, 23, 12, 6677–6683, 2007.
52. Jones, A., Rule, J., Moore, J., Sottos, N., White, S., Life extension of self-healing polymers with rapidly growing fatigue cracks. *J. R. Soc. Interface*, 4, 13, 395–403, 2006.

53. Granger, S., Loukili, A., Pijaudier-Cabot, G., Chanvillard, G., Experimental characterization of the self-healing of cracks in an ultra-high performance cementitious material: Mechanical tests and acoustic emission analysis. *Cement Concrete Res.*, 37, 4, 519–527, 2007.
54. Mauldin, T.C. and Kessler, M., Self-healing polymers and composites. *Int. Mater. Rev.*, 55, 6, 317–346, 2010.
55. Thies, C., Microencapsulation, in: *Encyclopedia of Polymer Science and Technology*, 2002.
56. Ghosh, S.K., Functional coatings and microencapsulation: A general perspective, in: *Functional Coatings*, pp. 1–28, 2006.
57. Gan, S.N. and Shahabudin, N., Applications of Microcapsules in Self-Healing Polymeric Materials. In book: *Microencapsulation - Processes, Technologies and Industrial Applications*. IntechOpen, 2019 (online). https://www.intechopen.com/books/microencapsulation-processes-technologies-and-industrial-applications
58. Grubbs, R.H. and Tumas, W., Polymer synthesis and organotransition metal chemistry. *Science*, 243, 4893, 907–915, 1989.
59. Schwab, P., Grubbs, R.H., Ziller, J.W., Synthesis and Applications of RuCl2(CHR‘)(PR3)$_2$: The Influence of the Alkylidene Moiety on Metathesis Activity. *J. Am. Chem. Soc.*, 118, 100–110, 1996.
60. Sanford, M.S., Henling, L.M., Grubbs, R.H. Synthesis and reactivity of neutral and cationic ruthenium(II) tris(pyrazolyl)borate alkylidenes *Organometallics*, 17, 5384–5389, 1998.
61. Rule, J.D., Sottos, N.R., White, S.R., Effect of microcapsule size on the performance of self-healing polymers. *Polymer*, 48, 13, 3520–3529, 2007.
62. Caruso, M.M., Delafuente, D.A., Ho, V., Sottos, N.R., Moore, J.S., White, S.R., Solvent-Promoted Self-Healing Epoxy Materials. *Macromolecules*, 40, 8830–8832, 2007.
63. Cho, S., Andersson, H.M., White, S.R., Sottos, N.R., Braun, P.V., Polydiniethylsiloxane-based self-healing materials. *Adv. Mater.*, 18, 997–1000, 2006.
64. Hoebe, K., Janssen, E. *et al.*, The interface between innate and adaptive immunity. *Nat. Immunol.*, 5, 10, 971, 2004.
65. Carroll, M.C., The complement system in regulation of adaptive immunity. *Nat. Immunol.*, 5, 10, 981, 2004.
66. Hickman, C.P., Roberts, L.S. *et al.*, *Integrated principles of zoology*, McGraw-Hill, New York, 2001.
67. Vivier, E., Raulet, D.H. *et al.*, Innate or adaptive immunity? The example of natural killer cells. *Science*, 331, 6013, 44–49, 2011.
68. Hia, I.L., Vahedi, V. *et al.*, Self-healing polymer composites: Prospects, challenges, and applications. *Polym. Rev.*, 56, 2, 225–261, 2016.

Index

Printed in the USA/Agawam, MA
September 16, 2026

910661.015